Y0-AAI-969

Communication Satellites for the 70's: Systems

University of Florida
1853

Progress in Astronautics and Aeronautics

Martin Summerfield, Series Editor
PRINCETON UNIVERSITY

VOLUMES	EDITORS
1. Solid Propellant Rocket Research. 1960	Martin Summerfield PRINCETON UNIVERSITY
2. Liquid Rockets and Propellants. 1960	Loren E. Bollinger THE OHIO STATE UNIVERSITY Martin Goldsmith THE RAND CORPORATION Alexis W. Lemmon Jr. BATTELLE MEMORIAL INSTITUTE
3. Energy Conversion for Space Power. 1961	Nathan W. Snyder INSTITUTE FOR DEFENSE ANALYSES
4. Space Power Systems. 1961	Nathan W. Snyder INSTITUTE FOR DEFENSE ANALYSES
5. Electrostatic Propulsion. 1961	David B. Langmuir SPACE TECHNOLOGY LABORATORIES, INC. Ernst Stuhlinger NASA GEORGE C. MARSHALL SPACE FLIGHT CENTER J. M. Sellen Jr. SPACE TECHNOLOGY LABORATORIES
6. Detonation and Two-Phase Flow. 1962	S. S. Penner CALIFORNIA INSTITUTE OF TECHNOLOGY F. A. Williams HARVARD UNIVERSITY
7. Hypersonic Flow Research. 1962	Frederick R. Riddell AVCO CORPORATION
8. Guidance and Control. 1962	Robert E. Roberson CONSULTANT James S. Farrior LOCKHEED MISSILES AND SPACE COMPANY
9. Electric Propulsion Development. 1963	Ernst Stuhlinger NASA GEORGE C. MARSHALL SPACE FLIGHT CENTER
10. Technology of Lunar Exploration. 1963	Clifford I. Cummings and Harold R. Lawrence JET PROPULSION LABORATORY

11. Power Systems for Space Flight. 1963
Morris A. Zipkin and Russell N. Edwards
GENERAL ELECTRIC COMPANY

12. Ionization in High-Temperature Gases. 1963
Kurt E. Shuler, Editor
NATIONAL BUREAU OF STANDARDS
John B. Fenn, Associate Editor
PRINCETON UNIVERSITY

13. Guidance and Control—II. 1964
Robert C. Langford
GENERAL PRECISION INC.
Charles J. Mundo
INSTITUTE OF NAVAL STUDIES

14. Celestial Mechanics and Astrodynamics. 1964
Victor G. Szebehely
YALE UNIVERSITY OBSERVATORY

15. Heterogeneous Combustion. 1964
Hans G. Wolfhard
INSTITUTE FOR DEFENSE ANALYSES
Irvin Glassman
PRINCETON UNIVERSITY
Leo Green Jr.
AIR FORCE SYSTEMS COMMAND

16. Space Power Systems Engineering. 1966
George C. Szego
INSTITUTE FOR DEFENSE ANALYSES
J. Edward Taylor
TRW INC.

17. Methods in Astrodynamics and Celestial Mechanics. 1966
Raynor L. Duncombe
U.S. NAVAL OBSERVATORY
Victor G. Szebehely
YALE UNIVERSITY OBSERVATORY

18. Thermophysics and Temperature Control of Spacecraft and Entry Vehicles. 1966
Gerhard B. Heller
NASA GEORGE C. MARSHALL SPACE FLIGHT CENTER

19. Communication Satellite Systems Technology. 1966
Richard B. Marsten
RADIO CORPORATION OF AMERICA

21. Thermal Design Principles of Spacecraft and Entry Bodies. 1969.	Jerry T. Bevans TRW SYSTEMS
22. Stratospheric Circulation. 1969	Willis L. Webb ATMOSPHERIC SCIENCES LABORATORY, WHITE SANDS, AND UNIVERSITY OF TEXAS AT EL PASO
23. Thermophysics: Applications to Thermal Design of Spacecraft. 1970	Jerry T. Bevans TRW SYSTEMS
24. Heat Transfer and Spacecraft Thermal Control. 1970	John W. Lucas JET PROPULSION LABORATORY
25. Communication Satellites for the 70's: Technology. 1971	Nathaniel E. Feldman THE RAND CORPORATION Charles M. Kelly THE AEROSPACE CORPORATION
26. Communication Satellites for the 70's: Systems. 1971	Nathaniel E. Feldman THE RAND CORPORATION Charles M. Kelly THE AEROSPACE CORPORATION

(Other volumes are planned.)

The MIT Press
Cambridge, Massachusetts,
and London, England

Progress in
Astronautics and Aeronautics

An American Institute of Aeronautics
and Astronautics Series

Martin Summerfield, Series Editor

Volume 26

Communication Satellites for the 70's: Systems

A collection of technical papers selected from the AIAA 3rd Communications Satellite Systems Conference, April 1970, subsequently revised for this volume.

Edited by

Nathaniel E. Feldman
THE RAND CORPORATION

Charles M. Kelly
THE AEROSPACE CORPORATION

Printed by The Alpine Press Inc.
and bound by The Colonial Press Inc.
in the United States of America.

ISBN 0 262 06045 0 (hardcover)

Library of Congress catalog card number:
74-165161

III.
Systems for the Emerging Nations

IV.
U.S. Domestic Systems

Preface

Little that has been written about communications satellites aids in understanding the problems that must be solved in order to overcome their present inadequacies and realize their full potential. This preface attempts to give an account of the nature and magnitude of the work that still remains to be done if satellites are to compete more effectively with other forms of electronic communications.

Point-to-point communications satellite systems have grown in size and complexity since the publication of Progress in Astronautics and Aeronautics, Volume 19, *Communication Satellite Systems Technology,* in 1966. The International Telecommunications Satellite Consortium (Intelsat) has put into service the Intelsat III family of satellites (with 1200 telephone circuits) and the Intelsat IV (with some 6000). Although a "domestic" communications satellite system is being constructed for Canada, no decision has yet been reached on a domestic system in the United States, although such major organizations as COMSAT, American Telephone and Telegraph Company, Western Union, Hughes Aircraft Company, and General Telephone and Electronics Corporation have submitted proposals for such a system to the Federal Communications Commission (FCC). Approvals for competing domestic satellite systems are, however, expected in the near future. Communication satellites for defense have advanced from the 100-lb Initial Defense Communication Satellite to 500-, 1000-, and 1500-lb spacecraft (United Kingdom Skynet, U.S. Defense Phase II, and Tactical Communications Satellite programs, respectively). European point-to-point communications satellites are imminent: Skynet II (United Kingdom), Symphonie (Franco-German), and SIRIO (Italian). The National Aeronautics and Space Administration (NASA) has made major contributions to data collection and aeronautical communications and navigation with its Applications Technology Satellite (ATS) programs, and the programed ATS-F and -G satellites continue to arouse interest. The latest ATS spacecraft, in the 2000-lb class, will carry equipment for a variety of experiments, such as (1) direct television broadcasting to 2000 isolated community receiving stations in Indian villages, (2) satellite tracking and data relay, (3) aircraft position location, and (4) propagation at millimeter wave frequencies. All of the satellite programs mentioned above (except for the tactical communications satellite) utilize only low-powered transmitters—some 10 watts of output. Past experience indicates that low-power satellite services will continue to multiply throughout the decade.

One may evaluate the work that yet remains to be done by examining the role of satellites in the so-called "communications revolution." To be described as revolutionary, a communications satellite either should provide significantly new service and/or capabilities or should reduce drastically the cost of an existing service. Present intercontinental television is just such a new service, but its $8000 cost per hour is high for extensive usage. Even if the cost of the satellite segment were reduced to zero, however, the cost of transmission from the originating studio to the broadcast transmitter on another continent would decrease by less than 15%, since the bulk of the cost is in the terrestrial facilities (such as the earth station and the microwave link to the central office or television operating center). Use of expensive ground stations far from urban centers reduces interference but raises over-all costs. Substantial economies in intercontinental television or telephone service seem unlikely with the present system. Moreover, the 7-year design life of Intelsat IV, its high investment costs, and the cost of the Atlas-Centaur launch tend to inhibit change; it thus seems likely that point-to-point intercontinental communications satellite operations will be virtually frozen for a decade, except for possible improvements in multiple-access and modulation techniques.

The Canadian Department of Communications, with assistance from NASA, will demonstrate communications at 12.5 gigahertz in a cooperative communications technology satellite scheduled for launch at the end of 1974. The reduced interference at this frequency makes it possible to locate satellite ground stations within the cities themselvés. More than the successful development of the technology is necessary, however, to make progress on this basic technological-economic problem. Intelsat nominally represents more than 75 countries. Although a large amount of publicity is given to the potential uses of satellites for extensive intercontinental television networking, data and facsimile transmission, and instruction, the broadcasting and educational interests and the press are not included on Intelsat, which represents solely the telephone and telegraph ministries (thus becoming a restricted commercial organization concerned solely with conventional telecommunications traffic). One or more new international consortia appear necessary to support satellite systems catering to these other interests.

In domestic systems, the use of satellites for television station networking may prove significant if it affects the constitution of the net or provides some new service or flexibility. If—in an area with developed terrestrial microwave facilities, such as the United States—it merely reduces the cost of national networking via long microwave links by 20 to 50% (as is generally predicted), its effect on broadcasting will be small. If the savings were applied to programing, for example, they would increase the total budget available for television programing by a mere 1 to 3%. In an undeveloped, remote area, however, satellite links may be the only economic means of providing network service.

Higher power satellites (i.e., hundreds of watts to several kilowatts per channel) are being developed very slowly because of the complex engineering required, the high research and development costs, and many other difficulties, such as reaching agreement among the diverse potential users. Although such high-power satellites may prove to be an extremely efficient means of distributing a dozen or more television channels to a large number of cable television systems using low-cost ground stations, the cost of generating many channels of new programing will be hundreds of millions of dollars, while wiring U.S. cities with cable will require an initial capital investment of between $5 and $25 billion. Thus the lack of both urban cable television systems and programing to fill large numbers of channels are limitations as significant as the deficiencies of the satellites. Of course, a direct broadcast satellite would make cable television distribution systems unnecessary and provide even larger population coverage. Since there exist no high-power satellites that would make low-cost (i.e., $25,000) multichannel community ground stations at cable television system head ends practical, it is obvious that the still higher power satellites needed for multichannel direct broadcast over large areas are even more remote.

Domestic point-to-point communications satellites for developed nations may provide a useful backup capability for peak loads and emergencies or may function as an alternative to conventional microwave and cable long-haul facilities, but in the latter role are not likely to have a great impact on terrestrial communications facilities in this decade. In undeveloped or remote areas, however, improved communications satellites may be the most economic means of providing television network and multichannel telephone service.

Defense communications satellite systems are expected to increase in size, capacity, and applications throughout the decade, primarily because of their flexibility, mobility, survivability, and cost-effectiveness (features which have not yet received much attention in domestic commercial satellite systems).

We also seek answers to the following questions: In the 1970s, what other technologies will satellites compete against or supplement? How are communications satellite systems likely to be integrated into intercity and international communication services and facilities? Telephone and television will continue to be the most important services, while videophone, data, and record communications may increase substantially in importance. It is expected that the most frequently used surface facilities will be cable and microwave links utilizing solid-state components, although a circular waveguide with some 400,000 voice circuits will be introduced and there will perhaps be limited use of short laser links. In the developed countries, the millimeter waveguide offers economies of scale over high traffic density routes, thus competing with microwave links while providing sufficient capacity to cope with the growth of demand for videophones. Long-haul laser links appear to offer a greater capacity (by a factor of ten) than the circular waveguide with much less than a factor-of-ten increase in cost. These may therefore serve a succeeding decade if sufficient demand develops.

Cable television may grow not only in North America but in western Europe as well. Between the creation of digital networks and the need to supply television signals to the head ends of cable television systems, terrestrial microwave point-to-point links will proliferate. Low-cost satellite ground stations (comparable in cost to a microwave terminal) will be necessary to assure that satellites will play a significant rather than a minor role in this service. In the less developed countries, video cassettes and inexpensive video playback equipment will compete with satellites in providing instructional programing to areas beyond the range of terrestrial broadcast transmitters and translator service. The video cassette, which can cope with a degree of linguistic and cultural diversity impossible for a single satellite channel, is suitable for controlled experimentation with new programing. In the more developed countries, video cassettes are uniquely suited to providing the home viewer the diversity of programing he is unable to obtain on broadcast television channels controlled by governmental administrators.

Space communication facilities will consist primarily of point-to-point satellite links with low-powered transmitters operating into large ground terminals unless a major spacecraft development occurs. New applications that exploit the unique line-of-sight capability of synchronous satellites (e.g., communications to aircraft and other mobile terminals, data collection using small antennas, and the availability of inexpensive ground terminals for voice, data, and television communication) may make only negligible progress without a national commitment.

To effect large reductions in the price of toll television or data service, higher power satellites are needed to work with small ground stations located within population centers (i.e., bypassing most of the present surface facilities). High-power satellites will determine whether satellite communications will play a substantial role in providing new services and competing with established terrestrial facilities.

Reliable, inexpensive ground terminals do not appear to be unrealizable, especially because of the availability of new solid-state components and the economies possible with mass production. In a satellite system, investment is concentrated in the ground terminals. Less developed countries would do well to focus on the production of ground terminals, shunning the development of spaceborne hardware, since the design of prototype satellite hardware and the demonstration of reliable space operations are costly and time-consuming.

If satellite systems are to provide new services in the 1970s, further improvements in multiple access are also essential. For all sizes of ground stations, the development of highly reliable unmanned equipment (using nontracking antennas with sidelobe level control to enhance orbit utilization) is necessary to achieve substantial economies.

Satellite antenna gains of 40 to 65 decibels may prove necessary, along with the development of the required accurate, long-lived stabilization, station-keeping, and multiple-beam-steering systems. For other applications, satellites with higher output power than currently developed (i.e., by a factor of 10 to 100) are essential. They in turn are dependent on the development of large, oriented solar arrays. The high-power satellites are essential not only for the less developed countries of the world, but also for the sparsely inhabited areas of developed countries, such as Alaska and much of Canada. The high cost of the ground stations in the initial Canadian domestic satellite system seriously handicaps the effort to integrate that country's remote northern areas into a single, cohesive unit. Interestingly, the types of satellites needed by advanced countries may be very similar to those needed by the less developed countries.

Under Congressional direction, NASA de-emphasized point-to-point communications satellite activity when COMSAT was established, but then proceeded to de-emphasize all satellite communications as well. By limiting the funds for all such activities, NASA opted for slow and modest progress. At present funding levels (1971), decades will be required to achieve the technical advances called for by the 1967 National Academy of Sciences study. Research and development supported by the United States created most of the technological base on which Intelsat rests; with exceedingly modest further support since the early 1960s, progress has been commensurate. Western Europe's communication satellite programs have demonstrated negligible results.

Research and development leading to more versatile high-power communications satellites must be emphasized if satellites are to strengthen their position in future world communications. Congressional support of a major international cooperative program is essential. Today there is no national or international commitment to such a research and development effort; without support on the order of $200 million per year to the development of the space segment, no dramatic new services or improvements in communications via satellite will occur in the 1970s. Among space activities, communications satellite research and development is the only area with the potential to become a multibillion dollar industry that can both make a profit and support world understanding as well.

Since scarce resources must always be allocated among several alternatives, long-term study and analysis of communications requirements, spectrum allocations, competing systems, and policy issues are needed to foster intelligent decision making and to justify necessary changes. A new international consortium appears particularly necessary to support the development of high-power satellites, since such a program appears beyond the means of any present communications organization.

The organizational aspect of the problem is expected to be difficult and time-consuming. If NASA were to go forward with the necessary technical developments while the potential users organize, the major goal—to implement a high-powered reliable communication satellite—would be advanced. Since bilateral agreements in the satellite communications area have proved consistenly more effective than multinational agreements, perhaps strong U.S. leadership in negotiating a series of bilateral agreements may be a means to evolve such an international program.

In conclusion, international and domestic point-to-point communications satellite systems are expected to multiply throughout the decade. In developed regions they will remain auxiliary to the over-all ground communication complex, even though significant in dollar volume. In many undeveloped regions, however, point-to-point systems could become the primary means of communication. If potential users of high-power satellites make the necessary plans for financing, regulation, and legislation (both national and international), many new applications with large growth potential can be developed for populated as well as undeveloped regions.

Volumes 25 and 26 of *Progress in Astronautics and Aeronautics* contain a sample of research in communications satellites drawn largely from the AIAA 3rd Communications Satellite Systems Conference held in Los Angeles, California, in April 1970. The uniqueness of these volumes is that they report on developments in all aspects of communications satellites, reflecting the wide range of disciplines within the AIAA. The papers were selected to emphasize analysis and performance data verified by laboratory or orbital testing and likely to prove applicable to future systems. The 73 papers in the two volumes are an impressive indication of the work that has been done and the solutions proposed to some of the technological problems described in this preface.

This volume was sponsored by the AIAA Technical Committee on Communication Systems (Richard B. Marsten, 1969 Chairman; Nathaniel E. Feldman, 1970 Chairman) and was an outgrowth of the AIAA 3rd Communications Satellite Systems Conference under the general chairmanship of Charles M. Kelly.

Nathaniel E. Feldman

Charles M. Kelly

April 1971

I. CANADIAN DOMESTIC SATELLITE COMMUNICATIONS SYSTEM

CANADIAN DOMESTIC SATELLITE COMMUNICATIONS POLICY

J.H. Chapman*

Department of Communications, Ottawa, Ontario, Canada

Abstract

The Canadian Government set out, in a White Paper dated 28 March, 1968, certain criteria under which the Domestic Satellite Communications system would be brought into being. A Canadian satellite corporation, established by Government legislation to design, develop and construct both the satellites and earth terminals of the system, came into being on 1 September, 1969. The Corporation, known as "Telesat Canada", involves a three way ownership scheme, between the general public, the telecommunications common carriers and the Government of Canada. The President and provisional directors were designated by the Governor-in-Council as set out in the Act establishing the Corporation. Subsequent appointments of directors, to include at least two members of the Public Service of Canada, will be made in accordance with the by-laws of the company. The Corporation will be financed through the sale of shares to the Government, the telecommunications common carriers and the public.

Canada was the third nation to have a satellite in space, and it is hoped that sometime in 1972 she will have become among the first in the Western world to have a domestic satellite communications system in operation. As early as July 1967, because of the great interest which had developed in a possible Canadian domestic satellite communications system, the Canadian government announced the creation of a Task Force to advise the Canadian government on satellite policy in general and, in particular, on the use of satellite technology for domestic communications.

Presented as Paper 70-428 at the AIAA 3rd Communications Satellite Systems Conference, Los Angeles, Calif., April 6-8, 1970.
*Assistant Deputy Minister (Research).

Canada is a large country of about 4 million square miles, with a population of 20 million. The country is larger than the U.S.A. but has about 1/10th of the population. Most of the people in Canada are distributed in fewer than a dozen large communities within 200 miles of the U.S./Canadian border. This ribbon of population is well served by communication systems. There are 48 telephones per hundred people, and television programs reach 95% of the people with at least one program and 75% with two programs.

There are two main microwave radio relay systems serving this population belt. One is provided by the Trans-Canada Telephone System, which is formed from the major telephone companies in Canada, and the second is provided jointly by two major railways, the Canadian National and the Canadian Pacific.
This network, with 1000 radio relay stations, is the heaviest development of microwave radio relay of any country in the world, and has a density of microwave radio relay about double that of the U.S.A. However, it is deficient in broadband capacity off the main trunk routes, that is, in the remoter or less densely populated areas, particularly in the North.

The capital cost to repair this deficiency would be measured in the hundreds of millions of dollars. Although it might be feasible to build up a terrestrial radio relay system in the Western Arctic, there is a combination of rugged terrain and wide stretches of open water in the Eastern Arctic that makes impossible the use of any terrestrial system, based on the present technology, to meet this need. In this area the major communications are provided by military systems, which were established on a basis that is, of course, other than that of commercial viability.

The telecommunications common carriers made a proposal to the government that they should be permitted to set up a domestic satellite communications system. A counter proposal of sorts was received from the broadcasters, suggesting that a satellite communications system should be owned by the broadcasters. Their basic argument was that the satellite could provide broadband facilities that the broadcasters needed for distributing their television programs. This is the well-known conflict between the telephone common carriers and broadcast interests that is not easily resolved in a free-enterprise economy.

The government set up a task force to examine the situation. The task force reported the various views to the government, which then decided on a policy and published a White Paper on Domestic Satellite Communications in March 1968.

The Government proposed that a mixed corporation, Telesat Canada, should be established by an Act of Parliament. It settled the ownership question by decreeing that ownership would be established through equity in the corporation, which the Government, the common carriers, and the public would share. The broadcasters would be represented on the one hand through the Government representation, because the Canadian Broadcasting Corporation is one of the major broadcasters in the country, and on the other, the independent broadcasters could be represented among the public owners. This would be sufficient to provide a representation of the broadcasters' interests on the Board of Directors. There was, in this proposal, an intentional built-in conflict of interest in the Board of Directors of Telesat Canada. It was the government's intent that the conflicts should be resolved within the Board between the parties concerned. The government foresaw that some problems might become intractable - for example, rates - and proposed that these would be referred to a regulatory agency that would be set up and given due authority in this area before the Telesat corporation became operational. The government also reserved for itself certain powers, such as the approval by the Board of Directors of the nomination of the president and approval of the Canadian content in satellite and ground station contracts.

The White Paper foresaw the Telesat Corporation providing a complete system. Both the satellites and the ground stations will be owned by the Corporation. It could under certain conditions compete with established common carriers. It is therefore a common carrier itself. The limitations on its powers are not spelled out in detail in the enabling legislation, but, in principle it acts as a wholesaler of complete radio-frequency channels, 6 MHz wide, suitable for broadband television distribution. It sells services directly to broadcasters who will undertake to use one or more rf channels. The microwave tails to Earth stations may be provided in any commercially or economically viable manner. In general the microwave tails from Earth stations to the customers' premises are usually not provided by Telesat and in most cases will be provided by the common carriers. Direct access to the earth stations is therefore available to other users, such as broadcasters.

There were several additional over-all objectives of the Canadian Government. The first was to maintain control of domestic communications within Canada in the hands of Canadians. Such control is essential to the fabric of a modern state. A second objective of the Government was to retain for Canada as much of the industrial development and manufacture as was feasible, with due regard for economic and technological realities. For example, there was no intention of developing launching systems for Canadian satellites. These would be bought from the best source - the source which provided the best product at the best price. Furthermore, there was no intention of redeveloping components such as traveling wave tubes or solar cells that are readily available on the market.

Technical study contracts were placed with industry; these studies lasted 6 months and were completed in November 1968. The main objectives of these studies were to define a satellite design suitable for a Canadian domestic satellite communications system, to provide an implementation plan for the development and fabrication of the segment that would have an optimum participation by the Canadian industry, to prepare realistic cost and schedule estimates for the full implementation of the space segment of the system, and to examine the tradeoffs between satellite and ground station parameters for the selection of the Canadian system.

At this stage in the program, no decisions had been made regarding the actual size of the Canadian system. The immediate cost of the total system -- that is of satellite or satellites, launching facilities, and ground stations -- was estimated to range from $65 million to $160 million. The higher cost would be incurred if three 12-television-channel satellites were built, two in orbit and the third in reserve on the ground, and an extensive terrestrial system was constructed immediately. The lower cost would be incurred if two 6-television-channel satellites were built, one in orbit and the second in reserve on the ground, and the terrestrial system was built over the years to match the demand. Subsequently, a decision in principle was made to develop a system based on 6-channel satellites and to install a modest sized earth station network.

Legislation was passed by the Canadian Parliament in June 1969 for the establishment of the Canadian Satellite Corporation. The Corporation, known as Telesat Canada, came into being on September 1, 1969. The objects of the Corporation are to establish satellite telecommunications systems providing, on a commercial basis, telecommunications services between locations in Canada. It will be a project that will help to knit Canada together more closely and to narrow the gap in the differences in living standards between those Canadians who live amid the rigours of the frontier and those who live amid the comforts of the cities.

NETWORK DISTRIBUTION OF TELEVISION BY SATELLITE

C.A. Siocos*

Canadian Broadcasting Corporation, Montreal, Quebec, Canada

Abstract

Television distribution by satellite is discussed as seen for Canadian applications and particularly for the networks of the Canadian Broadcasting Corporation (CBC). Emphasis is placed on broadcasting network system aspects, including performance.

I. Network Operations and Requirements

The CBC operates two distinct television networks, one in the English language and one in the French language. The English network is coast to coast. The French network, on the other hand, does not as yet extend from coast to coast. It exists, predominantly, east of Manitoba.

Coverage at present is such that 96% of the English-speaking population and 93% of French-speaking Canadians can receive CBC programs in their own language. Since by its mandate the CBC must strive, within its means, to reach all Canadians, considerable effort is expended each year to increase existing coverage. However, because further expansion at present is to areas where microwave links capable of carrying television are not presently available and new links cannot be provided at reasonable cost, broadcast stations in these areas are not electronically linked to the network center. Satellite transmission is expected to provide direct connection to the network for all these stations and for any new station in remote areas. The first requirement for the satellite, then, is to reach all parts of the country and provide a relatively uniform grade of service everywhere.

Presented as Paper 70-430 at the AIAA 3rd Communications Satellite Systems Conference, Los Angeles, Calif., April 6-8, 1970.

*Staff Engineer, Office of the Vice-President, Engineering.

Canada extends over roughly 90° of longitude. There are thus six time zones within the country. A seventh time zone exists in Newfoundland where, traditionally, time is ½ hr later than Atlantic time. It is the policy of the CBC to provide, insofar as possible, the same program at the same local time anywhere in Canada. Accordingly, at present, in the coast-to-coast English network, a five-time zone operation is provided (the two extreme time zones, Newfoundland and Western Yukon are merged operationally with their adjacent time zones.) Another requirement for the satellite, then, is that it should fit into this multiple time zone pattern.

In Canada, as in other very large geographical areas, regional differences in interest are important, but, because Canada is a single country, interest in centrally provided programs is also great. Accordingly, each of the two programs, the French and the English, is made up from central contributions and from regional or local contributions. In general, stations receive the packaged program, containing the network and regional items, from a regional center. Regional centers, in turn, receive the network contribution for packaging and redistribution from the two central stations, one for each network. There are six regional centers outside the two main stations. They all have production facilities and, in addition to local and regional material, they occasionally produce programs for all the network. There are, furthermore, two additional production centers, one in the federal capital and one in the capital of the Province of Quebec, contributing material to the main stations of both CBC networks. A further requirement for the satellite, accordingly, is that it should preserve regionality both in the configuration of the network and in program origination localities.

Technical quality should be such that the system will provide a favorable picture to a well-adjusted television home receiver within the service area of the transmitter (Figs. 1a and 1b).

II. Current Terrestrial Network Facilities

Coverage is provided by transmitters belonging to either CBC or affiliated private stations. At latest count, there were 87 CBC and 147 affiliated station transmitters in the English network and 19 CBC and 42 affiliated television station transmitters in the French network. Geographically, the distribution of main transmitters (excluding rebroadcast transmitters slaved to such main transmitters) is, at present, as shown in

Figs. 2 and 3. A large number of rebroadcast stations, utilizing high-quality receiving installations to pick up the main transmitter for immediate rebroadcast from the local transmitter, are used to extend coverage further. It is planned that, before the end of 1974, another 19 English and 9 French main transmitters will be installed by the CBC to serve remote areas.

Figures 2 and 3 show that, at present, not all of these stations are connected by microwave links. As a rule, in the southern part of the country, where telecommunications companies have adequate microwave facilities, direct connection to the network is universally provided. At far North locations, programs are received from the center on tape. Helical scan television tape reproducers are used. Of course, programs are shown with a delay of days and the tape stations operate for about one-quarter of the hours that connected stations do. Helical scan tapes for distribution to these stations are recorded at several of the regional centers.

To make it possible to show the same program at the same local time across the connected network, a large television tape recording center, strategically located on the backbone of the microwave network near the middle of the western provinces, provides three delayed repetitions of programs for consecutive distribution to the three western time zones. Furthermore, to accommodate the Atlantic time zone in this respect, a smaller tape recording center is operated in Halifax to provide 23 hr delay so that certain network programs are shown at the correct time on the following day.

Regional networks are operated at the Atlantic seaboard for the Maritime provinces and for Newfoundland. In the remainder of the country, however, regional networks are only set up for special occasions and local, rather than regional, content is provided normally.

The main trunk, or backbone, facility of each of the two networks has two-way capability. Each way, a single channel only is available, except that in accordance with distribution pattern requirements, two sections of this trunk now use an additional one-way channel.

As a rule, television audio is transmitted separately from its associated video, either by different facility or via the same facility. No diplexing of audio and video signals, either by frequency or time division, is normally used.

III. Utilization of Communications Satellites

In general, a multibeam distribution plan would be suitable in the case of a number of areas which usually carry independent programs and are only occasionally connected together as an over-all network. In the opposite situation, i.e. where the regions usually are connected into a national network and only less frequently act independently, a single beam from the satellite is more advisable. For the CBC network, it was concluded that an all Canada beam from the satellite would be the most advantageous.

At present, due to the high cost of a satellite TV channel, operations requiring many satellite rf channels are considered prohibitively expensive. Such operations would result from servicing time zones and regions by separate satellite channels. Accordingly, a judicious mixture of surface and space links must be found for best fulfillment of the set of economic and, in many cases, socio-political objectives prevailing. There are various factors peculiar to the satellite which must be taken into consideration, such as: possible outages resulting from eclipsing of the sun at the satellite and from intense solar noise coming into direct view of the Earth station antenna; length and cost of backhaul facilities possibly required between Earth station and broadcaster; possible vulnerability of the satellite and cost of maintaining terrestrial backup facilities; etc.

A simplified block diagram showing the use of 2 satellite rf channels for the Eastern and Western subarctic regions of Canada and for long-haul network connections, plus six regional networks, by surface links, to fulfill time-zone and regional distribution for the English language CBC network is shown in Fig. 4.

The French language network has no Northwestern stations and will employ, therefore, a single satellite channel used both for extension of coverage to the North and for long-haul connections in the South.

IV. Initial Satellite Distribution Plan

For the English network, the connection plan is shown on the map of Fig. 5. It follows the lines of the block diagram of Fig. 4. The dashed-line connections shown between adjacent regional networks represent terrestrial two-way links for occasional and emergency use. Per occasion use of these connections will be made for reverse-flow feeds from region to center, and of course they are essential to provide forward-flow service in case of satellite failure during the initial phase of implementation when one only satellite will be in orbit.

The connection plan for the French network is shown in Fig. 6.

A feature of the English network plan is that the Pacific Time Zone rf Channel of the satellite is fed from either the Network Center in Toronto or from the Regional Center in Vancouver. This arrangement is dictated by the requirement for regional control. Switching the uplink between these two locations is to be as rapid as an ordinary program switch and is to be controlled by the broadcaster. Should more rf channels be added to the distribution plan in the future, they in turn must have the ability to be fed either from the main or the appropriate regional center. Thus, the plan is conceived so that any distribution path can be implemented either by terrestrial or by space link as needed.

Operation during solar eclipses will be maintained by batteries. Should these fail, satellite position, at 109° W longitude, would make operation during solar eclipse at the satellite unnecessary for locations as far West as the Mountain Time Zone. However, for the Pacific Time Zone, such eclipses happen during the latter part of the broadcast network day and, therefore, ideally, the rf channel to be used for the Pacific Coast and the Northwest should have the ability to carry on operations during eclipse. In the case where even such partial eclipse operation is not available, the situation could be covered by rearrangement of programming but at the price of some loss of flexibility.

It is estimated that solar noise interference will produce a total outage of less than 30 min/yr. Given the cost involved in providing duplicate equipment required to eliminate such outages, it is not planned to have twin antennae at the receive-only TV stations of the North, even after the initial period when a protection satellite will be in orbit. The present state of the art does not allow us to think in terms of earth antennae capable of practically instant reorientation to the spare satellite.

The satellite link facilities also must provide flexibility in network operations. To accomplish this objective, Earth stations will be provided with channel selection ability. For unattended stations especially, selection should be remotely controlled. Furthermore, transmitting Earth stations shall have the ability to transmit, on demand, on all rf channels dedicated to broadcasting.

Planning for mobile transmitting Earth stations cannot be contemplated as yet mainly because of the problems of siting temporary stations in an environment of high microwave link density that is likely to prevail in most areas where such program originations may be desired. On the other hand, by siting the Regional Center receiving Earth stations so that they are free of interference problems in the 6 GHz band as well as the 4 GHz band, and by providing transmitting equipment for occasional use, these stations can be temporarily converted into transmitting stations when needed. This would go a long way toward providing additional flexibility. Of course, the possibility of greatly increased backhaul costs to achieve transmitting ability is a crucial factor in this case.

Television sound will be carried, by multiplexing, on the same satellite rf channel as the video signal. Although the method of multiplexing has not been finalized, one of the requirements will be that it shall not preclude the future addition of a second audio channel. In addition, a cue and control channel capable of providing a small number of signaling circuits also will be multiplexed in each television carrying satellite rf channel. It will be used for remote control network switching and for carrying printed information on scheduling and traffic changes.

V. Expected Picture Quality

Impairments fall into three main categories, i.e. linear distortion effects, non-linear distortion effects and interfering signal effects.

Usually, more than one such impairments are present at any given time and the over-all picture quality is affected by their simultaneous action. Customarily, estimation of picture quality in long paths is, however, made solely on the basis of random noise. It is well to remember, however, that such an estimate is optimistic in that it does not take into account all the other sources of picture degradation.

Signal-to-Noise Estimation

In the following, the usual convention for the expression of television signal-to-noise ratio will be adhered to, i.e. the ratio of the peak-to-peak luminance signal to the rms value of the random noise signal. Furthermore, to bring into line, from the subjective point of view, the various spectral distributions of the noise contributed from the diverse natures of the path links, weighting will be applied to noise. Noise will be limited to a bandwidth of 4.2 MHz, and the CCIR noise weighting for 525-line systems will be used. This will prove adequate for both monochrome and color television in the case of transmission paths but only for monochrome television in the case of transducers. In this latter case, color television can be handled by postulating, as a first approximation, an equivalent reduction of 1.5 dB to the transmitter power.[1]

With reference to Fig. 1, we should note that

1) For weighted signal-to-noise ratios of picture sources, Shadashige[2] gives the following: vidicon type cameras, 50 dB, image-orthicon cameras, 46 dB, and 3rd generation recording of vidicon signal, 46 dB; thus, 46 dB appears as a reasonable assumption for this case.

2) For in-studio transmission paths, minimum signal-to-noise ratio is specified as 66 dB (weighted) in the CBC.

3) Terrestrial intercity links and Earth station backhaul links, for the distances involved (much shorter than the coast-to-coast connection of the present Canadian specifications), can be expected to conform to the CCIR standard for a hypothetical reference circuit for the 525-line system, viz., 56 dB (weighted) minimum.

4) The tolerance limit for the space link signal-to-noise ratio is expected to be 54 dB (weighted) for the main network facilities and 48 dB for the remote connections.

5) A typical value for the signal-to-noise ratio of a broadcast transmitter is taken as 54 dB (weighted).

Accordingly, for the long chain of Fig. 1, addition to the noise powers as above results in an over-all signal-to-noise (weighted) ratio of 43.6 dB prior to broadcast propagation and reception. Likewise, for the shorter chain of Fig. 1, the resultant signal-to-noise ratio (weighted) is 43.2 dB taking into account the higher S/N likely to prevail in satellite links to remote areas.

The signal-to-noise ratio degradation resulting from broadcast propagation and reception depends on a) whether the receiver is in an area of high or low field strength, b) whether the receiver is in an area of high or low industrial and other ambient electric noise, and c) the receiving antenna gain and the receiver noise figure.

Near the transmitter, this degradation is negligible but nearer the limit of the service area it cannot be ignored. In Canada, as in the USA, at the limit of the service area the carrier-to-noise ratio (using home receivers and antennae of typical quality) goes down to 30 dB.

To relate baseband signal-to-noise ratios, as above, to the carrier-to-noise figures used in the well known TASO tests in the USA which, according to Dean[3], represented the ratio of radio-frequency rms signal during sync peaks divided by the rms noise voltage (unweighted white noise) over a 6 MHz channel, it is necessary to remember that a) following modulation, the video peak-to-peak picture corresponds to the rf peak picture signal; b) the ratio of the picture signal to picture-plus-sync signal is 0.714; and c) at reference white there is still 12.5% of peak carrier left.

Thus, for X dB of carrier-to-noise ratio, as used in the TASO evaluations, the signal-to-noise ratio will be as follows:

(C/N)		X dB
Peak-to-rms conversion		+3 dB
Picture-to-composite conversion		-2.93 dB
Modulation depth guard		-1.16 dB
(S/N)	=	X dB -1.1

The above S/N is not at the output of the receiver, and therefore the 6 dB loss resulting from carrier attenuation from the receiver is not included.

Accordingly, in high field strength areas, we may expect a C/N (weighted) of approximately 43.4 + 1.1 = 44.5 dB from the aforementioned chains. For white noise for which CCIR weighting is 6.1 dB, this would correspond to 38.4 dB unweighted and, according to the TASO results, it would make picture quality to appear as halfway between "fine" and "excellent" to at least 50% of the observers.

Of course, any additional complexity of the transmission path, such as double passage through the satellite or more tape recording, would result in a somewhat lower picture quality because of the additional noise.

Additional Subjective Impairments

To estimate over-all picture quality resulting from other types of impairment, it is necessary a) to know the law of addition for the distortion in question over tandem links and b) the relationship between picture quality grade and degree of signal degradation. These relationships as yet have not been adequately determined in all cases.

CCIR Recommendations 421-2 and 451-1 indicate laws of tolerance addition for long-distance links. However, the correctness of their application to all types of chains needs to be proved and the CCIR is now studying this matter along with the whole field of over-all picture quality.

The correspondence of picture quality to signal degradation has been investigated in some countries for a number of types of impairment. Among the most recent investigations are those, during the past 5 years, by the British Post Office team of Allnatt et al.* Also, prior to the adoption of a color television standard for Europe, the European Broadcasting Union had determined, in 1965, for a few types of distortion, the amount of signal degradation resulting in picture quality halfway between "rather poor" and "fairly good".

As an example, for differential gain the limit as above, determined by the EBU, was 40%. Using the CCIR law of addition (three-halves law for differential gain) as a rule-of-thumb for chains such as those of Fig. 1, we obtain, for the longer chain an over-all differential gain limit of nearly 40%. Individual differential gain values used were: 7% for 3rd generation tape recording of a camera signal, 5% for in-studio transmission, 10% for intercity and backhaul links, 10% for a modern broadcast transmitter (the 1961 EIA standard has 15%), and 2% for a receiver. It is rather probable that in this case the CCIR law of addition is pessimistic.

Multiple Impairments

Since final picture quality depends on the cumulative effect of all impairments present, it is necessary to know the law

* The bibliography of Ref. 5 lists the published results of these investigations.

governing the addition of subjective impairments of different natures. Lewis and Allnatt[4,5] have proposed a subjective impairment unit, termed "imp", which has been found empirically to have the property of summability, at least for the types of impairment investigated by the British Post Office. As Weaver[6] points out, these impairments were either the result of linear distortions or of interfering signals, and therefore strictly speaking, the empirical proof of summability of their imp measures cannot without further proof be taken as covering the effects of non linear distortions as well. The subject of multiple impairments is also presently under study by the CCIR.

VI. Concluding Remarks

Added coverage and improved operations such as outlined do not come without cost. It should, however, be stated that according to studies carried out by the CBC, this mixed method of space and terrestrial link utilization minimizes these additional costs. It also should be noted that satellite link costs would be appreciably lower if backhaul microwave links were unnecessary. This also would favor picture quality requirements. The cost of backhaul links is such that artificial site shielding should be investigated vigorously.

Since the necessity for backhaul links arises from the satellite having to share frequencies with terrestrial microwave circuits, and thus the Earth stations having to be sited far enough away from towns (terrestrial links on these frequencies being present there), clear frequencies are needed for the satellite. Availability of clear frequencies also would mean that no low ceiling to satellite signal flux density would be needed. Then the total television distribution system from space could be engineered to minimize costs by balancing satellite EIRP against Earth station sophistication.* To the extent that Earth station sophistication depends on antenna size, it will have a lower limit placed on it from orbital utilization considerations, but it is quite probable, given the ultimately large number of television receive-only stations, that systems costs could be lowered by clear frequencies.

*See, for instance, Ref. 7.

References

[1]Mertens, H., "Method of Specifying the Signal-to-Noise Ratio for a Satellite Television-distribution System," EBU Review, Pt.A, No. 103, June 1967, pp. 104-110.

[2]Shadashige, K., "Study of Noise in Television Broadcast Equipment," Journal of the Society of Motion Picture and Television Engineers, Vol. 78, Dec. 1969, pp. 1069-1076.

[3]Dean, C.E., "Measurements of the Subjective Effects of Interference in Television Reception," Proceedings of the Institute of Radio Engineers, Vol. 48, June 1960, pp. 1035-1049.

[4]Lewis, N.W. and Allnatt, J.W., "Subjective Quality of Television Pictures with Multiple Impairments," Electronics Letters, Vol. 1, Sep. 1965, pp. 187-188.

[5]Lewis, N.W. and Allnatt, J.W., "Subjective and Objective Impairments in Television Pictures," IEE Conference Publication, No. 46, 1968, pp. 2.1.1-2.1.5.

[6]Weaver, L.E., "The Quality Rating of Color Television Pictures," Journal of the Society of Motion Picture and Television Engineers, Vol. 77, June 1968, pp. 610-612.

[7]Talyzin, N.V. et al., "Optimum Parameters and Economic Efficiency of a Multistation Communication-Satellite System," Telecommunications and Radio Engineering, Pt.2, Vol. 24, Nov. 1969, pp. 59-65.

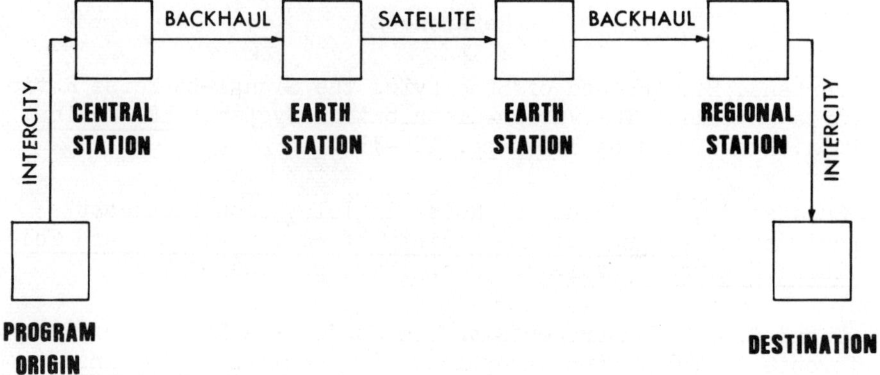

a) Long configuration.

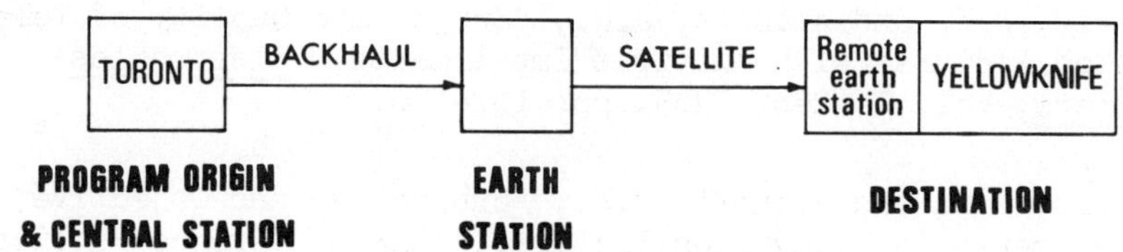

b) Short configuration.

Fig. 1 Usual broadcast chain example.

Fig. 2 CBC television network. English language.

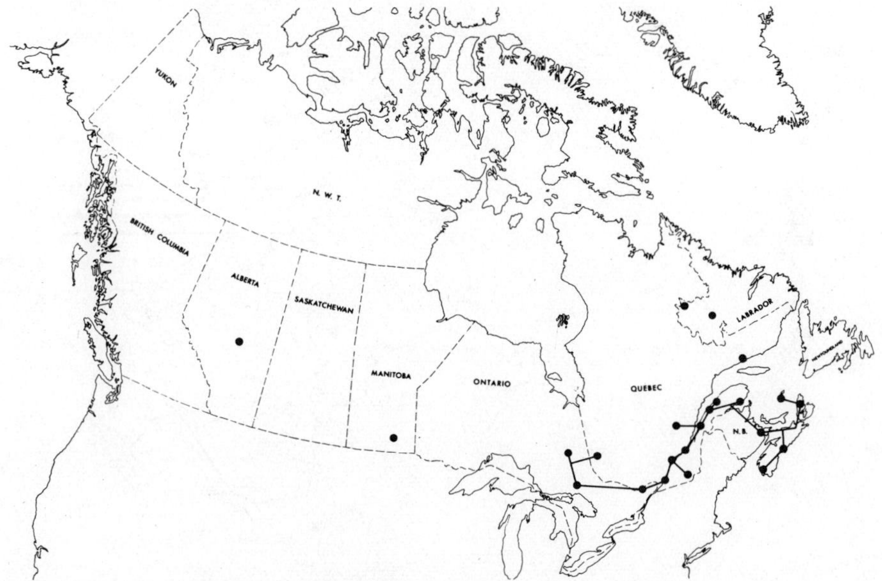

Fig. 3 CBC television network. French language.

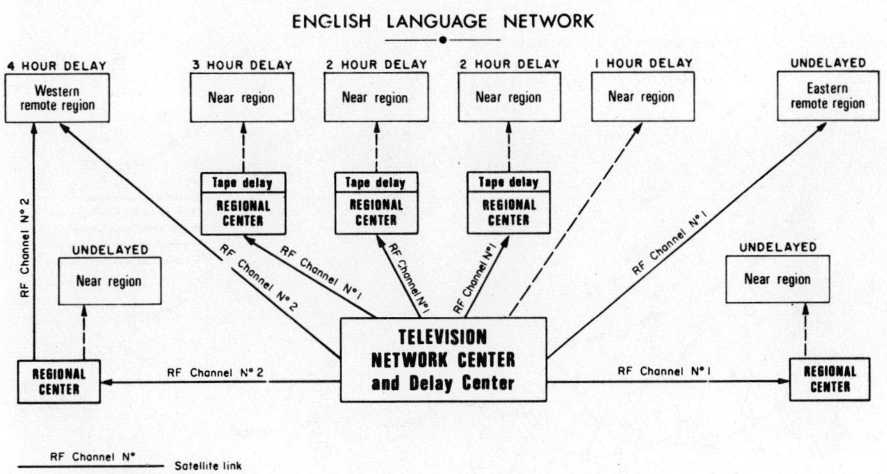

Fig. 4 Block diagram of a distribution plan utilizing two satellite rf channels.

Fig. 5 CBC distribution plan including satellite links. English language television network.

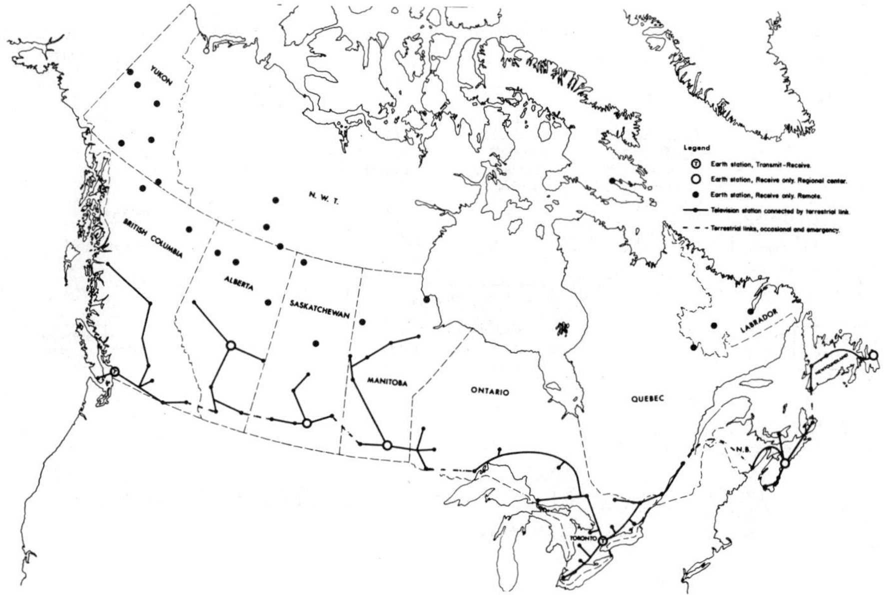

Fig. 6 CBC distribution plan including satellite links. French language television network.

II. EUROPEAN PROJECTS

THE FRANCO-GERMAN TELECOMMUNICATION SATELLITE SYMPHONIE

B.R.K. Pfeiffer* and P. Viellard†

Symphonie Satellite Project Group, Brétigny, France

Abstract

The primary mission objective of the Franco-German telecommunication Satellite Project Symphonie is the gain of technical knowledge and experience in the development and application of telecommunication satellites. Two Symphonie Satellites with an anticipated lifetime of 5 years will be located in a geostationary orbit at 15° West. The SHF coverage zones are given by a global receiving beam of 17.2° and two spot-beams for transmission to Europe-Africa and America. Each satellite has two transponders in the range 6-4 GHz, and 90 MHz bandwidth each. The minimum EIRP in the transmitting zone will be about 29 dbW. The satellite is three-axis stabilized, using a flywheel and a system of cold gas thrusters for attitude control. A bipropellant hot gas system provides proper position correction during the lifetime of the satellite. A liquid bipropellant apogee motor is integrated in the structure of the satellite. The electrical power is provided by three solar arrays spaced by 120°. The two flight models of the satellite will be launched by means of the ELDO Europa II launcher.

I. History

In June 1967 the French and German Governments agreed to the joint project of a telecommunication satellite. It is the purpose of this agreement between the two governments to build a prototype and to launch two flight models from French Guyana by means of the ELDO-launcher Europa II.

Presented as Paper 70-406 at the AIAA 3rd Communications Satellite Systems Conference, Los Angeles, Calif., April 6-8, 1970.

*Satellite Project Manager.

†Satellite Deputy Project Manager.

Before combining their efforts, both countries had performed preliminary planning for similar national projects, project "SAROS" in France and project "OLYMPIA" in the Federal Republic of Germany.

In 1968 a request for a proposal was released to industry in both countries. Two industrial groups, SYMCOSAT and CIFAS, submitted proposals. At the end of 1968, CIFAS was chosen as contractor.

Table 1 Telecommunication experiments

Type of Experiment	Possibilities	Number of carriers[a]
Television	1 picture - 3 voice channels (1 service channel)	1 (+1)
Television	1 picture - 1 voice, 9 commentary channels (1 service channel)	1 (+1)
Broadcasting	18 voice channels (1 service channel)	1 (+1)
Broadcasting	3 to 5 voice channels for each carrier (2 service channels)	2 (+2)
Telephone	132 telephone channels	1
	60 telephone channels	1
	12 telephone channels for each carrier	6
Broadcasting and telephone	10 voice channels for each carrier	3
	12 telephone channels for each carrier	6
Telephone	Time division multiplex-PCM	1

[a]The carriers are not related to one particular SHF-ground station.

II. Mission Objectives

In view of the increased application of telecommunication satellites for the transmission of TV, voice, telex, and other data the project has the following primary objectives: a) to gain technical knowledge and experience in the development of telecommunication satellites, as well as their application; b) to improve the experience of the European industry for further participation in similar international projects; and c) to perform an experimental program of TV, voice, and telephone transmission as defined in Table 1.

III. Telecommunication Requirements

The various required zones of SHF coverage are shown in Fig. 1.

Receiving zones

Global beam 17.2° with the satellite at 15° ± 0.5° West, 0° ± 0.5° North.

Transmitting zones

1) Spotbeam, Europe and Africa, 13° x 8° Center, 11° East, 11° North. Main axis, UK-Madagascar.
2) Spotbeam, America, 13° x 8° Center, 43° West, 3° North. Main axis, Montreal - Buenos Aires.

The frequency band of 500 MHz is divided into 4 subbands, each of 90 MHz, as shown in Table 2.

Table 2 Symphonie frequency bands

Transponder band	Receiving bands	Transmitting bands
a	5940 - 6030	3715 - 3805
b	6065 - 6155	3840 - 3930
c	6195 - 6285	3970 - 4060
d	6320 - 6410	4095 - 4185

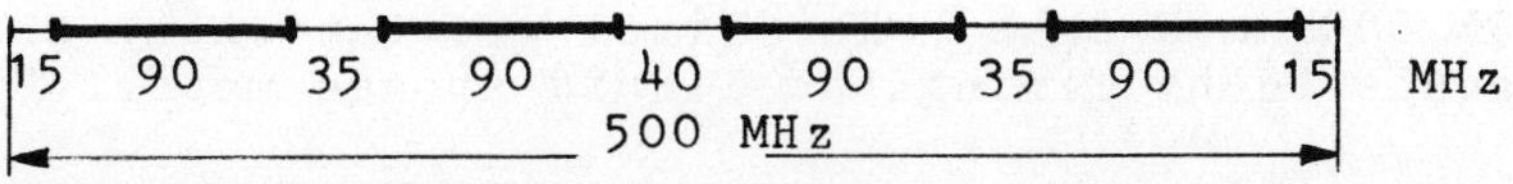

15 90 35 90 40 90 35 90 15 MHz
500 MHz

The bands a and c are used by the first satellite flight model, the bands b and d by the second flight model. The transmitting and receiving beams are orthogonal circular polarized.

The receiver characteristics are as follows:

Sensitivity

$g/t = -15$ db/^{o}K, where g is the receiving antenna gain and the noise temperature at the receiving input in space.

Variation of sensitivity

- 1 db during lifetime.

Power density

- 77 dbW/m^2 nominal value, - 87 dbW/m^2 minimal value, and - 57 dbW/m^2 maximum value.

The transmitter characteristics are as follows: a) EIRP, 29.0 dbW at the edge of the zones; b) variation of the EIRP during lifetime, $\pm$ 1 dbW. The EIRP permits the use of SHF-ground stations with 12-14 m diameter antennas.

IV. Mission Constraints and Launching Procedure

The main constraints for the satellite are the limited payload capability, the limited space under the shroud, and the vibration levels of the launcher EUROPA II, to be used in the Symphonie program.

The satellite will be launched from French Guyana into the geostationary orbit subsequently by the three stages of EUROPA II, a solid perigee stage and a liquid bipropellant apogee motor, which is fully integrated into the satellite.

The first three stages inject the satellite into a parking orbit of 300-km altitude, the solid perigee stage puts the satellite and apogee motor in the transfer orbit with about 300-km perigee and about 36,000-km apogee. The final geostationary orbit is achieved by firing the liquid apogee motor.

V. Technical Concept of the Satellite

a) Configuration and Structure

The satellite will be 3-axis stabilized, using a flywheel in combination with a cold gas system for attitude control.

Figure 2 shows the configuration of the satellite. It has a hexagonal main body of 1.7-m maximum diameter and 50-cm height. The central tube in this body fits the separation plane on one side and is connected on its other side (top side of the satellite) with the apogee motor. The electronic subsystems, as well as the tanks for the hot and cold gas systems, are contained in the three compartments of the main body. On the top side of the body (in its launching position) are the SHF-antenna-feeders and reflectors. On the lower side are the VHF antennas and most of the hot and cold gas nozzles. At three of the six corners of the main body are the solar panels, each of them consisting of 4 subpanels. Table 3 shows the mass breakdown of the satellite, including apogee motor.

Table 3 Mass breakdown of the satellite and apogee motor

Structure (incl. thermal control)	71.0 kg
Attitude control and orbit correction system (incl. fuel)	58.6 kg
Transponder	16.6 kg
Telemetry/Telecommand	9.2 kg
Solar Generator	17.5 kg
Power Supply	15.3 kg
Harness and pyrotechnics	15.2 kg
Antennas	5.9 kg
Apogee motor (structure, 16.0 kg; fuel 148,5 kg; He 0.6 kg)	165.1 kg
	374.4 kg
Margin	11.6 kg
Total weight	386.0 kg

b) Thermal Control

The thermal control of the satellite will be passive, using superisolation at the top and the bottom of the satellite and second surface mirrors at the side walls.

To protect the satellite, and particularly the SHF-antenna reflectors, from the heat irridiated by the apogee motor during its operation, a heat-shield surrounds the motor nozzle. A major problem is the thermal control of the solar panels. Because of self-shadowing, the outer subpanel can become extremely cold. Either by active heating or increased thermal capacity the temperature will be kept to technically acceptable limits.

c) Apogee Motor

A liquid bipropellant apogee motor is used for the final injection into the geostationary orbit. Aerozin 50 (AZ 50) and N_2O_4 are used as propellants for the 40 kp thruster. The specific impulse is 298 sec. The thruster already is space-qualified as control thruster for the third stage of the EUROPA II launcher.

The amount of fuel put into the tank before launching corresponds to the nominal velocity increment (1 sigma-case) of 1485 m/sec. In case of a major injection error (3 sigma-case), the missing velocity increment for injection into the geostationary orbit is provided by the hot gas system of the satellite, thus reducing the possibility of inclination corrections over the entire lifetime of 5 years. The cylindrical fuel tank is separated by a bellows. The fuel is pressurized by helium.

d) Attitude Control and Orbit Correction System

This subsystem of the Symphonie satellite provides orbit and attitude control in the transfer phase and the phase in the geostationary orbit. The most important elements of this system are the hot gas system, the cold gas system, the flywheel, and the various sensors.

The function of the subsystem is described in detail, following the different phases of the mission.

1) Transfer Phase

After the separation from the perigee stage, the satellite is in a transfer orbit of about 300-km perigee and 36,000-km apogee. The satellite rotates at 120 rpm. Using Earth and sun as a reference, the actual spin axis is determined and, if necessary, corrected for proper ignition of the apogee motor. For the corrections of the spin axis, one nozzle of the hot gas system is used. After burn-out of the apogee motor, the hot gas system provides the final orbit injection in the 3 sigma case.

2) Acquisition Phase

After the orbit corrections, following the injection, the satellite and apogee motor are despun by a yoyo-system. The remaining spin is reduced to zero, using the flywheel and the cold gas system. Reference for this procedure is the Earth.

3) Geostationary Phase

The attitude control of the satellite is a 3-axis control and is semipassive. The axis of the central tube of the satellite is perpendicular to the orbit plane and maintained by the flywheel. By means of a closed-loop, onboard operation, the pointing of the SHF-antennas is controlled. The control of the two remaining axes, which exchange each 6 hours, is performed by ground control. The proper signals are derived from onboard IR-sensors.

The attitude measurement is performed by the following sensors: 1 pencil-type IR-sensor, 2 static infrared sensors, and 4 solar sensors.

The pencil-type infrared sensor detects Earth's horizon by means of a CO_2-band during the spinning phase. The view angle of the sensor is $1.3^\circ \times 1.3^\circ$.

The static infrared sensors measure the attitude in the geostationary orbit. The precision of this sensor is $\pm 0.15^\circ$. The solar sensors are digital sensors.

The orbit corrections North-South and East-West are performed 1 to 2 times/yr by ground control by means of the hot gas system.

The hot gas system is a bipropellant system using AZ 50 and N_2O_4. Its velocity increment is as follows:

Spin axis correction during transfer phase and initial corrections in geostationary orbit:	173.7 m/sec
Orbit corrections, East-West and North-South for 5 years:	206.8 m/sec
Remaining fuel:	11.5 m/sec
Total:	392.0 m/sec

The cold gas system consists of 8 motors, each of 0.6 N thrust. The total impulse is as follows:

Attitude control, discharge of flywheel, spin corrections, reorientation:	900 N sec
Remaining fuel:	500 N sec
Total:	1.400 N sec

The 5.7-kg flywheel has the following characteristics: angular momentum 20 $\pm$ 2 N·m/sec., rotation 3,000 rpm, diameter 350 mm, height 110 mm, power consumption 9 Watts, total weight 9.0 kg (incl. electronics).

e) Transponder and SHF-Antennas

The frequency band, allocated to telecommunication satellites, is used for Symphonie as shown in Fig.3 and 4. For the realization of the transponder for telecommunication satellites, two concepts are possible: 1) RF-transponders with single frequency conversion and 2) IF-transponders with double frequency conversion.

The IF-transponder concept was chosen for Symphonie. The total amplification of about 120 db is achieved at the receiving frequency of 6 GHz, the intermediate frequencies in the range of 250-750 MHz and the transmitting frequency in the range of 4 GHz. The reason for this choice was that a great part of the amplification is achieved in the IF-range by transistor amplifiers. Because of the amplification in different frequency ranges, there is no serious gain limitation.

Only one TWT is needed for each subband of Table 2. Furthermore, the following telecommunication requirements are fulfilled more easily: 1) equalization of gain slope, thus minimizing crosstalk, 2) channel filtering, 3) equalization of group delay distortion, 4) extraction of telecommand signal, and 5) insertion of telemetry data.

In all Symphonie transponders the same local oscillator frequencies are used, thus yielding different IF bands. The down and up converter frequencies are 6675 MHz and 4450 MHz, respectively.

The signal path can be seen in the block diagram, Fig. 5. The input signal of the 17.2° receiving antenna (gain 16 db) is separated by the input diplexer into two transponder bands. A low noise (700°K) amplification of 26 db is accomplished by the two-stage tunnel diode amplifier (TDA). Following the TDA, a down converter with a conversion loss of 6 db shifts the input frequency range to the IF frequency range.

The bandwidth of these two stages is 500 MHz with the proper receive-out-of-band-response. After the IF pass-band filter, the IF signal is amplified by the 6-8 stage IF amplifier (36-db gain). Because of the degradation of the TWT, the amplification of the IF amplifier can be switched to 38, 40, and 48 db.

By a filtering network, the TC signal is separated, the gain slope corrected, and signals out of band suppressed. The following parametric up-converter (about 3-db gain) shifts the IF band to the transmitting band, which is finally amplified by the 13-w traveling wave tube. The SHF output of each transponder can be switched to each transmitting antenna.

f) Telemetry - Telecommand

Telemetry and telecommand can be transmitted either by the VHF or SHF link. However, during the injection procedure, only the VHF link is used.

The telecommand system is designed to the requirements of NASA - Tone Digital Command System.

Since the number of commands to be transmitted is greater than the possibilities given by this code, 2 different addresses are used. The VHF-Telecommand Frequency is 148.25 MHz.

Four different subcarriers between 7 and 12 KHz are used, 2 for each of the 2 satellites. The required false order probability is 10^{-5}. The telemetry system can be operated in two different modes, a low-power mode and a high power mode, with two different corresponding telemetry formats.

Because of the high amount of telemetry dates to be transmitted during the transfer phase, the high-power mode is used primarily during the injection procedure. Telemetry and Telecommand are handled by stations in Kourou, Brétigny, and Weilheim.

g) Power Supply

The electrical power for the various satellite subsystems is provided by the three solar panels. Each panel consists of 4 subpanels. A total of about 22,000 n/p 2 x 2 cm silicon solar cells provides a maximum power of about 300 w at the beginning of life.

Table 4 Power budget (Mean consumption)

Subsystem	Power consumption in transfer-phase (watts)	Power consumption in geostationary orbit (watts)	Power consumption during eclipse (watts)
Transponder	-	123.0	-
Attitude control system	7.8	18.5	14.5
Telecommand	1.91	1.91	1.91
Telemetry	8.7[a]/30.6[b]	8.7[a]	8.7[a]
Power conditioning	10.4	9.4	7.5
Total	28.7[a]/50.71[b]	161.51	32.61

[a]Low-power telemetry (small bitrate).
[b]High-power telemetry (high bitrate).

Because of the degradation by protons and electrons, the power at the end of life is 180 w.

The power consumption of the satellite subsystem is shown in Table 4. The block diagram of the power subsystem is shown in Fig. 6.

The regulated main bus of 27 v ± 1% is connected to the various converters.

These converters are synchronized by the central clock and operate at about 8 and 32 KHz in order to reduce the onboard RFE problems.

Two NiCd batteries, each 3.5 Ah and 14 cells in series, provide the peak power in the transfer phase and during eclipse. The battery is charged by a step-down regulator and discharged by a step-up converter.

VI. Integration

The first and second flight models will be integrated in France and the Federal Republic of Germany, respectively.

VII. Conclusion

The project is a challenging technical and management task for the European industry. In this respect it also should be mentioned that 4% of the development costs for the satellites will be given to Belgian firms.

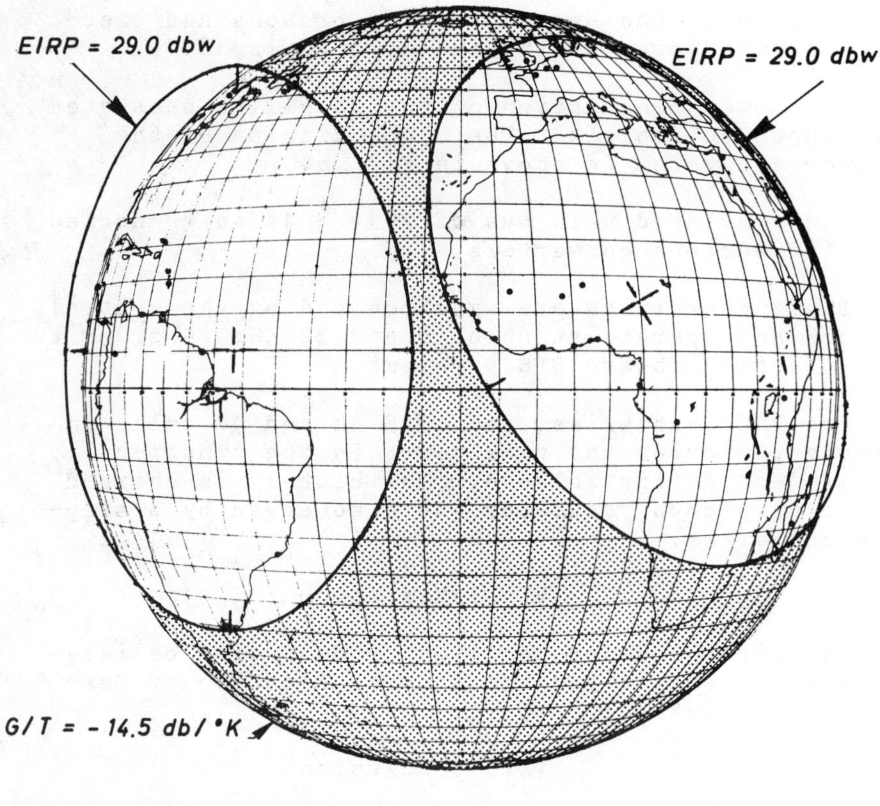

EIRP = 29.0 dbw

EIRP = 29.0 dbw

G/T = - 14.5 db/ °K

Fig. 1 The SHF-coverage zones.

Fig. 2 The Symphonie Satellite.

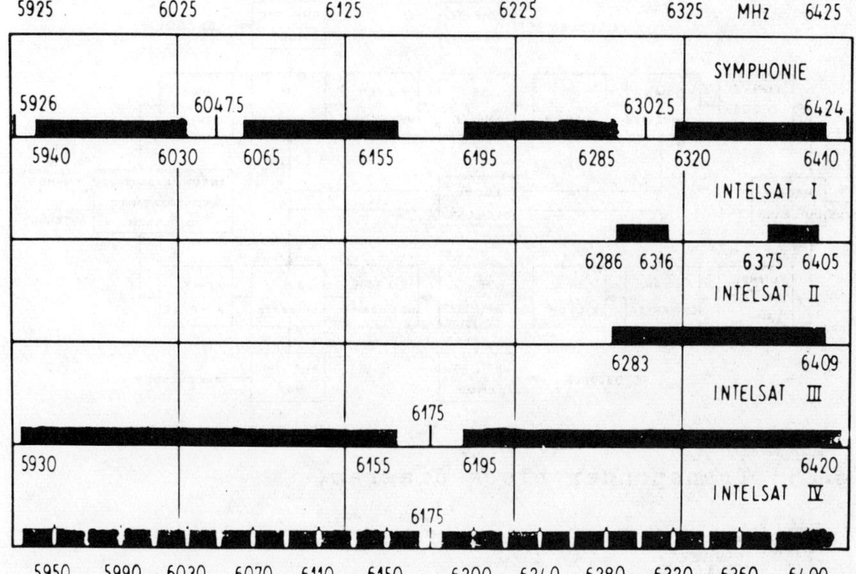

Fig. 3 Receiving bands of Symphonie.

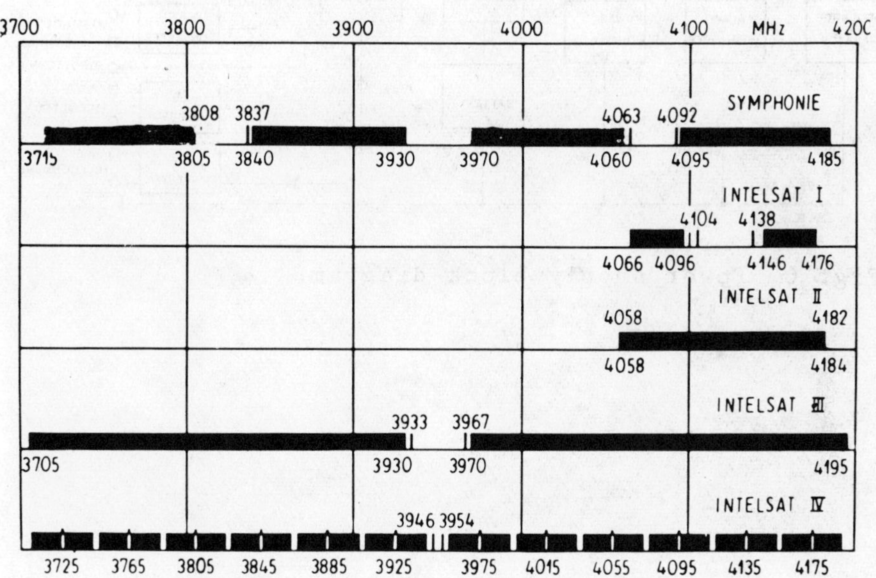

Fig. 4 Transmitting bands of Symphonie.

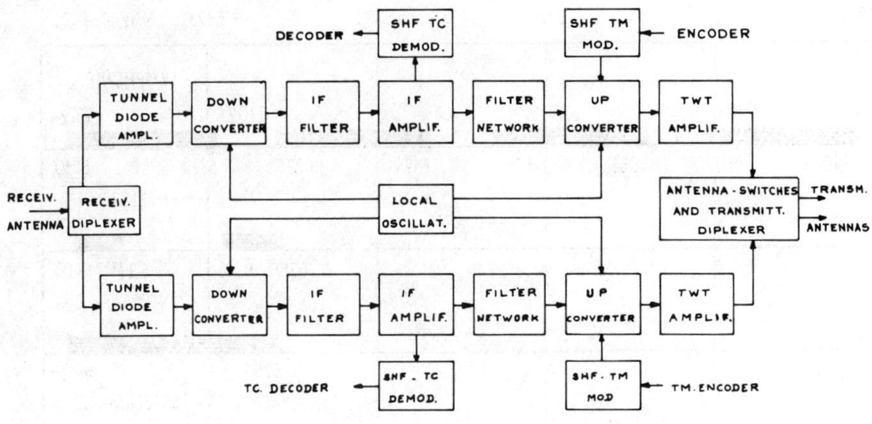

Fig. 5 Transponder block diagram.

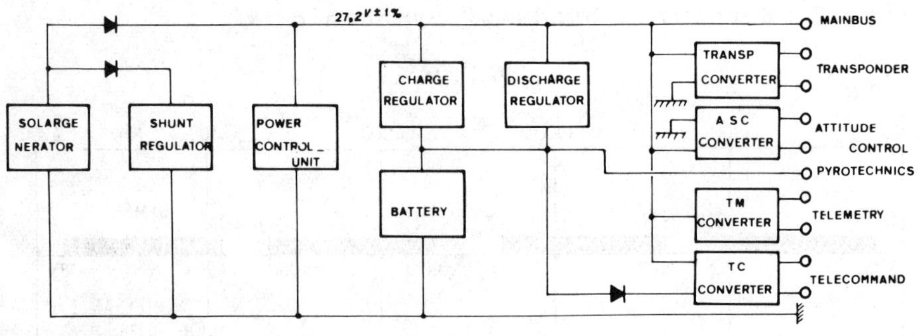

Fig. 6 Power supply block diagram.

AIR-AIR AND AIR-GROUND EXPERIMENTS FOR A NAVIGATION SATELLITE PROGRAM

Bertrand Manuali*

Centre National d'Etudes Spatiales, Bretigny-sur-Orge, France

Abstract

The necessity for aircraft experiments to provide a basis for a navigation satellite program has been emphasized for many years. The lack of a satellite with a UHF transponder in a 24-hour orbit forces one to substitute a simulation program. This paper is concerned with such a simulation program involving aircraft, a balloon, and an Earth station. In the first phase of the experiment (July 1970), the satellite is simulated by an aircraft flying at high altitude, and this aircraft transmits signals to another aircraft equipped with a high gain antenna (10 db) and a digital transceiver (18 kb/sec). In a second phase (September-October 1970), the satellite is simulated by a stratospheric balloon at an altitude of 40 km and equipped with two transponders (400 and 1600 MHz) and a circularly polarized antenna. The Earth station is in the 400 - 444 MHz band and the balloon-aircraft link is in the 1550 - 1650 MHz band. In this second phase, the aircraft is equipped with the same high gain antenna, but the 90 kb/sec transceiver uses multiplexing techniques.

1. Introduction

As a result of studies carried out in France since 1966 on the use of satellites for civil aviation purposes, the French space agency (CNES) and the French civil aviation agency (SGAC) have undertaken an experimental program on the basis of the technical choices made in the context of the Dioscures Project.

Presented as Paper 70-407 at the AIAA 3rd Communications Satellite Systems Conference, Los Angeles, Calif., April 6-8 1970.

*Chief, Navigation Program.

A technical development and test program was approved in January 1968. The purpose of this program was to simulate, in April 1970, an Earth-station/satellite/aircraft link using a stratospheric balloon at an altitude of 40 km to represent the satellite.

In December 1968, the United States (NASA and FAA) informed CNES of the possibility of using the ATS-E satellite equipped with a UHF L-band transponder. France then endeavored to adapt the aeronautical equipment of the Dioscures Project to the characteristics of this transponder - i.e., to an Equivalent Isotropic Radiating Power (EIRP) of approximately 27 dbw. A simplified model of the aircraft transmitter-receiver was built, and production of the aircraft antenna was accelerated. The equipment was to be ready in January 1970, but unfortunately the satellite went into a flat spin immediately after injection into geostationary orbit and could not be used for these experiments.

In November 1969, an attempt was made to establish a second experimental program on the basis of the first stratospheric balloon program. This led to the idea of two experiments, as follows: 1) an experiment involving two aircraft, A and B, in which aircraft B simulates the satellite; and 2) an experiment involving an Earth-station/balloon/aircraft link in which the satellite is simulated by a balloon.

2. Objectives of the Experiments - Planning

The purpose of the experiments is to prove the validity of the technical and technological choices made in connection with the study of the Dioscures system, particularly the choice of a 10-db net gain electronic scanning antenna for the aircraft, the choice of delta modulation, and the choice of the digital multiplex "voice + data + position-determination code."

The purpose of these experiments was to measure the following: 1) intelligibility of the voice channel in an operational environment, 2) data transmission error rates (1200 b/sec), 3) ranging accuracy, 4) multipath effect, and 5) effect of storms.

Even though an attempt will be made to achieve these five objectives in the course of both experiments, the aircraft/aircraft experiment will achieve only limited results with regard to objectives 3 and 4, while objective 5 will assume only secondary importance in the balloon/aircraft experiment.

Figure 1 shows the areas within which the experiments will be carried out on the following dates: 1) aircraft/aircraft experiment between June 1 and July 15, 1970; and 2) Earth-station-balloon/aircraft experiment between September 1 and October 15, 1970. The analysis of the results should be completed in December 1970.

3. Aircraft/Aircraft Experiment

This experiment will use the two simplified transmitter-receivers developed for the ATS-E experiment. This equipment is capable of handling a single channel with a bit rate of 18 kb/sec.

The channel is used in the following manner: 1) in the "communication" mode it is wholly occupied by a coded voice channel-delta controlled slope modulation; and 2) in the "position-determination" mode, the channel is divided into four subchannels, distributed as follows: a) subchannel 1-frame code, b) subchannel 2 - 150-b/sec data transmission, c) subchannel 3 - ranging code, and d) subchannel 4 - 150-b/sec data transmission. In the position-determination mode, the transmitter-receiver functions as a transponder, the telemetry code being retransmitted immediately.

The two aircraft will be equipped as follows: Aircraft A - simplified transmitter-receiver with a bit rate of 18 kb/sec and a 10-db electronic scanning antenna; Aircraft B - transmitter receiver modified to function as an Earth station transmitting the ranging code at regular intervals and measuring the intervals between reception and transmission of the code. The aircraft will have a low-gain antenna of about 3 to 5 db. Diagrams of the equipment are shown in Figs. 2 and 3.

3.1. Measurements Relating to Objectives 1, 2, and 3

The intelligibility of the voice channel, the data transmission error rates, and the ranging accuracy will be measured on the basis of the various C/N ratios (carrier noise ratio). The nominal C/N ratio is 44.5 dbHz; it will vary from 40.5 dbHz to 48.5 dbHz in steps of 2 db. A vocabulary of 250 words will be used to measure intelligibility. A pseudo-random code will be used for the data-transmission error-rate measurements.

The ranging measurements will be carried out in aircraft B, which will transmit the telemetry code and measure the interval between transmission of the signal and its return. Each measurement will be recorded.

Since the distance between the two aircraft is practically constant, the measurement will be affected both by rapid fluctuations stemming from thermal noise and the multipath phenomenon and by the slower fluctuations resulting from the movements of the two aircraft and from the drift of the equipment. Since the range measurement rate is 12.5 measurements/sec, these slower fluctuations will be of the order of 2 m between two measurements, which is negligible.

3.2. Objective 4: Multipath Effect

The amplitude and spectrum of the reflected signal will be evaluated by using a pure nonmodulated carrier and by decreasing as much as possible the discrimination resulting from the antenna. The received signal level will be 20-30 db higher than the nominal value so as to eliminate the influence of thermal noise.

Evaluation of fading amplitude. This will be done by detecting the envelope of the received signal, which will be recorded and then processed to measure the values exceeded during 5% and 1% of the total time.

Evaluation of the spectrum of the reflected signal. The reflected signal comprises a coherent part with the same frequency as the carrier and a noncoherent part whose spectrum spreads over a few hundred hertz. We shall use the sine and cosine outputs of the transmitter-receiver phase demodulator, the bandwidth of the loop having been reduced to 200 Hz.

Figure 4 shows the geometric configuration of the experiment. In the first phase the following values will be adopted: $\alpha = 0^\circ$ and $B = 6^\circ$, 8°, 12°, 16°, and 20°. The state of the sea and the height of the waves are known, the latter being measured by means of a laser.

3.3. Objective 5: Atmospheric Noise

Intelligibility and the error rate in data transmission will be measured when the aircraft carrying the receiver flies within the storm zone or close enough to it for the active zone of the perturbation to be seen by the antenna beam. These measurements will be made under link-computation conditions corresponding to the nominal value and to the nominal value +2 db -- i.e., 44.5 and 46.5 dbHz for comparison with the results obtained in calm weather. Measurements also will be made on the noise temperature of the system under the same conditions.

4. Earth-Station/Balloon/Aircraft Experiment

This experiment involves: 1) an Earth station called SITTEL, operating on 400 MHz, which was developed for the tracking of the stratospheric balloons launched by CNES (Fig. 5); 2) a 50,000-m^3 stratospheric balloon equipped with a repeater at 400 MHz and one at 1600 MHz; and 3) an aircraft equipped with the antenna and the complete transmitter-receiver developed for the Dioscures Project, which receives messages at a bit rate of 90 kb/sec and transmits at 36 kb/sec. The radio-circuit diagram is shown in Fig. 6. In addition, both the aircraft and the balloon will be tracked independently by means of radar, and, to this end, they will be fitted with a C-band transponder.

4.1. Characteristics of the Radio Signals

The aircraft on-board equipment will receive from the ground, via the balloon, a PCM/PM signal in the 1550-MHz frequency band. Time-multiplexed and transmitted at a rate of 90 kb/sec (NRZ-C code), this signal will comprise: 1) a frame synchronization code; 2) voice channel No. 1; 3) voice channel No. 2; 4) voice channel No. 3; and 5) a subcommutated channel comprising a) the 144-bit position-determination code (recurrence 25 Hz), b) a 300-b/sec data transmission circuit, c) a 150-b/sec telex circuit, d) a 70-b/sec data transmission circuit, and e) "circuit engaged" signals.

In regard to the aircraft-to-balloon link, the time multiplexed signal transmitted at the rate of 36 kb/sec (NRZ-C code) on a carrier frequency of 1650 MHz comprises either the position-determination multiplex consisting of a) the frame synchronization code, b) 300-b/sec data transmission (circuit 1), c) the 360-bit position-determination code, and d) 300-b/sec data transmission (circuit 2) or the communication multiplex consisting of a) the frame synchronization code, b) a voice channel, and c) a 150-b/sec telex channel.

4.2. Experiments Performed

The objectives of the experiments in the Earth-station/balloon/aircraft experiment were similar to aircraft/aircraft experiment objectives but with the additional advantage of making it possible to test the digital multiplexing of all data; test the call procedures and channel occupancy signals; measure the absolute distances, since the balloon and the aircraft are tracked independently by radar; and achieve perfect simulation of the multipath effect when a satellite is used. The measurements to be made will bear on voice channel intelligibility, data transmission error rates, ranging accuracy, and multipath effect.

The experiments will be performed over a period of $1\frac{1}{2}$ months (September and October 1970), with fifteen balloons being launched, each of which can perform at least two hours of experiment. This experimental program may be continued in April 1971 with the Concorde Aircraft.

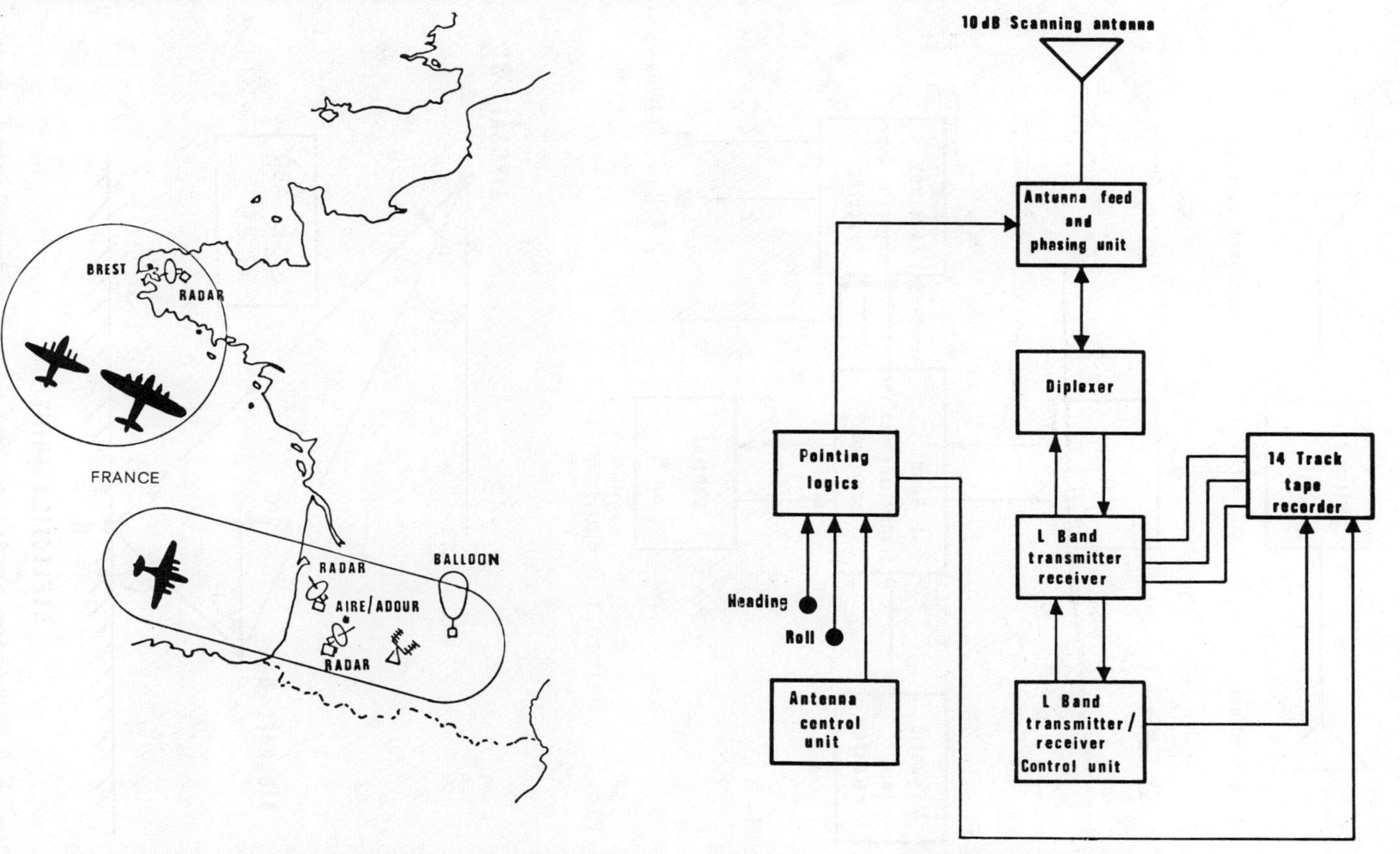

Fig. 1 Map of the area for the experiments.

Fig. 2 Aircraft A equipment.

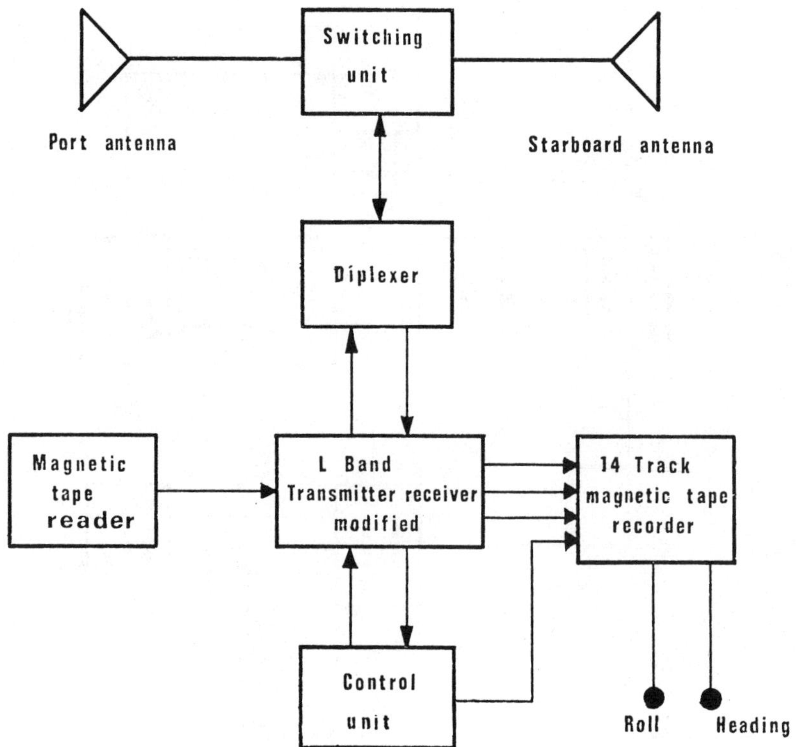

Fig. 3 Aircraft B equipment.

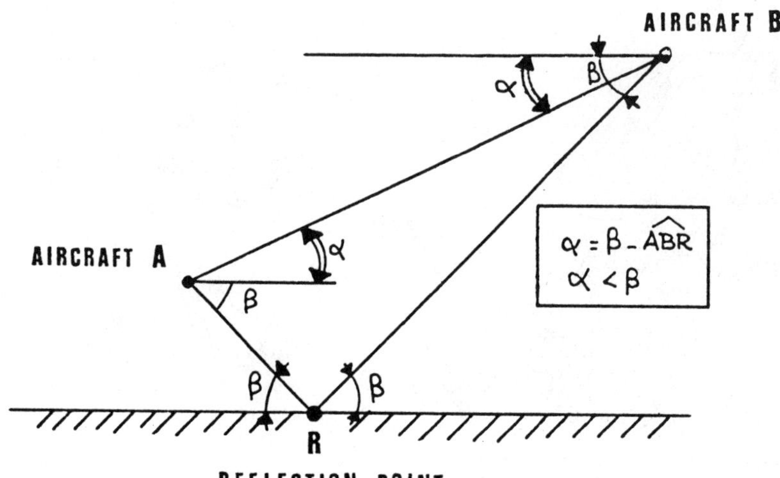

Fig. 4 Elevation angles of the direct and reflected beams.

Fig. 5 The 400-MHz SITTEL Earth station.

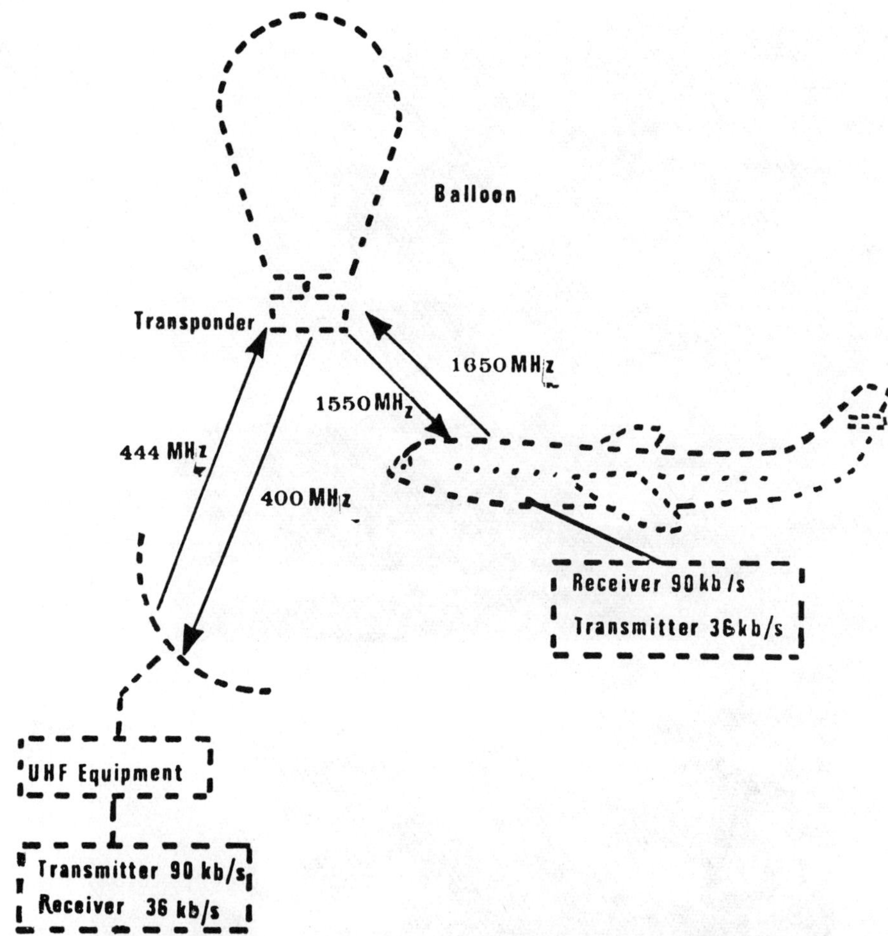

Fig. 6 Equipment for the balloon experiment.

THE ITALIAN SATELLITE SIRIO - SHF PROPAGATION AND COMMUNICATION EXPERIMENTS: SCIENTIFIC OBJECTIVE AND ORGANIZATION OF THE EXPERIMENTS

F. Carassa,* G. Drufuca,† and A. Paraboni ‡

Laboratorio di Comunicazioni Elettriche, Istituto di Elettrotecnica ed Elettronica del Politecnico di Milano, Milano, Italy

Abstract

The interest in future use of frequencies above 10 GHz for space and Earth communications has been continuously increasing in the last years. On the other hand at these frequencies atmospheric attenuation (and noise) become important and experiments are needed which allow us to collect data on these parameters. Of particular usefulness are the measurements made by using a geostationary satellite. To this purpose an Italian experimental satellite, named SIRIO, will be launched in the middle of 1972 and three ground stations will be installed in Italy.

SHF (Super High Frequency) propagation measurements will be performed by using a frequency of about 18 GHz from Earth to satellite and 12 GHz from satellite to Earth. Fading statistics, frequency dependence and possibility of path diversity will be investigated. In addition to propagation measurements, the satellite will allow communication experiments. Transmission of television or telephony with various types of modulation and multiple-access telephone transmission are foreseen. A broad international participation in the experiment is expected, thus increasing the amount of data which may be collected during storms in different regions.

Presented as Paper 70-501 at the AIAA 3rd Communications Satellite Systems Conference, Los Angeles, Calif., April 6-8, 1970.

*Professor of Electrical Communication.

†Research Scientist.

‡Research Scientist.

Introduction

SIRIO is an Italian experimental satellite which will be put in synchronous orbit (15°W longitude)at the beginning of 1972, in the framework of the space program of the CNR (Consiglio Nazionale delle Ricerche, i.e., National Research Council).

In Fig. 1 some of the main parts of the satellite are shown; the diameter and length of the main cylinder covered by solar cells are 1432 mm and 870 mm, respectively, and the mass (without the apogee motor) is 195 kg. The satellite, built by a group of Italian industries, is based on the previous development of ELDO-PAS satellite. The launch will be performed by NASA, by virtue of an agreement between CNR and NASA.

The main purpose of the SIRIO project is to investigate the communication applicability of frequencies above 10 GHz (SHF experiment). However, other scientific experiments will be carried out by SIRIO, on confined plasma and cosmic rays.

The SHF experiment will be performed by using a frequency of about 18 GHz from Earth to satellite and a frequency of about 12 GHz from satellite to Earth. Propagation and communication experiments are included in the program. The Communication Laboratory of the Politecnico di Milano, Istituto di Elettrotecnica ed Elettronica, is responsible for the general system outline and for the scientific work connected with the program [i.e., a) propagation measurements, data reduction, and related research; b) communication experiments].

Telespazio is responsible for the engineering and development coordination of the onboard and ground equipment and for the installation and operation of the Italian ground stations.

The Consiglio Nazionale delle Ricerche will control the program through a technical-administrative committee, which includes members of the Istituto Superiore delle Poste e Telecomunicazioni, and an executive secretariat, which includes the project manager of the entire SIRIO project.

Propagation Experiment

The propagation experiment aims to gather data on propagation to and from space at frequencies above 10 GHz, and in particular at 18 and 12 GHz. It is well known that at these frequencies atmospheric attenuation (and noise) become important, contrary to what happens at presently used frequencies; in particular, rain absorption and scattering become more and more important as the frequency increases. Yet congestion of presently used frequencies calls for consideration of these higher frequencies for space(and Earth) communications. Such frequencies are of particular interest for regional systems because of the quantity of information to be transmitted and the ease of obtaining highly directional onboard antennas. The attenuation produced by rain of known characteristics (e.g., diameter of drops and number of drops per unit volume) can be calculated accurately; very little is known, however, about actual distribution in space and time of such characteristics. Evidence exists that heavy storms frequently extend over only a few kilometers or much less. This means that detailed information on rain distribution would be required to calculate the over-all path attenuation. Statistics on rain distributions at ground level in terms of rain rate may be obtained by dense arrays of rain gages; statistics on rain distribution in space, in terms of rain reflectivity, may be obtained by meteorological radars. Exhaustive experiments may be carried out through actual propagation measurements possibly associated with radar observations and/or rain measurements on Earth. Although the sun can be used as the space source at the frequencies of interest, a coherent source on a satellite is better suited for propagation experiments.

The SIRIO SHF experiment aims to complement similar experiments programed by NASA (with ATS-E and ATS-F satellites), including frequencies to be tested and regions to be covered. By transmitting at 18 GHz from Earth to satellite, and at 12 GHz from satellite to Earth, the experiment will permit the following measurements and recordings:

a) At 12 GHz, all stations: a1) path attenuation (up to 10 samples/sec may be taken); a2) differential attenuation and phase on a frequency interval of 530 MHz[§]; and a3) improvement resulting from space diversity, using two receiving stations[≠].

b) At 18 GHz, only transmitting stations: b1)path attenuation; b2) differential path attenuation on a frequency interval of 770 MHz; and b3) improvement resulting from space diversity, using two transmitting stations.

c) At 12 GHz: antenna noise measurements.

We will review briefly the principles that have been adopted to perform the foregoing measurements. The basic idea has been to use a simple onboard repeater, i.e., all measurements performed by Earth stations. Figure 2 shows the spectra of the signals through the satellite. The spectra used for the propagation experiment, are labeled PROP; the spectra used for the communication experiments are labeled COM. In illustrating the various spectra, the frequency scale has been distorted severely to better show their relevant characteristics.

In the propagation experiment, two frequencies are transmitted from Earth, each ∿ 770 MHz apart from the other; these two frequency lines are obtained through amplitude modulation at ∿ 386 MHz of a carrier at ∿ 17.4 GHz (see RIC in Fig. 2). The satellite converts the above signal in frequency, using a local oscillator (LO) whose frequency differs by ∿ 10 kHz from the virtual carrier, so that an intermediate frequency signal (IF-PROP) is obtained, which is composed of two lines, centered at ∿ 386 MHz, having a frequency separation of 20 kHz and corresponding to the two received oscillations. At this point, a third oscillation,generated onboard and amplitude-calibrated, is added, to be used at Earth as reference for the amplitude measurement of the other two oscillations. After filtering and conversions, the above signal is

---

§This measure gives indications on the possibility of wideband transmission.

≠Space diversity will take advantage from the limited extension of rainstorms.

centered at ∿ 265 MHz (3rd IF-PROP) and thereafter modulates in amplitude the output carrier of ∿ 11.6 GHz (TRANSM). In addition to the sidebands, the carrier actually is transmitted at a calibrated value, so that, measuring its amplitude at Earth, the path attenuation from satellite to Earth is obtained. (This is the measurement listed above as a1). In addition, by comparing the two sidebands, the differential attenuation and the differential phase may be obtained (a2). Of course, all the above measurements may be performed by any receiving station on Earth; using two or more receiving stations at one site, space diversity efficiency can be tested (a3). All receiving stations also can perform antenna noise measurements (c) by measuring the noise in a band outside the received spectrum.

If, at any receiving station, the amplitude of each lateral line in each of the received sidebands is measured and compared with the center reference line of the same sideband, the path attenuation from Earth to satellite (b1), and the differential attenuation on the same path (b2) may be obtained. If the two lines from Earth to satellite, instead of originating at the same Earth transmitter, are produced by two different transmitters at the same site, diversity efficiency at 18 GHz may be measured (b3). If the two transmitters are in different regions, path attenuation at 18 GHz at two different regions may be measured simultaneously. The above possibility may be extended to more than two regions by using more than two lines. In the transmitted spectrum (TRANSM) from the satellite, a carrier also is indicated (TM) for telemetering transmission.

In Fig. 2, the spectra corresponding to the communication experiments also are reported. These spectra are located differently in frequency with respect to the propagation ones so as to be differently filtered and amplified. In particular, at the transmission from the satellite, one of the sidebands of the communication signal is eliminated by the transmitting rejection filter, whose attenuation characteristic is indicated as REJ FILT in TRANSM. It is easily seen that this filter has no effect on the spectrum used for the propagation experiments.

A simplified diagram of the onboard equipment is given in Fig.3. At the receiving side, the input converter is shown, followed by the summing device for the reference line and by the broad-band IF amplifier, the bandwidth of which (110 MHz) is capable of amplifying both the propagation signal and the communication signal. These signals thereafter are separated by filtering and are converted to a second IF, which is at 24 MHz, with a bandwidth of 40 kHz, for the propagation signal, and at 70 MHz, with a bandwidth of 2.5 or 35 MHz, for the communication signal. These bandwidths determine the noise performance of the equipment, which permits operation with a very small input signal (high path attenuation) in a propagation mode, and with a comparatively larger signal in the two communication modes.

After the second IF, the propagation signal undergoes a further conversion to a third IF at ∿ 265 MHz. The propagation signal and the communication signal finally are added together and joined to the telemetering signal (TM) to feed the balanced modulator that produces the spectrum TRANSM of Fig. 2, apart from the carrier, which is added after the rejection filter. A redundant traveling wave tube amplifier, having a saturation prower output of 10 w, provides the final amplification. The power output at the carrier is controlled by a PIN diode attenuator at the input of the traveling wave amplifier.

The despun antenna has the Earth coverages reported in Fig. 4 for 1, 2 and 3-db relative attenuation. As can be seen, two different pointings of the antenna are foreseen: one particularly suited for European coverage and the other permitting marginal coverage of North America.

A simplified diagram of the main Earth station to be used in Italy is shown in Fig. 5. The antenna is a Cassegrain-fed paraboloid having a diameter of 8.5 m and a gain of 60.5 db at 18 GHz. A cooled parametric amplifier having a bandwidth of 500 MHz is used at the receiver front end; consequently, the expected over-all G/T of the Earth station is 37.5 db. After the parametric amplifier, the signal is converted to a first intermediate frequency, centered at 750 MHz, at which an amplifier, having a bandwidth of 500 MHz, is used. The signal is then distributed to the various receivers (having 70-MHz center frequency), as indicated. On the transmission side, a modulator is fed by the modulating sinusoid at 386 MHz in the propagation mode and by the signals originating from the modulators in the communication mode.

The transmitting power amplifier has a saturated power output of 500 w. A controlled attenuator, driven by the measured attenuation on the up link, will be provided at the input of this amplifier to reduce the signal level variations at the satellite.

As shown in Fig. 5, a centralized frequency generator gives all needed frequencies and is locked to the carrier received from the satellite, to assure the required frequency precision at the satellite itself.

Detailed information on the equipment and on the obtainable performances in the various experiments are given in a compaion paper[1]. Path attenuation may be measured in the propagation experiments to approximately 35 db. For diversity efficiency measurements , a transportable Earth station is foreseen having a G/T that is 13 db lower with respect to the fixed stations described previously.

Communication Experiment

The communication experiment consists of the transmission of television or telephony with various types of modulation. Its possible finalization as an experiment of an interuniversitary link has been considered in choosing the communication parameters.

In clear-weather conditions, a standard 625-line television signal may be transmitted by frequency modulation, together with a subcarrier at 6 MHz for the audio signal. If the over-all frequency deviation is 17 MHz, a weighted signal-to-noise ratio of 45 db is obtained with a margin of 5 db against the threshold. This margin is expected to increase with the use of improved demodulation methods now under study. Also, experiments with PCM transmission of television signals possibly will be undertaken.

A second communication experiment to be performed consists of multiple-access telephone transmission; the multiple access is obtained by frequency division with PCM modulation of the carriers. The total number of carriers will be 12, each one capable of transmitting 70 kb/s, equivalent to one telephone channel or to high-speed data. In this experiment, using onboard the IF bandwidth of 2.5 MHz, the margin is 25 db on the up link, and 18 db on the down link.

Earth Stations to Be Installed in Italy

Three Earth stations will be built in Italy, two fixed and one transportable. One fixed station will be installed at the Politecnico di Milano, together with a meteorological radar. Propagation data and radar data will be recorded simultaneously to determine how accurately the rain attenuation can be evaluated from radar observation. The second fixed station will be installed at Fucino near the Telespazio Earth station for commercial satellite communication. This station will have at its disposal rain data at ground, given by a field of fast rain gages installed at Fucino mainly for other purposes (studies of propagation at frequencies above 10 GHz for Earth radio relay application). The transportable Earth station will be located normally at Fucino for investigation on space-diversity efficiency.

International Participation

A broad international participation in the experiment is expected, as discussed in some recent meetings in Milano. These meetings were attended by research managers from Belgium, Canada, Denmark, Finland, France, Germany, Netherlands, United Kingdom, U.S.A. A common method of data recordings will be discussed among the participants at the meetings to simplify the over-all evaluation of the results of the propagation experiment.

Conclusion

The propagation experiment through SIRIO satellite is likely to be carried out with many Earth terminals, thus increasing the amount of data which may be collected during storms in different regions. The communication experiment will permit testing of different modulation schemes for television and telephone transmission.

References

[1]Fanti, P. and Tirrò, S., "The Italian SIRIO Experiments: Satellite and Ground Equipment," Paper 70-507 (published elsewhere in this volume).

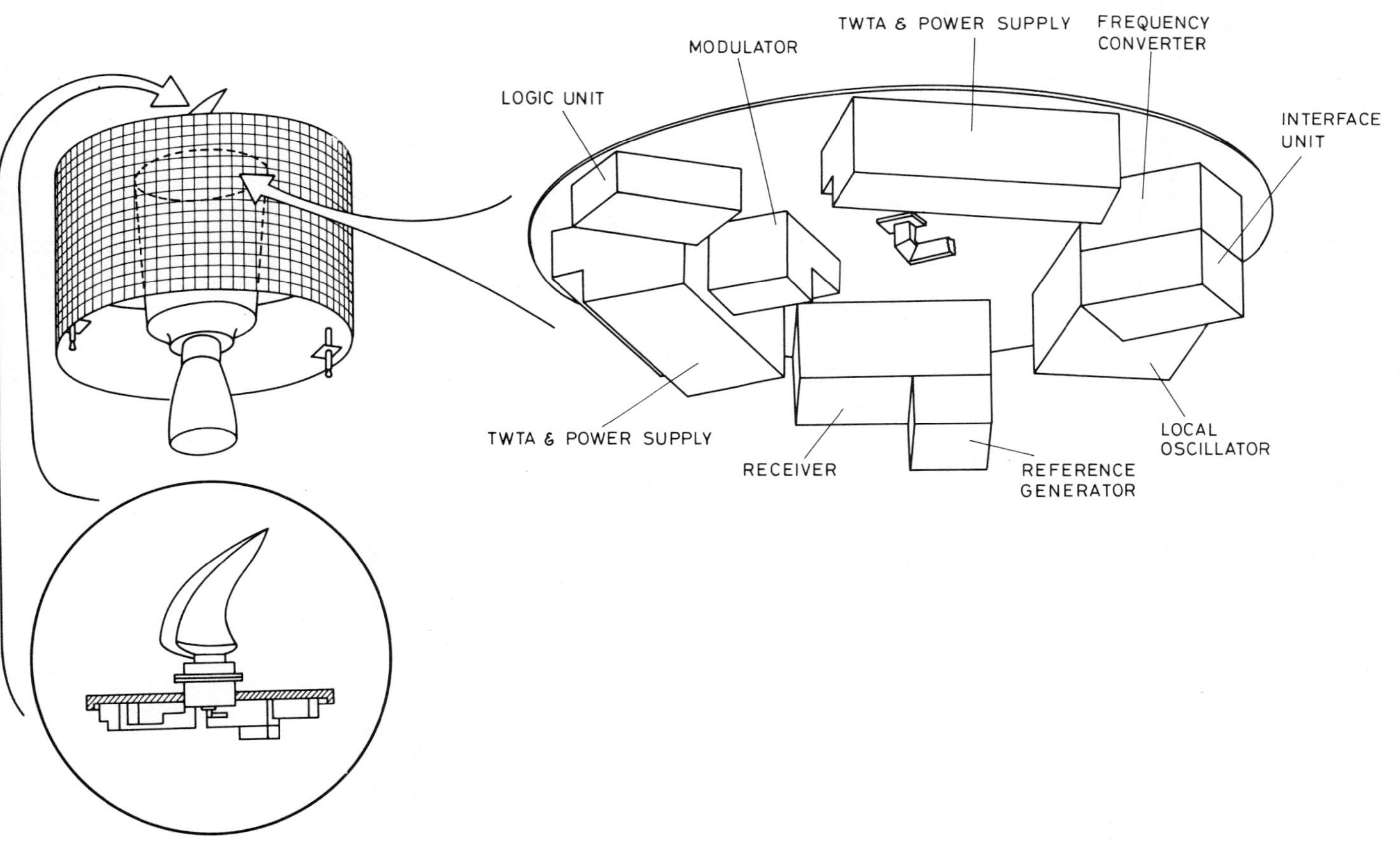

Fig. 1 SIRIO satellite.

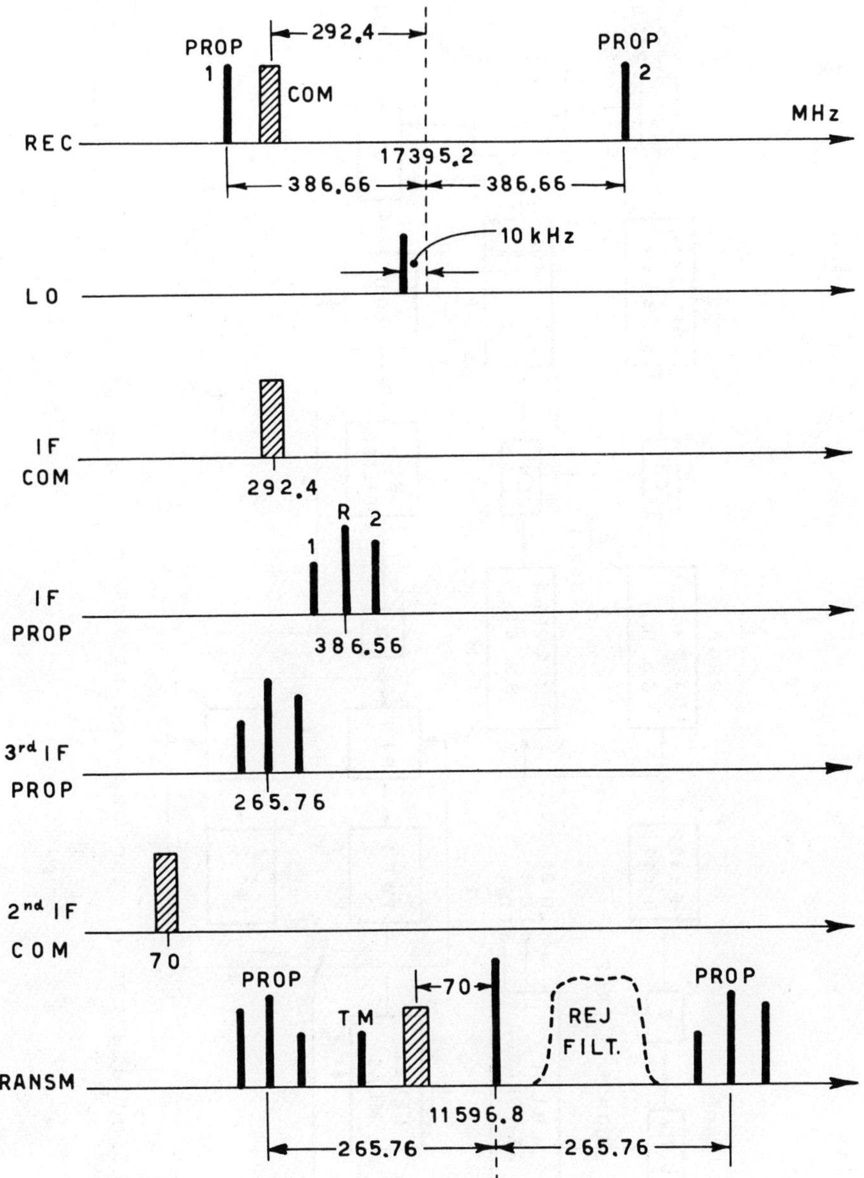

Fig. 2 Spectra of the signals through the satellite repeater.

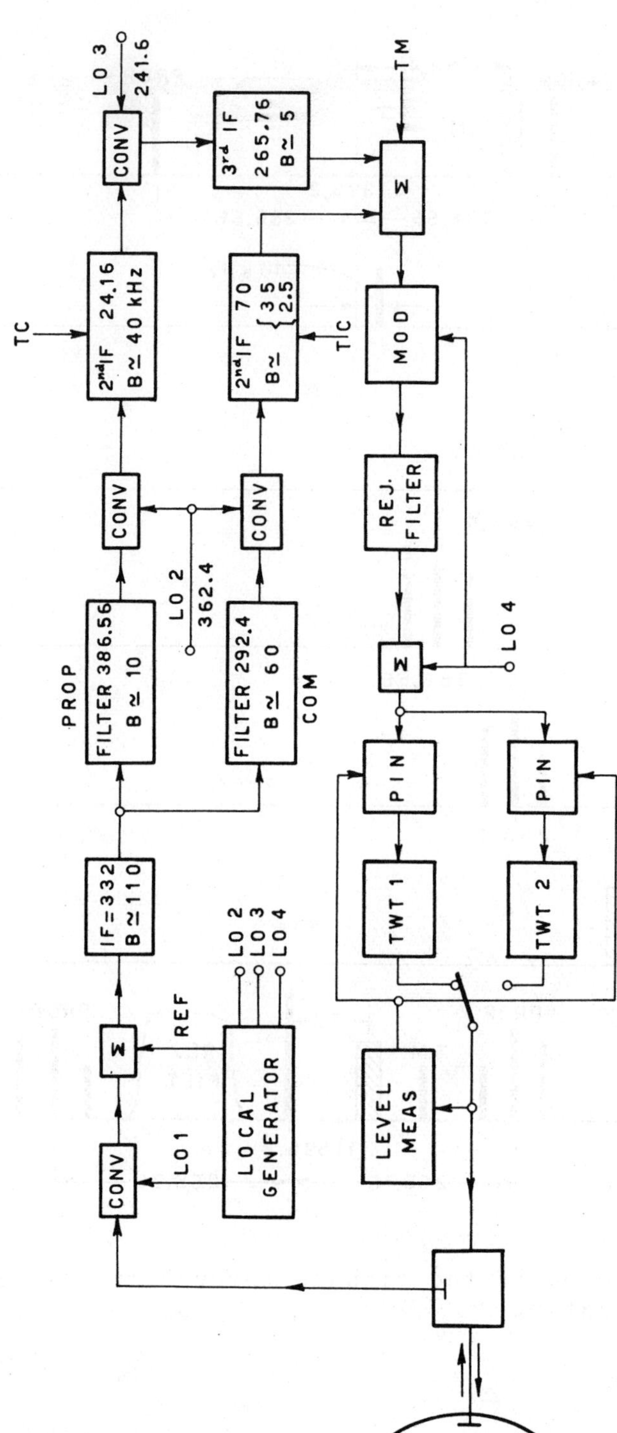

Fig. 3 Simplified diagram of the onboard repeater.

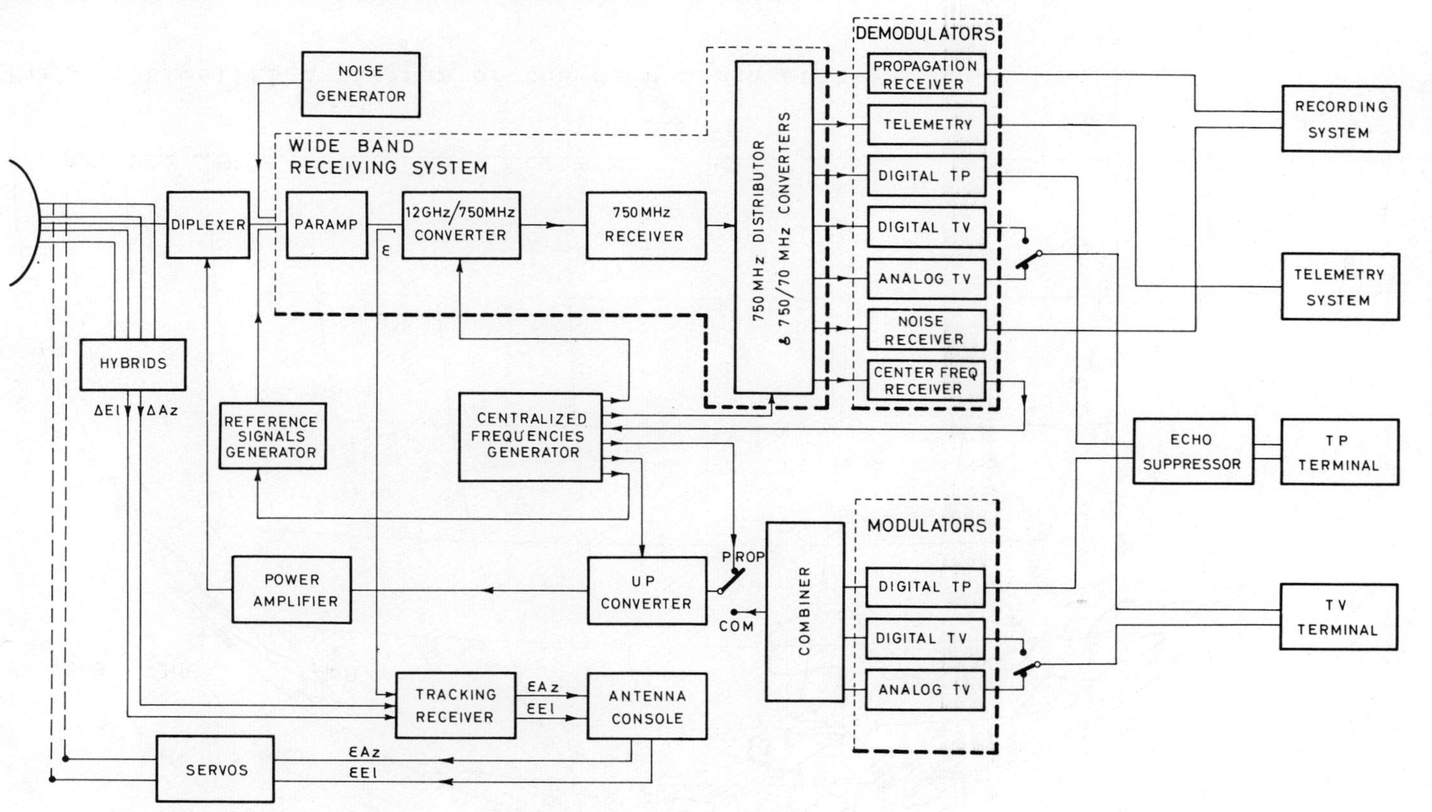

Fig. 4 Antenna coverage.

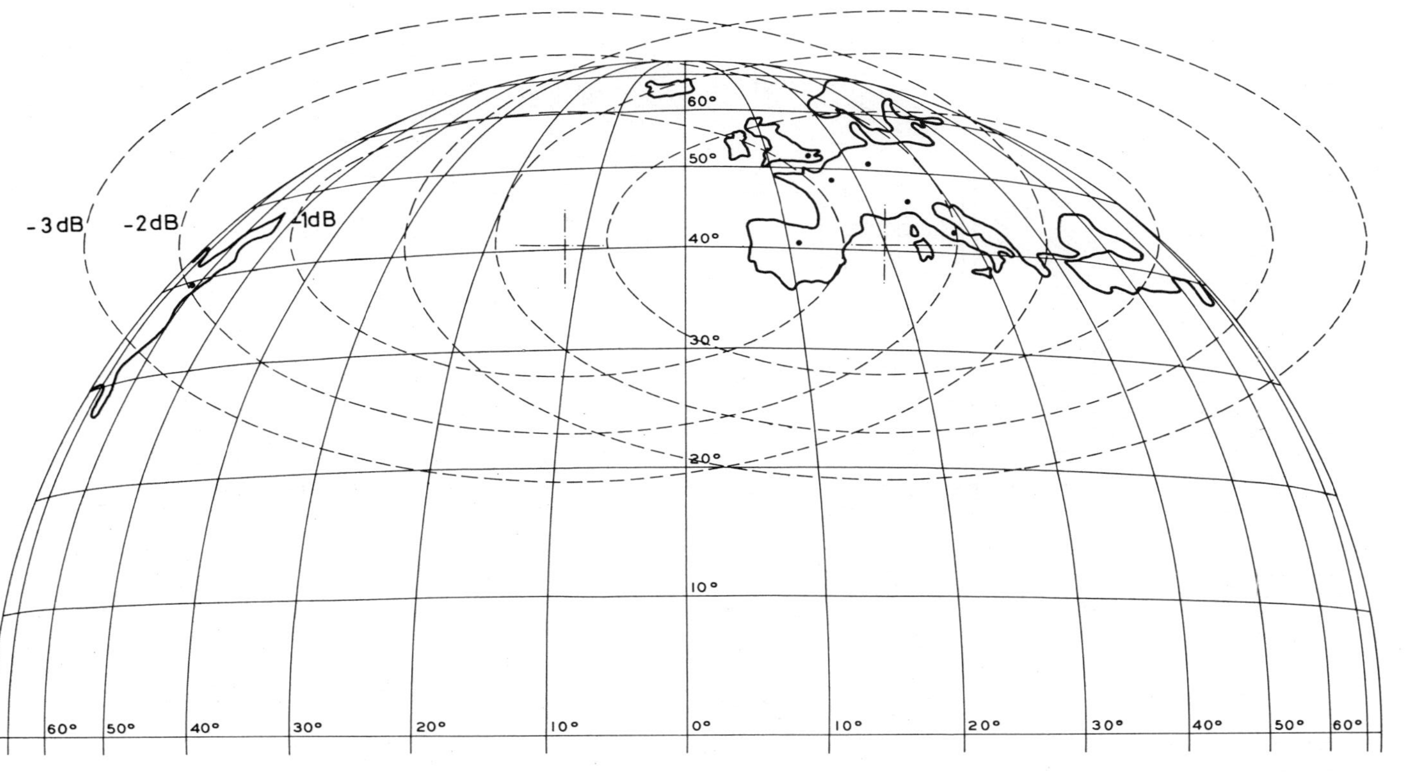

Fig. 5 Simplified diagram of the main Earth station.

THE ITALIAN SIRIO EXPERIMENT: SATELLITE AND GROUND EQUIPMENT

P. Fanti* and S. Tirro†
Telespazio S.p.A., Rome, Italy

Abstract

The propagation and communication experiments that will be performed using the Italian satellite SIRIO will provide valuable information on the behavior of the atmosphere at frequencies above 10 GHz. To perform these experiments, the construction of a satellite to be launched by a Thor Delta booster is described. The construction of three Earth stations (two fixed and one transportable) and an operational center for satellite control also is described. The available signal-to-noise ratio will permit the measurement, with acceptable accuracy, of the atmospheric attenuation and the propagation distortion in a 750-MHz band at 18 GHz and in a 550-MHz band at 12 GHz, and also the determination of the signal characteristics required for transmitting good quality television signals.

1. Introduction

The frequency band below 10 GHz today is about fully utilized in the most developed countries, and the coordination between the satellite system and the terrestrial radio links becomes more and more critical. The International Radiocommunications Consultative Committee (CCIR) also seems unlikely to permit the satellite systems to use frequencies below 10 GHz. Considerations relative to the satellite antenna size needed to cover small areas, and relative to the wider bands available, lead to an appreciation of the use of frequencies above 10 GHz. With the Super High Frequencies (SHF) propagation experiment performed with the Italian Satellite for Oriented Research (SIRIO), a measurement of the transmission characteristics of the atmosphere at these frequencies will be attempted. Such characteristics already

Presented as Paper 70-502 at the AIAA 3rd Communications Satellite Systems Conference, Los Angeles, Calif., April 6-8, 1970.

*General Manager (deceased, March 1970).

†SHF Experiment Project Manager.

have been measured on terrestrial radio links, but there is a substantial lack of knowledge as to satellite links.

Supplementary attenuation up to 15 db at 12 GHz and 25 db at 18 GHz can be expected for a very small percentage of the time; under the same limiting conditions, a differential attenuation of only a few decibels and a differential phase of less than 5° are expected in a band 500 to 800 MHz wide.

The SIRIO program is managed by the Italian National Research Council. The spacecraft will carry three experiments: two magnetic experiments and the SHF propagation and communications experiments. Telespazio, consultant to the I. N. R. C. for the SHF experiments (the subject of this paper), also will be responsible for contracts, technical specifications preparation, and program managing.

The flying hardware (transponder and despun antenna) will be manufactured by Selenia S.p.A. as a first contractor and a number of other Italian firms as subcontractors. The first contractor for the ground stations has not been chosen yet, so some degree of uncertainty still exists about the configuration of the Earth stations.

2. System Design

The SIRIO program foresees the completion and launch of a satellite, the development of which had already started under contract with the European Launcher Development Organization (ELDO). When the program became national, the spacecraft was optimized to obtain the maximum Equivalent Isotropically Radiated Power (e. i. r. p.) at SHF with coverage of Europe and the East Coast of the U.S.; so a maximum e. i. r. p. of 31.5 dbw at beam center was obtained (see Sec. 3.1). A brief analysis will justify the system configuration chosen by the experiment inventor, F. Carassa (see also Ref. 1). The purpose was to design a system to conduct propagation and communication experiments (not simultaneously). The following good weather G/T is needed at Earth stations, with the available e. i. r. p.: a) about -16 db ($1/^{o}K$) for amplitude measurement of a single frequency transmitted by the spacecraft; this result holds at 11.6 GHz with 20-db supplementary attenuation, coherent detection, and phase-lock noise filtering (with 150-Hz noise bandwidth and 8-db threshold); b) about 26 db ($1/^{o}K$) for propagation measurements at 11.6 GHz with 500-MHz swept bandwidth and some tens of Hz sweep rate; this result holds with 20-db supplementary attenuation and phase-lock filtering (a noise bandwidth of the order of

1 MHz is needed); c) about 26 db (1/ K) for FM television transmission, with 3-db supplementary attenuation, 10-db threshold signal-to-noise ratio, and 35-MHz bandwidth occupancy.

This result is based upon the use of gaseous helium cooled paramps. Therefore, G/T deterioration resulting from supplementary attenuation is significant. This G/T gives a substantial margin above threshold for a narrow-band communication experiment (2-MHz bandwidth).

The foregoing results show that a system designed to obtain the maximum communication efficiency (i.e., large e. i. r. p. at 11.6 GHz) cannot perform continuous exploration at both 11.6 and 17.4 GHz. In fact, if the 17.4-GHz measurement is performed onboard the spacecraft, it could require a very sophisticated receiving and measuring system plus the need for telemetry transmission of a large amount of measured data. On the other hand, if the measurement has to be performed at the Earth station, a large 17.4-GHz e.i.r.p. is needed, i.e., the spacecraft should give enough primary power to supply a 17.4-GHz transmitter and a 12-GHz transponder (which must also have the possibility of working as an 11.6-GHz transmitter for propagation experiment). The 17.4-GHz transmitter should be much more powerful than the 11.6-GHz one.

We could think of a system exploring a narrower bandwidth, but the amplitude and phase distortions are measurable only in a large bandwidth (see Table 1 for measurement precision). For the foregoing reasons, continuous explorations were discontinued, and single frequency measurements will be performed in the propagation experiment. It will be possible to measure the following parameters: a) absolute attenuation at 17.4 GHz; b) differential attenuation at 17.4 GHz ± 386

Table 1 Measurement precision for propagation characteristics

Measurement precision	Good weather (db)	Bad weather (db)
11.6-GHz absolute attenuation	+ 1.4	+2.2
Down-link differential attenuation	+ 0.8	+1.6
17.4-GHz absolute attenuation	+ 1.1	+ 1.9
Up-link differential attenuation	+ 0.5	+1.3
Down-link phase distortion (°)	+ 0.8	+1.8

MHz; c) absolute attenuation at 11.6 GHz; d) differential attenuation at 11.6 GHz ± 265 MHz; e) phase distortion at 11.6 GHz ± 265 MHz.

Two lines are transmitted from the Earth station, then down-converted onboard the satellite to frequencies only 10 KHz apart and narrow-band filtered. The satellite transmits a more complex spectrum, including the up-link information and a triplet of coherent frequencies for phase-distortion measurement. Therefore, it is possible to perform all the measurements in the Earth station with a small e.i.r.p./frequency transmitted by the satellite; telemetry data are needed only as a backup.

3. Onboard Equipment Description

The spacecraft is spin-stabilized with a 90-rpm spin velocity; the solar array is 954 mm high, and the spacecraft diameter is 1432 mm. Thus a primary power of 102 w is obtained at life end (2 years after launch) and at the summer solstice. The total spacecraft weight (excluding adapter and apogee motor propellant and case) is 182 kg. The size and weight are such that an improved long tank Thor Delta vehicle (with six strap-ons) is suitable for launch.

The available primary power permits the satellite to radiate 10 w at SHF. The power amplifier is a traveling wave tube (TWT). The despun antenna gain is 22.5 db at the beam center (covering Europe and the Atlantic Coast of the U.S.), so that a saturated e.i.r.p. of about 31.5 dbw can be obtained at the beam center.

3.1. SHF Transponder Description

A simplified block diagram is shown in Fig. 1. For simplicity, the redundant TWT is not shown. Three different configurations (corresponding to the three experiments) can be selected by ground command.

3.1.1. Propagation. The two received lines are converted at two frequencies only 10 KHz apart. A reference signal (whose level is stabilized within ± 0.2 db throughout the satellite life) is added after conversion to avoid up-link measurement errors resulting from transponder gain instability. The reference signal frequency is offset by 10 KHz from each line. The selectivity of the tracking filters used on the receive side permits reduction of this distance to only 5 KHz, so it will be possible, if desired, to perform the experiment with two simultaneously transmitting Earth stations.

It is important to keep the frequency spacing constant to permit an easy separation and recognition of each line in the propagation characteristics measurement system. This is obtained by phase locking the reference signal generator to the heterodyne base oscillator. The reference signal level can assume two different values, which can be selected by ground command. The best values will be assessed at a later stage in the program, depending on the attenuation range to be observed with the maximum precision.

In the receiver, the two lines plus the reference signal are down-converted at about 24 MHz, then passed through a crystal filter (40-KHz bandwidth) to optimize the signal-to-noise ratio at the TWT input. The total power is kept constant. After a first up-conversion to 265 MHz, the signal is again up-converted to the 12-GHz band. The carrier is suppressed at the converted output, and a residual carrier is added after filtering. The residual carrier level is well stabilized at the TWT output. The TWT nominal back-off is about 7 db, 600 mw being delivered on the residual carrier and on each sideband. So an e.i.r.p. of 49.5 dbm on the residual carrier and on each sideband is obtained.

3.1.2. Narrow-band communication. The intermediate frequency (IF) is 70 MHz, and the bandwidth is 2 MHz. Twelve one-channel carriers can be transmitted at one time, with the PCM-PSK-2Ø technique. The optimum TWT back-off is 4.5 db, so an e.i.r.p. of 46.6 dbm/carrier is obtained. The carrier reject filter at 11.6 GHz also filters out one sideband, so the entire available power is radiated in one sideband only. The automatic gain control (AGC) can be excluded by ground command.

3.1.3. Wide-band communication. The frequency transmitted by the Earth station remains the same, so the IF is still 70 MHz, but the available bandwidth is 35 MHz. The TWT back-off (resulting from the residual carrier and telemetry carrier presence) is 2 db, so a useful beam center e.i.r.p. of 29.5 dbw is obtained at TV frequency. A telemetry PCM-PSK-2 Ø carrier is transmitted in every configuration with a maximum bit rate of 1024 bps. The TWT output level is about 100 mw for this carrier in any configuration. The transponder weight and power consumption breakdown is shown in Table 2. The frequency plan is shown in Fig. 2.

Table 2 SHF transponder weight and power breakdown

Unit	Mass (kg)	Power consumption(w)	
Diplexer	0.60	...	
Down converter	0.60	0.65	
Local oscillator	1.20	8.0	
Reference signal generator	0.35	0.9	
Receiver	1.30	2.35	
Modulator	1.00	0.1	
Power amplifier no. 1	2.30	50.0	(η = 20%)
Power amplifier no. 2	2.20	...	
Output switch	0.70	3.0	
Interface unit	0.30	1.0	
Central PWM regulator	1.70	7.7	(η = 90%)
Cabling	1.90	0.5	
Growth	0.95	...	
Total	15.00	74.2	

3.2. Despun Antenna Description

Dual channel coaxial rotary joints have not yet been used onboard telecommunication satellites. A good rotary joint at 4 and 6 GHz has been developed recently by Philco-Ford under Intelsat research contract, but a 12-GHz and 18-GHz joint could be much more difficult to design and manufacture.

To avoid the need for such a joint, the SIRIO antenna is composed of a parabolic rotating reflector illuminated by an offset fixed conical horn. The parabolic sector is obtained by intersection of a paraboloid with a vertical axis cylinder. A simple layout of the despun antenna is hown in Fig. 3. The paraboloid axis is squinted 6.3° above the horizontal plane of the satellite in order to obtain a 6.3° squinted beam. Because of the reflector-feed relative motion, circular polarization must be used. The beam is equalized over the 12- to-18-GHz band by proper application of the phase error technique. Thus a 3-db coverage area of 6° (elevation) x 12° (azimuth) is obtained; taking into account ± 1° pointing error in azimuth and ± 0.75° in elevation, the coverage area reduces to 4.5° x 10°. The net coverage area is given in Fig. 4 (-3 db contour). Opposite polarizations are obtained in transmission and reception using an orthomode coupler followed by a circular waveguide iris polarizer. The beam center net gain of the antenna, including polarizer losses, is 22.5 db at 12 GHz and 23.5 db at 18 GHz.

Two pointing directions are possible in azimuth (both at 6.3° elevation). The first is toward the middle of the Atlantic Ocean (1.5°W azimuth), and the second is toward Western Europe (2.5°E azimuth).

The despun antenna weight and power consumption breakdown is shown in Table 3.

<u>3.2.1. Steering system design</u>. The d.c. motor is driven by an error signal giving velocity and position control. Velocity control is needed during each revolution to compensate for the d.c. motor torque variations. The position error is obtained by comparing the infrared sensor output with the position magnetic pick-up output as in Fig. 5. Another pick-up placed 180° apart from the first one resets the flip-flops after each comparison. The position pick-up gives 1 pulse/revolution. Dividing this frequency by two, a pulse 4° apart from the magnetic pick-up pulse can be obtained. This pulse can be utilized as the antenna reference pulse for the second pointing position. The over-all pointing error is foreseen to be ± 1° for the first pointing position. Contributions to this error are as follows: a) infrared sensor ± 0.3°; b) assembly tolerances ± 0.1°; c) motor position magnetic pick-up to ±0.1°; and d) servo system ± 0.5°. These values are intended to be conservative. A slightly larger error is foreseen for the second pointing position, which is

Table 3 SHF antenna weight and power breakdown

Unit	Mass (kg)	Power consumption(w)
Reflector and structure	0.700	...
Polarizer and feed	0.150	...
Rotation mechanism	1.760	3.5
Control circuits	1.350	2.3
Control circuits (stand-by)	1.350	...
Changeover circuits	0.450	0.4
Total	5.760	6.2

the result of greater uncertainty for the antenna reference pulse.

3.2.2. Mechanical drive assembly. A d.c. motor gives a control system intrinsically less precise than that of a synchronous motor because of motor torque variations during each revolution. Nevertheless, a d.c. motor has been selected because of its more flexible control. In fact, the maximum torque can be as high as the available electrical power, whereas in a synchronous motor a limit load angle and a maximum torque exist. Thus it is possible to consume power up to 20 w (nominal power consumption is 5.5 w) if a larger torque is needed. The torque vs speed characteristics are given in Fig. 6.

The motor drive assembly, manufactured by Ball Brothers Research Corp., has already successfully qualified for the Philco-Ford Skynet Satellite. It does not use brushes, so a very high reliability is achieved. Proper feeding of the two phases of the motor is obtained by using a rotating position resolver mounted on the motor axis (Fig. 7). The true motor is a synchronous one, but the synchronous motor plus resolver assembly is equivalent to a d.c. motor. The bearing is spring preloaded and has an axial bearing capability of about 2 lb. The classical Vackote lubrication system will be used.

4. Earth Stations Description

4.1. Fixed Stations

4.1.1. General description. A good TV link with frequency modulation requires at least 58-db antenna gain at 12 GHz, so a little margin above threshold is obtained in good weather conditions. This gain can be obtained by means of a Cassegrain type antenna having an 8.5-m diam dish, assuming the total antenna efficiency equal to 60%. The gain at 18 GHz is foreseen to be 60.5 db. The antenna noise temperature should vary from a minimum of 50°K (60° elevation) to 100° K at 5° elevation.

To avoid down-link absolute attenuation measurement errors, the antenna will be equipped with a complete autotracking system; an amplitude monopulse feed is adopted. Tracking accuracy is expected to reach 0.02° rms with 110/h steady wind and gusts up to 130 km/h. The antenna can track a geostationary satellite both manually and automatically. The beacon signal corresponds to the down-link residual carrier, with about 10-dbw beam center e.i.r.p. Both feed and structure detailed configurations have not been definitely decided upon yet. The possibility of using a larger diameter antenna (12 m) still exists to permit high rate data transmission, time division multiple access (50 Mbit/sec), or digital TV (65 Mbit/sec) transmission experiments. It is interesting to observe that the European Post and Telecommunication Administrations Conference foresee, for their stations at 12 GHz, antennas with 13- to 15-m diam.

A simplified block diagram of the over-all Earth station is shown in Fig. 8. The low-noise receiver is a three-stage parametric amplifier, gaseous-helium cooled, with 50° K noise temperature (including microwave receiver contribution) and 550-MHz bandwidth. To avoid measurement errors resulting from gain and phase instability, three reference frequencies very close to the frequencies received from the satellite are injected at the low-noise receiver input. It is also possible to inject a narrow-band chopped noise for a radiometric measurement of the antenna noise temperature in various weather conditions. The resulting signal spectrum at the input of the low-noise receiver is shown in Fig. 9. This spectrum is composed of three bands: a center band, including the residual carrier plus a reference signal 100 KHz away, and two side-bands, each including one side-band from the satellite and a reference signal 100 KHz away.

A wide-band receiver converts the 550-MHz band from 12-GHz down to 750 MHz (center frequency). Then three narrow-band receivers convert each group of frequencies down to 20 MHz (propagation experiment). The residual carrier is phase-locked, and a clean replica is obtained. This replica is then utilized to synthesize (in the centralized frequency generator) all the frequencies needed for up and down conversions in the Earth station. With this simple technique a good stability of the 20-MHz spectrum is assured on the receiving side, and the proper spacing between the satellite heterodyne and the two lines transmitted from ground is assured on the transmitting side.

The power amplifier can use a single wide-band TWT or two klystrons; although a final decision has not been made yet, the use of klystrons is suggested for reliability, life, and simplicity considerations. The klystrons are air-cooled and permanent-magnet focused, and their power supply is much simpler than the TWT's.

A possible configuration of the transmitting system is shown in Fig. 10. Each klystron is excited by both lines but amplifies only one of them. The output branching system is composed of two 3-db couplers and two filters tuned on the upper frequency. Each klystron has a saturated output power of 1 kw and 40-MHz bandwidth and is mechanically tunable in a 400-MHz band. Thus one of the two klystrons also can be used for communications experiments.

Equipment for narrow-band and wide-band communication experiments is shown in Fig. 8; the intermediate frequency for the communication signal is 70 MHz. Proper spacing of the twelve telephone carriers is assured by locking, in each Earth station, the transmitted frequencies to the satellite heterodyne frequency.

4.1.2. Description of propagation characteristics measurement system. A simple block diagram of the propagation measurements system is shown in Fig. 11. Each line is filtered by the phase-lock technique, using a very stable voltage controlled crystal oscillator that can be modulated in a very narrow range. The clean replica so obtained is utilized for amplitude detection. Figures 12 and 13 show, respectively, square law and coherent detections. The coherent detector gives a better signal-to-noise ratio, and no threshold caused by detection occurs. Thus the threshold is determined completely by the phase-lock filter.

The tracking filter must follow only Doppler variations and short-term variations of the satellite heterodyne. Practically, the equivalent noise bandwidth of the phase lock (PL) can be made as small as the short-term frequency spectrum of the satellite heterodyne times π - that is, about 150 Hz - so the PL works well above threshold in any foreseen operating condition. We can now conclude that the measurement error increases linearly with the noise-to-signal ratio. The detector output is processed in a logarithmic amplifier before comparison of the various levels. This comparison is then made by simple differential amplifiers, giving direct readings of absolute and differential attenuations in db. Very likely the comparison will be made by switching the outputs of two detectors and using only one logarithmic amplifier, followed by a synchronous detector, in order to avoid the errors caused by logarithmic amplifiers gaining instability. This possibility is still under evaluation.

As for phase-distortion measurement, it can be made easily enough by proper choice of the heterodyne frequencies. In fact, the satellite transmits the frequencies f_0, $f_0 + (K-1)f_0$, $f_0-(K-1)f_0$, i.e., f_0, Kf_0, $(2-K)f_0$, where $K = 491/480$. These frequencies can be converted to the same $f_c = K'f_0 = 20\text{MHz}$, using:

f_0 (1-K') heterodyne to convert f_0(phase ϕ_{x0})
$\{[1-K'] + [K-1]\} f_0 = f_0(K-K')$ heterodyne to convert Kf_0(phase ϕ_{x2})
$\{[1-K'] - [K-1]\} f_0 = f_0(2-K-K')$ heterodyne to convert (2-K) f_0 (phase ϕ_{x1})

The three frequencies transmitted from the satellite are obtained by amplitude modulation of a carrier, so they are coherent, and the phase function $(\phi_{x2} - \phi_{x0})-(\phi_{x0} - \phi_{x1}) = 2\phi_{x0} - (\phi_{x2} + \phi_{x1})$ is zero. In case there is a distortion caused by the propagation, and the phase characteristic is no longer linear, the phase function value will not be zero. The phase detector is shown in Fig. 14. The expected phase distortion amounts to only a few degrees, so particular care has been taken with the coherence of satellite transmitted frequencies and with Earth station distortion compensation by the introduction of reference signals.

The antenna noise is measured by a ratio radiometer, a block diagram of which is shown in Fig.15. The input precision attenuator is set manually to minimize the output reading. The IF amplifier AGC assures linear operation of the square law detector.

4.1.3. Data recording system. Although this subsystem is not yet well defined, we can foresee that it will be equipped with a compressor because the correlation time of the measured parameters is rather high. The registration probably will be made in digital form, with every elaboration done at the Milan Polytechnic. The following parameters will be recorded, with a maximum sampling frequency of about 10 Hz: a) propagation measurement system outputs; b) atmospheric data (rainfall intensity and wind velocity); c) Earth station characteristics (antenna pointing errors and radiated power); d) satellite ephemeris (from antenna axes encoders); e) Doppler shift (from range and range rate equipment); f) telemetry data (switch status, despun antenna position, TWT output power, AGC voltage, spin axis orientation, and spin velocity);and g) time signal.

4.2. Transportable Station

4.2.1. General configuration. This station will be used for propagation experiments only and will permit the following measurements to be performed: a) down-link absolute attenuation on a site different from the fixed station ones; b) down-link absolute attenuation in a site very close to a fixed station (space diversity experiment).

The measurement error resulting from thermal noise will be much larger than the one for fixed stations. In fact, the receiving system G/T at 5° elevation is only 24.7 db (1/°K) against the 36.2 db (1/ K) for fixed stations. These G/T are intended for good weather conditions. The transportable station will use a 4-m diam antenna with 50% efficiency at 11.6 GHz and a single-stage ambient temperature paramp with 300/°K receiving system noise temperature. The steering system will be very simple, including only a velocity servo with position display. The low-noise receiver bandwidth will be 400 MHz, so that receiving one sideband and measuring up-link absolute and differential attenuations also will be possible. This can give some 8.5-db advantage in the signal-to-noise ratio when the fixed transmitting station sees a 20-db rain attenuation at 12 GHz. The transportable station may also be equipped to transmit one of the two up-link frequencies to permit an 18-GHz space diversity experiment. During diversity experiments the station will be connected to the fixed station via teletype (TTY) and will receive a time signal from the fixed station.

4.2.2. Propagation characteristics measurement system. The transportable station measurement system will include only the equipment required for absolute down-link attenuation, i.e., residual carrier acquisition circuit, reference signal generator, and two coherent detectors followed by a radiometer and a logarithmic amplifier. The increase in complexity resulting from up-link measurements (Sec. 4.2.1) is very small.

4.3. VHF Station for Satellite Station Keeping

The satellite will be launched by NASA, and the station keeping will be effected by NASA for the first month. Nevertheless, as the foreseen life of the satellite is two years, the station keeping function will be needed much longer. Very likely, a VHF station will be erected for this purpose, with telemetry, command, range and range-rate operation capability. Possibly, this station could become an auxiliary to the NASA STADAN network by agreement between NASA and the Italian National Research Council.

The telemetry data will be received at SHF and/or at VHF, so the best location for the station will still be Fucino or Milan. In this way it will be possible to use the same demodulating and decoding equipment both for VHF and SHF telemetry.

5. System Performance

5.1. Propagation Experiment

A detailed analysis of the precision of the various measurements is not the purpose of this paper, so we will give here only the rough estimates already obtained from approximate analysis. These results hold in the following hypothesis:

a) coherent detectors are used for amplitude measurements;

b) a radiometric system is used to avoid errors caused by logarithmic amplifiers gain instability;

c) errors caused by wind gusts are compensated at ± 0.2 db;

d) an automatic control of Earth station transmitted power is used in order to keep the lines received on the satellite at the same level as the reference signal;

e) the noise error is 3 σ , and the peak errors are added.

The results are shown in Table 1 for good weather and bad weather (20-db attenuation at 11.6 GHz and 35-db attenuation at 17.4 GHz simultaneously) conditions. A substantial error percentage is due to long term variations (throughout the satellite life) of the measurement conditions (reference signal levels, sidebands balance, etc.). These error components therefore are substantially reduced by periodic system alignment in clear weather conditions, at least for differential attenuation and phase distortion measurement.

5.2. Narrow-band Communications Experiment

The link parameters are shown in Table 4. With 64-kbit/sec bit rate and 44.5-db unweighted signal quality, a margin of 15.8 db above threshold is obtained in good weather conditions.

The good weather system margin is a base line for the system described here; in fact this margin is computed on the hypothesis of twelve equal carriers being received onboard the satellite. This condition cannot be expected in bad weather conditions (bad weather at just one transmitting station would be sufficient); the base line applies only if each station e.i.r.p. is inversely proportional to the local supplementary attenuation. In this way the power flux density received on the satellite always will be kept constant, but the up-link signal-to-noise ratio always will be equal to the threshold value. In the second column of Table 4 the real situation of the Earth station transmitting its maximum available power in the limiting conditions is shown. If we assume that the supplementary attenuation in the up-link is 25 db, a deterioration of 15 db can be tolerated in the down-link. This deter-

Table 4 Narrowband communications experiment (PCM-PSK 2 ∅)

Up-link	Good weather	Bad weather (threshold)
Earth station e. i. r. p. per carrier (dbw)	86. 5	86. 5
Free space loss at 17. 4 GHz(db)	209	209
Satellite G/T (db/°K)(2 db off beam center)	-14. 7	-14. 7
Supplementary attenuation (db)	...	25
Boltzmann constant (dbw/°K MHz)	-168. 6	-168. 6
Up link (C/N)1MHz (db)	31. 4	6. 4
Down link		
Satellite e. i. r. p. per carrier (dbw) (2 db off beam center)	16. 6	16. 6
Free space loss at 11. 6 GHz (db)	205. 5	205. 5
Supplementary attenuation (db)	...	10
Earth station G/T (db/°K) (30° elevation)	37. 4	32. 7
Boltzmann constant (dbw/°K MHz)	-168. 6	-168. 6
Down link (C/N)1 MHz	17. 1	2. 4
Resulting (C/N)1 MHz	16. 9	1
Channel bit rate (K bit/sec)	64	64
Bit energy to noise density ratio(db)	28. 8	13
Margin above 13-db threshold (db)	15. 8	...
Unweighted signal to quantization noise ratio (db)	44. 5	44. 5

ioration, resulting from the antenna noise increase, corresponds to a supplementary attenuation in the down-link of about 10 db.

Table 5 Wideband communications experiment

Up-link	
Earth station e. i. r. p. (dbw)	90. 5
Free space loss at 17.4 GHz (db)	209
Satellite G/T (1 db off beam center) [db (1/°K)]	-13. 7
Boltzmann constant (dbw/°K MHz)	-168. 6
Up-link (C/N)1 MHz (db)	36. 4
Down-link	
Satellite e. i. r. p. [2-db back-off; 1-db off beam center (dbw)]	28. 5
Free space loss at 11. 6 GHz (db)	205. 5
Earth station G/T (db/°K) (good weather conditions; 30° elevation)	37. 4
Boltzmann constant (dbw/°K MHz)	-168. 6
Down-link (C/N)1 MHz (db)	29
Resulting (C/N)1 MHz (db)	28. 2
Maximum baseband frequency (MHz)	6
Peak-to-peak frequency deviation (MHz)	23
Carson's rule bandwidth (MHz)	35
Margin above 8-db threshold (db)	4. 8
Video signal quality (psophometrically weighted with CCIR emphasis) (db)	52

5.3. Wideband Communications Experiment

The link parameters are shown in Table 5. With 23-MHz peak-to-peak deviation, a margin of about 5 db above threshold and 52-db weighted video signal quality are obtained. A higher margin can be obtained, of course, by decreasing the peak-to-peak frequency deviation and considering also that the Carson rule is very pessimistic. A substantial margin increase also can be obtained by simultaneous use of threshold extension demodulation and of video signal processing after demodulation.

It is possible to perform digital TV and higher rate TDMA experiments with 12-m antennas.

6. Conclusions

The SHF experiment will permit us to determine the effect of the atmosphere on a satellite link at 12 and 18 GHz.

The ground stations and the basic spacecraft designed for this experiment also are suitable for use in a regional system for TV program transmissions covering a very small area. Furthermore, the TWT to be used has an output power that is well suited to the electrical power available on the satellite and to the regional system requirements.

The development of this critical component and of other critical components in the ground stations (the paramps and the power amplifiers, for instance) will be a substantial advance toward the use of 12- and 18-GHz frequency bands with satellite systems.

Reference

[1]Carassa, F., Drufuca, G., and Paraboni, A., "The Italian Satellite Sirio - SHF Propagation and Communication Experiments: Scientific Objective and Organization of the Experiments," preceding paper, this volume.

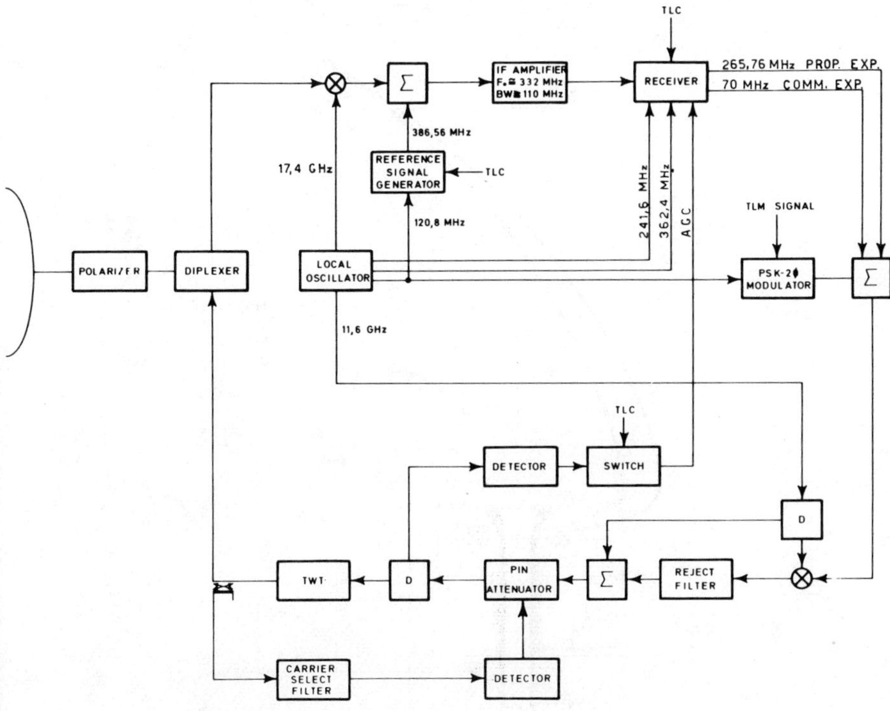

Fig. 1 SIRIO SHF transponder block diagram.

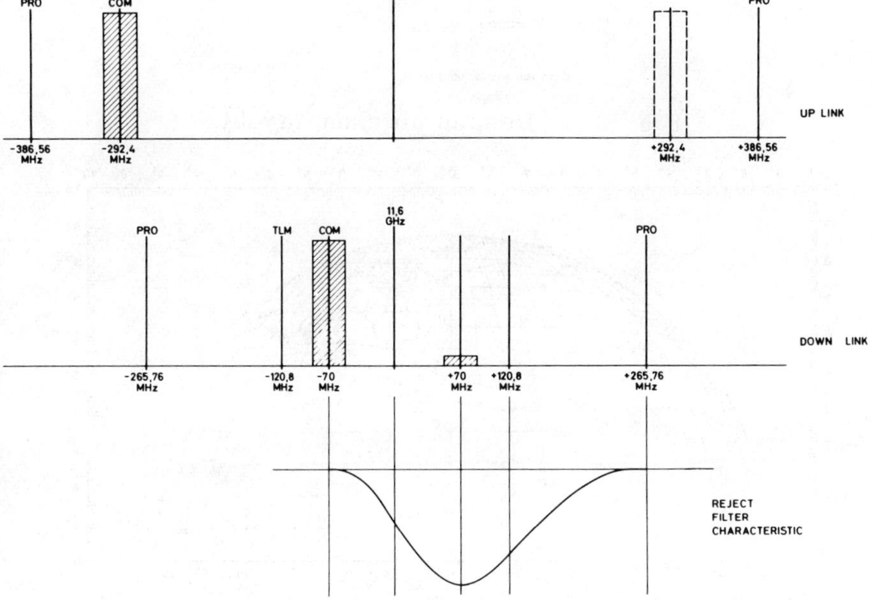

Fig. 2 SHF experiment frequency plan.

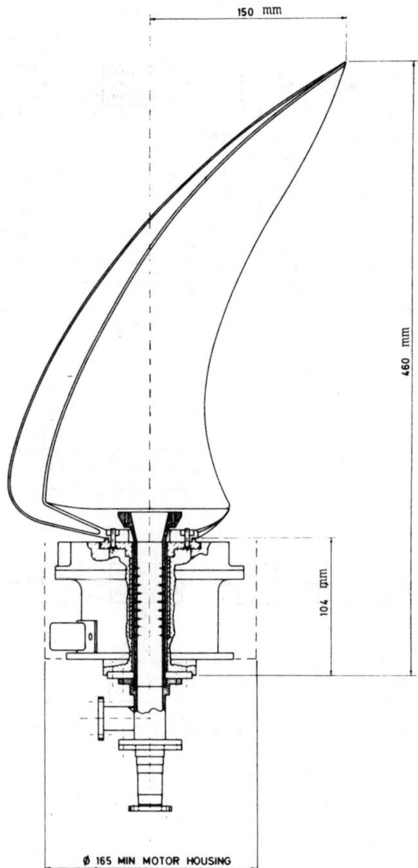

Fig. 3 Despun antenna layout.

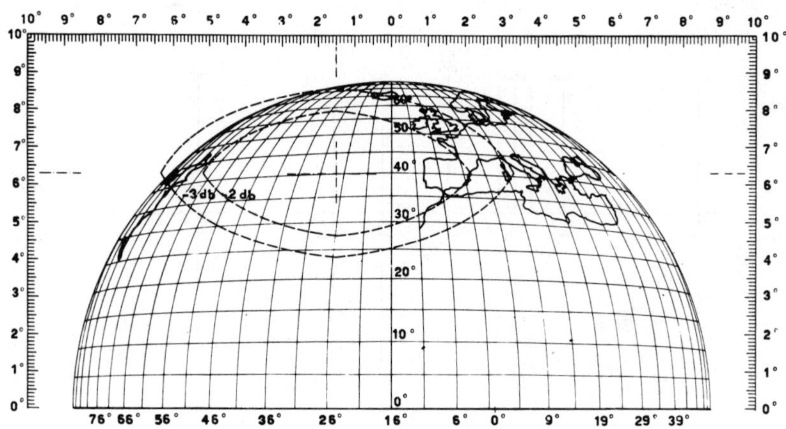

Fig. 4 Despun antenna coverage area.

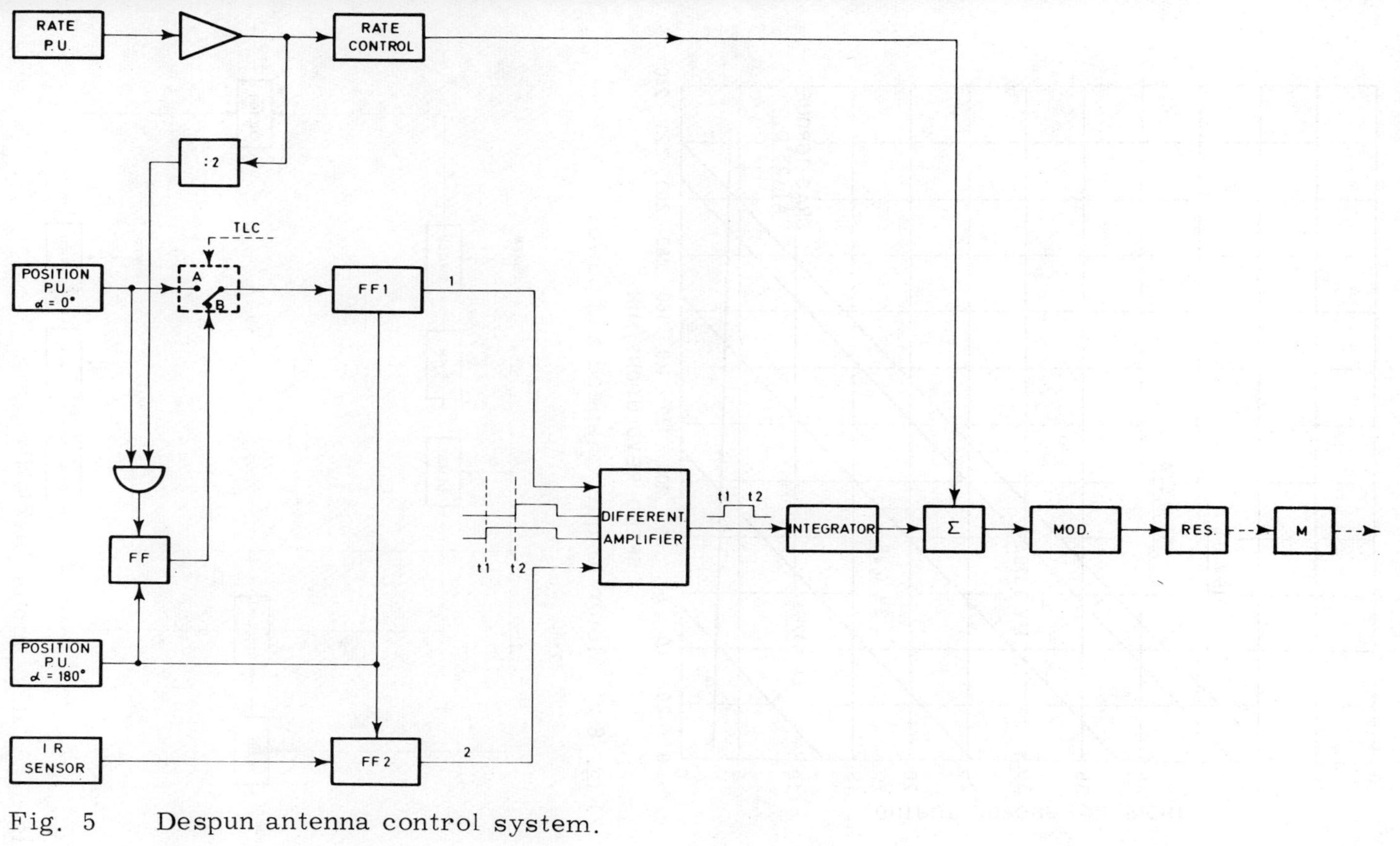

Fig. 5 Despun antenna control system.

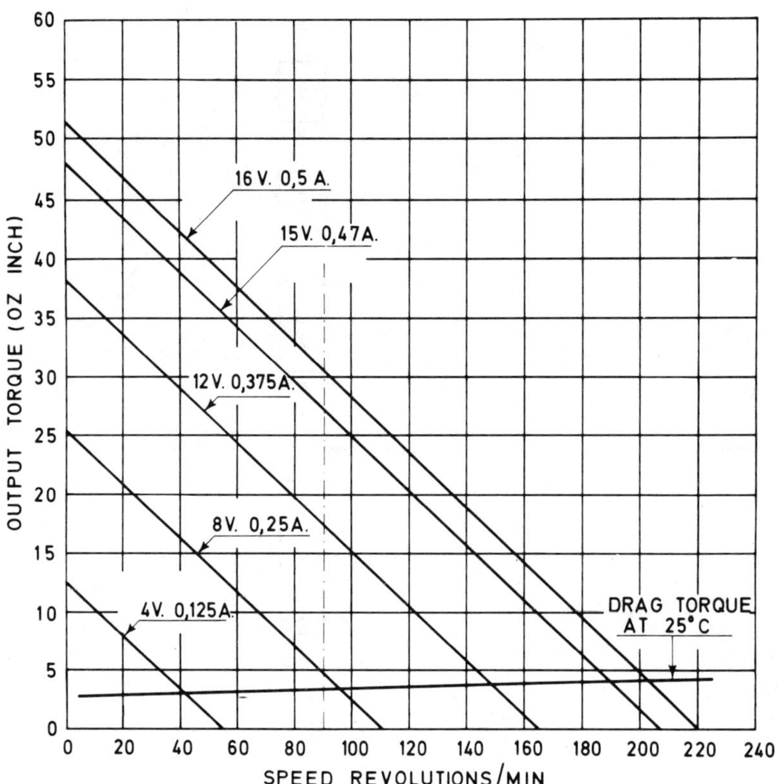

Fig. 6 Motor torque vs speed and power.

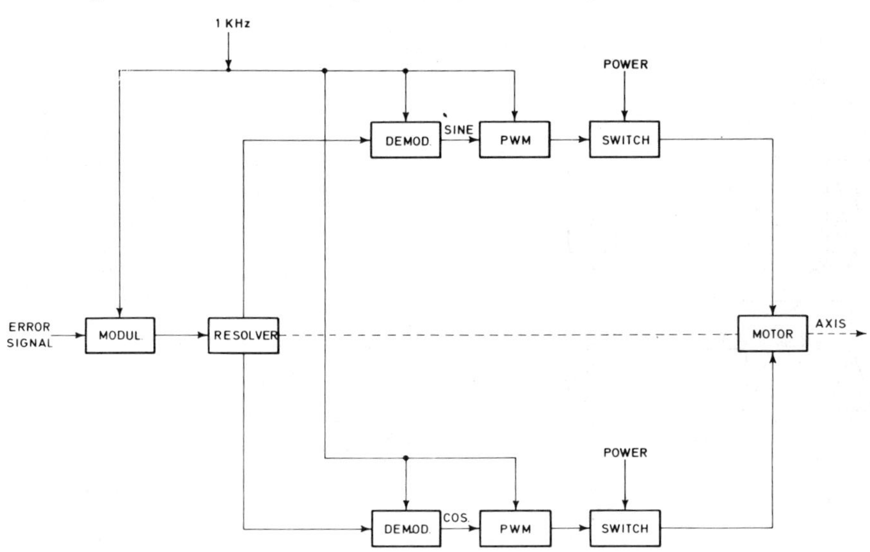

Fig. 7 d. c. motor block diagram.

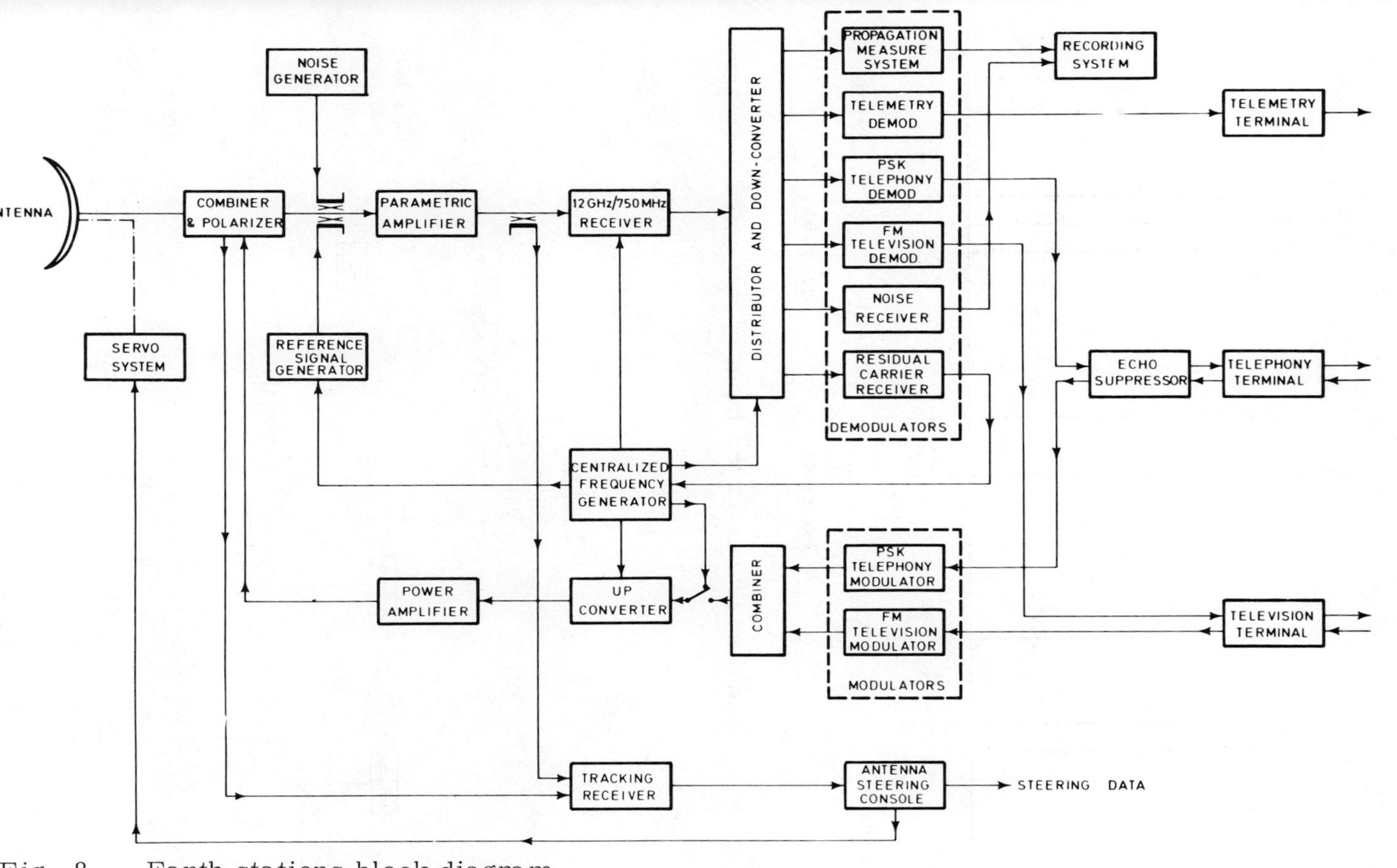

Fig. 8 Earth stations block diagram.

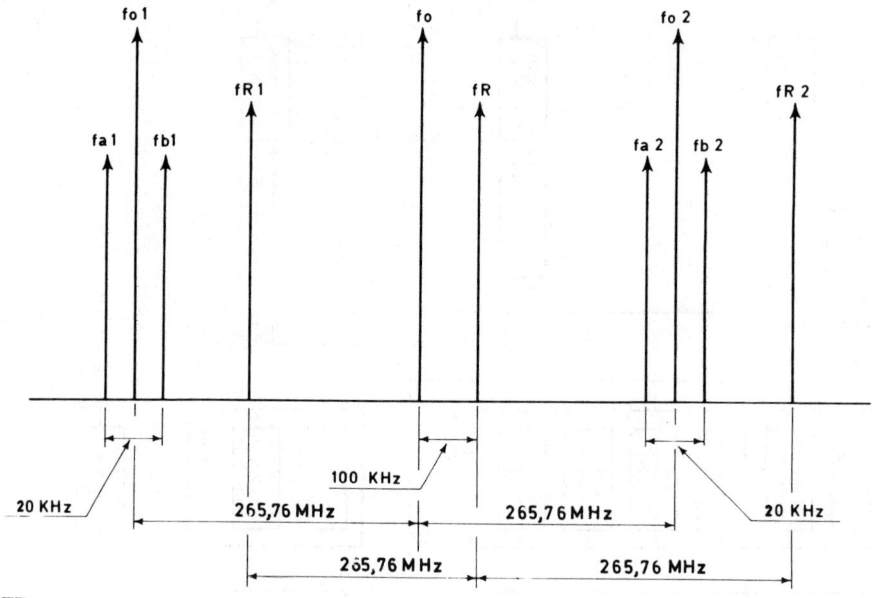

Fig. 9 Signal spectrum at the input of low noise receiver.

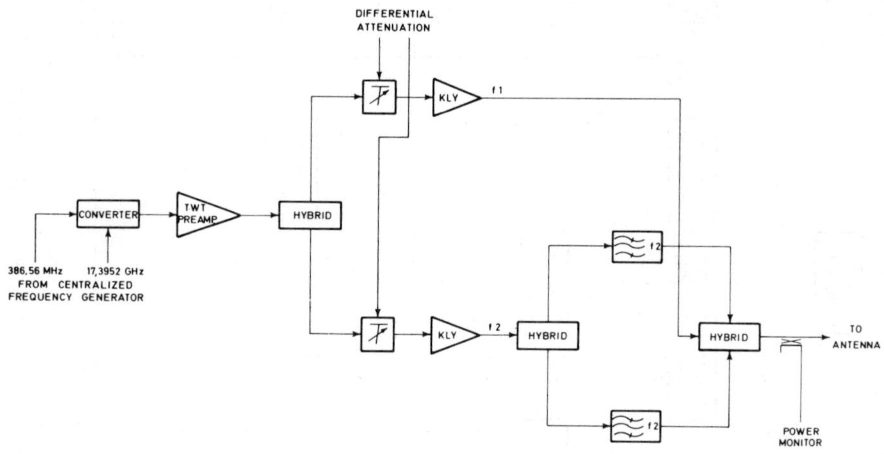

Fig. 10 Transmitting system block diagram (propagation experiment).

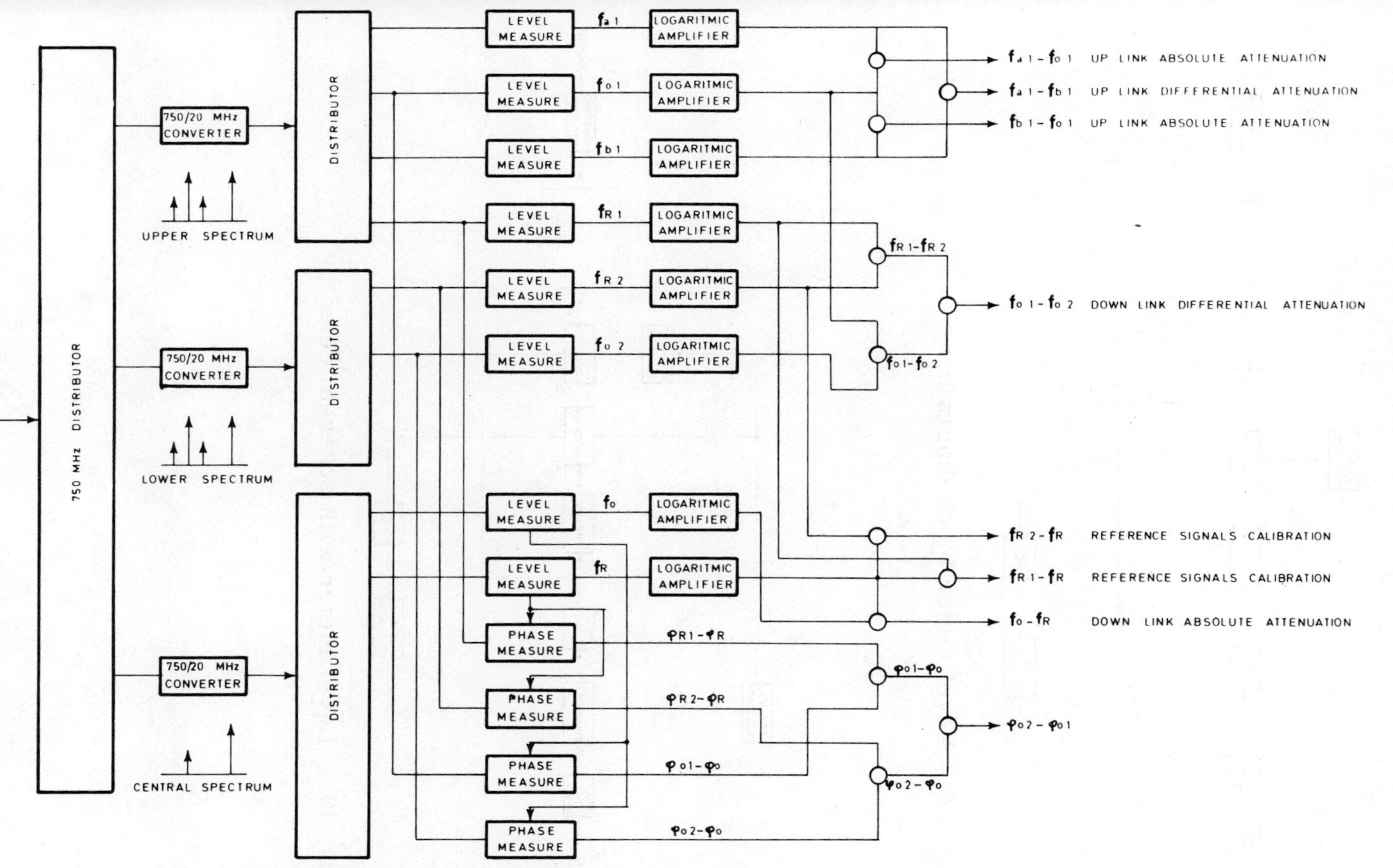

Fig. 11 Propagation measurements system .

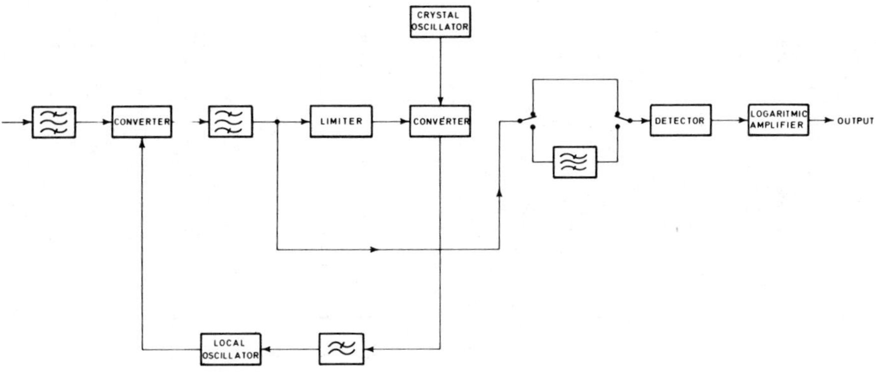

Fig. 12 Level meter block diagram.

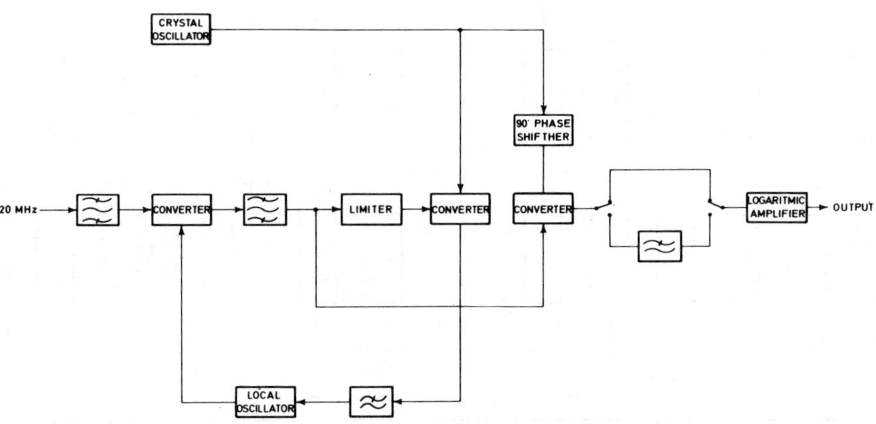

Fig. 13 Level meter with coherent detector.

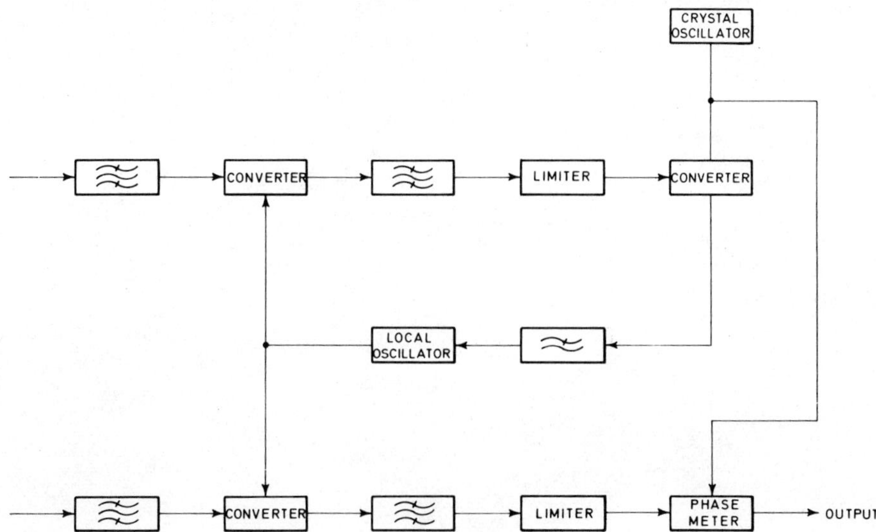

Fig. 14 Phase meter block diagram.

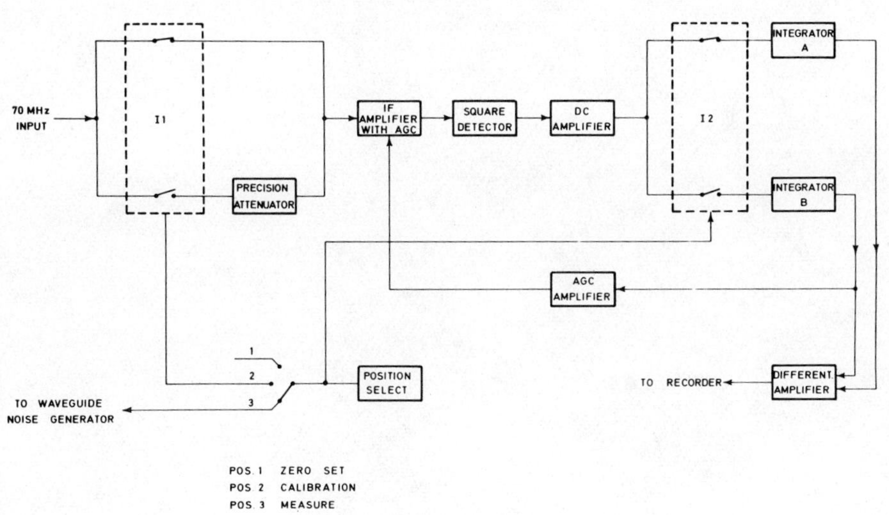

Fig. 15 Radiometer block diagram.

III. SYSTEMS FOR THE EMERGING NATIONS

SYSTEMS FOR EMERGING NATIONS: AN OVERVIEW

Robert P. Morgan*

Washington University, St. Louis, Mo.

and

Joseph B. Margolin+

George Washington University, Washington, D. C.

Abstract

This article provides an overview of papers presented at the 1970 AIAA Communications Satellite Systems Conference Session on Systems for Emerging Nations and of efforts by less-developed nations to utilize communication satellites for meeting educational and national development needs. The India/U.S. ITV Satellite Experiment Project will involve the broadcasting of family planning, agricultural and community development information direct to augmented television receivers in rural India via the ATS-F satellite. Project SACI is a Brazilian program in which plans are being made for the use of communication satellites to help meet that country's educational needs. Taken as a whole, the Conference papers include consideration not only of technical systems design but of educational software requirements and social, economic, legal and political factors as well.

Introduction. The India/U. S. ITV Satellite Experiment

The period 1969-1970 saw two developing countries, India and Brazil, take steps towards the utilization of communication satellites for educational and national development needs. On September 18, 1969, the Department of Atomic Energy (DAE) of India signed a Memorandum of Understanding with the National Aeronautics and Space Administration (NASA) of the United States to conduct the India/U. S. ITV Satellite Experiment Project. According to the basic purpose of the

---

Presented at the AIAA 3rd Communications Satellite Systems Conference, Los Angeles, Calif., April 6-8, 1970 (not preprinted).

*Director, International Development Technology Program.

+Director, Educational Policy Group, Program of Policy Studies in Science and Technology.

Memorandum[1], "The DAE and NASA agree to use their best efforts to conduct an experiment in the use of (the) ATS-F (Satellite) for direct broadcast to rural community receivers and limited rediffusion (rebroadcast) through VHF transmitters of Indian-developed instructional TV program materials."

The experiment calls for NASA to make available to the Indians the onboard ATS-F satellite UHF-FM transmitter that could be used to transmit one video and two audio channels to augmented conventional TV receivers. India will be totally responsible for implementing the ground hardware and software aspects of the experiment that is now anticipated to start in 1973 and last for approximately one year. The general experimental objectives are to:[1]

a) "Gain experience in the development, testing, and management of a satellite-based instructional television system, particularly in rural areas, and to determine optimal system parameters.

b) "Demonstrate the potential value of satellite technology in the rapid development of effective mass communications in developing countries.

c) "Demonstrate the potential value of satellite broadcast TV in the practical instruction of village inhabitants.

d) "Stimulate national development in India, with important managerial, economic, technological, and social applications."

Primary Indian instructional objectives in the NASA-DAE experiment include using television to contribute to family planning objectives, to improve agricultural practices, and to contribute to national integration. Four rural areas in the least-developed sections of India have been chosen as locations for placing a total of 2000 community television receivers in 2000 villages.[2,3] Urban areas also will receive and rebroadcast satellite television signals. At present, television in India is used to a very limited extent.[4] In early 1967, the only television broadcasts were in the New Delhi area with some 5000 receivers in operation and an estimated audience of 288,000 persons.[5]

Hoped-for side effects are the accelerated development of an Indian electronics industry and the creation of a spirit of national cohesiveness through participation in a national development project which might have a unifying effect, offsetting some of the centrifugal forces currently at work in India. The NASA-DAE experiment is viewed by leading Indian space scientists as the forerunner of a national satellite system which would reach many of India's 568,000 villages in the late 1970's.[6] One estimate indicates[2] that an Indian national satellite system would result in substantial cost savings over an equivalent terrestrial microwave net with conventional rebroadcast stations. However, some concern has been expressed about both the economic and social implications of the ITV experiment and subsequent rapid development of satellite-based television in India.[7,8] A key issue within India has been the question of which Indian agency or agencies should control the satellite television transmission.[9]

The moving force behind the ITV experiment is Dr. Vikram Sarabhai, Chairman of the Indian Atomic Energy Commission and Space Research Organization. Dr. Sarabhai sees in television a powerful tool to help India meet its staggering development problems. He closed a recent speech before the Society of International Development with the following remarks:[2] "It is quite clear that the problems of management and software for a television system to perform the role of an initiator of change and development are more formidable than the problem of hardware development. But when we have worked through our problems and learned through experience we would have gained something far more precious from the standpoint of development than the television system."

Brazil's Project SACI

Project SACI is a Brazilian program focusing on the use of satellite communication to help meet that country's educational development needs. Although Brazil has not yet signed an agreement with a country capable of launching a satellite, a sizable Brazilian planning program is underway, and ministerial decrees have been passed by the ruling military government to expedite the organization and implementation of a Brazilian educational satellite system.[10] According to a February 1970 report,[10] the SACI group has expanded from four engineers in 1968 to twenty-eight engineers and fifteen educators, sociologists, economists, and TV programming experts working full time. Consulting assistance is being received from United States and British space systems organizations.

The case for a Brazilian educational satellite system has been stated by Fernando de Mendonça, Scientific Director of Project SACI:[11]

"In 1967 roughly 40% of the Brazilian population could be classified as illiterate. Five million children of school age did not attend school. Brazil has one of the worst literacy records in South America. This is not only bad from humanistic and egalitarian standpoints, but constitutes a severe obstacle to economic progress."

"It seems unlikely that the quantitative and qualitative inadequacies of the Brazilian educational system can or should be tackled by an extension of the type of education presently in use...."

"Large-scale use of the new media, including a combination of satellite and ground radio and television transmission, seems to be the only solution...."

Brazil has some 4,000,000 urban television receivers now in use.[10] Terrestrial telecommunications infrastructure has been and is being developed, primarily along its Southeastern Atlantic coastline. Part of the rationale for a Brazilian satellite system would be to spur the development of remote sections of this vast country, especially the poorer Northeast region, by improving communication with and making use of the educational resources of the cities of the Southeast. According to Dr. de Mendonça, large-scale use of the new media via a combination of satellite and ground radio and television will result in "the establishment of a powerful new source for community development and the ability to reach any area of the country sooner and at a much lower cost than with surface communications."[11]

The 1970 AIAA Communication Satellite Systems Conference

At the conference session on "Systems for Emerging Nations" held on April 7, 1970, six papers[16-21] were presented on various aspects of communication satellite systems for emerging nations. Two of these papers (Jamison et al. and Prasada and Singh) dealt specifically with communication satellite systems planning for Brazil and India, respectively. One paper (Pay) proposed an ambitious system for providing technical aid to developing countries via a consulting services satellite. The other three papers (Sarkar, Dlugatch, Jordan) dealt in a general way with economic, social, and political aspects of satellite utilization.‡

The presentation by Jamison, Suppes, and Leme concentrated on plans for the development of software for the Brazilian

---

‡Because of space limitations, condensed versions of only those papers presented by Singh, Dlugatch, and Jordan have been included by the Editors in this Volume.

Advanced Educational Technology System (AETS/SACI). Included are estimates of the number of hours of programming required for an operational system. In contrast to many studies that ignore software aspects, this study indicates that some 143,000 distinct hours of programming are required from a staff of some 400 programmers, not including studio, graphic arts, and administrative support, at an initial cost of $72 million or about $25 million/year during the first three years. An additional $15 million/year is estimated to be required to expand and substitute software on a continuing basis. The authors also estimate that about 50 channels will be required in a 14-hour/day operation, which raises the possibility that some mix of radio and television (both full-motion and slow-scan or still-picture) will be utilized because of costs associated with all full-motion television.

The emphasis on software planning for the Brazilian system is in marked contrast to the general literature on satellite systems for education. As the summary of the paper distributed at the Conference makes clear, an approach which emphasizes software is valuable even if a satellite-based educational system does not come into being because "the software can be used with conventional radio and television, and the cadre of trained personnel will have basic skills transferable to programming in any required formats." The summary also describes a series of proposed educational experiments aimed mainly at the primary school level but with activity at the secondary and university levels as well. In general, the presentation indicated a serious effort on the part of Brazil to move towards implementation of an advanced educational technology system utilizing communications satellites.

The paper by Prasada and Singh proposes a hybrid airborne-satellite system for communication and mass education for India's development. This paper, which summarizes work performed in a system design course at the Indian Institute of Technology, Kanpur, and is not associated with the NASA-DAE experiment, outlines a system which is national in scope. Circling airplanes are used to broadcast to some fifteen regions in India in the appropriate regional languages. A synchronous satellite is employed for nationwide TV program distribution and communications. The authors believe that a uniscript, multi-lingual direct broadcast satellite is unsuited to the needs of a country with the ethnic and linguistic diversity of India and that satellites must take their place within a proper communications "mix."

The presentation by Sarkar assessed the role of new technology and, more specifically, communications satellites for bringing rapid socio-economic development to emerging nations. He concentrated not only on the benefits to be derived from satellites but on the possible dangers which might result if new technology is applied without a thorough understanding of the goals and values of people in those nations. The presentation, which was of a philosophical nature, also examined certain aspects of international development and the function of international organizations.

Pay's paper presented a preliminary feasibility study of a system, Project Delphi, for providing technical assistance via satellite from advanced technology centers to centers in developing areas. A satellite video-telephone system is proposed for serving 9000 experts by 1980. Such a system which, in effect, substitutes communications for transportation, could mean a reduction of U. S. technical assistance experts abroad, provided that communication via satellite proved to be a satisfactory alternative to the overseas expert. The ambitious plan calls for four synchronous satellites covering four regions and, in some locations, utilizing as much as 6.4 GHz of bandwidth and handling as many as 175 television channels. In view of the costs and spectrum resource utilization of such a project, it would seem appropriate to consider in more detail just how such communications links might be employed for technical assistance, what the software and programming would consist of, and what benefits could be expected for the costs to be incurred.

Dlugatch's paper, "Status Report on Communications Satellite Systems for Developing Nations," examines a number of proposed systems from the standpoint of the user. He concludes that much more attention must be paid to nontechnical aspects of satellite-based system proposals, such as local economic growth factors, cultural and educational needs, regionalism, manpower resources, industrial capacity, and national and international politics. Key questions which Dlugatch feels must be answered are: Are the nation's resources adequate to supply the required number of professionals which will be needed as the satellite reaches more pupils? Where will hardware maintainers and installers come from? How much teacher training and curriculum updating can the educational budget afford? How and what standards will be used to measure effectiveness?

In spite of assertions that the technology is moving too rapidly, Dlugatch lists a number of technical problems which he feels must be attacked to assure success. These include: 1) development of random multiple access satellites, 2) development of appropriate educational technology, 3) development of cheap, reliable, small receiving terminals, and 4) development of low-cost, efficient maintenance systems. The need for developed countries to debug thoroughly both hardware and software "at home" rather than after a "sale" is emphasized if trust and confidence on the part of emerging nations are to be developed and retained.

The final paper by Jordan constitutes a preliminary analysis, from a generalized economic point of view, of the use of educational television for improving the quality and quantity of education in developing countries. A simple model is presented which relates economic development in emerging nations to education and the level of applied technology. Television broadcast satellite systems appear sufficiently attractive to Jordan for aiding in economic development and nation building to warrant more detailed studies of costs and benefits. He feels that it is essential that such studies be carried out in view of the scarcity of resources in many countries and the many competing demands for use of those resources.

The Conference also featured a panel discussion on "Systems for Emerging Nations" held on April 7, 1970. Panelists included all of the speakers at the afternoon sessions and three additional members: 1) Mr. James R. DuMolin, Research Assistant, International Development Technology Program, Washington University, St. Louis, Mo.; 2) Dr. John Hanessian, Jr., Director, International Studies Group, Program of Policy Studies in Science and Technology, George Washington University, Washington, D. C.; and 3) Dr. Walter A. Radius, Consultant, National Aeronautics and Space Administration, Washington, D. C.

The discussion focused to a large extent on policy aspects and on nontechnical factors, mirroring the concerns of many of the speakers in the afternoon. Mr. DuMolin, who had spent two years in developmental rural education in the Peace Corps in Brazil, spoke of the need for good maintenance and of the problems that would have to be overcome because of unreliable electric power and other uncertainties in the rural infrastructure. Dr. Hanessian stressed the need for an innovative government which is determined to revolutionize the educational system of a developing country in order to bring about social change which in turn will facilitate economic development and modernization. He stated that one good way to do this is through educational television and direct broadcast satellites, but there are important international problems to be considered such as the question of intrusion into national sovereignty and the possibility of harmful propaganda. Dr. Radius reiterated the need for careful planning and a realistic approach to the costs of satellite systems. A lively discussion ensued with some members of the audience taking strenuous exception to the somewhat cautious views of the panelists.

Concluding Observations

In contrast to the technological emphasis of previous years, the 1970 AIAA Conference sessions on "Systems for Emerging Nations" examined some of the economic, social, cultural, and organizational obstacles to large-scale improvement of education in less-developed countries using communication satellites. Meetings in the previous years had viewed communication satellites as "an effective form of foreign aid."[12] A more cautious mood seemed to prevail this year, perhaps reflecting the fact that the United States had been exhibiting its lowest interest in foreign aid since World War II as a result of pressing domestic problems and its seemingly endless involvement in a tragic and an unpopular war. Since the meeting took place, Mainland China has successfully launched and orbited a communications satellite while, in the United States, budget cuts have forced NASA to delay the launching of the ATS-F satellite from 1972 to 1973.

Some people believe that the development of new and effective systems for education, making use of satellites and related technology, could result in substantial improvements in the educational capability of nations as well as better methods of international cooperation. Others feel that the rapid spread of new communications technology could result in the squandering of precious foreign exchange by less-developed countries, social upheaval, and the extension of centralized political control by repressive governments. A somewhat middle-of-the-road view holds that, "The dangers of communication satellites, in terms of propaganda, and political or cultural hegemony, appear to have been exaggerated, and so do the hopes vested in such development as a rapid breakthrough in educational television in the under-developed countries."[13] In general, political aspects of foreign aid and international communications were alluded to at the Conference but were not discussed to any extent.

In the future, greater attention will likely be paid to potential domestic applications of satellite technology, in competition with or complementary to cable TV, video recording, and other communications and educational innovations.[14] The role of the United States vis-a-vis the less-developed countries is tied to U.S. foreign aid policy which, in turn, is subject to current review and debate. Although this may not be the appropriate forum to enter that debate, it is clear that the gap between the "haves" and "have-nots" continues to be real and that the per capita GNP in the U.S. is thirty to forty times that of India; thus, the U.S., with 6% of the world's population, consumes 40-50% of the world's resources. Finding ways to make constructive use of technology to meet human needs within a framework of international peace, justice, and cooperation should become a major priority for the United States.[15]

According to the 1967 UNESCO Yearbook, there were 700,000,000 illiterate people over 15 years of age in the world in the year 1950. In the decade from 1950 to 1960, this figure rose by another 40 million people. Many countries are making great efforts to improve education but qualified teachers and financial resources are in short supply.

Two developing countries, India and Brazil, are moving forward to implement plans for utilizing communication satellites to aid in achieving their educational and development objectives. The results of these efforts may have important implications for other countries, including the United States, which is entering the 1970's with serious educational problems and with a need to re-examine its role in the world.

References

[1]"Memorandum of Understanding between the Department of Atomic Energy of the Government of India and the United States Aeronautics and Space Administration," September 18, 1969, NASA, Washington, D. C.

[2]Sarabhai, V., "Television for Development," speech delivered at meeting of Society for International Development, New Delhi, November 14-17, 1969.

[3]Lal, C., "Choice of Receiving Clusters for the National TV Satellite Project," Indian National Committee for Space Research, 1969.

[4]"A Quantitative Evaluation of the Effectiveness of a Pilot Rural Agriculture Television Project," Department of Adult Education, National Institute of Education, National Council of Educational Research and Training, New Delhi, India, July 1968.

[5]Owen, W., "Distance and Development: Transport and Communications in India," The Brookings Institution, Washington, D. C., 1968, pp. 110-111.

[6]Sarabhai, V. et al., "INSAT-I, The Case For a National Satellite", Science Today, Vol. 4, No. 11, May 1970, pp. 52-54

[7]Khanna, K. C., "Satellite TV Grid For India, Case for Keeping The Options Open," Times of India, December, 1969.

[8]Srirangan, T. V., "Space and Telecommunications in India--Some Doubts," Science Today, Vol. 4, No. 11, May 1970, No. 11, pp. 66-67

[9]"National Satellite Authority Suggested," Times of India, November 15, 1969.

[10]"Project SACI Report No. III," Tech. Rept. LAFE-104, São José dos Campos-São Paulo-Brasil, February 1970.

[11]Mendonça, F. de, in Astronautics & Aeronautics, Vol. 8, February 1970, pp. 37-41.

[12]Sielman, P., "Satellite Educational Television-An Effective Form of Foreign Aid," presented at AIAA 2nd Communications Satellite Systems Conference, San Francisco, Calif., April 8-10, 1968.

[13]"Communications Satellites," Stockholm Papers No. 1, Stockholm International Peace Research Institute, Almquist and Wiksell, Sweden, 1969, p. 52.

[14]"Progress Report: Program on Application of Communications Satellites to Educational Development," Washington Univ., St. Louis, Mo., November 12, 1970.

[15]Morgan, R. P., Walker, R. M., and Falconer, J. P. R., "Science, Technology and the United Nations," statement prepared for the St. Louis Hearings of the Presidential Commission for the Observance of the 25th Anniversary of the United Nations, Washington Univ., St. Louis, October, 1970.

[16]Prasada, B. and Singh, J.P., "ACME - A Hybrid Airborne Satellite Television and Communications System for India" (elsewhere in this volume).

[17]Dlugatch, I., "Status Report on Communications Satellite Systems for Developing Nations" (elsewhere in this volume).

[18]Jordan, P.L., "Communications Satellites, Technology Transfer, and Economic Development" (elsewhere in this volume).

[19]Pay, R.G., "Project Delphi - Technical Aid to the Developing Nations by Consulting Services Satellite," AIAA Paper 70-474, Los Angeles, Calif., April 1970.

[20]Jamison, M. and Leme, M.J., "Software Production for the Brazilian Advanced Educational Technology System" (obtainable from the authors).

[21]Sarkar, S.K., "Communication Satellite for Socio-Economic Development (obtainable from the author).

ACME - A HYBRID AIRBORNE SATELLITE TELEVISION AND COMMUNICATIONS SYSTEM FOR INDIA

Birendra Prasada* and Jai P. Singh+

Indian Institute of Technology, Kanpur, India

Abstract

A proposed television and communications system for India is described. It uses airborne television transmitters to provide both instant and delayed localized telecasts and a satellite in the geostationary orbit at 82°E for nationwide television distribution and long-distance inland voice, data, and facsimile communications among the primary and regional switching centers of the Indian telecommunications network. The economics, technical feasibility, and the effectiveness of a number of alternatives have been studied and compared. It is concluded that a hybrid airborne satellite system, as described in this paper, is the best choice for India: it is economical; it involves a technology which is quite compatible to that existing in India; its implementations may result in considerable spin-offs for the Indian electronic industry; and, most important, it can provide effective programing based on regional requirements. The importance of software considerations, such as the system organization, communication content, etc., is shown.

Presented as Paper 70-472 at the AIAA 3rd Communications Satellite Systems Conference, Los Angeles, Calif., April 6-8, 1970. The authors gratefully acknowledge the contributions of H. K. Kesavan (Professor of Electrical Engineering, IIT-K and Chairman, Department of Systems Design, University of Waterloo, Canada), L. S. Srinath (Professor of Mechanical Engineering, IIT-K), and R. C. Arora, R. Balasubramanian, S. Chatterjee, M. Krishnan, K. B. Ohri, P. R. K. Rao, V. S. Rathore, S. P. Thakur, and K. P. Vishwakarma--all members of the Systems Design Group.

*Professor and Head, Department of Electrical Engineering.

+Now Research Engineer, International Development Technology Program, Washington University, St. Louis, Mo.

Introduction

ACME system for TV and communications is an outcome of an interdisciplinary systems design project undertaken at IIT-Kanpur by a group of ten students and three faculty members. The problem for the Systems Design Group was to design a system for communication and mass education for India's development. A hybrid airborne satellite system was chosen as the best system. The functional design of the ground, airborne, and space segments of the system was undertaken, and the results form the main theme of this paper.

The nontechnical, i.e., socio-economic and political, considerations discussed in the first two sections of this paper have heavily influenced the choice and the design of the system. Like the design of any other meaningful mass communication and education system, the needs and the environment of the particular socio-political system have been considered concurrently along with purely technological considerations in evaluating various alternatives and synthesizing the final system.

System Identification

Everyone is aware of the dilemma that India currently faces--poverty; unsatisfactory dividends from the investments toward development projects; narrowing of loyalties to caste, region, religion, and languages; irrationality; intolerance, and lack of empathy among her citizens. There is an immediate and indispensible need for educating masses, disseminating to people information and skills necessary for development, effecting social change, and inculcating desired attitudes and attributes necessary for the development of the country. To make the fullest use of the new technology, a harmony is needed established between the social system and the technological environment. The common belief that economic change brings social change is not reasonable. All available evidences hold social customs, tradition, and attitude responsible for India's slow economic development.[1,2] To make any real breakthrough in developmental results, it is first necessary to make appropriate social changes. Special attention also should be paid to the recommendations of the Kothari Commission on Education[3] for reorienting the purposes of education and bringing excellence to it, for education is the main bearer of social change. An examination of the situation suggests that multiplication of the conventional schools and teachers does not provide an effective and efficient solution any longer for India's needs of revitalizing and reorienting the educational system and providing adequate-quality education to the increasing number of students.

In view of the needs, the following objectives were proposed for the system: 1) mass communication for information dissemination to create the social and political awareness necessary for the citizen's participation in the social, economic, and political transformation of the country and to inculcate desired attitudes and attributes; 2) quality school education up to the secondary level for the entire population in the age group of 5-14 years in a phased manner; 3) vocational training for an adequate percentage of the Indian population of all age groups to utilize human and material resources effectively; 4) literacy education for illiterate people above 14 years of age and continuing education for all; 5) civilian and military long-distance inland communication channels for data, voice, and facsimile communication; and 6) meaningful entertainment for the masses.

Television is chosen as the most appropriate instrument for achieving these. Other choices at this stage were that of conventional school system, radio, and film. Television is selected because it offers a unique communication process with speedy and simple distribution and is immune to the barrier of illiteracy. It has been found to be an intimate and persuasive medium. It has the potential to bring quality education to the massess. It can help to a great extent in reorienting the purposes of education and revitalizing it; it can be a teacher of students and teachers simultaneously and can help immensely in the professional growth of teachers.

After establishing that television is the best choice, the economics, technical feasibility, and effectiveness of five different proposals for a nationwide TV system were studied in detail. It is concluded that an airborne TV system, supplemented by a satellite in the geostationary orbit for long-distance inland voice, data, and facsimile communication, is best suited for our purpose. The five proposals were: 1) a terrestrial VHF system with microwave line-of-sight links for regional as well as national real-time program distribution (at least 2 channels are available in every part of India); 2) an airborne TV system without provision for real-time nationwide program distribution (2 channels available in every region); 3) a uniscript-multilingual direct broadcast satellite system (2 TV channels are available throughout the country, and each video channel is accompanied by several audio channels); 4) a satellite multiscript-multilingual direct broadcast system employing 8 satellites in the geostationary orbit; 24 TV channels available throughout the country in 12 major languages; and 5) a hybrid airborne-satellite system — airborne TV for regional coverage, satellite for nationwide TV program distribution and communication purposes; low-UHF band terrestrial transmitters employed to provide TV coverage to unsatisfactory reception areas below circling aircraft and to areas where airborne TV is not economical.

During the Fourth Five-Year Plan of India, the Post and Telegraphs Department proposes to link the state capitals and the major cities by a 6-GHz microwave communication system. A terrestrial TV system is prohibitively expensive, even if an assumption is made that the additional capacity can be built along the Post and Telegraphs Department links at the cost of $ 933/km. Since the Post and Telegraphs Department plan is for about 10,500 km of microwave network and the desired terrestrial system[1] requires 30,000 km of microwave network, at least 20,000 km must be erected at a cost of $ 6000/km.

The multilingual-multiscript direct satellite broadcast proposal is uneconomical. The available usable frequency spectrum limits the feasibility of this system to either SHF/VSB or UHF/VSB, both of which demand rf power beyond the capabilities of the present day or near-term (=1972) state-of-the-art. In uniscript-multilingual satellite direct broadcast systems problems are created by the Super High Frequency (SHF) reception; UHF/VSB systems are not possible for the 1972 time frame. The technology involved in SHF operation is incompatible to that existing in India. A supply analysis has shown that India would require a minimum of 1.14 million 23" TV receivers. Manufacture and maintenance of such a large number of receivers for SHF reception, whether direct or through a SHF to VHF and WBFM to VSB converter, would make India heavily dependent on the developed nations. A detailed economic study shows that the direct broadcast proposals involve large amounts of foreign exchange and would thus produce excessive strains on the fluid Indian economy.[1]

A study of the effectiveness of the uniscript-multilingual satellite direct broadcast systems shows that these are not suited to the needs of India. The regional and linguistic diversities of India cannot be ignored in the design of a system for communication and mass education. The effectiveness of the communication depends heavily on the narrowing of the credibility gap between the TV communicator and viewers. It is also dependent on how closely viewers identify themselves with the TV character. Regional language broadcasts are a necessity for India to maintain compatibility with the policy of providing education in the regional language and for reaching millions of illiterates who cannot be reached otherwise. Also, it is necessary to program the material so that it is relevant to the regional character of the viewers. Different needs of education and curricular patterns, differing contents of courses based on local interests and surroundings, and different standards of accomplishment among various regions require regional programing. It would not be realistic to telecast a single suggestion for cultivation of wheat to farmers in Punjab and Madras when the two regions have little in common in terms of soil characteristics, seasons, method of cultivation, and food habits. Providing several audio channels for each video channel is not satisfactory. Here the communication will be seriously affected by the lack of identification between the TV communicator and viewers, which widens the credibility gap.

In these circumstances, a hybrid airborne satellite system appears to be the best choice for India. Being regional in nature, it would provide effective and meaningful communication. Table 1 represents a summary of the costs of the five proposals. The hybrid system is slightly more expensive than the airborne and uniscript-multilingual satellite direct broadcast systems. The revenue from the commercial use of satellite communication channels makes the hybrid system more economical and attractive than the airborne system. The uniscript-multilingual direct broadcast satellite proposal is inadequate and unsuitable for our requirements, as discussed earlier.

System Description

The system proposed is based on the concept of airborne television broadcasting where the altitude of the airborne transmitter is responsible for the large area coverage resulting from the increased line-of-sight. A satellite in the geostationary orbit, 82°E of Greenwich, would provide a real-time nationwide television program distribution and long-distance inland voice, data, and facsimile communication. Low-power ground UHF transmitters would be used to provide TV coverage to zones of unsatisfactory reception beneath circling aircraft. This is the result of the irregular radiations off the ends of the airborne transmitting antenna, which is horizontally polarized. This irregular radiation is common to the ground-based TV antennae also, but, if the null extends out to 45° below the horizon, it is inconsequential for a ground transmitter because it extends over a radius of only 500 ft for a 500-ft tower. The null effect for a 22,000-ft airborne antenna is 4.2 miles, which is further aggravated by the fact that the air-craft moves anywhere within a 20-mile-diam circle, thus moving the null effect and extending the zone of unsatisfactory reception to a 42-mile diam. The geometry of Kerala does not favor airborne TV, and hence Kerala has been provided a terrestrial coverage woven around one high-power transmitter and two medium-power unattended "boosters." Figure 1 shows the TV coverage of India as envisioned in the system. It also shows the channel assignment and the language(s) to be broadcast in each region.

Table 1 Summary of costs of various proposals

System	Cost	
	Total first cost (in million $)	Total annual cost[a] (in million $)
1) Terrestrial	622	201.3
2) Airborne	305.3	144
3) Uniscript-multilingual satellite direct broadcast[b]	503	131
4) Multiscript-multilingual satellite direct broadcast[b]	627	180.6
5) Hybrid airborne-satellite	445.5	155

[a]Total Annual Cost (TAC) is the total annual expense of owning the system and keeping it in use. It has a first cost component in terms of depreciation and interest, and the other component is called operating cost, which includes maintenance, operation, programing, and overhead expenses. Comparison of the costs of the various proposals should be made on the basis of TAC. For methodology of costing and detailed calculations, see Ref. 1.

[b]Includes cost of satellite communication (receive-transmit) facilities to ten high-capacity communication traffic centers.

The ACME system provides at least two UHF channels in over 90% of the Indian territory. Some 80% of the Indian population receives programs which are suited to their needs in their own regional language.

A detailed study has shown that DC-6B four-engine prop aircraft, operating at an altitude of 22,000 ft, are the most suitable aircraft for our purposes of installing airborne TV transmitters which would cover an area with a radius of nearly 200 miles. Every such aircraft would have two TV transmitters, each with 3-kw rf power, and other necessary monitoring, navigational, and communication equipment. These aircraft would trace a tight figure 8 in a 10-mile radius, and the programs would be transmitted by a gyroscopically stabilized turnstile antenna. Aircraft trace a figure 8 to minimize the poor reception area on the ground at the center of the orbit area.

The continuous flight time of a DC-6B aircraft at 22,000-ft altitude, with 10,000 lb of payload and 45 min of reserve fuel for diversion to a distant airport in case of adverse weather, is slightly more than 5.5 hr. With two fully equipped aircraft, each region would have uninterrupted 11-hour service daily. In addition, the system plans a central pool of five aircraft. Aircraft from this pool will be assigned on demand to various regional setups during emergency and scheduled maintenance of the regular aircraft. Use of these backup aircraft would enhance the reliability of operation.

The system would have a ground programing, unlike the Midwest ETV program in the USA where programs originated in the aircraft itself from tapes and films. In the ACME system, programs will originate on ground and will be relayed to the airborne transmitters through 2-GHz microwave ground-to-aircraft radio-relay links. For 6 hours, during the daytime, programs for both TV channels will originate at their respective ground stations, either live or recorded. Ground transmitters, for filling the unsatisfactory reception zones beneath circling airborne transmitters, also will broadcast the same program on UHF frequencies different than the ones in use by the airborne transmitters in that particular region. In the later half of the day's operation, one of the two local channels would be substituted by a national TV hookup distributed by the satellite. In the beginning, only one central station would have facilities for distributing programs through the satellite, but at a later stage all sixteen regional installations can be provided with this facility at a small additional cost. The ground programing employed here has the advantages of live programs, reduction in the weight to be carried by the aircraft that makes feasible the addition of a third channel at the time of expansion, and the real-time national program distribution.

The satellite provides multiple access to ten high-capacity ground-communication traffic centers and has the capacity to relay 3600 one-way voice channels or a corresponding number of data and facsimile channels. When the nationwide TV program distribution is operative for five hours a day, this capacity is reduced to 3120.

Table 2 gives a detailed breakdown of the system components and cost.

Design of the System

Introduction

For the purposes of design, the system has been divided into three segments--ground, airborne, and space. The ground segment comprises all regional program originating setups, satellite communication Earth terminals, UHF transmitters, 1.14 million community TV receivers, and 800,000 windmill generators to power the TV receivers in unelectrified areas. The airborne segment includes 35 airborne transmitters, whereas the space segment consists of a lone dual-spin communications satellite in the geostationary orbit.

System Specifications

The airborne and terrestrial broadcasts will conform to CCIR 'B' specifications. It is a 625-line, 50-frame/sec., 5.0-MHz system which employs vestigial sideband modulation for video transmission and frequency modulation for audio. India has adopted this standard, and it appears to be the best compromise between the required resolution and economy in frequency spectrum usage. The low-UHF band (470-585 MHz) has been selected for airborne and terrestrial telecasts. The choice of VHF operation has been precluded as the radio noises generated within a prop aircraft have been found to occupy frequency spectrum up to 400 MHz. Economy in the use of the frequency spectrum has been obtained by the "use one, skip one" plan of frequency assignment as shown in Fig. 1. Worst-case co-channel separation is 540 miles whereas the worst-cast adjacent channel separation is 310 miles. Airborne broadcast links have been designed to provide a first-grade service in a 200-mile radius.

The satellite communication links, whether for communications or TV, employ wideband frequency modulation. Wideband FM can tolerate much greater interference levels than compact modulation techniques, such as AM, VSB, and SSB, with no substantial effect on the quality of users' signals. Use of FM also permits substantial reduction in satellite power. The reduction in satellite power is the result not only of the

Table 2 Cost breakdown of hybrid airborne satellite system

A. Total equipment and facilities cost	
a) Ground segment	
15 ground stations attached with airborne transmitters (with program origination, satellite TV reception and ground-to-airborne transmitter TV relay facilities), @$2 million	$ 30 million
3 highpower ground stations (with program origination and satellite TV reception facilities), @$2 million	$ 6 million
Satellite communication (transmit-receive) facilities for ten selected main and primary telecommunications switching centers, @$2 million	$ 20 million
Satellite telemetry, command, and control center (extended facilities to Experimental Satellite Communication Earth Station, Ahmedabad), @$3 million	$ 3 million
b) Airborne segment	
35 DC-6B aircraft with 3-5kw UHF transmitters, @$1.47 million	$ 51.5 million
c) Space segment	
ACME satellite in the geostationary orbit (cost of the satellite and orbital placement), @$17 million	$ 17 million
d) Receiving segment	
1.14 million UHF receivers, @$180.0 (includes antenna and installation costs)	$206 million
800,000 windmill generators for unelectrified areas, @$140.00	$112 million
Total first cost	$445.5 million
B. Total annual cost	
(Annual operating cost + interest (6%) + depreciation (for satellite 20% annual; for rest of the equipment 10% annual) + programing cost)	$155 million

usual improvement in signal-to-noise ratio (SNR) related to the FM modulation index, but also of the non-uniform distribution of the noise in the demodulated spectrum. In addition, the use of FM permits operation of the satellite power TWT in the most power efficient, i.e., saturated, region. In addition to the reduced power requirements, operation in the saturated region also compensates for short-term fluctuations in the signal level, provided the signal is strong enough to drive the TWT into saturation.

The system is designed to provide multiple access to the satellite to 10 heavy route Earth terminals (all main and 6 selected primary switching centers of the Indian telecommunications plant) on a Frequency Division (FDMA) basis. A multi-destination carrier concept has been employed. In this technique, a traffic center transmitts all the channels for the nine receiving centers using a single carrier and signal channels meant for them. This reduces the number of carriers to be handled by the satellite repeater to 10 (from a possible 90), with each center transmitting a separate carrier for each destination and hence reduces the intermodulation distortion increasing the capacity of the satellite repeater for the same ERP provided the operation is not bandwidth limited. It also permits a partial demand assignment of channels because the channel assignment for the nine destinations could be altered according to demand within the total channel allocated to a center for transmission. A study was made[4] to evaluate the performance of various multiple-access techniques in view of existing telecommunication plant, near-term future expansion already on the way or about to begin, and traffic pattern. It concluded that the main utility of the satellite relay for telecommunications in India lies in relaying bulk traffic among the widely separated main and primary switching centers. Time Division Multiple Access (TDMA) would have been the ideal choice if we did not have to worry about the compatibility of existing FDM terminals at the switching centers, and those planned for the proposed expansions, and the problem of network timing control. Demand assigned systems based on single-channel per carrier and a common channel-band concept did not seem to be at all attractive in view of the complexity and cost involved and little gain, since traffic among all stations will have about the same peak hours with a great degree of overlap because the same time-zone is involved. The modified FDMA concept already described therefore has been stated as the best choice for the Indian system under existing circumstances.

Satellite up- and down-links have been assigned 6.055-6.400 GHz and 3.855-4.200 GHz frequency bands, respectively. Frequency planning has been done so as to have a flat intermodulation spectrum with smoothed peaks. The highest third-order products given by $(2f_{tv} - f_{tf})$ and $(2f_{tf} - f_{tv})$, where f_{tv} and f_{tf} are, respectively, television and telephony downlink frequencies, have been kept out of the transponder band.

In satellite relay links, TV baseband is compressed by a ratio of nearly 4:1 by exploiting its redundancy through first-order interpolation to conserve the radio frequency spectrum and to permit a smaller reduction in the communication capacity of the satellite when the TV relay is operative, following the scheme proposed by Hochman.[5]

Design of the satellite TV as well as of the communications links conforms to CCIR recommendations on the output signal-to-noise ratio (SNR). Satellite communication Earth terminals have been characterised as 34-db G/T (Antenna Gain to System Temperature (^{o}K) ratio) for communications and 32-db G/T for TV relay-receive stations. These specifications are unlike those in current international practice for Intelsat satellites, as the ACME satellite has a considerably higher rf power output than Intelsat II or III. The weighted SNR for TV signal at the output of the Earth station receiver is 63 db, whereas that of the voice channel, in worst case, is 76 db. Syllabic companding has been employed in communications links to enhance the output SNR.

In satellite links, a carrier energy dispersal technique, based on the addition of a low-frequency waveform to the baseband, is employed to bring the satellite power flux density reaching the ground within CCIR recommendation of $-(152+\varnothing/15)$‡ dbW/m^2/ 4kHz by spreading the rf power in the entire rf band.

For TV receivers, electronic tuning and modular construction for high maintainability is recommended. Use of integrated circuits promises higher reliability, compactness, and lower power requirements. Most of the functional blocks of the TV receiver are within the reach of Medium Scale Integration (MSI) technology. Each block, such as rf tuner, vision IF, audio IF, discriminator, etc., could be used as separate modules. If some fault occurs, faulty modules can be pulled out and a new one inserted, thus cutting down the off-service time.

Tables 3-5 show the characteristics of airborne TV transmitters, satellite Earth stations, and TV receivers.

---

‡$\varnothing$ is the angle of arrival on Earth.

Table 3 Airborne TV transmitter features

Aircraft	DC-6B [modified hull and tail]
rf power[a]	
visual	3 kw
aural	0.5 kw
Power amplifier	Klystron
Transmitting Antenna	25 ft turnstile mast-gyroscopically stabilized and hydraulically operated.
Auxiliary power source	20 kw gasoline turbine

[a]For each TV channel.

ACME Satellite

The Systems Design Group also considered the design of a suitable satellite for the ACME system and decided upon a conceptual design based on the concepts and techniques developed independently by Landon and Stewart[6] of RCA, Billerbeck and Owens[7] of COMSAT, and Iorillo[8] of Hughes. The proposed dual-spin cylindrically shaped satellite (Fig. 2 and Table 6) is composed of an outer section spinning at a constant rate about the axis of symmetry (spin-axis), and a nonspinning (mechanically despun) parabolic antenna platform. Spinning-pulsed hydrazine jets have been used to control the satellite attitude along the lines of NASA's Advanced Technology Satellite-C (ATS-III). The rotor section contains three flexible and deployable thin-film cadmium sulfide (CdS) solar cell arrays, which unreel in space, on command, to power the communications, telemetry, command, and control subsystems. When unreeled, these arrays make the moment of inertia about the spin axis greater than that about either of the transverse axes, and provide conditions for good stability. Stability of the satellite, when arrays are not unreeled, is taken care of by a nutation damper on the top of the despun section[1,7,8] and the control mechanism. Though no detailed design of the control mechanism has been worked out, Porcelli's design[9] seems to be very attractive for the ACME satellite. It has original features of controlling, simultaneously, the attitude and the coning motion of the spin-axis, thus eliminating the requirement of a coning damper. The long-term effects of destabilizing torques are corrected by the fast-closed loop action of the control law, which is automatically applied anytime the spin-axis drifts away from the dead-zone. The effects of the limit cycle operation thus produced are reduced by the passive damper on the despun portion. The following requirements have been establi-

Table 4 Satellite Earth station characteristics

Item	Earth station type		
	TV relay-transmit	TV relay-transmit	Message-transmit receive
Antenna	50-ft Cassegrain feed; no tracking.	50-ft Cassegrain feed; no tracking.	50-ft Cassegrain feed; no tracking.
Transmit antenna gain	56.6 db	...	56.6 db
Receive antenna gain	...	53.4 db	53.4 db
Low-noise receiver	...	Uncooled parametric amplifier.	Uncooled parametric amplifier.
Receiving system noise temperature	...	115°K	80°K
Receiver bandwidth	...	30 MHz [4.17-4.20 GHz]	345 MHz [3.855-4.20 GHz]
Demodulator	...	Phase locked	Phase locked
Transmit erp	78.7 dbw	...	78.7 dbw
Transmit bandwidth	30 MHz	...	20 MHz [per 240 voice channels]
Modulation	Wideband frequency modulation[a][m=7; 4:1 video baseband compression]	Wideband frequency modulation[a][m=7; 4:1 video baseband compression]	Wideband frequency modulation [m=8]

[a] m=deviation index.

Table 5 Specifications for community TV receivers

General	
Size	23" screen
Frequency band	Ultra High (UHF)
Power supply tolerance (normal operation for voltage changes of %)	
a) Mains operated receivers	± 10
b) Windmill operated receivers in unelectrified areas	± 30
Available controls	Power switch, tuning (fine and coarse), contrast, brightness and loudness
Input characteristics[a]	
Input impedence at antenna terminal	300 Ω
Noise figure for the receiver	10-13 db
tuning	electronic
Output characteristics	
Minimum audio frequency range	150-5000 Hz
Minimum audio frequency output at 10% distortion	2 w
Minimum picture resolution	300 lines
Minimum interlace ratio	30/70
Reliability[a]	
Minimum mean time between failures requiring servicing, averaged over a production run	1000 hr

[a]Other specifications and characteristics should conform to those given in CCIR Document XI/53(Italy), 1966-69 (for Type B receivers).

shed for the control subsystem: $20^{\circ}30'$ tilt of beam from positive yaw-axis, 0.1° tolerance on the beam direction at 3 db points, and operational life of more than 5 years.

The Communication subsystem (Fig. 3) is much different than any in present use. The ACME satellite employs a direct rf-to-rf type transponder which receives in the 6.055-6.400 GHz band and transmits in the 3.855-4.200 GHz band with a direct frequency translation.

The design of the power amplification section of the transponder is quite unconventional. The incoming 345-MHz band, which contains 15 carriers, is broken into three 115-MHz bands through a branching network. This arrangement considerably reduces the

Table 6 Design parameters for ACME satellite

Diameter	5.5 ft
Length	7.42 ft
DC Power	360 w; three thin-film CdS solar cell arrays; silver cadmium batteries for eclipse operation
Weight	
Separated weight	858 lb
Apogee motor propellent	400 lb
Final orbit weight	458 lb
Communications	
Repeater	One direct rf-to-rf type.345MHz bandwidth; partial redundancy; multiple access capability
Antenna	3.12 ft dish; mechanically despun. 5°1' beam centered at 20° 30' aspect angle
Up-link frequency	6055 - 6400 MHz
Down-link frequency	3855 - 4200 MHz
Total effective radiated power	48 dbw
Capacity[a]	1560 two-way telephone circuits and 1 TV channel or 1800 two-way telephone circuits
Modulation	Wideband frequency modulation
Telemetry	One redundant 2.5 w VHF (136MHz) transmitter; omnidirectional antenna; PCM/PM modulation
Command	A 5 x 5 code with limited error-detecting capability; PCM/FM/PM modulation; 150 MHz operation
Tracking	140 MHz tracking signal on command

[a]When operating with Earth terminals specified in Table 4.

intermodulation noise and also avoids difficulties which would have been encountered by the use of a single TWT or several low-power TWTs in parallel with a 345-MHz bandwidth. For a five-carrier operation through a single power TWT, a 3-db backoff from saturation has been given to provide a 16-db carrier to intermodulation noise ratio sufficient for our purpose.

The ACME satellite command subsystem is capable of providing 252 command functions as compared to the 20-functional-command system used in Syncom and Relay I satellites. The ACME command system has no provision for stored commands, because it was considered unnecessary, but a command verification facility has been included in the design. A five-by-five code, i.e., a code consisting of a ten-bit word, five ones, and five zeros, has been used in our system. This code contains within itself a degree of error recognition. A single error in transmission is automatically detected and rejected since it does not fulfill the code requirements of five zeros and five ones. The command link uses a PCM/FM/PM modulation scheme. The digital control information is first used to frequency-modulate a sub-carrier which then phase modulates the 150-MHz carrier.

A novel telemetry scheme has been suggested. The slowly varying nature of a large number of parameters to be telemetered and the use of a proper threshold for selective filtering permit a considerably narrow bandwidth (≈500 Hz) than used either by Intelsat or Syncom communications satellites. Telemetry link employs PCM/PM scheme at a 136-MHz carrier frequency. One solid-state transmitter provides a 140-MHz tracking signal on command from ground. Redundant encoders have been used to improve the reliability of the telemetry scheme.

Some Important Considerations

1] The point-to-pint communication via satellite is no substitute for terrestrial facilities. At least in several areas, the surface links must grow so that a) outlying cities and villages gain access to domestic satellite earth terminals and b) connections between any place to Intelsat Earth terminals avoid two satellite hops in routing international telephone traffic.

2] Software aspects of the ETV part of the system should not be underemphasized. The hardware aspects of the system are essentially international in nature — equipment can be procured from other countries, and know-how can be obtained to manufacture them indigenously, but the system software aspects are truly local. The word "software" represents those aspects of the system which do not deal with the eqipment — the organizational structure of the system, the contents of the message and its effectiveness. As far as the organizational structure is concerned, there is a need for a universally accepted organization to administer the system, which could be provided only by an autonomous corporation with very limited government control. Of necessity, the government will be the major financier, but even then 5-year financial sanctions and a post audit system can severaly restrict the government influence.[1,10]

As far as the contents of the message and its effectiveness is concerned, communications research is required in India to determine the appropriateness of the symbols and vocabularies in the program, optimal load to be carried by the television show, and the like. Research is needed to study the propagation and diffusion of information in Indian society since research conducted in other countries has little or no meaning for India. Intra-culture difficulties also are foreseen. Most of the TV program producers will be from well-educated and urban populations and thus separated by social and educational distance from the masses of the people. This may lead to incommunicability unless communications research provides us with the essential information to frame techniques for successful communication.

3] A phased implementation of the system is necessary for various reasons — to avoid excessive strains over the fluid Indian economy and to provide a backlog of programs and trained personnel for subsequent expansions. It is necessary to start the implementation with a pilot project with one or two airborne transmitters for conducting communications research and field testing of equipment.

References

[1]Prasada, B. and Singh, J. P., eds.,"Project ACME (Advanced System for Communications and Mass Education) - Final Report", June 1969, Indian Institute of Technology Kanpur, India.

[2]Schramm, W. and Nelson, L. M., "Communications Satellites for Education and Development — The Case for India, Section 3,High Priority Tasks that call for Television", July 1968, Institute for Communications Research, Stanford University, Stanford, Calif.

[3]Report of the Education Commission (1964-66), 1967, Ministry of Education, Government of India, New Delhi, India.

[4]Singh, J. P., "Multiple Access Considerations for an Indian Domestic Telecommunications Satellite System", Tech. Rept. EE-COM-II, July 1969, Department of Electrical Engineering, Indian Institute of Technology Kanpur, India.

[5]Hochman, D. et al., "Application of Redundancy Reduction to Television Bandwidth Compression", Proceedings of the IEEE, Vol. 55, No. 3, March 1967, pp. 263-266.

[6]Landon, V. D. et al., "Nutational Stability of an Axisymmetric Body Containing a Rotor", Journal of Spacecraft and Rockets, Vol. 1, No. 6, Nov.-Dec. 1964, pp. 681-684.

[7]Billerbeck, W. J. et al., "Design of a Killowatt Power System for small Comsat's", AIAA Paper 68-464, San Francisco, Calif., 1968.

[8]Iorillo, A. J. et al., "Hughes Gyrostat System", Symposium on Attitude Stabilization and Control of Dual-Spin Spacecraft, August 1967, Aerospace Corporation, Los Angeles, Calif.

[9]Porcelli, G., "Optimal Attitude Control of a Dual-Spin Satellite", Journal of Spacecraft and Rockets, Vol. 5, No. 8, Aug. 1968, pp. 881-888.

[10]Prasada, B. and Singh, J. P., "Television, Education and Communication: Software Considerations", Paper 6, Nov. 14-17, 1969, Roundtable Conference D, 11th World Congress, Society for International Development Washington D.C., New Delhi, India.

STATION NO.	TV CHANNELS	LANGUAGES
1	14, 16	Urdu
2	18, 20	Punjabi & Hindi
3	14, 16	Hindi & Rajasthani
4	15, 22	Hindi
5	14, 16	Hindi
6	18, 20	Bengali
7	15, 22	Assamese & Bengali
8	15, 22	Oriya
9	24, 26	Hindi
10	28, 30	Hindi
11	18, 20	Gujrati
12	32, 34	Marathi
13	15, 22	Telugu
14	18, 20	Kanarese
15	32, 34	Tamil
16	14, 16	Malyalam

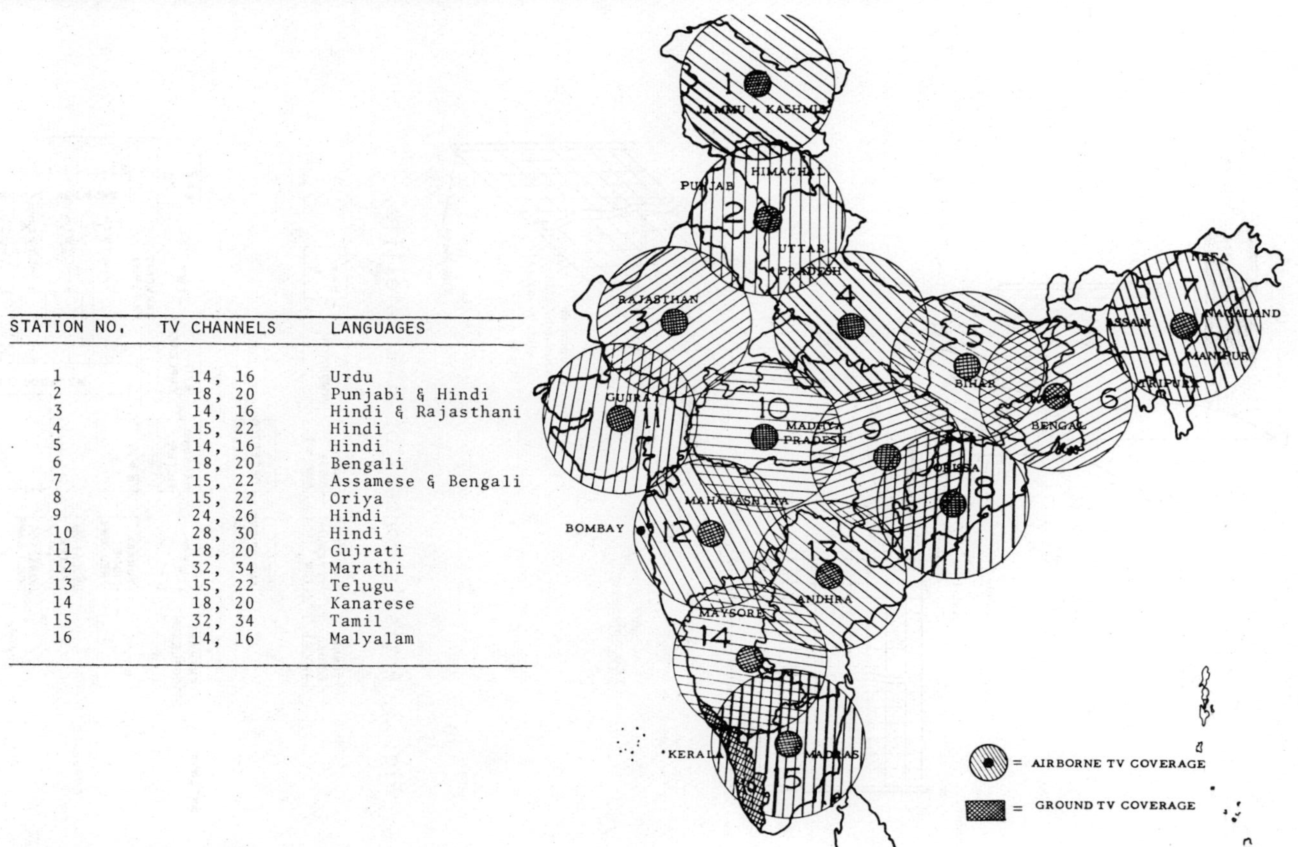

Fig. 1 TV coverage of India by the proposed system.

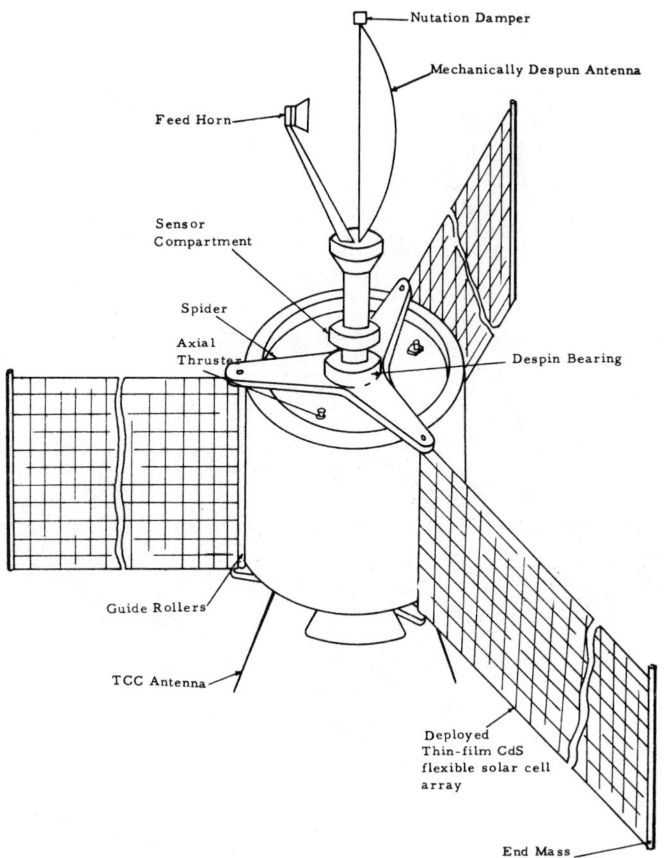

Fig. 2 Artist's concept of ACME satellite.

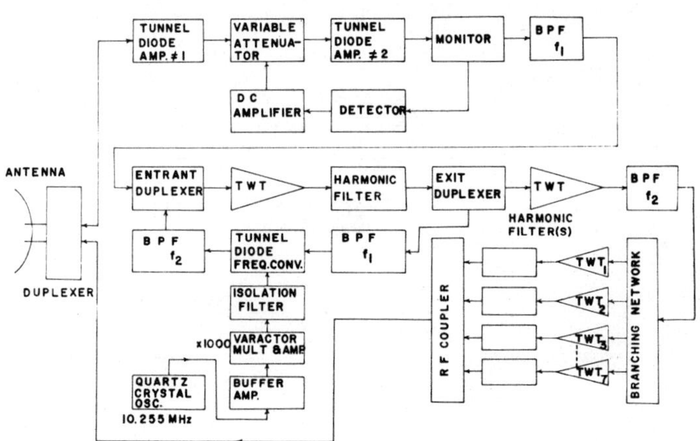

Fig. 3 ACME satellite transponder.

STATUS REPORT ON COMMUNICATIONS SATELLITE SYSTEMS FOR DEVELOPING NATIONS

Irving Dlugatch*

Hughes Aircraft Company, Culver City, California

Abstract

Many proposals have been offered for applications of communications satellites to the needs of developing countries. The purpose of this paper is to examine a few typical proposals and evaluate them from the standpoint of the user. The intent is not to deal with effectiveness but rather to analyze the reasons for the user's acceptance or rejection of the proposals, and determine whether there are deficiencies existing that prevent them from receiving wider acceptance. It is concluded that not enough consideration has been given to the non-technical aspects such as: economic growth factors peculiar to each country, cultural and educational needs, regionalism, manpower resources, industrial capacity, and national and international politics.

I. Introduction

In attempting to evaluate the many proposals for communications satellite systems for developing countries, we need to establish some basic principles for defining these systems from the user's viewpoint. That is, we need to examine the proposals through the eyes of the consumer who is not an engineer but is skilled in economics, education, and human communications. To this user: communications satellites are electronic bridges between distant communicators in competition with cables, radio, etc. They do not educate and do not entertain.

Presented as Paper 70-475 at the AIAA 3rd Communications Satellite Systems Conference, Los Angeles, Calif., April 6-8, 1970.

*Senior Project Engineer, Data Systems Division.

II. The Proposals

The first question to approach is: What is being proposed for the developing nations' application of communications satellites? Table 1 lists some of the proposed or planned systems. They range from paper proposals to operative systems. They are all remarkably alike technically, differing only in relatively minor implementation concepts. It is significant that the hardest selling pitch used by all proposers is based on economic and social advancement of the small nations. Table 2 compares the number of proposals that have communications alone for their purpose with the number that are for education alone. The same comparison is made for the technique employed. Observe that the overwhelming choice is for education and for broadcast satellites. Equal choice is the system with both applications and both techniques.

III. Satellite Justification

All of the proposals have in common the assumption that a satellite is necessary to the accelerated growth of the developing nations. From this it is an easy step to the rationale that huge benefits will be derived from applying the satellite to education or communications. It is significant that in the literature of economics, international communications is rarely seen as a bottleneck in a country's economic development. However, although it is accepted that the development will give rise to an increase in demand for international communications, it is not clear whether internal communications will benefit to the same extent.

Pursuing the international communications application further, we note that, as of January 1968, world communications included the following routings:

	Routed through
Rangoon to Djakarta	Tokyo
Dakar to Lagos	London and Paris
Samoa to Tahiti	San Francisco
South America to Central America	New York

These links are accomplished by radio and submarine cables. Note that, in competition with these media, satellites (at least, in global configurations) do not completely eliminate the routing bottlenecks and often provide links for which there is little demand.

Table 1 Proposed and/or Planned Systems [1, 2, 3, 4, 5, 6, 7]

System	Purpose	Technique	Receiving
Ascend	Communications and education	Relay and broadcast	Communications: 10 ft antenna 15 voice channels Education 5 ft antenna 1 TV channel
Intelsat	Communications	Relay	85 ft antenna 1500 voice channels
SACI	Communications and education	Relay and broadcast	6 ft antenna 1 TV channel (broadcast) 3 voice channels (relay)
ISRO	Education	Broadcast	10 ft antenna 1 TV channel 2 voice channels
Hughes	Education	Broadcast and relay	3 modes

Table 1 Proposed and/or Planned Systems[1, 2, 3, 4, 5, 6, 7] (Continued)

System	Purpose	Technique	Receiving
Hughes	Education	Broadcast	3 Modes 10 ft antenna 2 TV channels Broadcast 10 ft antenna 2 TV channels Relay 25 ft antenna 6 TV channels Relay
G. E.	Communications and education	Broadcast	10 ft antenna 1 TV channel 5 voice channels
A. I. L.	Education	Broadcast	15 ft endfire array 1 TV channel 1 voice channel

Table 2 Purpose and Technique Comparisons

		Index
Communications		0.142
Education		0.429
	Both	0.429
Relay		0.142
Broadcast		0.429
	Both	0.429

Statistics often are presented to show that a satellite system is cheaper than any other kind of information transmittal system. Unfortunately, the data are based on long-distance links and rarely indicate the relative costs for short range communications. Actually, long-distance transmission is the smallest part of the total system costs, and the satellite may suffer in any comparison with short-range cable and microwave systems. A more meaningful comparison would be on the basis of the solution of other than economic problems. For example, a very real problem for many countries is communications across natural barriers such as large bodies of water, mountains, etc. Another justification for the regional satellite could be the speeding up of the implementation of the internal communications where telephone and microwaves might not be installed throughout the nation for a long time to come.

Looking back at the list of proposals, it is no surprise to find most of them for regional types of systems. The one exception is the Intelsat system which, by its charter, is designed solely for international communications. A desire for world status, ambitions for the existing government of the nation, and a desire to enter the space age are poor reasons for taking on a heavy financial burden. It is a grave injustice to the developing nations to provide typical used-car salesman terms, "No money down. Pay for itself in five years." This does not mean that participation in Intelsat is not for any of the new nations. The point is that the characteristics of the Intelsat system are such that it may impose inordinate economic burdens on small nation members if the economic feasibility of this medium for <u>each individual case</u> is not carefully studied.

For example, the Intelsat standard system is based on the state-of-the-art and requires a receiving antenna and receiving and transmitting equipment on the ground that will insure signal quality that meets the CCIR standards. The cost of an 85-ft. antenna (required by the Intelsat standards) is estimated at about $750,000. A 40-ft. antenna can be purchased for approximately $260,000, a saving of roughly $500,000. Unfortunately, Intelsat estimates that the smaller antenna will require 6.5 times the share of the satellite power that the 85-ft. atenna commands. Using 1968 rates, the annual use charge for the system with a 40-ft. antenna is $260,000 compared to $40,000 for the 85-ft. system, an increase of over $200,000. Thus, in less than 3 years, the savings in capital investment would be wiped out and, in addition, the expected communications demand may be inhibited by the high tariff resulting from the increased costs. The solution lies in the use of true random multiple access with charges for actual use rather than on a full time basis. If large scale participation by small nations is to realize its promise, the switchboard-in-the-sky concept must become operational. Considering regional communications systems, random access is important to their success also because the situation is similar with many small stations accessing the satellite. With so many stations involved, the cost of the ground system is extremely critical.[8]

IV. Justifying the Education Application

The majority of the proposals stress education via satellite for developing countries. This would appear to be an area of great potential. Unfortunately, there seems to be considerable confusion as to the role satellites should play in education. About three years ago, a Los Angeles Times reporter in Nairobi wrote the following:

"Most studies show that highly developed countries have highly schooled populations. These studies have been used by most officials... to justify their heavy expenditures on secondary education. But the studies don't say which came first, the development or the education. Close examination shows that, in most cases, development came first. That certainly is true in the United States. The U.S. developed in the 19th Century on the strength of general primary schooling and a good apprenticeship system. Few people enjoyed a secondary or college education... Further statistics show that several African countries have as high a proportion of secondary school graduates today as some European countries did a generation ago. Yet Africa is far, far behind the economic level of Europe in the late 1940's."

It is evident that the educational improvements are needed at the primary level. However, the primary schools usually treated by the satellite system proposals are the urban ones. The curriculum and teaching methods are inherited from colonial masters who were trying to make the natives small copies of themselves. It is somewhat naive on the part of proposers of satellite systems to think that the urban teachers can be made over by "experts" via television. These education experts may have nothing to offer other than the same techniques that have been used for ages. If the "new" ways are so desirable, why are there intense battles in most American school boards over public school curricula?

As for the rural schools, these may have as few as a dozen children of different ages and stages of advancement. There is only one teacher, usually not well qualified because of the lack of funds for such work. Facilities are woefully limited (probably no electricity). These are children that cannot be spared to attend urban schools because they are needed to assist their parents in working the land. These are students who will probably never go beyond the 4th or 5th grade and most certainly have no need for the curricula used in the city. Peace Corps workers have reported that the teachers in such schools have developed programs that seem better suited to their community's needs than those designed by the outside educator.

The effectiveness of the satellite education system should not be measured by the communications efficiency but rather by the fulfillment of the specific educational goals and requirements of the user. For example, one should consider that:

1) Because the satellite will reach more pupils, a proportionate increase in the number of trained teachers will be required. Are the nation's resources adequate to supply the required number of professionals? If the educational system is poor (i.e., a low yield of trained teachers), what is the probability of ever achieving adequate teaching manpower? Note that the U.S. has a problem in obtaining sufficient competent teachers.

2) A need for mass education implies a low level of literacy and therefore implies technological and industrial backwardness. Where will the hardware installers and maintainers come from? Simplicity, ease of operation, and reliability of the equipment are

not the answer. Witness the overloading, malfunction, and inadequacy of maintenance staff suffered by the telephone systems of every large metropolis in the world.

3) Teacher training and curricula updating is bound to take up so much of the early capacity of the system that very little would be left for actual teaching. How much of this "setup time" can the educational budget afford (since, in effect, this implies an increased cost of educating each pupil until sufficient manpower is available)? Note that as the teacher population increases, its updating also heavily stresses the system.

4) What standards are to be used in measuring the effectiveness of the system? One suggestion is that the standard be that required for good citizenship within the specific nation in which the student lives.

V. Educational Television

Many of the workers in the field of application satellites when approaching education become involved in debates as to the merits of various communications media such as television. It should be stressed that communications is never an educational technique. It is no more than a medium for the distribution of the message. Both television and radio can be extremely effective, depending on the information to be transmitted. Thus, some messages are best sent by radio, others by television. It is difficult to make a decision on the basis of communications efficiency or economics until the education techniques that are best suited to a particular situation have been developed. However, since the proposals we are considering here place considerable emphasis on educational television, we should take a look at what has been and is being done in this field.

The paramount problem in applying new technological devices, such as television, in education is that new ideas are being forced to fit into old systems. The present system of education is oriented toward classroom teaching, time-bound instructions, teacher-direction, grades and examinations, etc. The teacher selects the lesson and directs the whole class in a formalized drill, exercise, note-taking, or question and response procedure. The emphasis is on covering and memorizing information. New technological devices such as television are not compatible with these constraints; hence there are conflicts.

It is significant that despite all that is claimed for educational television, its use in the United States is remarkably limited. Operational large systems include only: 1) Florida, connecting Gainsville, Cape Kennedy, Daytona Beach and Orlando; 2) Texas, SMU, TCU, University of Dallas and S.W. Center for Advanced Studies; 3) New York, Catholic Schools; 4) California, Stanford University. The latter system was initiated last year and eventually will include all the campuses of the University of California. A number of small scale systems are operational. In all cases, the experience is so narrow that the planning is more often the subject of technical papers on educational television than the actual engineering of the systems. The enthusiasm of hardware manufacturers has been dampened by the negative reaction the equipment receives from educators.[9] A typical experience is that of CBS Laboratories in selling a closed-circuit television system. In 1958, they arranged for an installation in the Stamford, Conn., public school system. It turned out that nobody wanted it. Everybody was afraid of it, particularly some teachers who felt it might replace them. The principal of Stamford's Riverbank elementary school salvaged the system, and, at present, very young children operate the system, presenting current events to the student body. In summation, the battle here at home is over acceptance and the hardware design of educational television. Little is being accomplished in the way of developing the educational techniques that are best suited to television. The technology is still being force fitted to an antiquated system and this approach can be blamed for the failure of systems such as The Midwest Program on Airborne Television Instruction. In this program, an airborne transmitter broadcast educational television over 150,000 sq. mile area. The average cost of receivers installed in particpating schools was $500. The program was discontinued because 1) it cost about $20 million and 2) it was incompatible with hundreds of independent school districts, differing curricula, and varying budgets.

In France and Spain systems of teleclaves are in operation.[10] Here television is used primarily for cultural education. Receivers are located in schools and churches in the villages and each teleclave is directed by a teacher trained in the techniques of monitoring cultural classes. One strikingly successful program has been operating in Italy for over 8 years.[7] Lessons aimed at illiterates are broadcast over the state-operated television network. It is estimated that about one million persons learned to read and write in the program.

Typical of most proposals for education television in an undeveloped country is the Colombian project. In this case, only the provisions for sending and receiving the programs came from outside. Educational technology was left entirely in the hands of the Colombian educators. One receiver was installed in each school in a room equipped for use as a small auditorium. This was not always the case, for space and equipment were scarce. Most of the problems with which the teachers were concerned were physical and included set operation, maintenance, and seating facilities. The chief effect was that teacher enthusiasm declined although some greater achievement by students could be attributed to television. In a report on the Peace Corps role in Colombia in achieving educational change with new media, George Comstock concludes, "...the point of reception requires assiduous attention... otherwise, there is only a show of education, although complete with stage settings of sending and receiving, before an empty theater."[11]

There are many technical and economic problems associated with educational television in underdeveloped countries. In Nigeria, there are frequent electrical power failures. Electrical connections are not available at all schools and cost five to six times as much to install as in Western countries. Set maintenance is a problem by reason of a lack of a system for repairs and a shortage of trained maintenance personnel. At the time of one particular study, three out of every four sets were inoperative in Nigeria.[7] In Colombia, 50% of the schools are within transmission range, but do not have electricity, and 30% have not joined because their local government cannot provide $8500 a year for operation and maintenance. The Colombian system was off the air for 2 months for lack of a video recording head.[7]

In summation, it is the conclusion of all who have been involved in educational television systems that the classroom teacher is the critical factor in any such system. Neither satellite nor ground-based systems can hope to succeed without fully integrating the teacher into the structure. It is not sufficient to bring the television set to the student. It is also necessary to create an educational system that truly ties the student and the teacher together with the television set and the producers of the programs. Being aware of this makes one worry that the satellite system might be blamed for education failures of the experimental Indian program. The literature is filled with technical and economic details but the characteristics of the planned educational programs are missing. One might note that the stated object of the project is to enable NASA to test the potential value of satellite technology in the rapid development of effective mass communications in developing countries. In line with this goal, India is initially planning to disseminate agricultural and home economics information. Success in the experimental stage thus would demonstrate the satellite's communications effectivity. Development effort on the operational satellite system should be accompanied by development of effective educational techniques for the masses.

VI. Broadcast Systems

The prime consideration in comparing and evaluating the various proposals for broadcast systems for developing nations is the estimated dollar costs. Table 3 is a collection of data obtained from various publications. A question mark indicates information that did not appear in the published article. The date after the name of the system is the year the data were published. All the figures are for operating rather than experimental systems.

The costs for the spacecraft and its launching include development of the hardware and the cost of spares to maintain the system for 5 years. Two ground relay systems are shown for comparison. The spacecraft costs vary widely from 23 to 90 million dollars. It is difficult to determine what is realistic since this cost depends on the payload requirements and the state-of-the-art at the time of the estimate.

The cost of transmitting stations depends on the number used in the system, and the cost per station is very close for the different proposals. This is because considerable experience with transmitters has been accumulated. Previous experience is not visible in the receiver costs with the prices ranging from 100 to 500 dollars with the mean at about $300. This cost includes the antenna which accounts for the deviations that exist. The most significant figure is the total cost per receiving terminal for a 5-yr. period. In particular, we note the highs of $1550 for the Hughes proposal and $3000 for the Spanish system. In general, the wide variations in terminal costs are caused by the differing quantities of terminals served. Thus an expensive spacecraft will show

Table 3 Broadcast System Estimated Costs ($)

System	Spacecraft and Launch	Transmission Station	Receiver terminals		5 yr total	5 yr. total $/receiver terminal
			Each	Total		
Ascend (1967)						
Brazil	23.7 M	3.1 M	288	50.7 M	94 M	470
India	23.7 M	3.1 M	288	376.2 M	470 M	420
Indonesia	23.7 M	3.1 M	288	76.8 M	125 M	500
A. I. L. (1967)	40 M	25 M	100	50 M	235 M	800
Hughes (1968)	27.5 M	5 M	500	5 M	725 M	1550
India (1969)	?	?	500	28 M	150 M	1000
SACI (1969)	98.7 M	?	350	52.5 M	?	
Spain (1966)	Ground relay	?	250	1.25 M	15 M	3000
Great Britain (1967)	CATV	?	224	0.164 M	?	

up as only a few dollars in the cost of operating one of a large number of receiving terminals. Nevertheless, the Hughes proposal is probably the most realistic estimate because it includes maintenance costs that are not considered in the others. This is pointed up by a comparison of the costs of a receiver and an antenna using the Spanish $250 as a standard. Observe that an underdeveloped nation 1) cannot produce at competitive prices in a newly started industry, 2) cannot import components or completed units without adding large freight costs and profits to foreign companies, and 3) must underwrite an expensive technician training program.

It is difficult to see how one can believe that a new and special receiver, a parabolic antenna, and a prime power generator can be supplied for very little more than Spain has paid for their receivers and a conventional VHF antenna. One might examine the cost of television reception in the United States. The American consumer does not have to pay directly for the transmission costs. The receiver costs about $250 and the owner will average $40/yr. for maintenance. Thus, over a 5-yr. period, it costs the American consumer about $450. Compare this with the estimates for the various proposals, and it becomes obvious that realistic figures cannot fall below $1000 and probably come fairly close to $3000. This would make the system with large numbers of receivers extremely expensive, i.e., SACI, $450 M instead of $53 M. The technical problems associated with broadcast systems are myriad.

VII. Conclusions

Some typical proposals for communications satellite systems for developing nations have been reviewed. The feasibility of the proposals has not been questioned. However, some problems have been uncovered that have not received adequate consideration. In general, the technology is seen to have advanced far too rapidly for the slower moving educators and economists. It is as though we were attempting to sell an experimental Cadillac to a farmer living in a remote, backward region where there are no gas stations or mechanics. Among the problems that need to be attacked to assure the success of these programs are development of 1) random multiple access satellites ("switchboard-in-the-sky"), 2) educational technology which can make effective use of communications satellites, and 3) low cost, reliable, easily maintainable, small receiving terminals.

References

[1]"ASCEND (Advanced System for Communications and Education in National Development)", Stanford University School of Engineering Report, June 1967.

[2]"COMSAT Shows INTELSAT A Going Global System", Astronautics and Aeronautics, December, 1969, pp. 11-12.

[3]Feldman, N. and Jamison, D., "An Instructional Broadcast Satellite System for Brazil", RAND Document D-19062, July 2, 1969.

[4]"Blackboard Satellite on a Passage to India", Business Week, September 27, 1969, pp. 34-35.

[5]Rosen, H., "A Satellite System for Educational Television", Astronautics and Aeronautics, April 1968, Vol. 6, No. 4, pp. 58-63.

[6]Steinmayer, A.G., "A Low Cost Communication System for a Developing Nation", Telecommunications, January, 1968, pp. 20-24.

[7]Sielman, P.F., "Satellite Educational Television — An Effective Form of Foreign Aid", AIAA Paper No. 67-963, October, 1967.

[8]Yeh, L.P., "Multiple-Access Tradeoff Study for Intra-South American Satellite Communication System", IEEE Transactions on Communications Technology, Vol. COM-16, No. 5, October 1968, pp. 721-730.

[9]"The Learning Circuit", Electronic News, October 27, 1969, pp. 4-5.

[10]Interview with Minister of Television, Madrid, Spain, October 1966.

[11]Comstock, G., "The Peace Corp Volunteer and Achieving Educational Change with New Media", Rand Document P-4174, August, 1969.

COMMUNICATIONS SATELLITES, TECHNOLOGY TRANSFER, AND ECONOMIC DEVELOPMENT

Paul L. Jordan*

The Rand Corporation, Santa Monica, Calif.

Abstract

The use of educational television to improve the quality and quantity of education in developing countries is examined. A simple model relating economic development in emerging countries to education and the level of applied technology is presented. The use of television broadcast satellites as a means for improving education systems in developing regions is discussed in the context of competition for scarce resources and the requirement to develop concurrently educational software and infrastructure.

Introduction

For developing as well as developed nations, economic growth may be directly associated with increased utilization of labor and capital to produce the industrial, agricultural, and service goods that make up their gross domestic products. The rapid advance of theoretical technology not only stimulates the flow of investment into high-productivity equipment and facilities, but also demands that increasing investments in education be made to improve the stock of human capital, because the potentially increased productivity promised by the new techniques and equipment concommitantly requires new or enlarged skills on the part of workers, managers, administrators, and teachers who make the socioeconomic system function.

Despite the lack of a well-developed theory of the economics of education, government officials and educational planners must apply reason and judgment to resource allocation decisions related to education. They must project trends, set

---

Presented as Paper 70-514 at the AIAA 3rd Communications Satellite Systems Conference, Los Angeles, Calif., April 6-8, 1970.

*Economist, System Sciences Department.

targets of achievement, and forecast system outputs, based, at best, on simple models having relatively little empirical verification. As experience lends credence to their forecasts, and fleshes out their models, the framework of a valid theory may become discernible. Of even greater importance, however, there must be a rationale available for today's decisions: those affecting current and near-term budgets; those affecting the current stock of human resources.

In growth economics the question of balance is ever present. Should industrial development take precedence over agricultural expansion? Should emphasis be placed on training teachers, farmers, scientists, or physicians? Should capital be allocated to steel mills or school buildings? Obviously, the answers to such questions will depend on the characteristics of the economy being studied.

In the area of human resources, emerging countries are characterized by low labor efficiency, low labor mobility, small numbers of occupational or trade specialists, few entrepreneurs, and social practices and institutions that minimize incentives for economic change. Consequently, a critical requirement in developing countries is the education of the populace in the sciences, technologies, and administrative skills on which desired advances will be based. Without the right numbers and types of trained indigenous personnel, foreigners must frequently be used to staff the high-skill positions in industrial and agricultural projects on which economic growth will depend, and this utilization of external manpower may serve more as a deterrent than as a stimulus to development.[†]

An important central economic requirement for development is the ability to produce savings. By foregoing current consumption, part of the national output can be invested in new capital--new resources from which greater production and, hopefully, greater productivity may be derived.

"... and one of the most difficult decisions on economic planning is to get the time cycles of different invest-

[†]Foreigners generally require higher salaries to work in developing countries than do the country's nationals. This may result in excess costs in producing items for home consumption or for export, thus rendering that portion of the national output less competitive internationally. Further, foreigners' salaries may be sufficiently out of accord with domestic salaries as to cause significant social and political problems.[1]

ments worked out so that the schemes with quick returns come in time to finance the larger projects whose ultimate productivity may be much higher but is much slower to achieve."[2]

There must be a balance among competing and complementary investments to achieve optimum growth rates within the existing and projected resource constraints. Piecemeal investments may help some sectors of a developing economy; however, only when all significant sectors are expanded--industry, agriculture, transportation, electrification, education, and so on--at mutually compatible rates, can an emerging country hope to reach the "takeoff" point from whence equilibrium growth may be maintained without large infusions of external assistance. Since each of these sectors must utilize trained manpower to expand and be competitively productive, the investment in human capital appears to take top priority.

Primary Education

In the emerging countries, primary education enrollments generally range from approximately 5 to 40% of the primary school age population.[1] These enrollment figures point to an immediate need to expand the primary school systems; however, they do not give the full picture. They do not fully describe the distribution of students by grade or by geographical location. Near urban centers, enrollments may be higher, since teachers are more likely to prefer the relative advantage of living in larger communities where some of the comforts and pleasures of city life are available, rather than in isolated, primitive areas where creature comforts are at a minimum. Even with compulsory primary education, the drop-out rates may be very high, with only a small fraction of those who enter first grade staying to complete the sixth grade.

These early drop-outs represent a vast waste of educational funds; frequently, they drop out in the first grade, re-enter, then drop out again, thereby increasing the costs of primary education. Furthermore, early drop-outs tend to lose whatever literacy they may have attained, which actually diminishes the value of their limited education to virtually zero.

The generally poor quality of primary education is as significant as the low enrollments and high drop-out rates. In many developing countries, the primary school teachers are predominantly "unqualified." They may have little more than a primary education themselves and no special training as teachers. Textbooks and other teaching materials, where used at all, are frequently poor, and curricula consist mostly of memorization of material read by the teacher. As a result of these deficiencies, those who attend may achieve barely sufficient skills in reading and writing to be considered literate, and efforts to expand the primary school systems rapidly by conventional means may tend to reduce even further the quality of primary education.

Secondary Education

As economic progress is made in developing regions, the demand for skilled manpower increases substantially, and technicians constitute a strategic group. They are required not only for industrialization, but also to improve the technical quality of the agricultural sector; and, where clerks, bookkeepers, secretaries, etc., can be considered technicians, the requirements for business managers, administrators, and general staff adds to the demand for technician level workers.

The supply of technicians arises mainly from secondary education, either academic or vocational. Thus, secondary education is the foundation of the high-level manpower required to achieve increases in the level of technology application and thereby in the level of national output in developing countries. The critical need for secondary-level personnel is demonstrated dramatically by the induced underemployment of physicians and engineers in India, where they are frequently forced to perform lower-level tasks because of the lack of skilled technicians. This underemployment further exacerbates the critical shortage of top-level professional personnel.

Many investigators have discussed the requirement to pursue an "unbalanced" educational policy by concentrating on secondary rather than primary education, in order to produce, as rapidly as possible, large numbers of secondary school graduates.[1,3] Since most nations are dedicated to universal education, and it is politically difficult to openly adopt a policy of not expanding primary schools, the most likely method of expansion of secondary schools in economically constrained countries will be by using new tools and techniques in conjunction with provision of more teachers, more buildings, etc.

Higher Education

Higher education is achieved in underdeveloped countries in two ways: either by establishing universities within the country or by sending the students to other countries. The establishment of a college or university facility within a country frequently is not an economic decision but is related to political and social motives. Since most "factories" have optimum plant sizes, it is almost axiomatic that, in the beginning, education of nationals at home will be more expensive than providing for their education in established foreign universities. Independent of size, the operation of indigenous universities is also more expensive, since the teaching and administrative staff frequently must be obtained, in large part, from outside the country. In some circumstances these personnel receive higher salaries, by far, than teachers or high government officials within the country. However, once this higher-priced teaching talent is employed, their replacement is difficult. Those students who graduate from the university tend to go on into high government office or, if they remain in the university as faculty, demand the higher wage given the extranational faculty. The latter circumstance tends to distort the salary structure of the country so that it is sometimes inadvisable that they be paid at the same rate as the extranational personnel; this either removes incentive for obtaining the higher education or, if the incentive is maintained, causes the graduate to go into fields other than teaching in order to make as much money as possible. Despite the problems in staffing the colleges and universities, expansion of higher education systems of developing countries is critical to the production of the business managers, administrators, lawyers, educators, and other professionally trained persons who make an economy function.

Educational Television as an Expansion Tool

Educational television (ETV) has, in the 1960's, shown great promise as a means of upgrading the quality and quantity of education, both in developed and developing countries. Its advantages have been lauded by educators, communications experts, budget directors, and others interested in cost-effective pedagogy. It has been the subject of many research and experimental programs and is a line item in the educational budgets of many countries, states, and municipalities. The proponents of ETV sometimes make it appear to be the panacea for education. Its detractors frequently see it as a misuse of scarce resources, a gimmick that may be overemphasized, to the detriment of established systems and goals, or a threat to the tenure of the teachers and administrators in current educational systems. Based on our review of literature in this field, it can be said that there is some merit in both views.

Educational television has been used, at least on an experimental basis, in hundreds of schools, by thousands of students in many countries. The data from these programs are predominantly qualitative and subjective, but from the research of Godwin Chu and Wilbur Schramm, of the Stanford University Institute for Communication Research, who surveyed more than 300 research papers on the use of audiovisual teaching aids for the United States Office of Education[4] and of Joseph Kanner who evaluated the instructional effectiveness of color in television,[5] the overall emphatic but qualitative conclusion is that ETV, either color or black and white, can be an effective educational tool if properly used.

Potential Effects of New Educational Techniques

To understand the potential effects of new educational techniques on a developing economy, we may postulate a simplistic model economy in which the current output Y(t) may be expressed by

$$Y(t) = A(t)f[K(t), L(t)] \tag{1}$$

where K(t) and L(t) denote the currently existing stocks of capital and labor, and A(t) is the current level of technical knowledge applied in the model economy. We assume constant returns to scale such that

$$\lambda Y = f[\lambda K, \lambda L] \tag{2}$$

By differentiating (1) with respect to time, we have

$$\dot{Y}(t) = A(t)\{[f(\cdot)\dot{K}/K] + [f(\cdot)\dot{L}/K]\} + \dot{A}(t)f(\cdot) \tag{3}$$

Thus the available means for achieving an increase in output is by increasing the stocks of capital or labor, or by increasing the level of technical knowledge applied in producing the national output. Increases in the stock of capital arise from investment of savings which, typically, are a small fraction of the national income of a developing country. Except for the discovery of vast new reserves of minerals which may attract large foreign investments, rapid increases in national output are more likely to derive from changes in the labor force and the applied level of technology. Even where large foreign or domestic investment funds are available, they may not be effectively used without a suitable number and quality of workers.

Nelson and Phelps[6] express the rate of change of the level of application of technical knowledge in a given period as

$$\dot{A}(t) = \phi(h)[T(t) - A(t)] \tag{4}$$

where $\phi(h)$ is a function of the average level of education $T(t) = T_0e^{\lambda t}$ and denotes the current theoretical level of technology associated with some exponential growth factor λ over a period t, and A(t) is the actual level of technical knowledge applied in period t. They assume that h is constant, or essentially so, over a short period. This assumption is reasonable for industrialized, developed countries with relatively high average levels of education. But for those developing countries which have low average educational levels, one might expect that the average level of education could be raised at a sufficient rate to significantly affect $\dot{A}(t)$.

In simple form, the average educational level $\bar{h}$ of a country at any instant may be given by

$$\bar{h} = \left[\sum_{j=1}^{P} (py)_j/P\right] \qquad j = 1, 2, 3, \ldots P-1, P \qquad (5)$$

where the (py) are the student-years of formal schooling of the jth person of the population P. However, considering the emerging nations, a simple statement of student-years neglects the quality of the educational programs. Even when differences in wage rates and living costs are taken into consideration, an expenditure of $500/student-year in one country vs $50/student-year in another, may imply a considerable difference in the quality of school years. Introduction of improved teaching methods, teaching materials, or teacher-training programs can improve significantly the output of the educational system, quite possibly at the same per-student expenditure rate. Of equal importance there may be a sizable portion of the population existing outside of the money economy and therefore not technically considered part of the labor force. In such cases, the labor force can be made to grow at a very much greater rate than the population growth rate, at least for some short period of time, merely by bringing these "outsiders" into the labor force. Again, the education system is the major means to prepare the new entrants for productive activity.

Higgins[7] has presented some of the problems in optimal expansion of education systems in developing countries. Considering the opportunity costs of allocating labor (including students who might be in the labor force if they were not students) and capital to education or other activities that contribute to the gross domestic product, different countries may, ideally, allocate these factors in different proportions, depending upon the existing qualities of their education systems and the availability of opportunities for expanding national output by conversion of natural resources or utilization of special national capabilities. Thus, given increasing levels of national output, an "expansion path" for a particular country's education system may be derived. Resource-rich countries such as Kuwait and Libya may find it virtually impossible to expand their education systems as rapidly as their income rises. Countries such as Greece and Japan, with comparatively high education levels but few natural resources may find that national output growth is achieved more readily by investment in capital stocks.

Satellite-Based ETV

The growing consensus on the efficacy of educational television and the new-found ability to illuminate an entire country with television signals from synchronous satellites have stimulated considerable discussion and support of satellite-based educational television as a shining hope for solving the educational problems of developing countries; and it may be that this is indeed true. But in the profusion of papers on satellite-based ETV, engineering design and costs of the space and terrestrial hardware required to transmit and receive the television signals appear to be the dominant topics.[8-20] In some cases comparisons of the costs and time required to cover a region by a microwave network or by a broadcast satellite are used to justify or support the implementation of satellite-based ETV. A major caveat for organizations and individuals interested in developing space systems for ETV is that dissemination of the television signal is only a small part of an educational television system.

Satellite-based ETV in a developing country must be considered as part of the overall education system. Not only must the country be financially and administratively capable of implementing the space and terrestrial portions of the ETV communication system, but it also must be able to provide the proper environment and peripheral equipment required to permit effective use of ETV. In American Samoa, it was found that grass huts previously used as school buildings were totally inadequate as ETV classrooms. Therefore, along with the expense of installing transmitters and receivers, and producing programing, it was necessary to provide for modern classrooms, complete with blackboards, desks, proper lighting, and so on.[21] Even in areas of cheapest labor, the costs of a modern schoolroom undoubtedly will exceed by far the cost of a television receiver and antenna. If textbooks, workbooks, and similar software are required to support ETV programs, these items too may have aggregate costs that equal or exceed the reception, transmission, and program production costs. Thus, cost analyses which deal only with specific ETV hardware and program production do not provide adequate information for planning the expansion of a developing country's educational system. Granted the efficacy of ETV as a teaching tool, how large an ETV system can a country afford in a given period? Despite the technical feasibility of complete coverage of a nation or region by satellite ETV transmission, can this ubiquitous signal be used effectively? These questions can be answered only by thorough analyses on a country-by-country or region-by-region basis.

Conclusions

The magnitude of the problems of population growth, food production, illiteracy, and other social conditions are of great importance to the political and economic stability of many developing countries. The possibilities of using a TV broadcast satellite for alleviating these problems and helping to achieve stable, growing economies, requiring little or no external economic aid, appear sufficiently attractive to warrant more thorough studies of the costs and benefits associated with TV broadcast satellites: not one-channel, 1-year experiments, but satellites configured to fit specific needs and capabilities. It is not enough to say that one TV set in each village or schoolroom could be served by a satellite--such a system might entail building tens of thousands of buildings to house the sets and their audiences, or thousands of miles of roads to service them. To estimate costs or impute benefits without regard to the total system requirements may indeed delay rather than accelerate implementation of ETV, by satellite or otherwise. Until firm data are made available, responsible authorities cannot objectively establish the priority or urgency of an ETV project and should rationally allocate their scarce resources to projects for which benefits and costs are clearly set forth.

References

[1] Harbison, F. and Myers, C. A., Education, Manpower, and Economic Growth, McGraw-Hill Book Company, New York, 1964.

[2] Ward, Barbara, The Lopsided World, W. W. Norton and Company, Inc., New York, 1968.

[3] Coombs, P. H., The World Educational Crisis, Oxford University Press, New York, 1968.

[4]Chu, G. C. and Schramm, W., "Learning from Television: What the Research Says," ERIC at Stanford, ERIC Clearinghouse on Educational Media and Technology at the Institute for Communication Research, Stanford University, Aug. 1968, Stanford, Calif.

[5]Kanner, J. H., "The Instructional Effectiveness of Color in Television: A Review of the Evidence," ERIC Clearinghouse on Educational Media and Technology, Stanford University, Jan. 1968, Stanford, Calif.

[6]Nelson, R. R. and Phelps, E. S., "Investment in Humans, Technological Diffusion, and Economic Growth," American Economic Review, Vol. 56, May 1966, pp. 69-75.

[7]Higgins, B., Economic Development Problems, Principles, and Policies, W. W. Norton and Company, Inc., New York, 1968.

[8]Burns, R. F. et al., "An Econometric Analysis of an Educational TV Distribution System," AIAA Progress in Astronautics and Aeronautics: Communication Satellite Systems Technology, edited by R. B. Marsten, Vol. 19, Academic Press, New York, 1966, pp. 983-1009.

[9]Dlugatch, I., "A Low Cost Communications Satellite Educational System," SP-2493, presented at the XVII International Astronautical Congress, Madrid, Spain, Oct. 1966.

[10]"Educational Television Via Satellite," Communications Satellite Corporation, Washington, D.C., April 1967.

[11]Haviland, R. P., "Broadcast Satellites: Their Applications and Implementation Impediments," AIAA Paper 68-1064, Philadelphia, Pa., 1968.

[12]Haviland, R. P., "Space Broadcasting - How, When, and Why," unpublished paper, General Electric Company, Philadelphia, Pa.

[13]Haviland, R. P., "Why Space Broadcasting?" AIAA Paper 68-422, San Francisco, Calif., 1968.

[14]Hesselbacher, R. W., "An Evaluation of Voice Broadcast Satellite Systems," AIAA Paper 68-423, San Francisco, Calif., 1968.

[15]Jansen, J., "Communication Satellites for Educational Television," Proceedings of the 18th International Astronautical Congress, Pergamon Press, Oxford, England, 1968.

[16]Jansen, J. et al., "Television Broadcast Satellite Study," Report CR-72510, Oct. 24, 1969, NASA.

[17]"Preparatory Study of a Pilot Project in the Use of Satellite Communication for National Development Purposes in India," prepared for United Nations Educational, Scientific and Cultural Organization by a Unesco Expert Mission in cooperation with a national counterpart team appointed by the Government of India, Nov. 18 to Dec. 7, 1967.

[18]Rao, B. S. et al., "Satellite Television: A System Proposal for India," presented at United Nations Conference on the Exploration and Peaceful Uses of Outer Space, Thematic Session I/India, A/Conf. 34/I.1., June 24, 1968.

[19]Rosen, H. A., "Satellite System for Educational Television," A/Conf. 34/1.5, Thematic Session II, United Nations Conference on the Exploration and Peaceful Uses of Outer Space, United States, June 20, 1968.

[20]Rubin, P. A., "Satellites for Education," presented at XIXth Congress of the International Astronautical Federation, New York City, Oct. 13-19, 1968.

[21]Feedler, Martha L., "ETV Goes Way Out and Brings the World to Samoa...," American Education, Vol. 3, March 1967, pp. 14-17.

IV. U.S. DOMESTIC SYSTEMS

DOMESTIC SATELLITE SYSTEMS OF THE FUTURE

Robert D. Briskman*

Communications Satellite Corporation, Washington D.C.

Abstract

This paper discusses possible domestic communications satellite systems of the future. The technical characteristics, capabilities and applications of such systems are described.

Introduction

It is a certainty that communications satellites will be implemented in this decade for providing various domestic services and, because of their introduction, innovations in communications networks may be possible. A domestic satellite system is currently in operation throughout the Soviet Union, and development of one for Canada has commenced. Several proposals for United States domestic systems have been advanced. The discussion herein concerns itself solely with "geostationary" satellites (i.e., those in near synchronous equatorial orbits). During this decade it is highly improbable that satellites in other types of orbits would be considered for providing United States internal commercial communications services.†

Presented at the National Electronics Conference, Chicago, Ill., December 8, 1970, and published in this volume by permission.

*Manager, Domestic and Special Project Office.

†It is interesting to note that the Soviet Union currently uses a nonsynchronous domestic satellite system. The orbit chosen provides improved coverage of its northernmost extremes; however, recent indications show a renewed interest in geostationary satellites.

Capabilities

Communications satellites are another form of communications media, others being microwave radio relay, coaxial cable and high-frequency radio. Like all these other media, communications satellites have unique advantages and disadvantages. The main advantages are the ability of all service points within view of the satellite to communicate directly with each other and to receive simultaneously the same broadcast transmission.

In the first instance, this means that communications between two distant locations (such as New York and Los Angeles) need pass through only a single satellite-borne electronic amplifier.≠ It is also feasible to shift the communications load through the satellite among several service points located throughout the United States in response to actual communications demand or to unequal communications loading caused by time differentials across the country. The second instance allows the same broadcast transmission, such as television, radio, or press wireless to be received everywhere in the satellite coverage area. If we assume a modest-cost Earth station, additional points of service for such broadcasts can be provided at an incrementally low cost since no additional satellite power is required as the number of Earth stations is increased.

It is interesting to speculate on how the unique capabilities of such satellites might be employed in domestic communications systems during this decade. A three-stage evolution might be envisioned with, because of diminishing foresight, the initial stage most clearly seen.

Initial Implementation

The first stage would be reflective of those satellite systems currently pending approval by the Federal Communications Commission. These are essentially wide-area bulk transmission distribution

≠Over 100 microwave radio relay stations or over 1000 coaxial cable in-series amplifiers would be required with these other media.

systems that would supplement the current long-haul microwave radio relay communications system. The most advantageous uses appear to be for broadcast distribution (network, CATV, and educational/instructional) and certain private-line data services. These systems also could be used in the domestic switched telephone network for providing trunks over the longer routes.

The bulk of the broadcast services mentioned in the previous paragraph would be provided to Earth stations located close to the metropolitan areas. These Earth stations would have small-diameter antennas (e.g., 32-ft.) and simple electronics. The received transmissions would be relayed from the Earth stations to the broadcast transmitters in the metropolitan area by conventional communications means.

The satellite for this first stage could be based on an Atlas-Centaur launch weight. Studies have shown that it is feasible to beam at the 48 contiguous United States[§] as many as 24 television transmissions simultaneously through a satellite of this weight class. Sufficient redundancy would be provided for a 7-yr life expectancy. The satellite could be held in orbital position precisely so that it would not be necessary to move the small-diameter antennas at the Earth stations or to autotrack the incoming signal. Besides the economic savings, the fixed antenna also allows such Earth stations to operate unattended. It is believed that this type of satellite could be implemented and orbited rapidly since two similar types already have been developed (TACSAT and INTELSAT IV). It is also possible to implement smaller satellites, suitable for an advanced Delta launch vehicle, which would have approximately half the capacity of the larger one.

§Beams could also be provided for other states, such as Alaska.

Subsequent Implementation

The second phase envisioned would be incremental and would be the addition of an extremely high-volume point-to-point capability suitable from both a technical and economic viewpoint for providing numerous internodal trunks in the domestic switched network. Satellite capacities of 100,000 or more circuits would be envisioned. To accomplish this would require the concentration of the satellite radiated power into very narrow beams accurately directed at terrestrial communication hubs and the availability of wide-frequency bands not limited by current service sharing restrictions with radio-relay facilities. Such new frequency bands will not be found at the favorable propagation frequencies below 10 GHz because of present frequency spectrum congestion. Proposals for obtaining sufficient spectrum above 10 GHz for the services just described will be proposed at the World Radio Administrative Conference in June 1971. Because of the unfavorable propagation conditions through the atmosphere at radio frequencies above 10 GHz, dual Earth stations will be employed at prime service points and separated so as to provide space diversity. This would overcome the propagation loss without incurring large spacecraft penalties.

The satellite envisioned would radiate broadcast transmissions, as mentioned in phase one but also would radiate several very narrow beams toward the major domestic communication hubs. The satellite electronics would be arranged so that transmissions coming up to the satellite on one beam would be routed automatically by ground command to go down on another beam.# Also, the electronics would be rearrangable by ground command so that the total capacity of the satellite could be quickly varied among the individual beams to match the actual traffic demand. It may be that, on future satellite versions, all (or at least a few) beams could be redirected by ground command to different service points in the United States. One possible use of

#The elementary beginnings of a "switchboard in the sky."

this capability could be for emergency communications during a local disaster. It is envisioned that a satellite of the type just described would require a Titan III C or Titan III D/Centaur launcher. With sufficient investment now in research and development, such a satellite could be orbited in 5 or 6 years.

The Earth stations envisioned would have modest size antennas with automatic tracking capabilities. The complexity and costs would be individually comparable to current Earth stations, and no equipment redundancy would be employed (i.e., dual stations are used for space diversity). One of the Earth stations in each pair might be located right at the terrestrial communications terminal. This would be possible under the assumption that exclusive (or favorably shared) satellite frequencies are allocated in the future. The other station would be located many miles away (e.g., typically 10-20 miles) and interconnected by high-capacity coaxial cable or other means.

Future Possibilities

The last stage of evolution would be the addition of direct broadcast satellites. Although there may be various unsolved legal, political, and economic problems, there is no question from a technical point of view that these satellites can be developed. It also is believed that such systems are expensive, and in the United States there may be few or no existing services that would be more economically provided by these satellites than by existing facilities. Nevertheless, a start in this decade on implementing such a system for domestic use is not out of the question, especially for new services. The system envisioned would radiate a few television and radio channels from a satellite directly to individual home receivers (adapted to receive the transmission by a simple, low-cost modification kit). It has been speculated that the use of such a capability for public service broadcasting might be of interest.

A direct broadcast satellite of this type would require a high in-orbit weight. Studies of suitable launchers generally has been in the Titan III D to Saturn weight classes. Several technological areas, such as thermal dissipation from the high-powered transmitters, must be developed further, as well as an appropriate frequency cleared for this use.

Summary

The previous discussion is brief and not totally inclusive. Projections into the future in rapidly advancing technologies have often been inaccurate. It would appear, however, that the development of domestic satellites may follow along the lines indicated.

AN INSTRUCTIONAL COMMUNICATION SATELLITE SYSTEM FOR THE UNITED STATES

E. M. Sheppard*

University of Maine, Orono, Maine

Abstract

An instructional communication system for the United States is proposed using seven synchronous satellites. The country is divided into fifteen regions, each containing from one to six states and a combined primary and secondary school population of 2-5.5 million pupils. Each region would have seven 40-MHz channels capable of carrying either data or one TV channel which would be enough to allow each state to have a minimum of one TV channel plus sufficient data capacity to allow for one Computer Aided Instruction (CAI) terminal for every 25 students. There would be a 2-MHz return link from each school back to the regional centers for the CAI system. Up and down link frequencies of 30 and 18 GHz are proposed in view of the substantial bandwidth requirement of 2.3 GHz for the system.

Introduction

At the present time in the United States there is a substantial interest in the problems of education and its application in society. The total enrollment in U.S. educational institutions (Fig. 1) has risen from 36 million in 1954 to 57 million in 1967 and is expected to hit 63 million by 1975.[1]

Presented as Paper 70-450 at the AIAA 3rd Communications Satellite Systems Conference, Los Angeles, Calif., April 6-8, 1970. This study was conducted during the author's Ford Foundation residency at the RCA Astro-Electronics Div., Princeton, N.J. 1967-68.

*Professor, Electrical Engineering.

Demands for more and better education have brought on a re-evaluation of teaching methods and a search for new techniques. Among the most promising of these are Programed Learning,[2,3] Computer Aided Instruction (CAI),[4] and Educational Television (ETV).[5] In essence, programed learning is the automation of segments of the teaching process. It is not a substitute for the teacher but rather a significant tool in the hands of a competent educator to be used in increasing the effectiveness and efficiency of the learning process. The application of programed learning ranges from the use of programed texts, workbooks, and simple teaching machines to its most sophisticated form, CAI. In CAI, a computer is programed to dispense information in small increments to the student and then to test him by his responses to related questions, reinforcing him when he is correct, correcting him, and giving an occasional hint when he is wrong; if necessary, the computer will lead him through the whole process again. In such a system, the student typically converses with the computer by means of a terminal that includes a cathode ray tube display, a keyboard, and a "light" pen for calling up and responding to information.

ETV has been widely used during the past decade, and although its merits are somewhat controversial, it has been repeatedly shown to be an effective teaching method.[6] Today, the trend is to use ETV with class and teacher participation. The class first watches some TV and then discusses the material presented or works problems, returning to the TV for further information. The classroom teacher is an important part of the process, in fact, the most essential part. The concept of the master teacher via TV thus is giving way to the use of TV as a supplement to the classroom instructor.

A key factor in the use of the ETV techniques is that the program material cost per student is relatively high. Therefore, the programs have not been of the highest quality, which has prevented attainment of the full potential of the techniques. Consider, then, the possibility of constructing a large scale two-way communication network to serve the educational community. The impact of such a network on the use of the aforementioned teaching techniques would be very substantial; the network would allow for the widespread use of the proposed methods on a very favorable economic basis. If we assume a demand for such a network to service the needs of ETV, CAI, and data retrieval, the following paragraphs examine the system requirements and implementation. Note that the study is based on grades Kindergarten through 12, thus representing 88% of the total student population.

An Educational Communication Network

It would be essential that the network be capable of highly individualized control by state and local school districts in terms of selecting services (subject matter, teaching methods, and teachers) while simultaneously offering the economic advantages of large-scale systems. It would be unreasonable to expect that any such system would be national in character, from a technical or political viewpoint. Thus, it seems reasonable and practical to divide the country into regions, the most obvious partition being state-by-state. However, while this might be appropriate in the case of the larger states, such as New York and California, this partition fails to provide a suitably large student population base for a distribution center in the less populated states, such as Idaho and Nebraska. Therefore, as a compromise solution, the continental United States could be divided into 15 regions (Fig. 2 and Table 1) each tending to be geographically compact and representing roughly equal student populations. (It is easy to see how this educational system could also be applied to other than secondary and primary school groups; for example, it could be used by professional groups to give updated information in evening courses.) Consider now the regional requirements for such a system.

One CAI Terminal for Every 25 Students

This requirement allows for approximately 15 minutes of CAI time/student/day. Although this time might at first seem too short to be worthwhile, the process is very efficient and should allow ample time for the students; some will require much less time than others, and it is not necessary for all students to use the computer every day.

Table 1 Educational regions

Region	Primary and secondary school population in millions 1966	1975
1. Maine, N.H., Vt., Mass., R.I., Conn.	2.5	2.7
2. N.Y.	3.8	4.1
3. Pa., N.J.	4.0	4.3
4. Del., Md., D.C., Va., N.C.	3.7	4.0
5. S.C., Ga., Fla.	3.5	3.8
6. Ala., Miss., La., Ark.	3.1	3.4
7. Ohio, W. Va.	3.2	3.5
8. Tenn., Ky., Ind.	3.1	3.4
9. Wis., Mich.	3.4	3.7
10. Texas, Okla.	3.7	4.0
11. Ill., Mo.	3.6	3:9
12. Ia., Minn., N.D., S.D., Nebr., Kansas	3.0	3.2
13. Mont., Idaho, Wyo., Ore., Wash.	1.9	2.0
14. Nev., Ariz., N.M., Colo., Utah	1.9	2.0
15. Calif.	5.0	5.5
Total	49.4	53.5

Six TV Channels

This is based on having a minimum of at least one channel/state given the regional divisions shown in Fig. 2.

It is necessary to translate the channel requirement into data and bandwidth requirements for an effective communication system. A requirement which is based on an advanced version of the RCA instructional system has been assumed. In this system, a 2400-bit line services 14 student terminals, yielding a maximum data rate of 171 bps/terminal. A maximum rate of 200 bps is assumed to be reasonable. Note that these data rate calculations do not make any allowance for transmission of voice or pictures, highly desirable inputs to the student. A method of overcoming this situation would be the use of local storage with periodic updating (covered later in the paper). For the return link from the schools to regional facilities, the rate requirement chosen is based on the ability of an individual to put information into a machine using a keyboard device. Assuming 100 words/min. as the peak rate of an expert typist, this translates into about 600 characters/min., or, at 5 bits/character, a rate of 50 bps. This would represent a maximum rate and is probably far above the actual rates required.

Finally, it would be necessary to determine a duty cycle for the use of channels to individual student terminals. For communication from the computer to the student terminal, a duty cycle of 50% is assumed, and from terminal back to computer, 10%. Thus, the effective data requirements/terminal are 100 bps and 5 bps respectively. Translated into regional requirements using the 25 students/terminal assumption yields 1) regional facility to schools--$8X10^6$ to $22X10^6$ bps and 2) schools to regional facility--$4X10^5$ to $1.1X10^6$ bps.

Possible solutions to realizing such a communication system might be by 1) cable and terrestial microwave links, 2) an airborne system such as the Midwest Program for Airborne Television Instruction (MPATI), 3) satellite, and 4) distribution of taped programs.

All are considered in terms of the 1975-80 time frame. Although any system actually built would probably employ some elements of the first, third, and fourth solutions, primary emphasis is placed on the use of a satellite system.

Satellite System

Figure 3 shows a map of the continental United States, as viewed from a synchronous orbit, and of the 15 educational regions previously outlined. The use of a synchronous satellite is dictated by the large number of antennas required for the system, one for every school or over 100,000. To consider anything but fixed antennas is impractical; the problem of correctly aiming an antenna is bad enough without having to make it track a moving satellite. Figure 4 shows a frequency plan for such a system and indicates that at least four different frequency allocations must be used to prevent overlap. Figure 5 illustrates a possible solution to the coverage of the different regions using 20 beams (1 to 2/region) varying from $2/3^{\circ}$ to $1\text{-}2/3^{\circ}$ in size. Although the states of Alaska and Hawaii have not been included in this system, it would not be difficult to add them.

The bandwidth requirement for such a system would be 1) six TV channels at 40 MHz each (FM would be highly desirable in this case) and 2) one data channel at 40 MHz, or, for the four different allocations, $(6+1)(40\text{X}10^{6})(4) = 1.12$ GHz + return link. The return link would be very narrow by comparison (less than 2 MHz). A bandwidth, therefore, of 2.3 GHz would be required for the bandwidth, considering both up and down links.

To find the room needed for a system with such very substantial bandwidth requirements, it is necessary to move up in frequency from the conventional bands. Figure 6 is a plot of atmospheric attenuation as a function of frequency. Frequencies in the region of 18 and 30 GHz (for down and up links, respectively) have been chosen for this system since they fall into windows in the higher frequency bands which are not presently in widespread use. It should be noted that the ATS-E satellite will have an experiment to test the use of 15 and 30 GHz for satellite communication. There are, in addition, several other programs aimed at the use of these general frequency bands.[7,8] Other frequency choices are also practical.

Table 2 shows a link analysis for a 40-MHz channel at 18 GHz. This particular example would represent a worst-case analysis since the area to be covered is the maximum encountered (a $1\text{-}2/3^{\circ}$ beam for the southwestern United States Region 14). A 3.9-ft. dish is assumed at the schoolhouse giving a 1° beam at 18 GHz. This same dish is also used for the return link, yielding a 0.6° beam at the uplink frequency. This means that there is a substantial problem in keeping the antenna correctly aimed and also in keeping the satellite in a fixed position. Given these conditions, 150 W of transmitter power yields a margin of 7.0 db in clear weather. For the data return channel down link, 10 W is sufficient to provide a margin of over 8 db.

Table 2 Down link (1-40 MHz channel at 18 GHz)

Transmitter power	+21.7 dbw	150 w
Transmitting system losses	-1.5 db	
Transmitting antenna gain	+39.9 db	$\theta = 1.67^{\circ}$, D = 2.34 ft
Space loss	-208.6 db	
Atmospheric loss	-0.5 db	
Receiving antenna gain	+44.3 db	$\theta = 1^{\circ}$, D = 3.9 ft
Receiving system losses	-1.0 db	
Received power	-105.6 dbw	
Required C/N	10.0 db	
Noise (BW=40 MHz, T=1000°K)	-122.6 dbw	

For data return channel

P_R = -117.4 dbw	BW = 2 MHz
P_T = 10 w	C/N = 10.0 db
Noise = -135.6 dbw	Margin = 8.2 db

Table 3 shows an analysis of a 40-MHz channel up link, such as would be used from regional centers to the satellite. Here, a 0.5° transmitting antenna is assumed and should not represent too great a problem since there would only be a few of these to aim. Again, a worst case is assumed in that a 1.67° receiving antenna is used. This corresponds to Region 14 and the largest area to be covered by a single beam. Note that a larger antenna may not be used on the satellite in this case due to the fact that the same antenna must also pick up return signals (at the same frequency of 30 GHz) from local schools throughout the region. 100 w of transmitter power yields a margin of 5 db under favorable weather conditions. Because so few transmitters are required it is proposed that space diversity be used at regional centers to overcome weather problems. Preliminary indications suggest that placing the transmitters and receivers about 10 miles apart should improve to an acceptable level[9] the reliability of the link between satellite and regional center. Studies of diversity techniques to combat the effects of rain at these frequencies are underway at several locations and it appears that heavy rain storms are confined to fairly small areas at any given time.

Table 3 Up link (1-40 MHz channel at 30 GHz)

Transmitter power	+20.0 dbw	100 w
Transmitting system losses	-2.0 db	$\theta = 0.5^{o}$
Transmitting antenna gain	+50.3 db	D = 4.7 ft
Space loss	-213.0 db	
Atmospheric loss	-0.6 db	$\theta = 1.67^{o}$
Receiving antenna gain	+39.8 db	D = 1.4 ft
Receiving system losses	-2.0 db	
Received power	-107.5 dbw	
Noise (BW=40 MHz, T=1000°K)	-122.6 dbw	
Required C/N	10.0 db	
Margin	5.1 db	

Table 4 shows an analysis of the up-link portion of the data return channel. Fairly low power levels (10 w for worst case of Region 14, $\theta = 1.67^{o}$) produce margins of 6db or better. The very narrow beam of the transmitting antennas at this frequency would represent a major problem in terms of pointing, as mentioned previously.

Table 4 Up link-date return (1-2 MHz channel at 30 GHz)

Transmitter power	+10.0 dbw	10 w
Transmitting system losses	-2.0 db	$\theta = 0.6^{o}$
Transmitting antenna gain	+48.7 db	D = 3.9 ft
Space loss	-213.0 db	
Atmospheric loss	-0.6 db	$\theta = 1.67^{o}$
Receiving antenna gain	+39.8 db	D = 1.4 ft
Receiving system losses	-2.0 db	
Received power	-119.1 dbw	
Noise (BW=2 MHz, T=1000°K)	-135.6 dbw	
Required C/N	10.0 db	
Margin	6.5 db	

Figure 7 shows the current state-of-the-art for both tubes and solid-state devices. In the time frame being considered, devices should be available.

A heavy rainstorm will wipe out communications in these bands, but the problem is not a frequent one, as is shown by Fig. 8. Even in heavy rainfall areas, the outages which may be expected amount to about 10 hr /yr. In terms of data communication, this amount should present no problem since, the recipient, aware of the problem, can wait until the situation improves to resume again. In the case of communication from regional center to school, such as in CAI, this arrangement would be a tolerable situation. The reliability of such a system would have to be balanced by a somewhat flexible programing concept. This would not seem to be a very high price to pay for the services gained. (It should be noted that highly reliable links between satellite and regional centers are very desirable and can be realized by the use of space diversity, as already mentioned.) Another point for consideration is that the worst case for transmission, Region 14, is also one of the driest regions and hence a point where the margins required for weather are small.

Because of the fairly substantial power requirements for the satellites and the need to keep down the number of frequency allocations, the use of several satellites is necessary. Table 5 shows the satellite breakdown; seven satellites are needed to serve populations from 2 to 12 million students each. The primary factor in determining the particular assignment of region to each satellite are the power requirements, which are shown in Table 6.

The power required to run the TWT's in the transmitters is so much greater than any other need for power in the satellite that this calculation essentially determines the over-all requirement. Power levels of 6-10kw should not be too difficult to obtain in the time frame considered.

Table 5 Satellite system

Satellite	Regions	Student Population
A	13 Mont.,Idaho, Wyo., Ore., Wash.	$2.0X10^6$
B	14 Nev.,Ariz.,N.M., Colo., Utah	2.0
C	3 Pa., N.J.	8.3
	10 Texas, Okla.	
D	2 N.Y.	
	12 Ia., Minn., N.D., S.D., Nebr., Kansas	7.3
	4 Del., Md., D.C.,Va.,N.C.	
E	6 Ala.,Miss., La.,Ark.	10.8
	8 Tenn., Ky.,Ind.	
	5 S.C., Ga.,Fla.	
F	7 Ohio. W.Va.	11.2
	11 Ill., Mo.	
	1 Maine,N.H.,Vt.,Mass., R.I., Conn.	11.9
G	9 Wis.,Mich.	
	15 Calif.	

Table 6 Satellite power requirements

Satellite	Power(kw)	η_{TWT} = 20% assumed
A	8.0	
B	9.0	
C	7.7	
D	7.6	
E	7.6	
F	6.9	
G	6.3	

References

[1]U.S. Office of Education, "Projections of Educational Statistics to 1975-76," Superintendent of Documents Catalog No. FS 5.210:10030-66, U.S. Government Printing Office.

[2]Schram, W., "Programed Instruction," Fund for the Advancement of Education (Ford Foundation).

[3]Skinner, B.F., "Teaching Machines," Scientific American, Nov. 1961, pp. 90-102.

[4]Suppes, P., "The Uses of Computers in Education," Scientific American, Sept., 1966, pp. 206-223.

[5]U.S. Office of Education, "Educational Television, The Next Ten Years," U.S. Government Printing Office.

[6]U.S. Office of Education, "Educational Television, The Next Ten Years," U.S. Government Printing Office, p. 52.

[7]Altshuler, E. E., Falcone, V. J., and Wulfsberg, K. N., "Atmospheric Effects on Propagation at Millimeter Wavelengths," IEEE Spectrum, July, 1968, pp. 83-90.

[8]Hathaway, S. D., "Possible Terrestial Common Carrier Applications Above 10 GHz," 1968 IEEE Convention Digest.

[9]Hogg, D.C., "Millimeter-Wave Communication Through the Atmosphere," Science, Vol. 159, No. 3810, Jan., 1968, pp. 39-46.

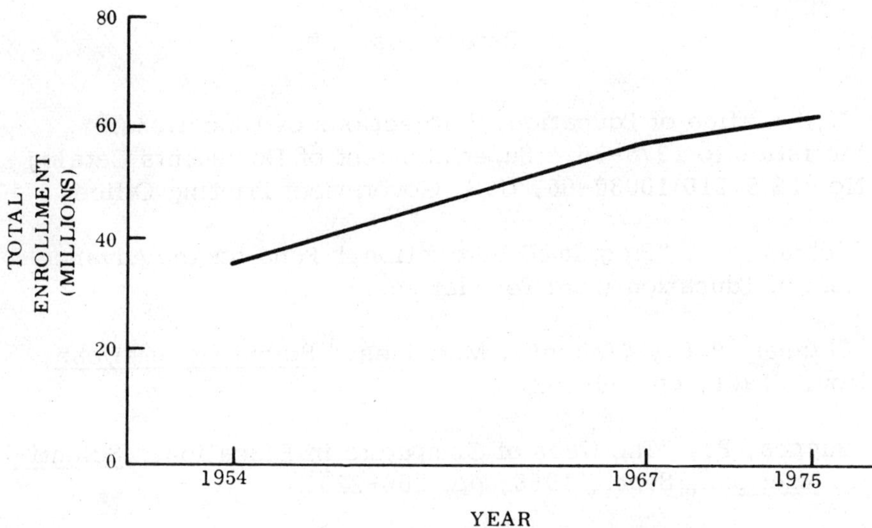

Fig. 1 Total enrollment in United States educational institutions.

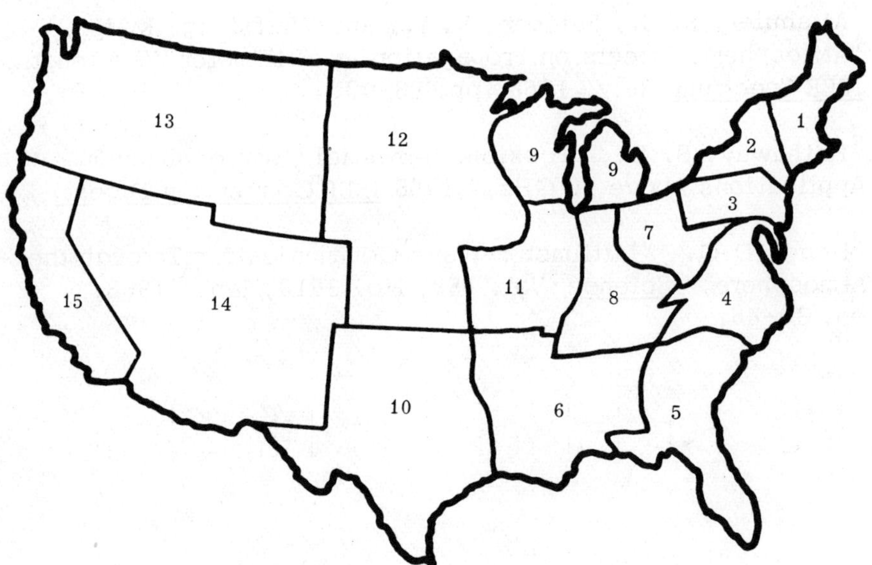

Fig. 2 Fifteen U. S. educational regions.

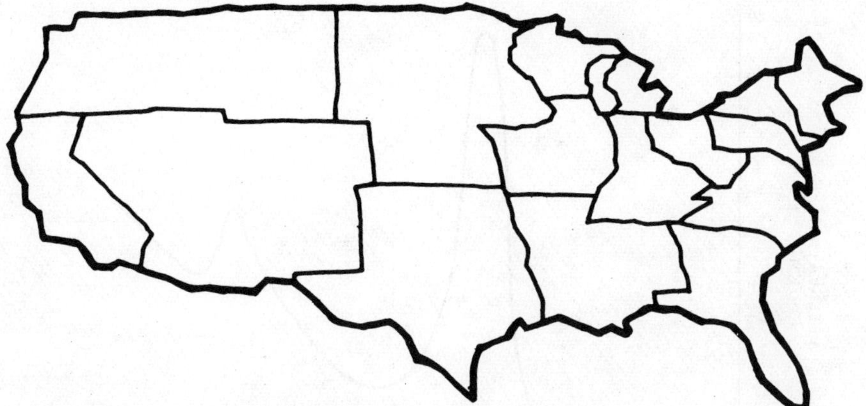

Fig. 3 Educational regions as seen from synchronous orbit.

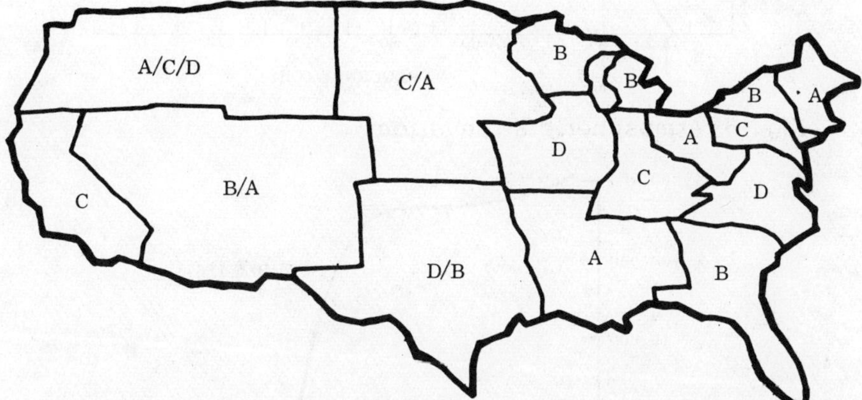

Fig. 4 Frequency plan for synchronous satellite system.

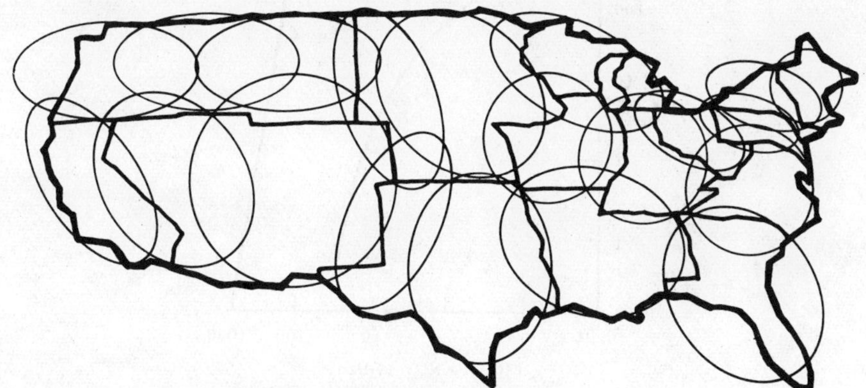

Fig. 5 Twenty-beam configuration.

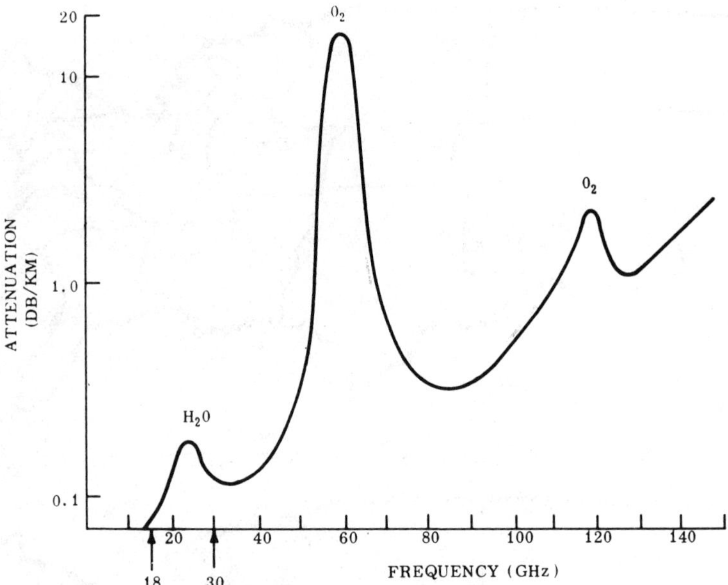

Fig. 6 Atmospheric attenuation.

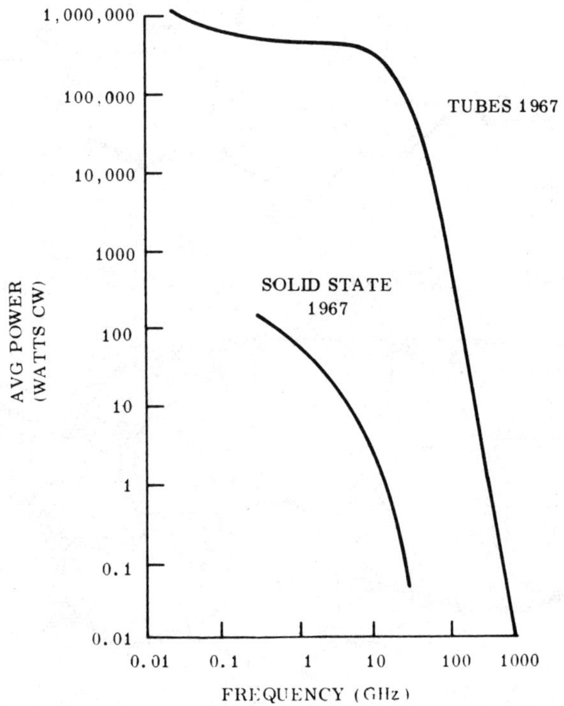

Fig. 7 Microwave power generation 1967.

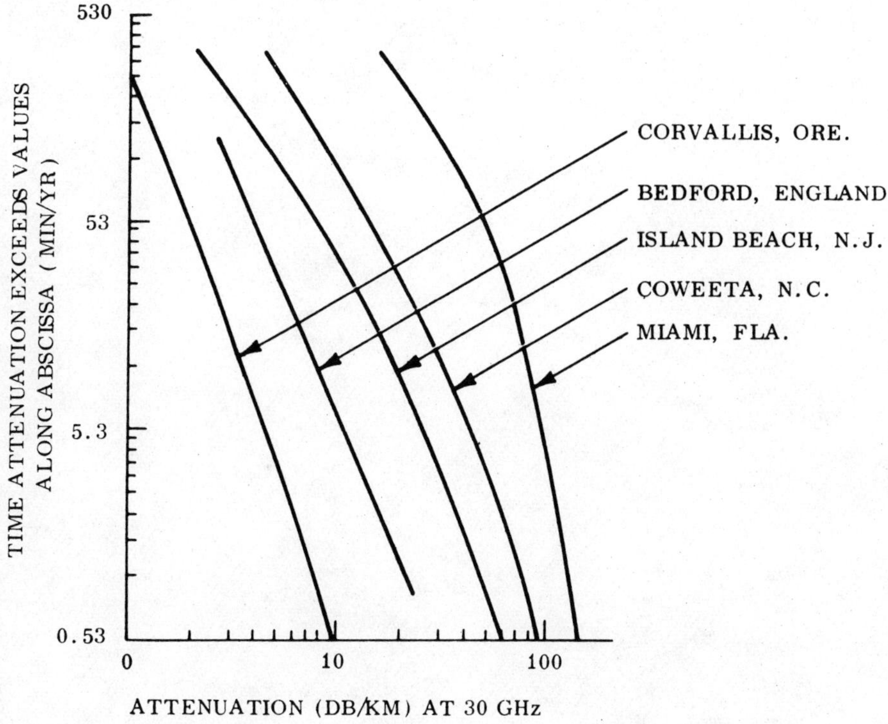

Fig. 8 Attenuation due to rain.

TV SATELLITE DISTRIBUTION AT FREQUENCIES ABOVE 10 GHz

John D. Kiesling* and Henry J. Meyerhoff/

COMSAT Laboratories, Clarksburg, Md.

Abstract

Satellite distribution of TV over an area on the Earth's surface provides a more flexible, more reliable, and less expensive alternative to terrestrial linkups. Severe attenuation of frequencies > 10 GHz is caused by heavy rain, usually localized and of short duration. A high-power spot-beam capability in the satellite, which allows propagation at high intensity to any Earth station subject to instantaneous severe fading, improves the reliability of the system without requiring excessive satellite-power margins. Based on limited statistical data, parametric studies are made to derive optimum values for antenna-gain margins, satellite e.i.r.p., battery duty cycles, and number of Earth stations per spot beam onboard the satellite, for a predetermined degree of system reliability.

I. Introduction

Communication distribution schemes, such as TV, can be implemented with geostationary satellites at frequencies above 10 GHz. However, frequencies above 10 GHz are subject to severe attenuation by heavy rainfall, usually localized and of short duration.

To combat these fades, spot beams from a larger satellite antenna (using alternate frequency assignments to avoid interference) propagate to any region in the distribution area,

---

Presented as Paper 70-454 at the AIAA 3rd Communications Satellite Systems Conference, Los Angeles, Calif., April 6-8, 1970. This paper is based upon work performed in COMSAT Laboratories under Corporate Sponsorship.

*Branch Manager, Systems Integration Laboratory.

/Member of the Technical Staff, Systems Integration Laboratory.

thereby insuring the specified reliability of continuous operation without requiring excessive power margins.

Other methods of overcoming severe rain attenuation along the satellite Earth-station link will not be discussed in this paper.

II. Measurements and Assumptions

Measurements have been made of rain-attenuation statistics for a representative location in the proposed distribution area.[1] It may be assumed, for want of data to the contrary, that beyond some rainfall rate, these statistics are identical and independent at all Earth stations distributed approximately uniformly over the zone. Figure 1 illustrates the probability that the rain attenuation at a representative station exceeds a certain level (in db). For example, x db of rain attenuation is exceeded at station A for q_x% of the time. The above assumptions imply that the same result holds at station B, and that the probability that attenuation due to rain at both stations A and B surpasses x db simultaneously is q_x^2%. It is to be remembered that these rain-attenuation statistics are also frequency dependent (Fig. 1).

Link reliability measures the probability of interrupted operation between the satellite and any Earth station. To insure a reliability of 0.01% would require a tremendous power margin onboard the satellite, where the distribution antenna gain is set by the distribution area. The peak satellite e.i.r.p. is called for over short intervals of broadcasting time, and within a small region in the distribution area. The spot beam, which is commanded over the normal command channel by the Earth station experiencing fades greater than the area-coverage antenna-gain margin threshold, satisfies these requirements with acceptable limits on satellite power. Spot-beam positioning with 3-db crossover is achieved by multiple feeds (Fig. 2) over the area distribution.

Power supplies for the spot beam may come directly from solar cells or via batteries. The battery specifications are based on statistical estimates for the length of, and intervals between, fades as a function of their magnitude. Again, since this data is not available, reasonable assumptions are made that consider the maximum anticipated length of fade inversely proportional to, and the minimum anticipated interval between fades independent of, the fade magnitude for deep fades (Fig. 3). This combined assumption is in accordance with measured rain attenuation statistics of Fig. 1.

III. Parametric Studies

Parametric TV-distribution studies considering a specified area and number of Earth stations having a certain reliability of continuous operation are used to determine values for spot-beam antenna size, and solar-cell capacity (with or without batteries) to minimize satellite weight, which is assumed roughly proportional to cost.

Consider a satellite area distribution communication scheme at some frequency $f_0 > 10$ GHz. Specify N Earth stations, where the independent identical rain-attenuation statistics are available in the form given in Fig. 1. The required reliability of continuous link operation between satellite and Earth stations is r%.

For no spot-beam capability, the required gain margin to ensure r% reliability is taken from Fig. 1 as x_0 db, where

$$r = q_{x_0} \tag{1}$$

and q_{x_0} is the probability that the rain attenuation at the Earth station exceeds x_0 db.

The e.i.r.p. for the area-coverage antenna in the absence of rain u_0, depends on the Earth-station parameters and the minimum tolerable signal-to-noise ratio at the receiver. The total e.i.r.p. e_0 for the area-coverage antenna is given by

$$e_0 = u_0 + x_0 \text{ (in db)} = 10 \log P_{u0} P_{x0} \tag{2}$$

where

$$P_{e0} = P_{u0} P_{x0} \text{ (in rf watts)} \tag{3}$$

Link reliability of r% may be achieved using a spot-beam antenna with rain-gain margin $y > x_0$ db, where the corresponding area-coverage rain margin has been considerably reduced to $x << x_0$ db.

There are M positions for the spot beam, which when taken together would cover the distribution area (Fig. 2). Hence, the gain of the spot-beam antenna is M times that of the zone antenna. Positioning of the spot beam may be accomplished by M feeds to the spot-beam antenna. For one spot-beam transmitter, the link reliability equation (described in

Appendix A) is given by

$$r_1 \simeq q_y + (N/M)(M-1/2)q_x^2, \quad q_y << q_x << 1/N \tag{4}$$

where q_y is the probability that the attenuation exceeds y db at any independent Earth station (Fig. 1). For two spot-beam transmitters, both with y db of rain-gain margin and both operating at slightly different frequencies to avoid interference, the reliability equation becomes (as in Appendix A)

$$r_2 \simeq q_y + (N^2/M^2)[(M-1)(M-2)/6]q_x^3, \ q_y << q_x << 1/N \tag{5}$$

For any specified reliability, r%, the greatest improvement derived from the spot-beam antenna occurs when link outage is equally attributable to spot and zone failures (Appendix A); i.e., for one spot beam

$$q_{y_{m1}} = (N/M)(M-1/2)q_x^2 \simeq r_1/2 \tag{6}$$

and for two spot beams

$$q_{y_{m2}} = (N^2/M^2)[(M-1)(M-2)/3]q_x^3 \simeq r_2/1.5 \tag{7}$$

The spot-beam gain over the gain for the area-coverage antenna is given in terms of the number of positions M, on which they must focus within the distribution area. Hence,

$$y = 10 \log MP_y \tag{8}$$

where $P_{u0}P_y$, $P_{u0}P_x$ is the power supplied to a spot-beam and area-coverage antenna, respectively. Solar-cell power supplied to the spot beam is required for only q_x% of the broadcasting time. A two-power-level device for the area-distribution antenna makes more efficient use of the solar cell's capacity by taking spot-beam power when the spot beam is not operating. For the additional complexity, the saving is not appreciable (Appendix B). Since the spot-beam e.i.r.p. is transmitted for relatively short intervals, a battery source of power with low duty cycle may prove very suitable.

t_y is the maximum anticipated time interval for a strong fade, y > x db. The required battery capacity to provide power to one spot beam to overcome such a fade is

$$B = P_{u0}P_y t_y \text{ (in rf watt-hours)} \tag{9}$$

where the battery weight is given by

$$w_B = k_3 B \text{ (in lb)} \tag{10}$$

Assuming that these deep fades occur in the thunderstorm season (approximately 10% of the year), and that the minimum anticipated time interval T between y db fades is independent of their magnitude, then t_y and y are inversely related in accordance with the straight-line relationship between fade depths and their probability of occurrence (Fig. 1); i.e.,

$$MP_y t_y = k_4 \tag{11}$$

The required solar-cell charging capacity S for the battery depends on the duty cycle t_y/T and the conversion efficiency factor η according to

$$S = P_{u0}P_y t_y/\eta T \text{ (in rf watts)} \tag{12}$$

The maximum anticipated time interval for simultaneous fades of y > x db amplitude in two regions is t_{yy}; hence, the battery capacity required for the two-spot-beam transmitters is

$$B = P_{u0}P_y[t_y + t_{yy}] \text{ (in rf watt-hours)} \tag{13}$$

where

$$t_{yy} << t_y$$

Similarly, the solar-cell capacity to supply the two-spot-beam transmitters is given by

$$S = (P_{u0}P_y/\eta T)[t_y + t_{yy}] \text{ (in rf watts)} \tag{14}$$

where the minimum anticipated time between simultaneous fades in two regions is also T hours (a daily phenomenon).

An inverse relationship between simultaneous fade intervals and their levels may again be assumed, where

$$MP_y t_{yy} = k_5 \tag{15}$$

The solar-cell capacity, in the absence of any battery facility for the one- and two-spot-beam transmitters is given by

$$\begin{aligned} S_1 &= P_{u0}(P_{x1} + P_{y1}) \quad \text{(in rf watts)} \\ S_2 &= P_{u0}(P_{x2} + 2P_{y2}) \quad \text{(in rf watts)} \end{aligned} \tag{16}$$

respectively, and weighs

$$w_c = k_1 S \text{ (in lb)} \tag{17}$$

Weight of the spot-beam antenna is proportional to its area, and becomes

$$w_s = k_2 A_0 M \quad \text{(in lb)} \tag{18}$$

where A_0 is the area of the area-coverage antenna and k_1, k_2 are constants.

For any spot-beam gain margin y db, minimum weight consideration relates antenna gain and solar-cell power for one- and two-spot-beam transmitters according to

$$M_1 = \sqrt{(k_1 P_{u0}/k_2 A_0)\ 10^{.1y}} \tag{19}$$

$$M_2 = \sqrt{2}\ M_1 \tag{20}$$

For the battery case, on the other hand, these gains are independent of the spot-beam gain margins and are given by

$$M_{1B} = \sqrt{(k_4 P_{u0}/k_2 A_0)\,[k_3 + (k_1/\eta T)]} \tag{21}$$

$$M_{2B} = \sqrt{[1 + (k_5/k_4)]} \cdot M_{1B} \tag{22}$$

The combined weight of solar cells for spot- and area-coverage antennas, spot-beam antenna, and battery (where required) is

$$w = k_1 P_{u0} P_x + 2k_2 A_0 M \quad \text{(in lb)} \tag{23}$$

There is a minimum value for the area-coverage gain margin, below which the assumption of statistical independence for rain-attenuation curves does not hold. This factor may limit the advantages to be gained with the spot-beam schemes outlined above.

In comparing the twin spot-beam transmitters with the single-transmitter case, the weight of the additional transmitter has not been considered.

IV. Example

Consider a TV satellite distribution scheme at 12 GHz with 50 Earth stations located over an area spanned by a 1.5-ft satellite antenna. Radio frequency transmitter power of 4 w satisfies signal-to-noise specifications at the ground receivers under clear weather conditions. Link reliability is set at 0.01%, and rain-attenuation statistics for identical independent Earth stations are shown in Fig. 1, where the zone antenna-gain margin must exceed 2.5 db to satisfy the preceding statistical assumptions.

Values for the constants derived in the previous section are as follows: $k_1 = 0.4$ lb/rfw (relates to solar-cell capacity); $k_2 = 0.25$ lb/ft^2 (relates to spot-beam antenna size); $k_3 = 0.67$ lb/rfw-hr (relates to battery capacity); $k_4 = 100$ hr (relates to fade-duration constant); $k_5 = 20$ hr (relates to double-fade duration constant); $\eta = 85\%$ (relates to battery-charging efficiency); T = 25 hr (relates to daily thunderstorm activity); and $A_0 = 1.8$ ft^2 (relates to area-coverage antenna area).

The area-coverage antenna e.i.r.p., would have to be increased by $x_0 = 18.5$ db to correspond to a link reliability of $r = q_{x_0} = 0.01\%$ in Fig. 1. Since the antenna gain is fixed by coverage considerations, the satellite transmitters would have to have an rf capacity of 285 w, as compared to the 4 w for clear weather conditions, where

$$18.5 = 10 \log P_{x0}$$

and

$$P_{e0} = 4\, P_{x0} \quad \text{(in rf watts)}$$

Now consider the addition of a spot-beam antenna with M feeds that can position the spot coverage over the distribution area onboard the satellite. (M refers to the relative gain of the spot- over the area-coverage antenna.) One or two additional transmitters feed into the spot-beam antenna at slightly different frequencies.

A range of values for the spot-beam gain $y > x_0$ db may be selected. For no battery capability, the minimum weight required for solar cell and spot antenna to yield the required e.i.r.p. determines spot-beam antenna gain when

$$M_1 = \sqrt{(k_1 P_{u0}/k_2 A_0) 10^{.1y}} \quad \text{for one spot transmitter}$$

$$M_2 = \sqrt{2}\, M_1 \quad \text{for two spot transmitters}$$

Hence,

y (db)	19	20	21	22	23
M_1	17	19	21	24	27
M_2	24	27	30	34	38

Associated with spot-beam gain y is the probability q_y that the rain attenuation at any Earth station will exceed y db (Fig. 1). Combined with a knowledge of the spot-beam antenna gain M, the area-coverage gain margin x db is read off Fig. 1 to correspond to the probability q_x, where

$$q_{x1} = \sqrt{\frac{(r - q_y)}{(N/2)[1 - (1/M_1)]}} \quad \text{for one spot transmitter}$$

and

$$q_x^2 = \sqrt[3]{\frac{(r - q_y)}{(N^2/6)[1 - (1/M_2)][1 - (2/M_2)]}}$$

for two spot transmitters

Hence,

y(db)	19	20	21	22	23
q_y(%)	0.009	0.0075	0.006	0.005	0.004
q_{x1}(%)	0.065	0.103	0.13	0.144	0.158
x_1(db)	7.1	5.2	4.5	4.2	4
q_{x2}(%)	0.3	0.41	0.475	0.51	0.54
x_2(db)	2.4≠	1.9≠	1.8≠	1.7≠	1.65≠

The rf-transmitter output power for area coverage and spot beams, together with the weight of power supplies and spot-beam antenna, are given by

$$P_{u0}P_x = 4.10^{.1x} \quad \text{(in rf watts)}$$

$$P_{u0}P_y = 4\ (1/M)10^{.1y} \quad \text{(in rf watts)}$$

$$w = k_1 P_{u0} P_x + 2k_2 A_0 M \quad \text{(in lb)}$$

Hence,

y	(db)	19	20	21	22	23
$P_{u0}P_{y1}$	(w)	19	21	24	27	30
$P_{u0}P_{x1}$	(w)	20.5	13	11.5	10.5	10
P_{e1}	(w)	39.5	34	35.5	37.5	40
w_1	(lb)	23.5	22.5	23.5	26	28
$P_{u0}P_{y2}$	(w)	13.5	15	17	19	21
$P_{u0}P_{x2}$	[w≠ (min)]	7	7	7	7	7
P_{e2}	(w)	34	37	41	45	49
w_2	(lb)	28	30.5	33.5	37	40.5

≠The area-coverage gain margins for the two-spot-beam capability are below the 2.5-db threshold, suggesting that the advantages of this scheme cannot be exploited fully.

There is a value of spot-beam gain that minimizes the weight of solar cells and spot-beam antenna for the preceding examples:

$$y = 20 \text{ db},\ M_1 = 19,\ P_{u0}P_{y1} = 21 \text{ w}$$
$$y = 19 \text{ db},\ M_2 = 24,\ P_{u0}P_{y2} = 13.5 \text{ w}$$

$$P_{u0}P_{x1} = 13 \text{ w},\ w_1 = 22.5 \text{ lb}$$
$$P_{u0}P_{x2} = 7 \text{ w},\ w_2 = 28 \text{ lb}$$

(It should be remembered that the weight of the additional spot-beam transmitter is not included in w_2.)

Now consider the two-power-level zone transmitter. The spot-beam gains are as before, which implies no change in the spot-transmitter power. The two power levels for the area-coverage antenna are such that

$$P_{u0}(P_{x1} + P_{y1}) = P_{u0}P_{z1} = P_{e1} \text{ for one spot beam}$$

and

$$P_{u0}(P_{x2} + 2P_{y2}) = P_{u0}P_{z2} = P_{e2} \text{ for two spot beams}$$

The product of the probabilities is given by

$$q_{x1}q_{z1} = \frac{(r - q_y)}{(N/2)[1 - (1/M_1)]}$$

for one spot beam

$$q_{x2}^2 q_{z2} = \frac{(r - q_y)}{(N^2/6)[1 - (1/M_1)][1 - (1/M_2)]}$$

for two spot beams

By iteration of Fig. 1,

y (db)	q_{x1} (%)	q_{z1} (%)	x_1 (db)	z_1 (db)	q_{x2} (%)	q_{z2} (%)	x_2 (db)	z_2 (db)
19	0.09	0.047	5.6	9.2	0.76	0.047	1.5§	9.2
20	0.212	0.05	3.2	8.7	1.23	0.046	0.8§	9.3
21	0.347	0.049	2.1§	8.8				

The power and weight parameters for the one spot-beam capability are given by

y (db)	$P_{u0}P_{x1}$ (w)	$P_{u0}P_{y1}$ (w)	$P_{u0}P_{z1}$ (w)	w_1 (lb)
19	14.5	19	33.5	21
20	8.5	21	29.5	20.5
21	7§	24	30.5	22

The weights of solar cell and spot-beam antenna are less than in the previous case. There is, of course, the additional complexity of a two-power-level transmitter, where full power $P_{u0}P_{z1}$ for the zone antenna requires that the transmitter have this increased capacity.

Now consider supplying the spot-beam antenna with batteries. The spot-beam gain factor M_B for minimum weight consideration is given by

$$M_{1B} = \sqrt{(k_4 P_{u0}/k_2 A_0)[k_3 + (k_1/\eta T)]} \quad \text{for one spot beam}$$

$$= 25$$

$$M_{2B} = \sqrt{1 + (k_5/k_4)M_{1B}} \quad \text{for two spot beams}$$

$$= 28$$

The overriding factor in determining these gain factors was the battery weight, not the component due to solar-cell capacity needed to charge them, where

$$k_3 >> k_1/\eta T$$

§The area-coverage gain minimum margin has been set at 2.5 db.

The area-coverage gain margins corresponding to any spot-beam gain margin y db were calculated previously, but now the spot-beam transmitter capacities are different. Hence,

y	$P_{u0}P_{x1}$	$P_{u0}P_{y1}$	w_1
(db)	(w)	(w)	(lb)
19	20.5	13	30
20	13	16	27.5
21	11.5	20	27
22	10.5	25.5	26.5
23	10	32	26.5

and

y	$P_{u0}P_{x2}$	$P_{u0}P_{y2}$	w_2
(db)	(w)	(w)	(lb)
19	7	11.5	28
20	7	14	28
21	7	18	28
22	7	23	28
23	7	28.5	28

The satellite-payload component for the battery case compares unfavorably with the previous examples. It is necessary that other uses be found for the batteries. If the capacity requirement for the battery may be pooled, constants k_4, k_5 are reduced: k_4 = 25 hr; k_5 = 5 hr. Hence,

M_{1B} = 13 for one spot beam

M_{2B} = 14 for two spot beams

and

y (db)	$P_{u0}P_{x1}$ (w)	$P_{u0}P_{y1}$ (w)	w_1 (1b)
19	20.5	25	20
20	13	31	17
21	11.5	38.5	16
22	10.5	49	16
23	10	61.5	15.5

and

y (db)	$P_{u0}P_{x2}$ (w)	$P_{u0}P_{y2}$ (w)	w_2 (1b)
19	7	23	15.5
20	7	28.5	15.5
21	7	36	15.5
22	7	46	15.5
23	7	57	15.5

For this case, the two-spot-transmitters scheme has an advantage over the one-spot-beam facility in payload, but this must be modified by the weight of the second transmitter.

V. Conclusions

Rain attenuation is an important consideration in TV satellite-distribution schemes for frequencies greater than 10 GHz. To maintain a high-reliability measure of continuous operation (≃0.01%) would require an excessive gain margin (≃18.5 db for 12-GHz transmission) over the power requirements for clear-sky operation. To provide an area-coverage antenna, whose gain was fixed by coverage considerations, with solar cells and transmitter capacity to meet these localized short-duration attenuation peaks would be most uneconomical. With the use of an additional spot-beam antenna and ancilliary transmitters, which can be switched on by command from the Earth station experiencing heavy rain attenuation, it is possible to achieve the desired measure of reliability. Parametric studies indicated values for spot-antenna gain and transmitter power that minimized their combined weight onboard satellite, where the transmitter and solar cell capacities were much below the previous case. Other parameters appearing in the system equations were the rain-attenuation statistics, which were assumed to be identical and independent beyond some minimum gain-margin threshold for all Earth stations uniformly distributed over the TV distribution area. Instead of solar cells supplying the spot beam directly for the short intervals required, batteries with low duty cycles, which reduced the satellite payload when the batteries could be shared with other systems, were considered. The possibility of a two-level zone transmitter was also considered as a means of efficiently using the solar-cell capacity without batteries onboard.

Penalties associated with the spot-beam scheme include additional Earth station capability and cost to handle at least two different frequencies, and additional satellite antenna with multiple feeds, transmitters, switches and batteries. On the positive side is the reliability measure attainable for reasonable-size satellite transmitters and power supplies at frequencies greater than 10 GHz.

Appendix A

Link Reliability Equations

N Earth stations are uniformly distributed over the area. There are M locations for spot beam (Fig. 2) implying that there are $m = N/M$ Earth stations within one spot beam.

The probability that attenuation at any station does not exceed x db (Fig. 1) is p_x. ($x > 2.5$ db to allow the assumption that attenuation statistics at all stations are independent and identical.) The probability that attenuation at all m stations within one spot-beam location is less than x db, is p_x^m.

The probability that attenuation at one or more of m stations exceeds x db is $1 - p_x^m$. Hence, the probability that attenuation is less than x db at station A is p_x.

Probability that attenuation lies between x db and y db at station A within one spot-beam location, and that the attenuation is less than x db at all remaining stations within the other spot-beam locations is

$$(P_y - P_x) \quad [p_x^m]^{M - 1}$$

and that the attenuation is less than x db at all stations within M-2 spot-beam locations, and greater than x db at one or more of the stations within the other spot beam location is

$$(P_y - P_x) \quad (M - 1)[P_x^m]^{M-2} \quad [1 - P_x^m]$$

The link reliability for station A depends on: 1) attenuation at station A less than x db - P_x; 2) attenuation at station A between x db and y db and, a) the availability of one spot beam with gain margin y db for station A, implying that the spot beam is not required at any other location

$$(P_y - P_x) \quad [P_x^m]^{M-1}$$

and b) the requirement for one spot beam at one other location

$$[P_y - P_x][P_x^m]^{M-2}[1 - P_x^m](M - 1)/2$$

The link reliability for the one spot-beam capability is given by

$$\begin{aligned} r_1 &= 1 - P_x - (P_y - P_x)\Big[[P_x^m]^{M-1} \\ &\quad + [P_x^m]^{M-2}\,[1 - P_x^m](M - 1)/2 + \ldots\Big] \\ &= q_x - (q_x - q_y)\Big[(1 - q_x)^{m \cdot M-1} \\ &\quad + (1 - q_x)^{m \cdot M-2}\,[1 - (1 - q_x)^m]\,(M - 1)/2 + \ldots\Big] \\ &\simeq q_y + m[(M - 1)/2]q_x^2, \quad q_y << q_x << 1/N \end{aligned}$$

where

$$q_x = 1 - P_x, \; q_y = 1 - P_y$$

If two spot beams are available, the link reliability is given by

$$r_2 = q_x - (q_x - q_y)\Big[(1 - q_x)^{m \cdot M-1}$$

$$+ (1 - q_x)^{m \cdot M-2}\,[1 - (1 - q_x)^m]\quad (M - 1)$$

$$+ (1 - q_x)^{m \cdot M-3}\,[1 - (1 - q_x)^m]^2[(M - 1)(M - 2)/3] + \ldots\Big]$$

$$\simeq q_y + m^2[(M - 1)(M - 2)/6]q_x^3,\ q_y << q_x << 1/N$$

Now link reliability r and probability q_x are measured on log scales. With no spot beam available, reliability is given by (Fig. 1)

$$r_0 = q_x$$

With one spot beam available, the greatest difference between $r_0 = q_x$ and $r_1 = q_y + [m(M - 1)/2](q_x^2)$ occurs when

$$(d\ \log r_1)/(d\ \log r_0) = 1$$

$$\simeq \frac{m(M - 1)q_x^2}{q_y + m(M - 1/2)q_x^2},\ q_y << q_x << 1/N$$

$$\therefore q_{y_{m1}} = m(M - 1/2)q_x^2$$

For the two spot beams

$$(d\ \log r_2)/(d\ \log r_0) = 1$$

$$\simeq \frac{M^2[(M - 1)(M - 2)]/2\ q_x^3}{q_y + m^2[(M - 1)(M - 2)/6]q_x^3},$$

$$q_y << q_x << 1/N$$

$$\therefore q_{y_{m2}} = [m^2(M - 1)(M - 2)/3]q_x^3$$

For the two-power-level device, the reliability is given by

$$r_1 = q_z - (q_z - q_y)\Big[(1 - q_x)^{m \cdot M-1}$$

$$+ [(M - 1)/2](1 - q_x)^{m \cdot M-2}\,[1 - (1 - q_x)^m]$$

$$+ [(M - 1)/2][(M - 2)/3](1 - q_x)^{m \cdot M-3}[1 - (1 - q_x)^m]^2 + \ldots\Big]$$

$$\simeq q_y + m[(M - 1)/2]q_x q_z, \quad q_y << q_z < q_x << 1/N$$

for one spot beam, and

$$r_2 = q_z - (q_z - q_y)\Big[(1 - q_x)^{m \cdot M-1}$$

$$+ (M - 1)(1 - q_x)^{m \cdot M-2}\,[1 - (1 - q_x)^m]$$

$$+ [(M - 1)(M - 2)/3]\,(1 - q_x)^{m \cdot M-3}[1 - (1 - q_x)^m]^2 + \ldots\Big]$$

$$\simeq q_y + m^2\,[(M-1)\,(M-2)/6]q_x^2 q_z, \quad q_y << q_z < q_x << 1/N$$

for two spot beams.

Appendix B

Two-Power-Level Device for Area-Distribution Antenna

Full power (P_e watts) is applied to the area-coverage antenna when the spot beam is not required. At other times, full power is shared between the area-coverage antenna $P_{u0}P_x$ and the spot antenna $P_{u0}P_y$. The reliability equation becomes, for one spot-beam transmitter (Appendix A)

$$r_1 \simeq q_y + (N/M)[(M - 1)/2]q_x q_z,$$

$$q_y << q_z < q_x << 1/N$$

where q_z is the probability that the attenuation at any Earth station is greater than

$$z = 10 \log(P_x + P_y) \text{ db}$$

and

$$P_{e1} = P_{u0}(P_x + P_y) \quad \text{(in rf watts)}$$

For two spot-beam transmitters

$$r_2 \simeq q_y + (N^2/M^2)[(M-1)(M-2)/6]q_z q_x^2$$

$$q_y << q_z < q_x << 1/N$$

where

$$z = 10 \log(P_x + 2P_y)\,\text{db}$$

A three-power-level device would drop the area-coverage rain-gain margins to one level when only one spot-beam transmitter is in operation, and to a lower level when both spot-beam transmissions occur simultaneously.

Reference

[1]Wilson, R.W., "Sun Tracker Measurements of Attenuation by Rain at 16 and 30 GHz," Bell System Technical Journal, Vol. 48, No. 5, May-June 1969, pp. 1383-1404.

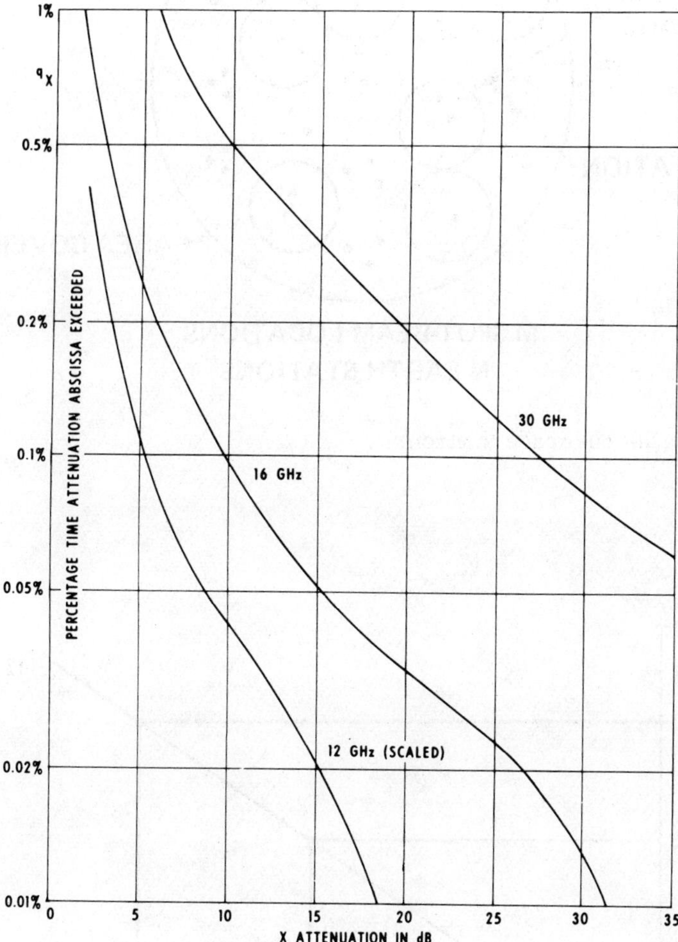

Fig. 1 Attenuation statistics for millimeter-wave link for satellite to any Earth station.

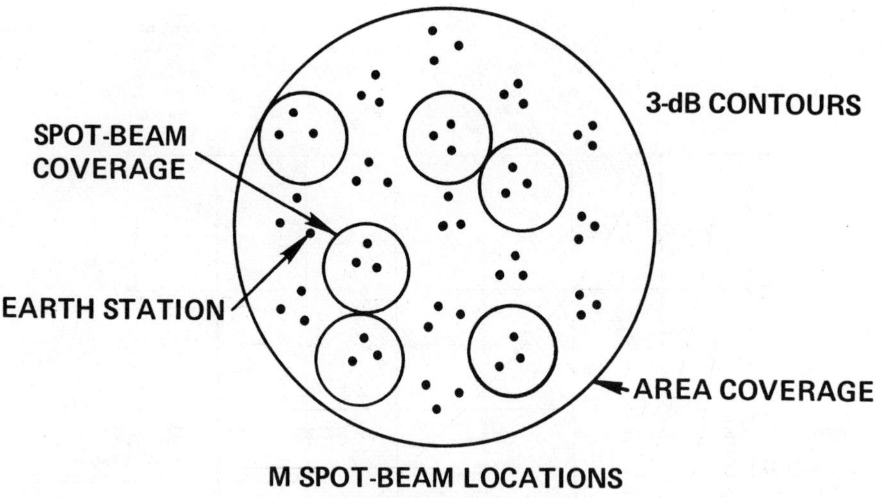

Fig. 2 3-db coverage contours.

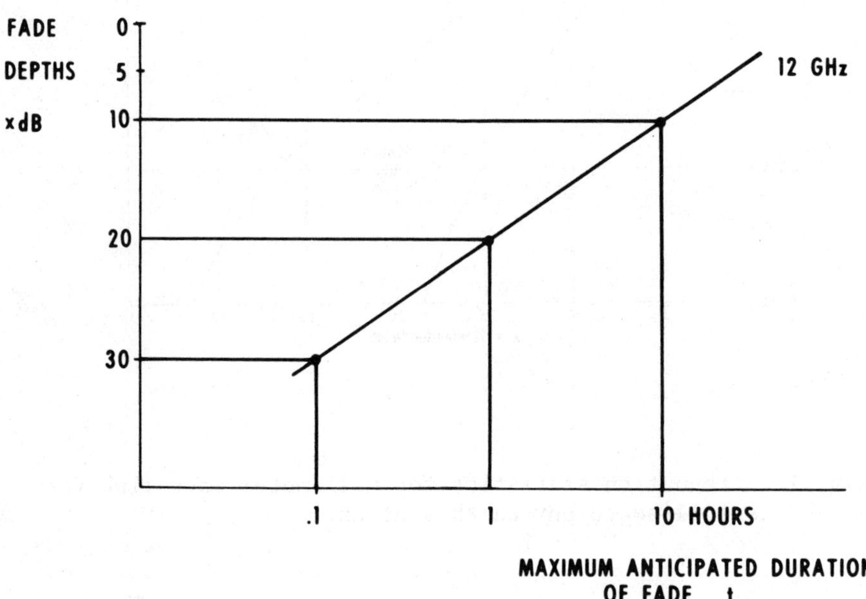

Fig. 3 Maximum anticipated duration t_X of fade of x db.

ATS-F AND-G SYSTEM SUMMARY

Harry L. Gerwin*

NASA Goddard Space Flight Center, Greenbelt, Md.

Abstract

The objective of these satellites is to demonstrate a 30-ft deployable antenna, provide a three-axis stabilized spacecraft with a 0.1° pointing capability, and provide an oriented platform at synchronous altitude for advanced experiments. The spacecraft will be launched from Cape Kennedy by a Titan III-C vehicle and will be placed at about 22,300 miles altitude.

At the top of the reflector support truss of the spacecraft are the 30-ft self-deploying reflector, the science platform and the two solar array panels and their booms. At the bottom of the truss is the Earth-viewing module carrying all of the necessary experiments, communications, control, propulsion, and sensor subsystems. Four frequency bands are provided: C, S L, and UHF. A multifrequency, single-conversion, ground-controlled transponder provides the receiving, processing, and transmitting functions. Sixteen experiments are planned such as evaluating new concepts in meterological data collection, defining the Earth's space and near-Earth propagation characteristics above the conventional rf spectrum, studying data on rf interference, demonstrating the relay by satellite of instructional TV conducted by India, relaying communications from low-orbiting satellites to ground stations, and scientific experiments designed to study the environment at synchronous altitude and to gain information on electromagnetic-ionospheric interactions.

Introduction

Dramatic achievements by the manned space effort have been widely presented, but the less-publicized, equally important

---

Presented as Paper 70-1307 at the AIAA 7th Annual Meeting and Technical Display, Houston, Texas, October 19-22, 1970.

*ATS-F and G Project Manager.

unmanned scientific and engineering efforts have not been widely dramatized. It should be noted that these unmanned space efforts have provided many new concepts and creative ideas for the benefit of man in his everyday environment. One of these unmanned efforts, the ATS Program, was conceived and initiated by the National Aeronautics and Space Administration early in the 1960's to advance space technology for the benefit of mankind.

The SYNCOM and EARLYBIRD series of satellites demonstrated the feasibility of relaying both wide- and narrow-band communications via synchronous satellites. The earlier ATS series demonstrated the spin scan camera that provided pictures of cloud cover. These pictures uncovered wind patterns and cloud movements previously unknown by the meteorological community. They showed cloud movements across the equator—it had been generally believed, before these pictures were taken, that cloud movement across the equator would not occur. These cloud patterns and movements now can be included when the factors that go into a weather prediction are considered. Timely warnings of severe storms and other weather hazards save lives and property. In 1969, the path of hurricane Camille was tracked and predicted accurately enough to permit evacuation of 70,000 people from the affected Gulf Coast area. Without this warning, ESSA estimated that thousands of people might have lost their lives.

The ATS-III has and still provides a hurricane and tornado watch. By virtue of its position in synchronous orbit it has an overview of the hurricane spawning ground. Because of its ability to provide a picture as often as every 5 min, a continuous picture sequence is available. This provides a history and a track of the hurricane. Thus the meteorologist can provide the general public with an up-to-date assessment of the hurricane position, based on this unique ATS capability, and is able to warn the public when danger is imminent. The Chairman of the U.S. Atomic Energy Commission, Dr. Glenn Seaborg, estimates that if weather could be predicted accurately for even three days in advance, man could save $60 billion a year, plus millions of lives and an untold amount of grief.

Another achievement of the earlier series of ATS was the first ground-to-satellite-to-aircraft communications link over the Atlantic Ocean.

The ATS-F and -G program will expand the capabilities of preceding satellites and provide advanced technologies in communications, navigation, and meteorology. A major innovation of the new series of ATS spacecraft is the high gain, 30-ft parabolic antenna. A second major innovation is the achievement of a tenth of a degree angle tracking accuracy—a capability required with the 30-ft antenna.

The general objectives of the ATS missions are to develop technology commonly required for a variety of applications and to provide an orbital testing platform for advanced concepts in space technology. Within the general objectives, there are more specific objectives assigned to each spacecraft. ATS-F and -G has the following specific mission objectives: a) to demonstrate a 30-ft deployable antenna in near geostationary orbit; b) to provide a three-axis stabilized spacecraft with 0.1° pointing capability in all three axes; c) and to provide an oriented platform at synchronous altitude for advanced technology and scientific experiments.

The ATS-F and -G program has proceeded on a predetermined project plan basis. In accordance with this plan, anticipated launch schedules for ATS-F and -G are early 1973 and early 1975. The experiments that will be carried on board ATS-F were selected in 1969. Delivery of prototype models of the experiment equipment is scheduled for late 1971; delivery of flight models is scheduled for early 1972.

ATS-F and -G will be launched from the Eastern Test Range, Cape Kennedy, Fla., by a Titan III-C launch vehicle. The satellites will be placed at an altitude of approximately 22,300 miles with an orbital period of 24 hrs. The insertion inclination will be about 1°. This orbital period matches the Earth's rotational period; thus, the satellite seems to remain at a fixed position in the sky. This synchronous type orbit provides full coverage of almost one-half of the Earth's surface.

The Spacecraft

The ATS-F and -G spacecraft, shown in the orbital configuration in Fig. 1, consists of five major assemblies: Earth-viewing module, reflector support truss, 30-ft parabolic reflector, solar arrays, and the science platform.

The Earth-viewing module (EVM) consists of three separate modules that when completely assembled form a structural cube approximately 54 in. along each edge. The module nearest the parabolic reflector contains all of the communications subsystem components and experiments associated with communications.

The top of this module supports the prime focus feed assembly. The center module contains the attitude control, power, telemetry and command, and auxiliary propulsion subsystems. The remaining module is the Earth viewing portion of the EVM, which contains the remaining experiments, the batteries, and the battery chargers. The interferometer, Earth sensors, sun sensors, star trackers, and auxiliary propulsion tankage also are located in this module.

The reflector support truss connects the EVM and the reflector assemblies. It is designed to maintain the position and alignment of the reflector/EVM axis.

The reflector antenna is a self-deploying 30-ft parabola. The ribs are flexible, semilenticular, aluminum members, parabolically contoured and chemically milled into a web beam pattern. The rf mesh, fabricated from tapered segments bonded together, is stitched to the ribs. The ribs are attached to a center hub through spring-loaded hinges connected to adjustment mechanisms for fine contouring. The hub assembly is torus with a channel cross section. Spring-loaded doors provide closure for the stored rib and mesh assembly, with door restraint provided by a hoop cable separated by a cable cutter at deployment.

The reflector is stowed for launch by rotating each rib 90°, flattening the rib camber, and bending the ribs around the hub wall. The flexible mesh is folded between adjacent ribs. The resulting wrap is contained by closure doors and cable. At cable separation command, the doors open, and stored energy unwinds the ribs tangentially. Stretching of the contoured mesh is provided when the ribs reach their final position. Figure 2 shows the reflector deployment sequence. After the reflector is deployed, each feed element illuminates the reflector, focusing the rf energy into a highly directive beam.

The two solar panel arrays are arranged to intercept the maximum sun illumination over a 24-hr period. Initial power provided by the panels is 500 w, reducing to 400 w near the end of the satellite life. A power supply regulator system is designed to provide maximum solar power output to the spacecraft and experiments throughout the entire life of the solar arrays.

The science platform is mounted on the hub behind the reflector where it can meet the scientific experiment requirements for viewing space.

Spacecraft Subsystems

The Command and Telemetry Subsystems provide ground control of the spacecraft functions and the capability of monitoring its health and status. Since the spacecraft is stationary with respect to ground stations, a continuous monitoring and command capability exists with few automatic functions required. The systems are compatible with and use the existing NASA network of ground stations. Up to 512 commands can be sent, and more than 600 data points can be monitored by the ground stations.

The Auxiliary Propulsion Subsystem is a self-contained module installed in the EVM. It contains a pressurized fuel tank and attitude and station-keeping valve-controlled rocket engines. The engines are located around the spacecraft to provide both attitude and orbital control.

The Communications Subsystem provides four frequency bands—C, S, L, and UHF. Monopulse attitude data are developed at C- and S-band frequencies to permit precise pointing of the antenna beam at a cooperating station. The S-band feed array is designed to permit electronic scanning to shift the radiated rf beam sequentially to follow a low orbiting satellite across the Earth's surface. The L-band antenna pattern as shown in Fig. 3 is uniquely shaped to provide coverage of the North Atlantic air traffic corridor from New York to Shannon, Ireland. The UHF feed provides a high-gain beam for transmission of TV signals to small ground terminals for educational TV programming. C-band is used as the main ground-spacecraft link for all communications operations.

An integrated, multifrequency, single-conversion transponder provides receiving, processing, and transmitting functions. It provides a capability to receive three different frequency bands and transmit four assigned transmission bands. Ground control of the transponder allows selection of frequency, bandwidth, and processing to perform operational and experimental functions. Phase lockup of the transponder local oscillator to a receive carrier is provided. The communications capability of the spacecraft is summarized in Table 1.

Table 1 Communications Performance

Frequency	Mode	Beamwidth (Degrees)	ERP (DBW)	G/T (DB/°K)
6 GHz (C)	REC	0.4	---	10.5
4 GHz (C)	XMT	0.6	54	---
2.25 GHz (S)	REC	1.1 (Scanned)	---	7.0
1.8 GHz (S)	XMT	1.35 (Scanned)	48	---
1.65 GHz (L)	REC	7.5/1.5 (Shaped)	---	-5.0
		1.5	---	2.5
1.55 GHz (L)	XMT	7.5/1.5 (Shaped)	42	---
		1.5	48	---
850 MHz (UHF)	XMT	2.6	48	---

The Attitude Control System (ACS) maintains requisite spacecraft attitude in orbit by using sensing systems that reference to Earth and polaris, a servo system, and an auxiliary propulsion system. The ACS maintains a pointing capability of 1/10° or better. The basis of this system is an onboard computer that converts sensed data to commands required to maintain proper spacecraft attitude.

The inertial reference assembly provides rate damping signals. The coarse and fine sun-sensor signals are used to acquire the sun, and the Earth sensor is used for Earth acquisition and to provide the desired local vertical reference (roll/pitch) for subsequent operational control modes. The polaris sensor is used for polaris acquisition and subsequently provides a reference for control of the yaw axis. The operational controller (Fig. 4) serves as the prime controller. The inertia wheels are the prime means for torquing the spacecraft with the auxiliary propulsion system, providing torques for unloading the wheels. The acquisition modes use the auxiliary propulsion subsystem for torquing the spacecraft.

Backup control modes at reduced power levels include an analog backup controller and ground command of the auxiliary propulsion jets based on telemetered sensor data. Backup experimental pitch and roll reference is provided by an interferometer which, when operated with a single ground station, provides an over-all accuracy of ±0.03° over a field of view of ±12.5°. The monopulse at S- and C-band provide an additional roll and pitch backup angle measuring capability.

The Experiments

Since the experiments for ATS-G have not been selected at this time, the following descriptions include only those experiments to be flown on ATS-F. These experiments are catagorized and listed in Table 2.

Table 2 ATS-F experiments

Meteorology
Very high resolution radiometer
Communications
Instructional television (ITV) relay
Millimeter wave
Radio frequency interference
Radio beacon
Spacecraft Applications
Data relay satellite system (DRSS)
Cesium bombardment ion engine
SAMOC and SAPPSAC
Navigation
PLACE
Scientific Experiments
Low-energy proton-electron
Low-energy protons
Solar cosmic ray
Auroral particles
Particle acceleration
Magnetometer
Solar cell radiation damage

Very High Resolution Radiometer Experiment

This experiment is directed toward evaluating the feasibility of new concepts in meteorological data collection and distribution, and new camera systems on board stabilized spacecraft.

The radiometer is designed to give both day and night cloud coverage information for the determination of cloud motions, tropical storm life cycles, extratropical storm life cycles, mesoscale phenomena, and cloud climatology studies. It is capable of operating in the infrared frequency between 10.5 and 12.5μ and in the visible spectrum frequency between 0.55 and 0.75μ. Operating through an 8-inch Cassegrainian telescope, the instrument can generate a picture in 20 min with a resolution of 10 km at the subsatellite point. In addition, the radiometer system supplies information used for Earth albedo measurements, Earth resources studies, and ocean temperature studies.

The principal investigator plans to reduce and analyze the radiometer data by a combination of digital computer and photo-imaging techniques. In addition to recording all radiometer data while the spacecraft is within view of the United States, the investigator will be provided with a near real time (within 1 hr) readout of selected frames of radiometer data. There are no present plans to provide this quick-look capability while the spacecraft is over Africa. As with most of the experiments aboard the ATS satellites, this experiment is available to anyone desiring to participate in its activities.

Instructional Television Relay

The objectives of the Instructional Television (ITV) experiment are to advance the state-of-the-art in wideband communications by demonstrating satellite relay of CCIR quality television to small ground receiving stations and to study the propagation characteristics of satellite to ground radio transmission.

Of prime importance to the rapid development of nations is the ability to economically disseminate information to the population. Information designed to improve occupational skills, increase food production, contribute to family planning, improve health and hygiene, and train teachers can be most efficiently disseminated by television. In areas of the world which do not have existing television distribution networks, such as India, this can best be accomplished by the use of a satellite to relay the broadcast from the central transmission site to community receivers as illustrated in Fig. 5. In densely populated areas the broadcast can be relayed locally to increase coverage.

This experiment, sometimes referred to as a "schoolhouse in the sky", will use the ATS-F as the satellite relay in a test to determine the feasibility of using broadcast satellites for national development. Information will be relayed to 2000 community receivers, serving up to 400,000 people, and to four redistribution centers reaching an additional 500,000 people. The ground sites, receivers and programming will be provided by the Indian government. An experiment duration of one year is planned with service to be provided for four hours per day.

Millimeter Wave Experiment

This experiment is the second in a series designed to define the Earth's space and near-Earth propagation path characteristics in a region extending above the conventional radiofrequency spectrum. Investigation in this region is necessary to reassign the overcrowded communications frequency bands currently available, thus accommodating communications growth needs for the 1970's and beyond. Data regarding propagation characteristics at 20 and 30 GHz, over a very wide bandwidth, will be obtained and analyzed. The knowledge gained will permit the design of operational systems utilizing frequencies in this range, which fall on either side of the 22-GHz water-vapor absorption line.

Radio-Frequency Interference (RFI) Experiment

The purpose of the radio-frequency interference experiment is to obtain data for the creation and improvement of a mathematical model of the electromagnetic environment of satellites in synchronous orbit. The study concentrates on the 6-GHz communication frequency band. Transmission sources working within this frequency band include microwave relays and the present COMSAT satellite communications relays.

The technical objectives of this experiment are to 1) determine the 6-GHz interference power at the satellite, 2) establish practical gain-noise limits for spacecraft, 3) establish a minimum satellite protection ratio, and 4) determine both geographical and frequency distribution of the rf noise. The data obtained will be analyzed and made available for spectrum utilization planning and for radio transmission regulatory determining purposes.

Radio Beacon Experiment

The radio beacon experiment provides a multifrequency spacecraft-to-ground radio link for investigating particles that affect radio propagation beyond the atmosphere. The beacon is designed to measure Faraday rotation, differential phase (doppler), phase and amplitude scintillation, and signal amplitude (absorption). The planned mode of operation allows for continuous emission on all frequencies. This experiment will provide a better understanding of the effect particles beyond the atmosphere have on radio propagation.

Data Relay Satellite System Experiment

The value of satellites in near-Earth orbit has been amply demonstrated in the areas of meteorological and geophysical data collection. There are two problems in the use of low orbiting data collection spacecraft. These are transferring the acquired data to the ground and the precise determination of the orbit parameters. The use of a synchronous satellite as a data relay is the most promising technique for solving both problems. The 30-ft parabolic antenna system on ATS-F and -G provides a high degree of rf energy focusing. This high-gain antenna and the capability to follow a low orbiting satellite as it passes around the Earth will be used in the data relay experiment.

As the data collection satellite orbits the Earth, it is within sight of any one ground station for only a short time. It is economically impractical to provide enough ground stations for continuous coverage. Thus, the satellite must either operate intermittently or be provided with a complex recording system to store data for delayed transmission to the ground. Figure 6 illustrates the improved coverage available with a synchronous data relay. The relay satellite can communicate with a low orbiting satellite for at least half an orbit and continuously relay data to a ground station in real time. This greatly reduces the need for onboard recording capability and makes the data immediately available for analysis and dissemination. Experimental data from the NIMBUS spacecraft will be transmitted to the ground at 4,000 bits per second. The link performance will be measured at rates up to 500,00 bits per second.

Orbit determination using data from several ground stations, each with a short observation time, is more costly than orbit determination using a synchronous satellite with an observation time of at least one-half orbit of the low orbiting satellite.

Since the data can be transmitted directly to a computer center in real time, the satellite orbit is available to the user on a more timely basis.

Cesium Bombardment Ion Engine Experiment

The primary objective of this experiment is to verify and obtain operational data on the utilization of an ion microthruster electric propulsion system. Interest in this area is expanding because of the high value of specific impulse that is achievable compared to chemical propulsion systems. The specific impulse expected of the ion engine experiment to be flown on ATS-F is 2500 sec. This compares with specific impulses of 125-200 sec for the typical chemical system used in a similar role today. An important element in this experiment is to examine the technology of preventing cesium from contaminating the spacecraft or experiments as a result of the ion engine firing.

Spacecraft Attitude Maneuvering Optimal Control (SAMOC) and Self-Adaptive Precision Pointing Spacecraft Attitude Control (SAPPSAC) Experiment

The SAMOC and SAPPSAC experiments are designed to demonstrate the use of a ground-based controller for attitude maneuvering and precision pointing of the ATS-F spacecraft using onboard torquers and sensors with coupling through command and telemetry links.

The objectives of the experiments are to demonstrate the ability to: 1) perform a single attitude slewing maneuver between two large-angle reference points, using optimal criteria of prescribed maneuvering time, propellant or electrical energy expenditure constraints, allowable body rates, or combinations thereof; 2) hold a precise attitude in the presence of disturbing inputs for an extended time, using adaptive control of spacecraft torquers; 3) generate a ground track or mapping pattern using optimal performance criteria; 4) track another inflight object, using optimal or suboptimal criteria for propellant, rates, and electrical power consumption; and 5) improve spacecraft control performance by updating the dynamics model used in the ground controller.

PLACE Experiment

Aircraft traffic density, the world over, is increasing year by year. If present safety standards are to be maintained in the future and costs held to a reasonable level, improved methods for aircraft position location and two-way communication between the ground and many aircraft must be developed. The Position Location and Aircraft Communications Experiment (PLACE), illustrated in Fig. 7, will evaluate new technical approaches to solving this problem. It will provide engineering data demonstrating two-way communication between ground terminals and many aircraft simultaneously, and practical experience applicable to air traffic control systems operating at the aeronautical L-band frequencies. The experiment will investigate the absolute and relative accuracies of several position location techniques to aid in the selection of the one most effective. Experiments, demonstrating the capability to communicate with up to 200 aircraft simultaneously will be conducted in close cooperation with the Federal Aviation Administration. It is anticipated that a position accuracy of 1 nautical mile will be achieved.

Scientific Experiments

The integrated scientific experiments package on board ATS-F is a group of seven experiments that are designed to study the spacecraft environment at synchronous altitude and to gain information on electromagnetic-ionospheric interactions. These experiments are: 1) low-energy proton-electron, 2) low energy protons, 3) solar cosmic ray, 4) auroral particles, 5) particle acceleration measurements, 6) magnetometer, and 7) solar cell radiation damage.

Five of the experiments are designed to obtain data on charged particles of several different types and over wide energy ranges. The sixth experiment will provide magnetic field data to be used in conjunction with the charged particle measurements to determine the dynamic processes which take place in the synchronous orbit environment. The seventh experiment is a continuation of previous ATS studies into solar cell degradation.

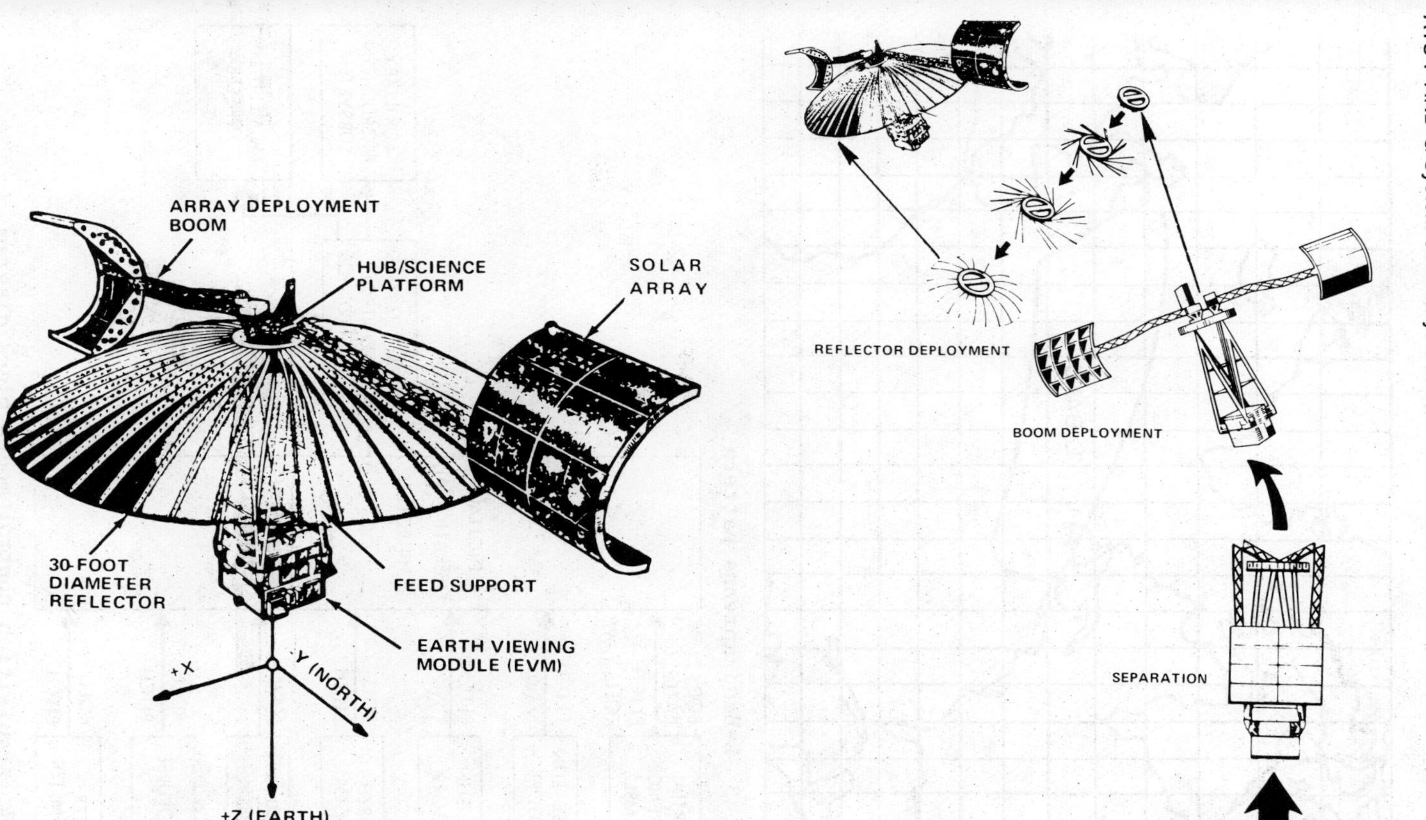

Fig. 1 ATS-F and -G orbital configuration.

Fig. 2 Reflector deployment sequence.

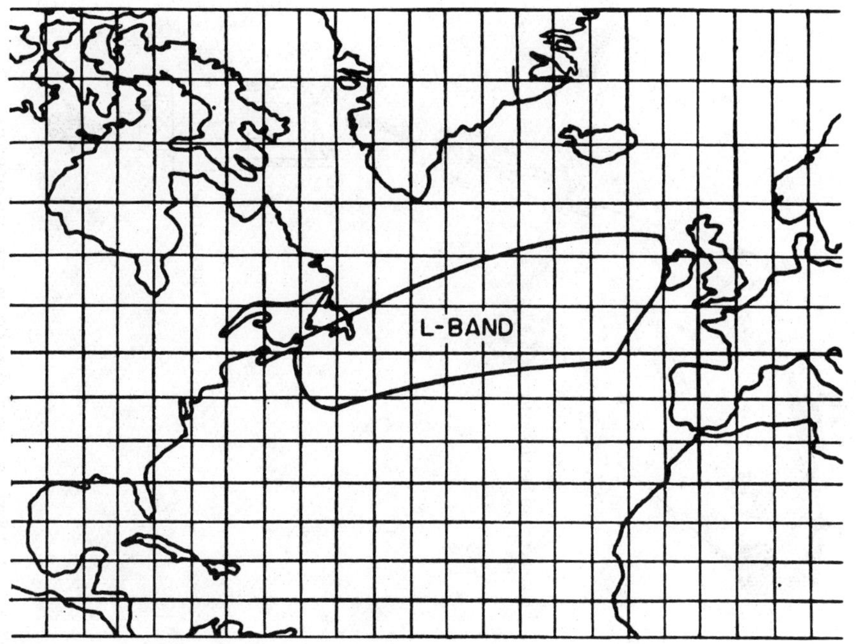

Fig. 3 L-Band antenna pattern.

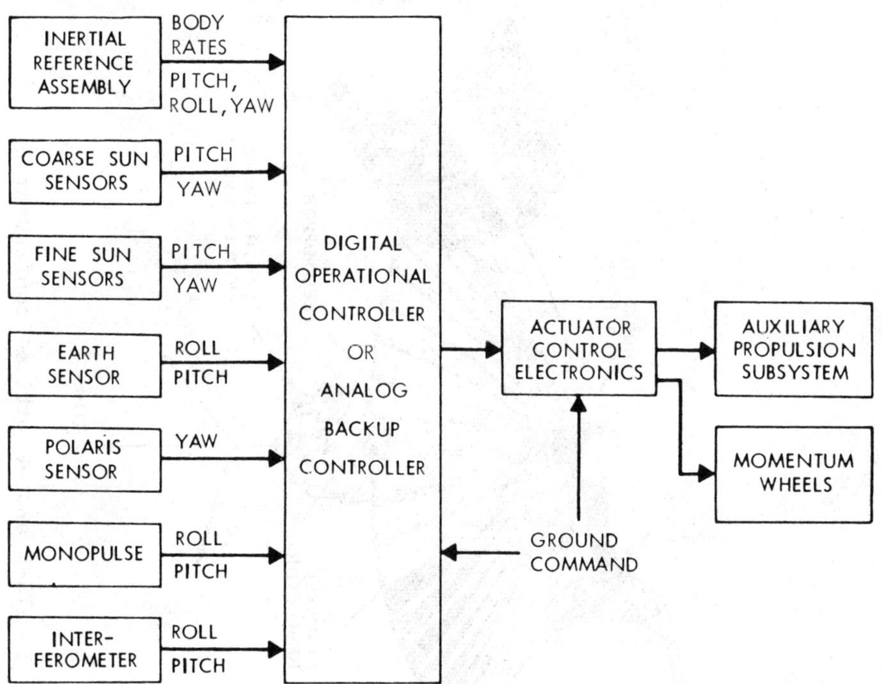

Fig. 4 Acquisition control modes block diagram.

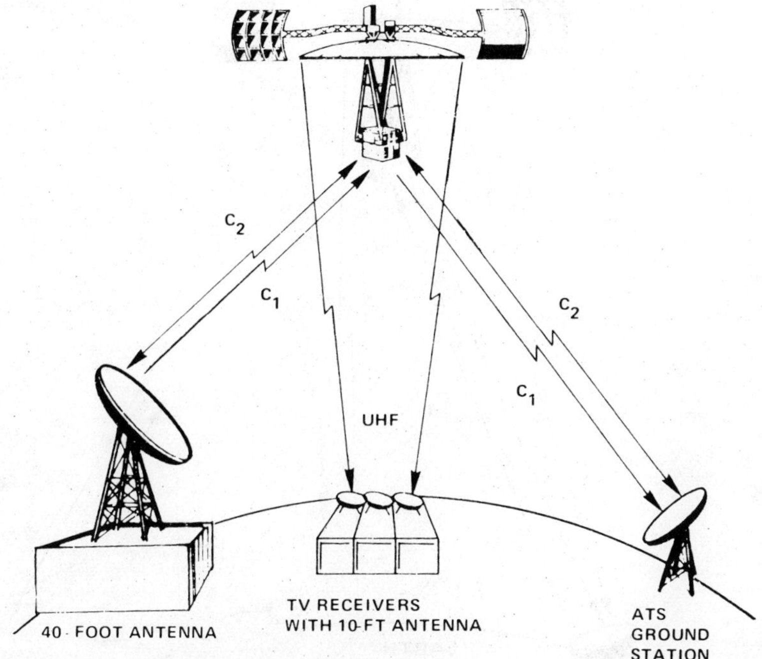

Fig. 5 ITV link geometry.

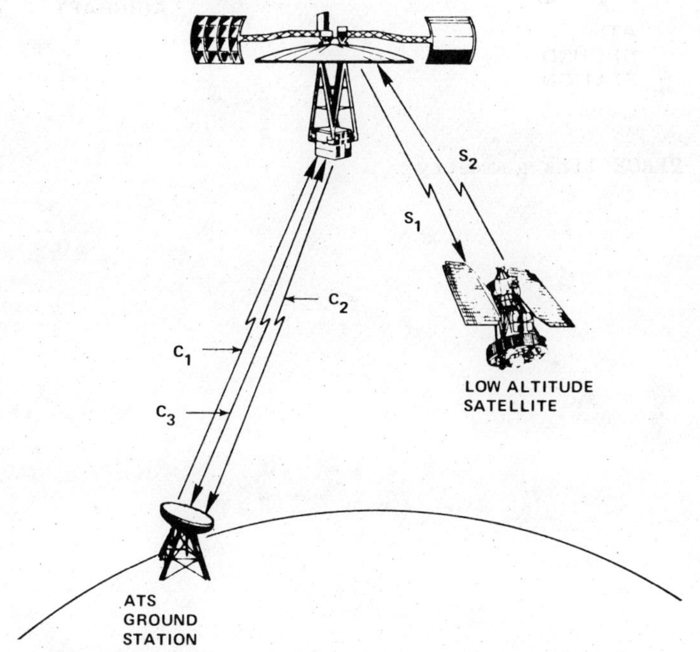

Fig. 6 Data relay experiment link geometry.

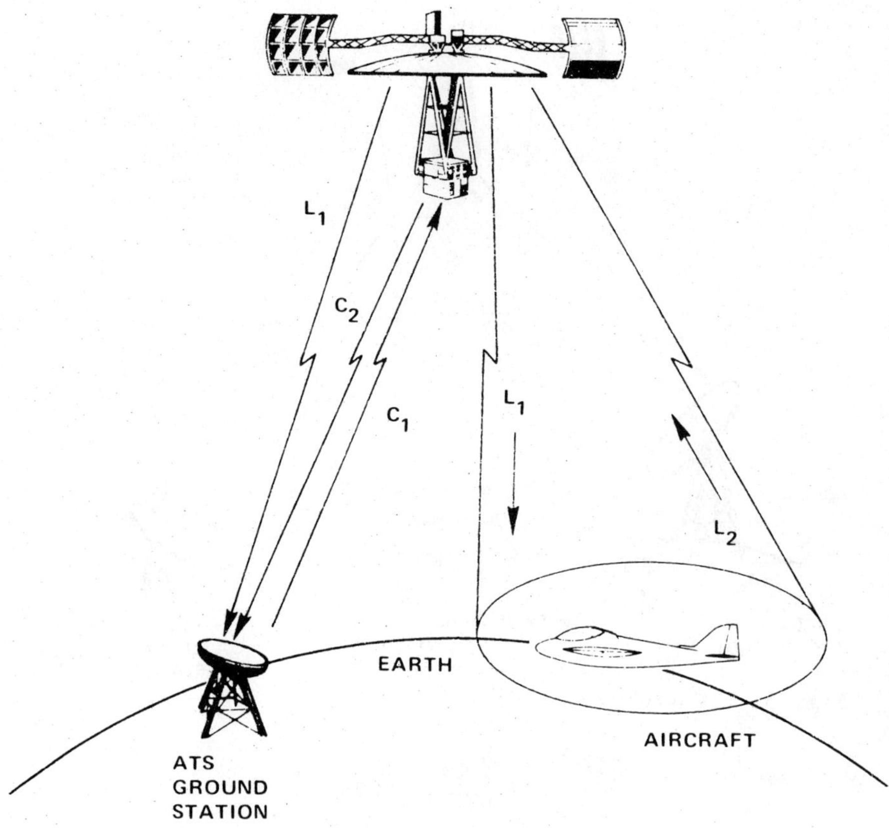

Fig. 7 PLACE link geometry.

V. AERONAUTICAL SERVICE SYSTEMS

SATELLITES FOR DOMESTIC AIR TRAFFIC CONTROL

D. D. Otten,* J. H. Craigie,† A. Garabedian,† and D. D. Morrison§

TRW Systems Group, Redondo Beach, Calif.

Abstract

The long-term domestic air transportation picture can be enhanced substantially by the application of satellite technology, although the problems of the next several years must clearly be resolved by other means. This paper briefly examines air transportation and air traffic control factors which relate to the application of satellites to surveillance, navigation, and communications over the contiguous United States. The surveillance function is likely to be the most important application; thus, this paper includes discussions of some of the design considerations which are significant in that application. The system described involves satellites, ground stations, and electronic hardware for a wide range of aircraft. All users will have, as a minimum, a VHF communications transceiver identical to that which aircraft now have, and a Location/Identification Transmitter (LIT) and antenna for surveillance which is unique to the satellite system proposed. This low-cost, lightweight transmitter and antenna will permit the identification and three-dimensional location of all airborne aircraft every second. The higher cost aircraft will have an option of satellite navigation equipment that can provide highly accurate three-dimensional position and velocity information. The use of satellites for rapid and reliable communication from any place in the country to an aircraft anywhere over the country also is discussed.

Presented as Paper 70-488 at the AIAA 3rd Communications Satellite Systems Conference, Los Angeles, Calif., April 6-8, 1970. Authors would like to acknolwledge A. J. Mallinckrodt, D. Brown, T. P. Nosek, G. E. Clark, G. B. Jordan, R. N. Miller, I. Zipper, and others at TRW who have contributed materially to the contents of this paper.

*Advanced Systems Manager, Control/Navigation Satellites.

†Senior Systems Engineer.

§Senior Staff Mathematician.

I. Introduction

During the 1960's air traffic has grown at a phenomenal pace and to such an extent that the air traffic control system is currently under a significant strain. During the same time frame, satellite technology has matured to the state that there is now promise of significant operational and/or economic advantage to using them for certain transportation services. The first such usage was the Transit Satellite System (first launched in 1960) which now finds limited application for maritime use. The next application will probably be oceanic Air Traffic Control (ATC). Such systems have been studied by many organizations (see for example Refs. 1-5), and are not the subject of this paper. Instead, this paper describes certain aspects of use of satellites for ATC over the contiguous United States (CONUS). Specifically, a moderate cost network of satellites can be used to provide: 1) identification and very accurate three-dimensional surveillance of every airborne aircraft once per second with minimal aircraft hardware costs; 2) completely effective collision avoidance and improved navigation and communications services for the higher cost aircraft which include commercial, military, and a percentage of general aviation; 3) a basic framework which may eventually provide economical navigation and collision avoidance service to all aircraft.

We believe that the implementation of this system will result in significant improvements over the present system in terms of safety, improved traffic flow, and reduced costs. In this paper, the system operation is described and some design tradeoffs are presented.

In the remainder of this introduction, a brief discussion of air traffic management is presented to provide a background for the ensuing discussion of the satellite system. As such, this section does not present a "balanced" set of ATC system requirements, but rather treats only those parameters which are germane to the satellite-based air traffic control concept.

The following description of basic operational services provided by the Federal Aviation Administration is taken from the "Design for the National Airspace Utilization System Summary"[6]:

"Air Traffic Rules. The rules governing the flight of aircraft serve the pilot by allowing him to anticipate the actions of others. Where control service is provided, rules offer a better means for the ground-based system to predict the future actions of the pilot.

Flight Planning and Inflight Information. Through Flight Service Stations, Notices to Airmen, and regular publications, the pilot is aided in selecting routes and airports most suitable for the safe and effective performance of his flight. While in flight, pilots receive timely notification of hazardous weather and significant status reports on navigation facilities and airports.

Aviation Weather Information. Current and forecast meteorologic information is required for preflight planning. Pilots in flight need current meteorologic data to continue flight in a safe and efficient manner.

Search and Rescue Alerting. The filing of a flight plan and subsequent followup action afford a means for assuring that search and rescue services will be alerted whenever a flight is overdue.

Ground-Based Navigation Facilities. Navigation facilities provide the pilot with accurate guidance and establish a common system of reference for the pilot and the controller.

Air Traffic Control. The ground-based ATC subsystem offers increased safety by keeping controlled aircraft separated from each other and from hazardous areas."

Although satellites have made contributions to meteorology and could likewise prove very useful for some of the inflight information services provided by FAA, this paper concentrates on the use of satellites for the last two services - navigation and air traffic control. Given an understanding of the capabilities afforded by the use of satellites, the application of the system to improving search and rescue services will be clear to those currently involved with such operations.

The mission of air traffic control is the safe orderly movement of traffic. Three primary tools are used: 1) surveillance (or data acquisition) - that process whereby air traffic controllers on the ground determine the location of aircraft in flight; 2) navigation - that process whereby aircraft determine their position and - in most cases of interest - adhere to a flight plan; 3) communications - voice or data.

The potential satellite ATC roles which will be addressed in this paper involve those three basic functions: surveillance, navigation, and communications. Overlayed on these three basic ATC functions are the management aspects: the regulations and procedures, computation, data processing and displays, and the people involved. Of these, only computation is discussed in this paper.

Air traffic projections considered as fundamental to the design are shown in Table 1. Aircraft fleet size is important because it gives an indication of the size of the avionics market and points out the fact that the vast majority of the aviation fleet is, and will continue to be, general aviation. The total number of annual flights is of interest for two reasons: first, about two-thirds of all flights are itinerant; second, the number of IFR flights will increase from less than one-third of the current itinerant flights to almost half of a much larger number of future flights. Thus, the Air Traffic Control System load (which is related primarily to controlled or instrument operations) will grow to an even greater extent than the increase in aircraft population itself would imply.

Of considerable significance in the description of air traffic as it relates to a satellite-based system is the peak instantaneous airborne aircraft count. Assuming an initial operational capability for the system of 1980, and a 30-yr design life for the system, it can be inferred from Table 1 that the system, if it were to service the majority of airborne aircraft, should be capable of handling about 100,000 airplanes at one time. It will be seen that the satellite system can readily do this.

Other guidelines used in the preliminary synthesis of the satellite system include:

1) <u>User Base and Costs.</u> The projected number of fatalities associated with air carrier/general aviation midair collisions,[7] the high general aviation growth rate and the difficulty in resolving the question of airspace organization make it virtually mandatory that the surveillance portion of this concept be economically available to all aircraft. This means that the price of the associated avionics should be lower than the projected price of the ATC radar transponder with digital altitude readout.

2) <u>Performance.</u> Surveillance - Reference 7 states that parallel runway separation standards for simultaneous IFR operations may be reduced to approximately 2500 ft. This would infer a 1σ position determination uncertainty requirement of less than 250 ft and, if any sort of blunder detection and warning is to be possible, a surveillance fix repetition period on the order of 1-4 sec.

The subject of altimetry and altitude determination as they relate to surveillance needs further analysis; but it appears that even if altitude separation standards remain the same as they are today, the 1σ surveillance error should be

Table 1 Aviation projections (adapted from Ref. 7)

	Typical lifetime of a major air traffic control system concept			
	1968	1980	1995	2010 (Extrapolation)
Aircraft fleet size				
Air carrier	2,500	3,500	6,500	
General aviation	114,000	214,000	500,000	
Military	20,000	20,000	20,000	
All users	136,500	237,500	526,500	~1,000,000
Annual flights (in millions)				
Itinerant	23.6 (29% IFR)	46.4 (43% IFR)	109.5 (49% IFR)	
Local	12.1	20.7	47.9	
Total	35.7	66.6	157.4	~300
Peak airborne aircraft				
Air carrier	1,300	2,100	4,600	
General	8,000	16,800	46,300	
Military	3,500	3,300	3,500	
All users	12,800	22,200	54,400	~100,000
Itinerant			42,400 (~31,800 enroute)	
Local			12,000 (~22,600 terminal)	

less than 300 ft. Given a surveillance system which can identify and locate all airborne aircraft in three dimensions, the accuracy and data rate desired for collision avoidance usage can be determined as a function of the volume of air space protected for each aircraft, aircraft dynamics, the probability of collision, total traffic, and the degree to which the traffic can be controlled. Such a study has not yet been done; however, preliminary estimates indicate that accuracy requirements and data rates are probably close to the numbers given previously. Navigation - The same 2500 ft runway separation criterion[7] requires 1σ track keeping accuracies of some 300 ft. This would infer that if satellite navigation were to be used in this area, the satellite error contribution should be less than that.

3) Availability. Full-time coverage over the entire United States must be available. Further, the ground system should be made with enough redundancy and ease of maintainability that ground station reliability should have an insignificant effect on over-all system reliability. Finally, since the majority of the users of the system will be in a competitive commercial situation, or on a limited budget in the case of general aviation, high reliability and maintainability must be balanced against initial and operating costs, and the traffic control system must be able to tolerate a few aircraft with inflight equipment failures.

4) Adaptability. This characteristic provides the ability on the part of the ultimate operator or consumer to use the system in a number of ways, not all of which may be evident at the time the system was conceived. As the needs of the traveler and the modes of travel change, the requirement for the ability of such a system as this to adapt to those changes becomes extremely important. If the system has operational flexibility the air traffic controllers and pilots will find ways to adapt the system to its environment, thereby enhancing its over-all usefulness and prolonging its life.

5) Growth Potential. The system must be capable of being phased into the domestic ATC environment such that it need not displace present systems overnight. Furthermore, the system must be capable of providing growth in the services provided just as the customer base itself will certainly grow. It is simply not economical nor feasible for commercial carriers, the military, or general aviation to replace equipment which is operating adequately before they have recovered their investment on that equipment. Further, they will not make such a change until it is clearly demonstrated that it is both beneficial and safe for them to make such a change. The old low-frequency radio ranges served the aviation community in the United States for a long time, and present vhf omniranges are expected to do the same. If correctly designed, this modern satellite-based ATC system will likewise serve the country for decades.

6) Communications. Although some aeronautical SATCOM will be desirable, economic considerations dictate that the vast majority of air-ground communications over the United States for the next decade or so will be accomplished with the extensive network of existing direct air-ground links. More use of digital transmissions, as opposed to voice, can be expected and will be useful in cutting down the pilot and controller work load, and in conserving rf spectrum bandwidth. In addition, the use of data link will reduce satellite power requirements by about an order of magnitude over that needed for voice for the same message. The advantage which satellites can offer is extremely reliable communications from any part of the country to any aircraft in the country that can justify the expense of the service.

Satellites could be designed to provide all communication, navigation, and surveillance services for all aircraft over the United States. However, with today's satellite technology the implementation of such a system would result in high costs for the satellite and user hardware. Such a system was the only one considered in Ref. 7 for implementation over the United States, and design details are available therein in Appendix F, Subsection 2.

The approach at TRW has been to concentrate upon designing a system which permits accurate three-dimensional surveillance of all airborne aircraft utilizing very low cost aircraft hardware and a modest cost network of satellites.*

In Sec. II the resultant surveillance system concept is described. Certain design tradeoffs related to the surveillance system are then discussed in Section III.

The same network of satellites is used to provide a high quality navigation system (see Sec. IV) and augmented communications (Section V), for high performance aircraft (commercial, military, and a portion of general aviation.) As mentioned in Sec. V, the communication of navigation data to lower cost aircraft via existing ground-based facilities is feasible, but since we have not yet studied the economic or operational aspects of this in any depth, we are not proposing it at this time. Computation and display are briefly considered in Sec. VI.

II. Surveillance System Description

The surveillance system to be described represents a refinement of a technique first proposed by R. L. Garwin[8] in 1967. In that paper, he stated:

"There will be small commercial grade transmitters on each aircraft. No aircraft without such beacons or without a communication radio would be allowed to fly within the United States .... The urgency to initiate the interim system is to implement Air Traffic Control and Collision Avoidance throughout the domestic airspace and to save, as much as possible, the present FAA operating budget by providing that agency with better and cheaper tools to do its job."

In this system concept all aircraft will be equipped with a small lightweight antenna and transmitter to operate through a satellite network. The system features the following characteristics: 1) All airborne aircraft locations will be precisely

---

*Each satellite required for the surveillance, navigation, and communications can be launched by the Atlas/Agena class of vehicles; that is, the satellite finally orbited will weigh approximately 1000 lb.

known in three dimensions (50 - 300 ft - see Sec. III) with updating about once/sec. 2) All airborne aircraft will be identified. 3) The system saturation point can be kept well above projected traffic forecasts for this century with a relatively modest use of rf spectrum. 4) Very low cost avionics (target price of $400).

The foregoing can be contrasted with the current radar system which by itself yields no altitude information and has position errors close to two orders of magnitude larger. The minimum cost of a certified transponder to operate with the radar is about $1000. At today's prices, adding the capability to digitally readout barometric altitude, will probably double this figure. Uniquely identifying all aircraft would be difficult. The current total number of air traffic control radars in use is 204; this number results in high replacement maintenance, and operations costs.

The operation of the satellite surveillance system is indicated in Fig. 1. The consideration of minimizing the user avionics price has led to a system design in which the aircraft's role in the surveillance scheme is simply limited to the autonomous, asynchronous transmission of a simply generated code burst.

Each satellite in the network relays the signals that it receives from aircraft to a ground station (ATC center). The time of arrival of each signal is then measured at the ground station. The time difference from a pair of satellites defines the range difference between the transmitter and the two satellite locations. The locations (or orbits) of the satellites are necessary to these computations and are known through tracking data. Thus, a hyperbolic surface is defined in much the same way as with the LORAN, DECCA, and OMEGA navigation systems. Measuring the range differences between three pairs of satellites (a total of four satellites is needed), the ATC center computers can determine the location and altitude of each aircraft; i.e., the aircraft is located at the intersection of three hyperbolic surfaces.

Although the foregoing method of computing position is readily visualized and understood (particularly for those familiar with hyperbolic navigation systems), there is an alternate way to compute position which is preferable from both the viewpoint of the computer requirements and the information yielded. In order to keep user hardware costs at a minimum, a low cost oscillator (1 ppm) is all that is needed and desired onboard the user. Thus, since time of a given broadcast is not known a priori by the ground station, we can consider the four quantities need to be computed; specifically, three components of position and time of broadcast. Using four satellites with good geometry, these computations can be made. The accuracy of the broadcast time as measured in nanoseconds is roughly equivalent to the position accuracy as measured in feet. Typical equations for such computations are available.[9] Thus, inherent in the ground systems computation is the calibration of the user's clock such that at the ground station that clock accuracy is known to less than 1 μsec. The possible utilization of this knowledge will be discussed in Sec. V.

In addition to the four operational satellites, at least two in-orbit spares will be used to provide the necessary high level of availability of at least four operating satellites (Sec. III).

The transmitted signals consist of 51.1-μsec bursts* which are biphase-modulated by a 511-bit code. Each code bit is therefore 0.1 μsec long. Each aircraft is assigned a specific pulse repetition period. The periods vary in 10-μsec increments between 1.0 and 1.1 sec such that 10,000 different pulse repetition periods are assignable. If, for example, 25 different codes are used, there will be a total of 250,000 assignable aircraft identifications. Therefore, in addition to measuring the times of arrival of the pulses from each satellite, the ATC ground station also identifies the aircraft associated with each pulse by determining the pulse code and the pulse repetition period for that pulse. This is done by using matched code filtering in the receiver which compress the 51.1-μsec pulses to 0.1-μsec widths and by a pulse sorting algorithm in the computer. It is possible to add a further dimension of identification by using frequency separation; however, in this paper only a single frequency band of approximately 25 MHz is considered. The significant feature of identification by these techniques is that no additional data need be radiated by the user other than the timing or ranging signal. This keeps the signal overlaps manageable as will be seen.

A simplified block diagram of the required ground station surveillance equipment is shown in Fig. 2. This equipment is duplicated for each of the four operational satellites in the network, except for the station clock and position computer. The satellite to ground links operate at C band (5000 MHz). Since a separate narrow beam (0.5°) ground station antenna is tracking each satellite, all satellite-to-ground links can be

*As discussed in Sec. III, 204-μsec pulses are also under consideration.

at the same frequency to conserve spectrum. The LIT coded pulses received from the satellites are translated to IF and compressed to a pulse width of 0.1 μsec by a multitap delay line which is matched to the pulse code. When the incoming pulse fills the delay line, the summation of the delay line taps by the code matrix results in a maximum amplitude pulse of 0.1-μsec width out of the code matrix. If N different codes are used, the code matrix consists of N different matched summations of the delay line taps. This results in a 511 x N code matrix with N outputs. An envelope detector and threshold circuit on each of these outputs detects the occurrence of the compressed pulses and a pulse time digitizer measures their times of arrival with reference to the station master clock.

The PRP correlator identifies the particular one of the 10,000 possible PRP's associated with each pulse time. When an unidentified pulse appears at the PRP correlator, the correlator looks at the next 1- to 1.1-sec time frame and associates all unidentified pulses occurring in that frame with the pulse to determine the possible PRP's. Unidentified pulses in the following time frame are then examined to reduce the number of possible PRP's. This process continues until only one possibility remains. Once the PRP is identified, the associated pulses are removed from this search mode and placed in the track mode. In the track mode, the computer predicts the time of arrival of subsequent pulses based on the known PRP's and if pulses arrive at the proper times (within a tolerance band), the track is maintained. If pulses are not detected at the proper times due to misses, interference, or the aircraft transmissions stopping, the track is still maintained for three frames. If no proper pulses occur during these three frames, the PRP track is dropped and a controller is notified.

Analysis has shown that the PRP sorting process converges to a solution surprisingly fast. For instance, for 1000 untracked pulses, on the average only two frames would be required for successful identification. Assuming that 100,000 aircraft are being tracked simultaneously and that there are at least 25 codes used (see the following section for further discussion on codes), the maximum number of pulses into the PRP correlators will be 4000/frames. In the unlikely event that all 4000 pulses must be sorted from scratch, about three frames on the average would be required. If all of these pulses are to be sorted simultaneously, the data storage requirements become large. For example, with 1000 pulses the storage requirements are calculated to be 10^5 words. In this case, the problem is attacked in small bits; i.e., pick an arbitrary 100 pulses, identify their PRP's, allocate them to the track mode and proceed with the next hundred. Therefore, in the extreme case of 4000 untracked pulses, identification of the first 100 would take 3 sec. The process would speed up as pulses are assigned to the track mode and the whole process would take on the average less than one minute. It should be noted that once the PRP's are identified from one satellite, this knowledge will aid in simplifying the PRP sorting of the other satellite signals.

Finally, the position computer computes the location of all tracked aircraft from the times of arrival of the identified pulses from each satellite and stores this information with updating approximately once/sec.

The Location/Identification Transmitter (LIT) is small, lightweight, and inexpensive as discussed in the following section. The transmitter could be turned on by a pilot switch but more likely would be activated by a switch tied into the ignition system. With quite simple facilities, the operation of the LIT could be quickly and easily verified at an airport prior to granting a clearance to takeoff.

The aircraft antenna will also be small, unswitched, and have a pattern which radiates (roughly) in the upper hemisphere. Antenna size is related to the carrier frequency selected (see next section), but dimensions are, in all cases, modest for the frequency range considered.

III. Some Surveillance System Design Tradeoffs

Carrier Frequency Selection

The best choice for the carrier frequency depends on two conflicting factors. At the higher frequencies, propagation errors (principally the ionosphere) are reduced; while at the lower frequencies the required aircraft transmitter output power is reduced, resulting in lower-cost avionics. The increased transmitter power at the higher frequencies occurs because of the system antenna requirements which result in the increasing space loss with frequency. These requirements are for a simple low gain aircraft antenna which is in view of all satellites simultaneously, and a CONUS coverage satellite antenna.

Because of the added effects of noise at the satellite, the minimum aircraft transmitter power occurs at about 250 MHz. Below this, man-made noise* begins to overcome the space loss advantages. In addition, the satellite antenna dimensions increase as frequency decreases in order to maintain CONUS coverage. For example, if the maximum permissible dimension is 30 ft, which is a reasonable size, then the minimum aircraft transmitter power occurs at about 400 MHz. Plots of transmitter power and noise variations as a function of frequency are given in Fig. 3.

Ionospheric refraction errors on range measurements vary inversely with the square of frequency as shown in Fig. 4. Both the ionospheric and the total errors (relative and absolute) for a system subject to random errors of 50 ft are also shown in this figure. Absolute errors are essentially constant for frequencies greater than about 800-900 MHz. Satellite ephemeris errors are not included, but can be made small in either case by simple calibration procedures and, hence, can be left out of this argument. Note that total relative navigation errors are nearly independent of frequency down to about 400 MHz. The relative error shown is for the case of two users 5 miles apart. As the users get farther apart, the curve becomes closer to that curve for absolute error. In practice LIT ground beacons will be placed at various locations (such as airports) to serve as calibration points. When used for collision avoidance, the effects of the ionosphere tend to become self-calibrating as its effect decreases as aircraft get closer. Although some further analyses need to be performed, it appears that if the carrier is kept above 400 MHz, the effects of the ionosphere will be small.

Figure 5 is a scatter plot showing the maximum dimensions of several upper hemisphere coverage aircraft antennas as a function of frequency. All other things being equal, dimensions should scale as f^{-1}. However, there is also a tendency as dimensions become small in inches (as in the 6-in. NAVSTAR L-band curved arm turnstile) to exceed the scaling law dimension both because of lack of incentive and difficulty of manufacture. It does not appear that size will pose major limitations on the antenna for subsonic aircraft at frequencies

---

*Man-made noise levels looking toward earth from a satellite are based on a population density extrapolation of measured urban noise levels for an antenna field of view encompassing CONUS. The expected man-made equivalent noise temperature varies approximately as $f^{-2.2}$ and equals 290°K at about 400 MHz.

above 400 MHz. However, at the lower frequencies, good circular polarized designs may result in excessive size. Therefore, at lower frequencies linearly polarized antennas are likely to be used.

From the previous results, three frequency bands are being considered for the LIT system. These three bands are in the regions of 450, 900, and 1600 MHz. Table 2 shows the results of a cost study made on LIT avionics designs at these three frequencies. Two pulse widths were considered at each frequency - the basic pulse width of 51.1 μsec and an expanded 204.4-μsec pulse width which will be described in more detail in the following paragraph. The required power outputs and the estimated relative costs for each design are depicted in the table.

Table 2 LIT avionics relative costs

Design	Frequency, MHz	Pulse width μsec	Peak power output w	Relative cost
1a	450	204	100	1.0
1b	450	51	400	1.2
2a	900	204	250	1.2
2b	900	51	850	1.3
3a	1600	204	900	1.4
3b	1600	51	3500	1.7

Figure 6 depicts the LIT avionics box and a 1600-MHz turnstile antenna. The dimensions are about 6-1/2 x 7-1/2 x 2 in. and the average power consumptions at 1600 MHz is approximately 25 w which is principally heater power for the output tubes. At the other frequencies, these parameters will not change significantly. The 1600 MHz antenna is about 6 in. diam and 2-1/2 in. high. At the other frequencies the antenna will increase in size as indicated by Fig. 5.

To summarize, the frequency tradeoff considerations show that for minimum avionics costs the lower frequencies are favored while for system accuracy the higher frequencies are slightly favored. It is recognized of course that the usual problems of spectrum availability and assignment play the major role in the choice of a frequency. No attempt has been made to factor this intangible into the present discussion. At 1600 MHz, the presently largely unused frequency allocation for aeronautical radio services from 1535 to 1660 MHz could be available; at 900 MHz, the top end of the uhf TV band or lower end of the TACAN band are possibilities; at 450 MHz, the DOD/FAA radar band may be a possibility. In conjunction with the previous frequency bands, it should be borne in mind that the single carrier LIT surveillance system as discussed here will require about 20 - 25 MHz of rf spectrum on the aircraft-to-satellite links. The satellite-to-ground links have been assumed to be located at 5100 MHz but can be placed in any convenient band above 2000 MHz.

LIT Modulation Parameters

The basic problem in designing the LIT pulse modulation parameters is to provide the capacity to locate and identify up to 100,000 aircraft using the same signal space. Therefore, the modulation should provide for a signal processing gain (PG) equivalent to

$$PG = 10^5 \text{ (S/N) required} \tag{1}$$

which is on the order of 60 - 70 db. In addition, identification of all existing aircraft must be provided by some means of coding each aircraft transmission to give approximately 500,000 unique ID's by 1995.

The signal processing gain is provided in two parts — by using a combination of low duty cycle transmissions (pulses) and biphase coded pulse modulation for pulse compression at the receiver. Identification codes are provided by using a combination of different biphase codes and pulse repetition periods (PRP's) for each aircraft transmission. The PRP's should be on the order of one to two seconds to provide position updating approximately once every 1 or 2 sec. These considerations have led to the LIT modulation parameters shown in Table 3. The compressed pulse width of 0.1 μsec results not only from a processing gain requirement but also from an accuracy requirement on pulse arrival time measurements for position location.

Using these parameters, the resulting processing gain is about 65 db computed as follows: 1) duty cycle of 51.1×10^{-6} gives 43 db; 2) pulse compression of 511 ideally equals 27 db but calculations indicate a gain of about 22 db can be achieved. The sum of the two parts then equals about 65 db.

The number of identification codes available is equal to the product of the number of biphase codes and discrete PRP's or 500,000. More ID's are possible by using more biphase codes

Table 3 LIT modulation parameters

Transmitted pulse width	51.1 μsec
Pulse repetition period (PRP)	Between 1.0 and 1.1 sec/pulse
Number of discrete PRP's	10,000-space 10 μsec apart
Biphase code length	511 bits
Bit length	0.1 μsec (compressed pulse width)
Number of biphase codes	50 or more

and/or PRP's. Increasing the number of PRP's beyond 10,000 is unattractive since it can place an excessive load on the ground station pulse sorting computers. Therefore more biphase codes would be used. System growth beyond 100,000 aircraft airborne capacity could be provided by adding an additional LIT carrier channel with duplicate codes and PRP's.

The PRP spacing of 10 μsec will be large enough so that a combination of doppler and transmitter PRP error will not cause confusion between adjacent PRP assignments. For less than Mach one aircraft, doppler effects including satellite motion will not exceed 1-1/2 ppm. Oscillator stability in the aircraft transmitter can be held to 1 or 2 ppm with low cost so that the maximum PRP jitter will be ±2-1/2 to 3-1/2 μsec. For high performance aircraft, better oscillator stability (0.2 - 0.3 ppm) will be provided resulting in a higher cost to these aircraft.

For the pulse compression filter in the ground station, the total processing gain (G) can be shown to be equal to

$$G = L\left[\sin(\pi L \Delta f \tau)/\pi L \Delta f \tau\right]^2 \tag{2}$$

where L - biphase code length in bits, Δf = frequency error in pulse due to doppler and system oscillator errors, and τ = code bit length or compressed pulse width (0.1 μsec). The gain can be maximized for a particular frequency error by picking an optimum code length L. The optimum code length and corresponding gain are equal to

$$L_{opt} = 0.372/\Delta f \tau, \quad G_{max} = 0.23/\Delta f \tau \tag{3}$$

Taking into account all error sources, the frequency error will not exceed 3-1/2 ppm as discussed for the PRP jitter requirements. Using this value, the frequency errors at the three bands of interest equal 5.6 kHz at 1600 MHz, 3.2 kHz at 900 MHz, and 1.6 kHz at 450 MHz. The corresponding optimum code lengths L are 665 bits, 1160 bits, and 2320 bits for the three frequencies.

Two code lengths have been investigated to date, 511 bits (close to optimum for 1600 MHz) and 2044 bits (close to optimum for 450 MHz). The attractive feature of the longer code is the increased processing gain available which reduces avionics costs as shown in Table 2. The disadvantage of the longer code is the increased pulse overlaps occurring from two or more aircraft which can cause more difficult design constraints for the satellite transponders and perhaps increase the interference or crosstalk between codes in the ground receiver.

The 2044 bit code cannot be used with the top two carrier frequencies unless postdetection integration is used because of excessive loss of matched filter coherency across the pulse width from the frequency error. For example, at 1600 MHz the 511-bit code results in 25.9-db processing gain, while the 2044-bit code results in 18.2-db processing gain. Therefore, for the long code the matched filter must be segmented into four 511-bit sections and the output of each section incoherently summed to provide postdetection integration of four 51-μsec pulses. The processing gain (taking into account 0.6-db postdetection integration loss) is now equal to 31.3 db. At 900 MHz only two postdetection integrations would be necessary because of the smaller frequency error.

The exact choice of a code length or pulsewidth will depend on the choice of carrier frequency, further analysis of pulse interference problems, and the auto and cross correlation properties of the codes themselves. A brief discussion of these areas follows:

There are two potential problems from pulse overlaps. These are at the compressed 0.1-μsec pulsewidth level and at the original transmitted pulsewidth level of 51 or 204 μsec. Discussing first the 0.1-μsec pulses, if two or more pulses occur in the same tolerance slot (equal to about 1 μsec maximum as predicted by the ground tracking computer) it may not be possible to make an accurate measurement of pulse arrival time and, therefore, position information could be lost for that pulse frame. The probability of this occurring can be estimated as follows. Assuming pulses from all aircraft are uniformly distributed over the pulse frame, the probability of interference on a particular pulse in a particular frame is

$$P_i = \delta N_c / T \tag{4}$$

where N_c = number of aircraft using the same biphase pulse code (4000), δ = tolerance slot (1 μsec), and T = pulse frame (≈1 sec), or about 0.004. Therefore, an average of four measurements in 1000 would be lost from a satellite. The probability of obtaining at least four measurements from four satellites for a three-dimensional position fix during any frame would be $(0.996)^4$ or 0.984. Thus, only 16 measurements out of 1000 would be lost and they will be randomly distributed. Considering the fix rate of once per second, this is quite acceptable. With five or six satellites available, the data loss will be even less.

The other overlap problem is the cross-talk interference between overlapping pulses using different codes. For example, if an aircraft transmits a pulse consisting of L code bits, the signal power at the receiver matched filter output will be L^2E_s where E_s is the signal energy per bit. Assuming the interfering pulses, like thermal noise, add incoherently,* the expected output signal-to-noise ratio (including interference) is

$$(S/N)_o = L^2 E_s / (LE_n + \overline{m} LE_s) \tag{5}$$

where E_n = thermal noise spectral density and $\overline{m}$ = expected number of interfering or overlapping pulses. The expected number of interferers $\overline{m}$ is equal to the number of aircraft in the system N_c times the pulse duty cycle, i.e., $\overline{m} = N_c \tau_p / T$. If we take the pulse period T = 1 and the pulse duration τ_p as L/W, where W is the signal bandwidth and 1/W is the bit duration, we have $\overline{m} = N_c L / WT$.

Substituting this expression in Eq. (5) yields

$$(S/N)_o = (S/N)_i / (1/L) + (N_c/WT)(S/N)_i \tag{6}$$

where $(S/N)_i$ is defined as the input signal-to-noise ratio E_s/E_n.

*This assumption appears to be pessimistic for well chosen codes.

From this expression, a number of conclusions can be drawn. When there are few aircraft so that N_c/WT is negligible the output signal-to-noise ratio, $(S/N)_o = L(S/N)_i$; thus, the signal-to-noise ratio increases linearly with the signal duration as expected for coherent detection. For high input signal-to-noise ratio $(S/N)_i \rightarrow \infty$, we have $(S/N)_o \rightarrow WT/N_c$. This is the interference limited case where the output signal to interference ratio depends only on the ratio of the bandwidth to the number of aircraft and is independent of the length L. Therefore, increasing L improves the performance against noise, but has no effect on the interference. In principle, it would be desirable to make the signal duration as long as possible in order to achieve a high signal-to-noise ratio with limited transmitter peak power. In practice, the duration of the signal is limited by range resolution and coherence interval considerations. The interference problem remains essentially independent of the pulse interval.

The selection of codes for minimal interference assuming asynchronous nonperiodic operation has centered on two types of pseudo-random sequences known to have favorable correlation properties. At present, a computer program is being written to determine the probability distributions obtained when the various codes are auto or cross correlated. It is hoped that this will provide the information necessary to determine the detection and false alarm probabilities. In addition, the relationship between the number of distinct code words and the false alarm rate should be indicated.

Satellite Network - Coverage, Accuracy, and Reliability Considerations

The design tradeoffs involving the satellite network configuration involve the availability, quality of service, and cost in providing the three principal services of communication, navigation, and surveillance. As indicated in a number of other studies,[2, 9-12] there are a number of useful satellite constellations to choose from. Although the needs are not yet quantified (a necessary prerequisite to the optimization process), enough is understood to discuss some of the trades between coverage, availability, accuracy, and cost; and to identify an attractive candidate system. Preliminary reliability and replenishment analyses have been performed to determine the requirements on satellite reliability and lifetime, the number of spares, launch rate requirements, and launch strategies which are needed to provide various levels of over-all system availability.

Availability

A total system breakdown must be an extremely rare event. Thus, since it could be inferred intuitively that at least one orbiting spare would be required, an analysis was performed for both 5 and 6 satellite constellations, i.e., one and two orbiting spares. This preliminary analysis treated the effects of satellite MTBF, satellite design, life, and booster reliability on satellite system availability and replenishment rates, and is summarized in Table 4. Exponential distribution assumptions were used throughout the analysis because in this case there are insufficient data available to describe any nonexponential satellite failure distribution, and since previous efforts had shown that a system of nonexponential segments yielded similar results. Furthermore, it was assumed that some dedicated launch facility was available which would be capable of launching a satellite six weeks after the need for it had been identified, but that two launches could not take place within a six-week period.

Table 4 System availability and annual downtime averages

		μ = 3 yr[b] T = 5 yr		μ = 5 yr T = 7 yr		μ = 7 yr T = 9 yr	
N[a]	S[a]	R_B = 0.90	R_B = 0.95	R_B = 0.90	R_B = 0.95	R_B = 0.90	R_B = 0.95
		System Availability					
6	4	0.9982	0.9987	0.9^350	0.9^366	0.9^377	0.9^386
5	4	0.9789	0.9824	0.9921	0.9936	0.9960	0.9967
N	S	Annual average downtime (days/yr)					
6	4	0.65	0.47	0.18	0.12	0.08	0.05
5	4	7.7	6.4	2.9	2.3	1.5	1.2

[a] N = number of satellites in system; S - number needed for success.

[b] The reliability of each booster — satellite combination is taken as $R(t) = R_B e^{-t/\mu}$ for $0 < t < T$, M = MTBF, $R(t) = 0$ for $t > T$, T = design life, R_B = booster reliability; note $0.9^386 = 0.99986$.

A satellite with a 5-yr MTBF and a 7-year design life (reasonable figures for current satellite designs), launched from a booster with a 0.9 launch reliability, produces a system availability of 0.99950 in the six satellite configuration tentatively selected. This in turn means that the probability of a system outage occurring during a 30-yr projected system life span would be 0.194. The probability of system outage would not reach 0.5 until the system had been operational 99 years! The expected outage time for such an event would be approximately 5-1/2 weeks. For such a system a maintenance launch rate of approximately one launch each seven months would be required. This would amount to 24 launches over a 10-yr span including the initial startup.

Coverage

While synchronous equatorial satellites are visible from all latitudes within the U.S., it will be seen that if only these orbits are used accuracy degrades seriously because of the poor geometry of the ranging measurements. Placing some satellites in inclined orbits enhances accuracy since those satellites when in the northern hemisphere, are at higher elevation angles. On the other hand, the satellites may now drop completely out of sight when they move to the southern hemisphere. Thus, with higher inclined satellite orbits, the advantages of better accuracy are paid for with requirements for more satellites in orbit to cope with loss of visibility during part of the orbit. These satellites would be in addition to whatever on-orbit spares may be required to protect against satellite failure. Examples of these coverage/accuracy tradeoffs are given in the following paragraphs.

The constellation initially chosen places one satellite (and a spare) in equatorial synchronous orbit ("central satellite") with four* others moving in inclined, elliptical orbits with their ground tracks circling the central satellite's subsatellite point. It is termed a "Rotating X" configuration. Placing perigee at the most southern latitude reached (by setting the argument of perigee at 270°), gives the circular ground track and also increases the percentage of time the satellite spends above the equator. The reduced perigee altitude, however, tends to cause the satellite to disappear below the horizon near perigee, for which the increased speed at perigee may only partially compensate. Figure 7 shows approximate ground tracks for these different "Rotating X" configurations. It can be seen from the figure that 30° inclinations are not suitable for

* To obtain these tracks, set $e = \tan(i/2)$, e.g., for $i = 20°$ inclination, $e = 0.176$.

full view coverage of the U.S. In fact, something less than 20° is required. A 12° orbit on the other hand takes in much of southern Canada.

Very high inclination and eccentricities, on the other hand, provide orbits yielding full coverage of the North American continent as long as enough satellites are in view. Furthermore, the elevation angles tend to be substantially higher which generally improves the user antenna coverage pattern picture. Ground traces for system with satellite inclinations of 63° and eccentricities of 0.52 are shown in Fig. 8. Placing 4 satellites in the "ring" and 2 as shown on the equator yields a very attractive system. The coverage in the U. S. is 100%; and the accuracies, as will be pointed out in a later paragraph, are excellent. In the event of the failure of 2 satellites in the ring, however, the system will be inoperative (i.e., less than 4 satellites will be in view) ten hours of every day. Going to even higher inclination and eccentricities will reduce this outage time, but the availability of this system can only approach that of the low inclination system as a limit. In the event that the two geostationary satellites were to fail, there would be only approximately 4 hours in the day when the remaining four inclined satellites would all be visible from the United States.

Accuracy

Error analyses for a number of satellite constellations have been performed for constellations with a central satellite at 100° W longitude, and with the user at 40° N latitude, 100° W longitude. The following assumptions were used in the analysis: 1) a random ranging error (1σ) of 15 ft; 2) ionospheric error of less than 10 ft (consistent with 1600 MHz); 3) no a priori altitude information; 4) satellite tracking errors not included (note: This yields relative rather than absolute position determination uncertainties. For reasonable satellite tracking errors and the constellations in question, the absolute would exceed the relative uncertainties by, typically, 20%); 5) range differencing or pseudoranging techniques (recursive estimation of user clock bias) are assumed, requiring a minimum of 4 satellites in view.

A number of significant points can be drawn from Table 5. First, comparison of the three 12° inclination cases indicates that the loss of a central satellite in a 5-satellite constellation is significantly more serious in terms of accuracy degradation than loss of a satellite on the ring. This led to the

Table 5 Surveillance system accuracies (1600 MHz)

X-constellation, user at 40°N, 100°W, except in case 6

Case	Inclination of rotating satellites	Remarks	Satellites in system	Satellites in view	Horizontal CEP, ft	1σ altitude, ft
1	12°		5	5	240	300
2	12°	North Satellite out	4	4	260	310
3	12°	Central Satellite out	4	4	390	500
4	20°		5	5	110	170
5	30°		5	4	80	170
6	30°	User at 30° N	5	5	40	80
7	63°		6	5	50	220

conclusion that a 6-satellite constellation should have two geostationary and four rotating satellites. Second, the 20° inclination offers substantial reduction in CEP's. The operational significance of these reductions is not clear at this time, although it does appear that the 1σ altitude difference may well be the more significant of two, if satellite-determined altitude for surveillance purposes is used in the air traffic control conflict prediction, and resolution process. Third, it is interesting to note that when the inclination is increased to 30° (Case 5), although the horizontal CEP goes down, the 1σ altitude does not improve because one of the satellites has dropped out of view. Moving the user southward (Case 6) picks up the other satellite and provides the expected accuracy. Finally, with the 6-satellite, high inclination constellation shown in Fig. 8, although the horizontal CEP continues to drop, the 1σ altitude uncertainty has actually increased. Although this at first may seem surprising, the reasoning lies in the fact that good altitude determination from a satellite is obtained when at least one satellite is very nearly directly overhead; whereas, in this case the elevation angle of the satellite as it moves all the way from the Atlantic Ocean to Hawaii — a full 12 hr — remains relatively constant at about 65°.

In summary, a 5-satellite rotating X constellation with inclined satellites at about 16° appears to offer adequate (typically 180 ft CEP, 250 ft 1σ altitude) determination accuracies even in the presence of a failed satellite (typically about 300 ft CEP, 375 ft 1σ altitude), but two on-orbit spares provided by a 6-satellite rotating X are required to attain desirable availability levels. Thus, an attractive candidate is a 6-satellite system with 5-yr MTBF and 7-yr design life satellites which results in a system outage probability of less than 20% for a 30-yr system life span. Because of the somewhat preferred location of the center satellite, the 6-satellite rotating X should probably have two geostationary satellites at a near-center location. Their optimum displacement is not yet determined.

IV. Navigation — Navstar System

The navigation system briefly described in this section has been under study for several years and is discussed in considerable further detail in Refs. 9, 10, and 13. Hence the discussion in this section will be brief.

The same network of satellites used for surveillance will also radiate independent navigation signals. The navigation process is the reverse of the LIT surveillance process in that in the navigation system the satellites radiate to users instead of users radiating to the satellites (Fig. 9). Similar position location principles apply, however, in both cases. The navigation user measures the differences in the times of arrival of signals from four satellites and computes his position in a similar manner to the ATC ground station. His task is, however, much more simple than the ground station task since he obviously does not have to go through the pulse sorting and identification task of the surveillance system ground station.

In the navigation system, each of four satellites transmits in time sequence (satellite transmissions are time division multiplexed) a ranging signal plus data on its orbital parameters and its onboard oscillator time error. This latter data are required since the satellite transmissions are time synchronized by individual high-precision crystal oscillators onboard each satellite. A ground tracking network determines the relative time error between the oscillators on each satellite by measuring the time of arrival of the navigation signals from the satellites. This oscillator correction data plus satellite ephemeris data are transmitted to the satellites where they are stored in the satellite memory and read out for transmission to users as part of the navigation signal transmissions. The data in the satellite memory are updated periodically by the master station in the tracking network.

The elements of one class of user hardware are indicated in Fig. 10. As indicated in Ref. 10, the frequency range of 1 - 2 GHz (within the L band) is optimum for the navigation service. The satellite signals are received by a simple, light-weight unswitched antenna (approximately 2-1/2 in. high and 6 in. diam) and processed by the receiver which decodes the satellite data and forms a digital number representing the time of arrival of the signal.

For the configuration of Fig. 10, the output of the processor interfaces with a small digital computer for automatic computation of position. The display indicated is one similar to a conventional inertial or Doppler navigation system; however, the NAVSTAR system is capable of driving any conventional or planned navigation display (e.g., moving map).

Table 6 summarizes the projected cost and weight of the navigation hardware for three out of many possible classes of users. Although in all cases, these costs are low for the service provided, it is clear that the very low cost general aviation user will not be able to afford this service. The cost is that projected for antennas, receivers, computers, and displays.

Table 6 Summary of navigation user hardware characteristics

Configuration	Weight, lb	Cost, $
Low cost	15	5,000
Moderate cost	15	9,600
High performance	40	30,000

In summary, using the same network of satellites the navigation system provides the following features: 1) unsaturable, all-weather operation; 2) common grid for all airspace; 3) accurate position and altitude (50 - 300 ft) determined onboard the aircraft; 4) accurate velocity and rate of climb information also available onboard the aircraft; 5) modest cost aircraft hardware (upwards of $5000, depending on user).

The need for extremely precise navigation and altimetry is called for in many sources, typically Ref. 14. At the user's option, the NAVSTAR system can provide this.

V. Communications System

Today, practically all air ground communications for aircraft in the U.S. are conducted on very high and ultra high frequencies (vhf, uhf), the latter being reserved primarily for governmental and military use. The communication frequencies of interest to air carrier and general aviation lie between 118 MHz and 136 MHz. This band presently contains 360 channels (50 kHz spacing) with various segments allocated for particular usage such as airport towers, enroute ATC, approach control, flight service stations, aeronautical enroute stations (air carrier), and the emergency frequency of 121.5 MHz. Navigation aids in the vhf and uhf region such as the VOR and ILS localizer facilities (108.1 - 117.9 MHz) also contain voice capability for ground-air communications of general interest and flight service station response when an aircraft cannot establish contact on one of the foregoing simplex frequencies.

Since two-way communications are required at all airports with FAA control towers, practically all aircraft are so equipped. Some minimal type equipments, consisting of tunable receivers and limited channel transmitters still exist today in general aviation; but these are rapidly becoming obsolete. Most aircraft today have 90, 180, or 360 channel discretely tuned sets, with transmitter powers ranging from 5 to 50 w, and with corresponding prices ranging from $500 to several thousand dollars.

The FAA operates two systems for the purpose of controlling enroute air traffic. These are the Air Route Traffic Control Centers (ARTCC's) controlling IFR traffic, and Flight Service Stations which perform a supporting role in the ATC function.

There are presently 21 ARTCC's in the contiguous U.S., each maintaining cognizance over a number of geographical sectors the size of which is determined by traffic flow, airway complexity and area of radio and radar coverage. Each sector is manned by from 2 to 5 controllers depending upon volume of traffic. Direct controller-pilot communications are maintained on an assigned frequency while the aircraft is within a given sector. When the aircraft approaches a sector boundary, it is "handed over" on another frequency to another controller responsible for the adjoining sector. The communication channels available for this purpose are implemented through local and remotely operated air-ground facilities as shown in Fig. 11. Indirect communications for ATC purposes is accomplished through ARTCC's ground links with other Centers, airport towers, Flight Service Stations, airline* and military operations, and other essential facilities.

With all the present communication facilities, there are still geographic areas in the U.S. for which airspace cannot be controlled due to the absence of communication and navigation

*The airlines maintain communications with their flights, principally through the ARINC network, for operational information. ATC functions may be relayed through this facility.

aid coverage. The Owens Valley in California is one example. With the advent of area navigation, this will become less and less tolerable. Reference 15 indicates the FAA's intent to expand the capacity and reliability of its ATC communication facilities, with additional and redundant remote-controlled air-ground sites.

En route traffic communications is by no means the whole story. The terminal air traffic control system requires a system of strategically located transmitter and receivers to provide high quality, full-time communications with aircraft adjacent to Approach Control and within Control Tower and airport traffic control area. In a manner similar to the ARTCC's, Approach (Departure) Control maintains direct communications with IFR aircraft until handed over to a tower operator or to the ARTCC. The tower, of course, must integrate these arrivals with other VFR traffic.

In addition to ATC, there are a host of other aeronautical activities requiring communication frequencies in the rapidly saturating vhf band. FAA Flight Service Stations provide the principal communication link with VFR aircraft as well as important weather and other advisories. Airport towers require additional frequencies for control of ground traffic. Military and private operators require frequencies for their local training and other activities. In order to relieve the growing vhf communication congestion, the FAA is pursuing two avenues of relief. The feasibility of increasing the number of channels by closer frequency spacing (25 kHz) is being studied, and automatic ground/air/ground communications are planned for increased channel efficiency.

Given the cost of satellite communications and the current investment in ground based facilities, it is likely that the present air-ground vhf communications system or a modified form of it will continue into the foreseeable future. Therefore, the role of satellite communications would be that of supplementing the present system (Fig. 12).

Data rather than voice communications are the most promising modulation form for satellite-to-user links because data communications make much more efficient use of the satellite links than voice. Automatic data link techniques, if used correctly, can significantly reduce pilot and controller workload. Emergency voice links may still be useful for the system if for no other reason than unforeseen circumstances not provided for in a data link.

Detailed studies of the optimum usage of satellites remain to be made. Nonetheless, with the satellite design considered previously for the surveillance and navigation functions, there could be communications capacity available of on the order of 20 - 40 kilobits of data on the ground-to-satellite-to-aircraft link and significantly higher capacity on the reverse link assuming operation in the 0.5- to 1.5-GHz bands and efficient design of the aircraft terminal. The exact data rate will be dependent upon detailed link design, more refined satellite design, and selection of specific carrier frequency, aircraft antenna and avionics.

The SatCom usage which has received appreciable attention to date is related to the overocean application (see, e.g., Refs. 2-4). For aircraft which would use this type of link the resulting price of the equipment is on the order of $10,000 or more. Although this would preclude most general aviation users, it is not beyond the means of commercial transports and many military users. Since these aircraft fly mostly different portions of the airspace than the other segments of aviation, their routes can be controlled with a relatively modest data flow from the ATC center via satellites to the users. Thus, these aircraft will generally be commanded only when they deviate from an IFR flight plan, or when the plan must be changed due to weather, collision situations, etc. Such users can readily afford the cost of the highly reliable and real-time satellite communications link.

Thus, should a collision situation be imminent between a light aircraft and a commercial transport, the light aircraft could be notified by ground link with the assurance that even if he cannot be contacted, or does not respond, the collision situation can be avoided by satellite communications to the commercial aircraft.

The possibility of instant communications with commercial aircraft may have important applications for airlines in terms of relay of manifest information and real time telemetry of important aircraft parameters. Such potential usage is clearly dependent on the economics involved and remains to be investigated fully.

The communication of collision avoidance and other traffic control commands and possibly even navigation data to low cost users by the ATC centers might well be handled by the present vhf direct air-ground system. The ATC generated commands could be routed to the ARTCC station within line-of-sight of the affected aircraft for transmission. Very simple digital receivers and displays could be developed for this use. A detailed investigation of the economics involved remains to be performed, however. We merely note here that the system has such growth potential.

Another possible communications scheme likely to be very suitable for Intermittent Positive Control was just suggested by R. L. Garwin. This involves using the ATC center's knowledge of the time and the code of the LIT transmissions from each aircraft to send addressed commands back to a unique aircraft via the satellites.

As an example, such a scheme could operate as follows. The ATC center could send to an aircraft a LIT pulse using that aircraft's biphase code and timed to arrive at the aircraft, say 1 msec (arbitrary) after aircraft transmission of its own LIT pulse. The aircraft's code generator could be triggered 1 msec after each LIT transmission and used as a correlator (but not transmitted) to detect pulses addressed to it by the ATC center. To uniquely address a single aircraft, pulses would be staggered from 3 satellites to be received precisely, say 1000 μsec, 1100 μsec, and 1200 μsec after aircraft transmission. Receiving this group of three pulses would constitute a bit of information. The command information can be encoded by the presence or absence of these pulse groups in subsequent 100 μsec time slots. Since the idea of a command link using LIT coding and timing is a new one only a preliminary look has been made to date on the feasibility of the concept. Further studies in this direction are planned, but the potential exists to design and use a low cost SATCOM receiver.

VI. Information Processing; Automated ATC

With the exception of a few centers (New York Common IFR Room, Atlanta IFR Room, Jacksonville Center), the current ATC system has virtually no automation. The systems planned for the next few years (NAS system for enroute, ARTS system for terminals) have only limited automation of clerical tasks associated with flight strips and automatic aircraft tagging of alphanumeric displays on display scopes. The real control functions performed by a controller have not been automated and are not likely to be automated in the next few years. Possible exceptions may be the conflict detection and resolution task and the automated flow control task which may appear in later stages of the NAS system.

It is generally recognized that the system of the 1980's will involve more automation, including automation of the control functions of a controller. There is, however, no general agreement on exactly how this will be done. One proposed approach was studied by the Department of Transportation Air Traffic Control Advisory Committee. The major system functions and system blocks are identified fairly well.

The basic assumptions used by the Computer Sizing Subgroup (Ref. 7, Appendix 1) of the Committee are considered generally appropriate here: 1) The controller will mainly act as a monitor. Most control functions will be taken over by computers. 2) Data link will be used to give control information to all pilots. 3) All aircraft are equipped with a transponder used by a ground trilateration system to determine position. Since we are studying satellite systems, this is replaced by the assumption that every aircraft carries a LIT transmitter. 4) All aircraft will have a display which gives control information. (note: We feel that this may have to be modified in later studies.) 5) The present physical division of ATC functions will remain. That is, there will be terminals and enroute centers and their duties will be about the same as in the present system. 6) There will be a national flow control center.

Using these assumptions, the subgroup committee estimated the size of the computer effort. Some of the results are given in Table 7. The figures are for a terminal control center, an enroute center, and the national flow control center. The enroute center handled an area of about 400 x 800 miles with about 4000 simultaneously airborne aircraft controlled by the center. The terminal system controlled about 2000 aircraft. The results in Table 7 are expressed in instructions per second required for the data processing of each of the major ATC functions.

Table 7 Computer sizing estimates (from Ref. 7, Appendix D)

Function	Enroute (IPS)[a]	Terminal (IPS)	National (IPS)
Conflict prediction and resolution (mixed)	2.56×10^5	15.1×10^5	-
Conflict prediction and resolution (high density)	9.2×10^5	(in command and control)	-
Data acquisition	27.3×10^5	48.9×10^5	-
Command and control	5.6×10^5	14.6×10^5	-
Additional enroute (displays, etc.)	29.2×10^5	-	-
Additional terminal	-	10.0×10^5	-
Central flow control	0.03×10^5	0.4×10^5	0.4×10^6
	7.22×10^6	8.90×10^6	0.4×10^6

[a]IPS = instructions/sec

The results were obtained by a number of people in a number of ways. In some cases, the instructions were counted in existing programs and the results extrapolated to include the predicted traffic densities. In other cases new programs were written and the instructions counted. After all the estimates were made, the results in Table 7 were multiplied by a factor of 12 to produce a final estimate. The multiplication accounts for the assumptions that 1) the program will be written in an inefficient high-order language, 2) there will be a large "executive overhead," and 3) the manufacturers estimates of computer speed always refer to the fastest computer instructions. The final estimate indicates that each busy terminal will need the computing power of about 100 present-day computers (i.e., computers which do about 10^6 instructions-1-sec). This was concluded to be quite feasible since it was estimated that 1975 computers will do about 20 x 10^6 instructions-1-sec and hence a multiprocessing system will do the job. The cost of the computers was also predicted and the resulting cost (even after multiplying by another safety factor of 8) was found to be reasonable.

We happen to believe that the computer size in the report was overestimated by at least an order of magnitude; but, even if we are wrong, there will be no problem provided the computer technology predictions are valid.

Now consider the impact of the LIT system on the automation problem. First, it is clear that a large part of the automation computer program is not affected. The basic command and control equations, for example, are not particularly affected by the fact that aircraft positions come from satellites instead of radars. There may be a favorable effect on traffic flow because of the higher accuracy of satellites, but this is not the subject of this paper. When the LIT system is used, the main impact is felt in the data acquisition scheme. The data acquisition scheme assumed by the computer sizing subgroup of the Alexander Committee was a radar multilateration scheme. Three or four radars measure ranges to an aircraft. The aircraft sends its own altitude via data link. These measurements are sufficient to determine the aircraft position.

With the satellite system, the basic multilateration method is very nearly the same mathematically. From the standpoint of computing effort, it makes little difference whether the multilateration stations are on the ground or in satellites.

An additional set of computations must be done in the LIT system to identify the aircraft. Since this computation was not included in the computer sizing report, a rough estimate of the size of the task was made. The results indicate that the additional computation load is normally a very small fraction (about 1%) of the multilateration load.

One major effect of the LIT system on the automation problem is the fact that every receiving station will receive signals from all the aircraft in the country. This is contrasted with a radar system in which the radar only receives signals from aircraft in its line of sight. This is a major difference which seriously affects the automation problem. To analyze the effects, consider the following possibilities: 1) Keep the present terminal and enroute structure. Each terminal and each enroute center will have its own receiving station. 2) Keep the present terminal and enroute structure. Add new centers which receive and sort LIT data, calculate aircraft positions, and send the aircraft positions to the appropriate enroute centers and terminals. 3) Reduce the number of air traffic control centers. This would involve eliminating traditional boundaries between centers and between centers and terminals. One might even reduce the number of centers drastical, e.g., to three.

There are some serious tradeoffs to be considered before selecting one of the previous possibilities. Configuration 1 requires more receiving equipment and a greater total computing load. Configuration 2 eases the computing load and uses less receiving equipment but imposes a communication load in order to send position data to the appropriate centers. Configuration 3 can ease the communication load. The system still needs long land-line communication to relay commands to the pilots via RCAG stations. However, it requires fewer bits to send occasional commands than continual position data and the total communication load is smaller.) On the other hand, configuration 3 introduces a number of problems associated with eliminating centers and traditional control boundaries.

From the standpoint of computations, the first case is probably the worst. That is, a center controlling traffic in Los Angeles would like to be able to throw away signals from aircraft over, say, New York. It would be possible for every center and terminal to compute the position of every aircraft over the U.S., but this is obviously wasteful. Fortunately, this is not necessary; in the tracking mode of operation, the center would have a list of all the aircraft under its control and a list of the PRP codes and time of previous pulse for each aircraft. The computer consequently knows what time to expect each pulse for each aircraft and can ignore those pulses which do not arrive at the expected time. This is a fairly small computing load — about 1% of the total multilateration problem. When a new aircraft appears in the center's jurisdiction, there will be no trouble, provided that the PRP and code are also sent to the center and provided to the computer (which they should be anyway). For a new aircraft, the computer does not know the time of the previous pulse. It must then search among all the pulses received for the appropriate code in order to extract a chain of pulses of the appropriate PRP. It can be shown that this is a light computing load compared to the rest of the problem.

A more serious problem develops if the computer must suddenly start from scratch to identify all of the aircraft in the area (the "search mode"). Suppose, for example, that power is lost to the computer for a period of time and it must restart. It knows the PRP's and codes of all the aircraft, but does not know where to start each train of pulses — it does not know the "previous pulse time" for any of the aircraft. A rough estimate of the time required to solve this problem was made. The estimate indicates that a computer with an instruction rate of 10^6 IPS (the enroute center of Table 7) should recover in less than 1 min.

Another problem would occur if the computer lost the list of PRP's and codes for the aircraft in its jurisdiction. It could sort out the pulses as indicated in the last paragraph but then it must compute the positions of all the aircraft in the country in order to determine which aircraft are in the immediate area. For 10^5 aircraft, the computer must perform 2×10^7 instructions (assuming 200 instructions for a multilateration computation). A 10^6 IPS computer could do this in about 20 sec, and a 7×10^6 IPS multiprocessor in about 3 sec, which indicates again that the problem is not serious. After the aircraft in the immediate area are identified, the computer returns to the tracking mode.

The previous section indicates that even if the present enroute and terminal structure is maintained, the computational problems will be reasonable. One can make the computational problem still easier by using options 1 and 3 previous. When identifying and sorting all of the aircraft over the country, the total work is performed more efficiently by doing the computations at a small number of centers.

If the present structure of terminals and centers is maintained, there is a disadvantage in the fact that the positions of all the aircraft in the country are available. If the traditional structure is changed, the disadvantage turns to a considerable advantage. Consider what could be done if a center knows the location of all the aircraft over the U.S., the flight plans of these aircraft, the current and predicted weather, and the current and predicted airport capacities.

This center could simultaneously consider the problems of flow control (adjusting the flow of aircraft to prevent overloading parts of the system) enroute control (maintaining spacing of enroute aircraft and terminal control (sequencing, spacing, and metering aircraft approaching the terminal). These functions are now performed separately with preassigned and rigid dividing lines between the different parts of the problem. By combining the functions in a central location it should be possible to perform the functions more effectively. For example, it should be possible to react more rapidly to unexpected decreases in an airport's capacity — the program can respond by slowing aircraft enroute to the airport instead of waiting until the aircraft get near the airports and putting them in a holding pattern.

We have not yet examined all the possibilities for air traffic control by a center which always knows where all of the aircraft are, nor do we at this time advocate highly centralized control. It is clear, however, that the possibilities are extremely interesting and promising.

References

[1]"Report on Studies Carried Out in 1969 in the Field of Traffic Control Satellite Systems and Proposal for the Initiation of a Preparatory Phase Towards the Development, in Collaboration with NASA, of an Experimental/Preoperational Traffic Control Satellite System," Jan. 15, 1970, European Space Research Organization.

[2]"Navigation/Traffic Control Satellite Mission Study," NASA Contract NAS 12-595, Vol. I-III., CR-86166, CR-86167, and CR-86168, Final Rept., June 1969, TRW Systems Group.

[3]Systems Engineering Study of Aeronautical Satellite Services," TR-DA1583(1), Final Rept., Dec. 15, 1967, Philco-Ford Corp.

[4]"Global Satellite System for Telecommunications, Air Traffic Control and Navigation," Final Rpt., DIOSCURES Project 1972-1985, July 1969, Centre National D'Etudes Spatiales (CNES), Paris, France.

[5]"Satellite Systems of Telecommunications Air Traffic Control and Navigation, Economic Study," Aug. 1, 1969, Centre National D'Etudes Spatiales (CNES), Paris, France.

[6]"Design for the National Airspace Utilization System," Summary of 1st ed., Sept. 1962, Federal Aviation Agency, Systems Research and Development Service.

[7]Report of Department of Transportation Air Traffic Control Advisory Committee, Dec. 1969.

[8]Garwin, R. L., "Satellite Systems for Navigation, Air Traffic Control, Collision Avoidance, etc.," Aug. 23, 1967; also Supplement April 18, 1968.

[9]"Satellite System for Precise Navigation, System 621B(U), Final Report, Vols. 1-7, AD 395 212L, AD 393 237L, AD 395 238L, AD 395 193L, AD 395 213L, AD 395 194L, AD 395 314L, respectively, Jan. 31, 1969, TRW Systems Group.

[10]"Constellation Study for Satellite Navigation with Passive Ranging," Final Report, TRW No. 11409-6001-R000; Navy No. X68-18630, May 10, 1968, TRW Systems Group.

[11]"Recommendations for a National Air Traffic Management System," May 15, 1969, ATA Air Traffic Control System Planning Group.

[12]The National Aviation System Plan, 1971 - 1980, U.S. Department of Commerce, Federal Aviation Administration.

[13]"Study of a Navigation and Traffic Control Technique Employing Satellites," Final Report, Vol. 1, Summary, TRW 08710-6012-R000, NASA No. N68-31277, Dec. 1969, TRW Systems Group.

[14]"Constellation Study for Satellite Navigation with Passive Ranging," Final Report, TRW No. 11409-6001-R000; Navy No. X68-18630, TRW Systems Group.

[15]Otten, D. D., "A Satellite System for Radio Navigation," AIAA Paper 68-1063, Philadelphia, Pa., Oct. 1968.

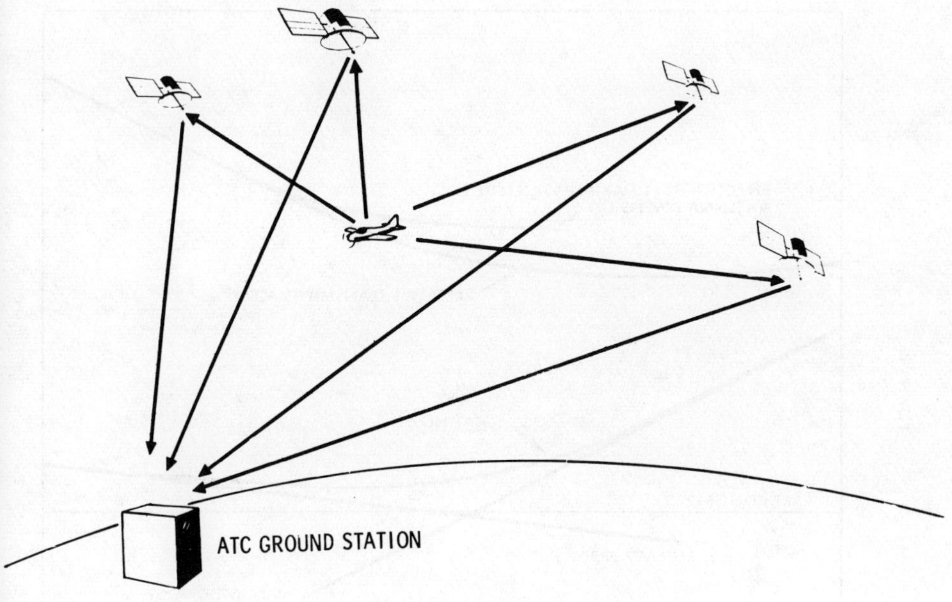

Fig. 1 Satellite surveillance system.

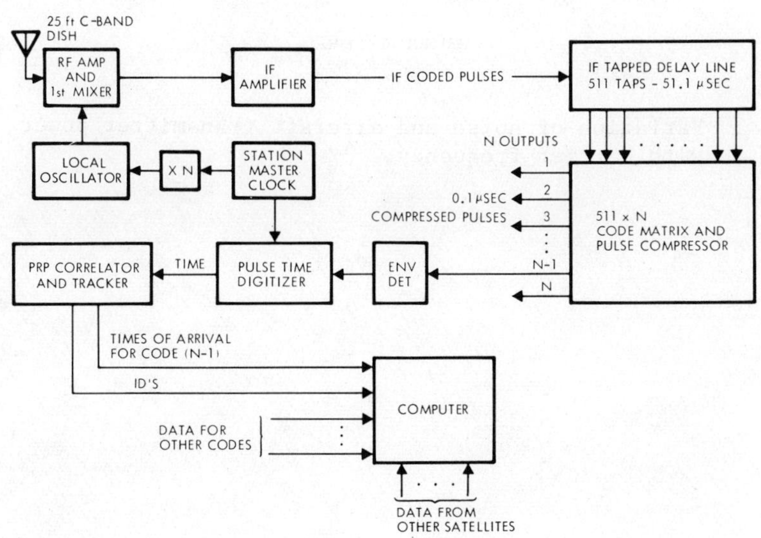

Fig. 2 ATC ground surveillance equipment for LIT.

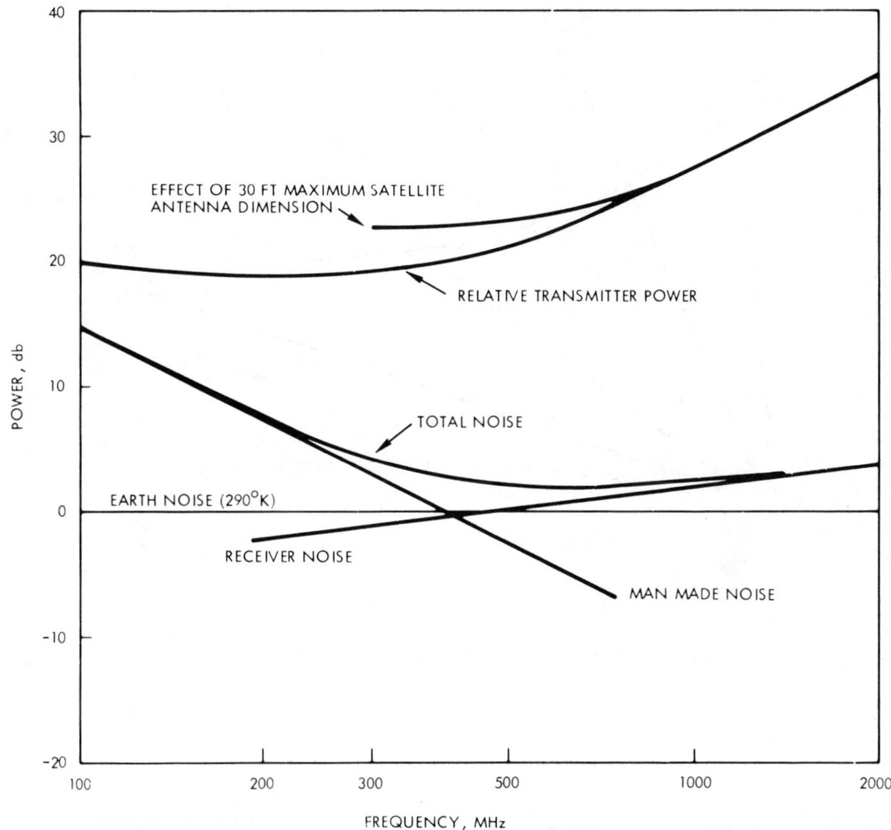

Fig. 3 Variation of noise and aircraft transmitter power with carrier frequency.

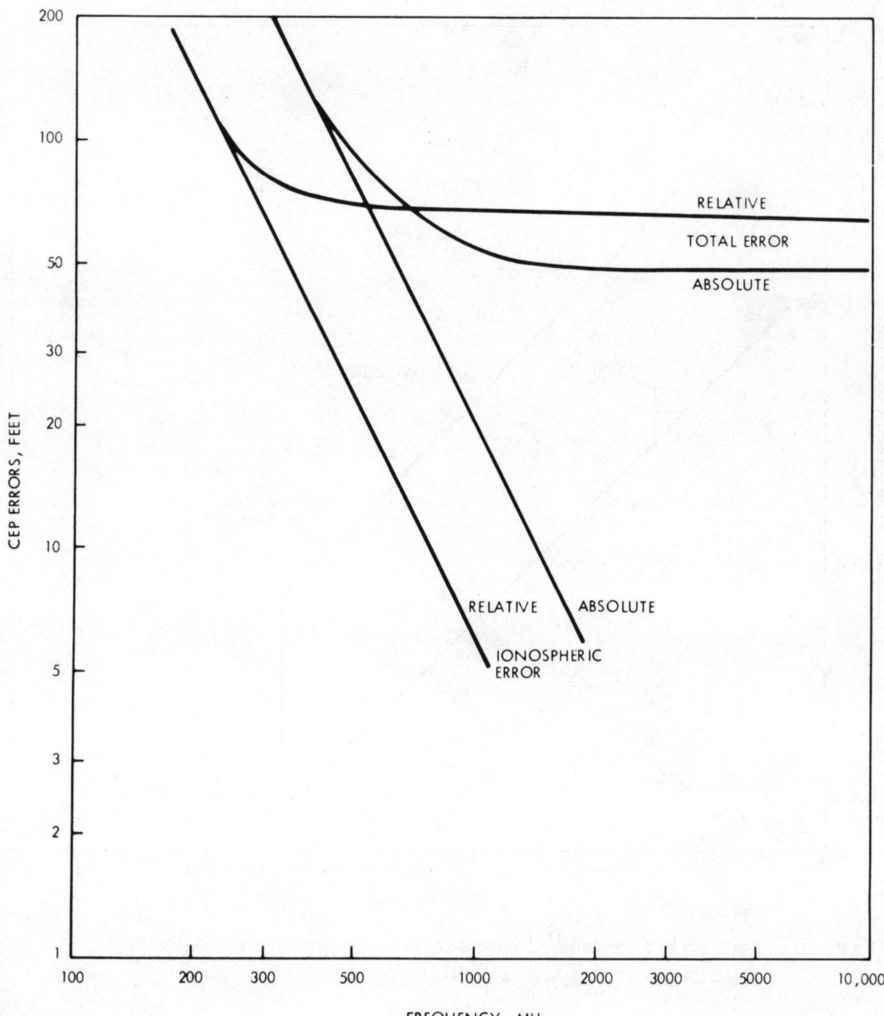

Fig. 4 Variation of ionospheric and total errors, (CEP) in a LIT system having 50-ft random errors; relative errors are for 5-mile distance, satellite ephemeris errors excluded.

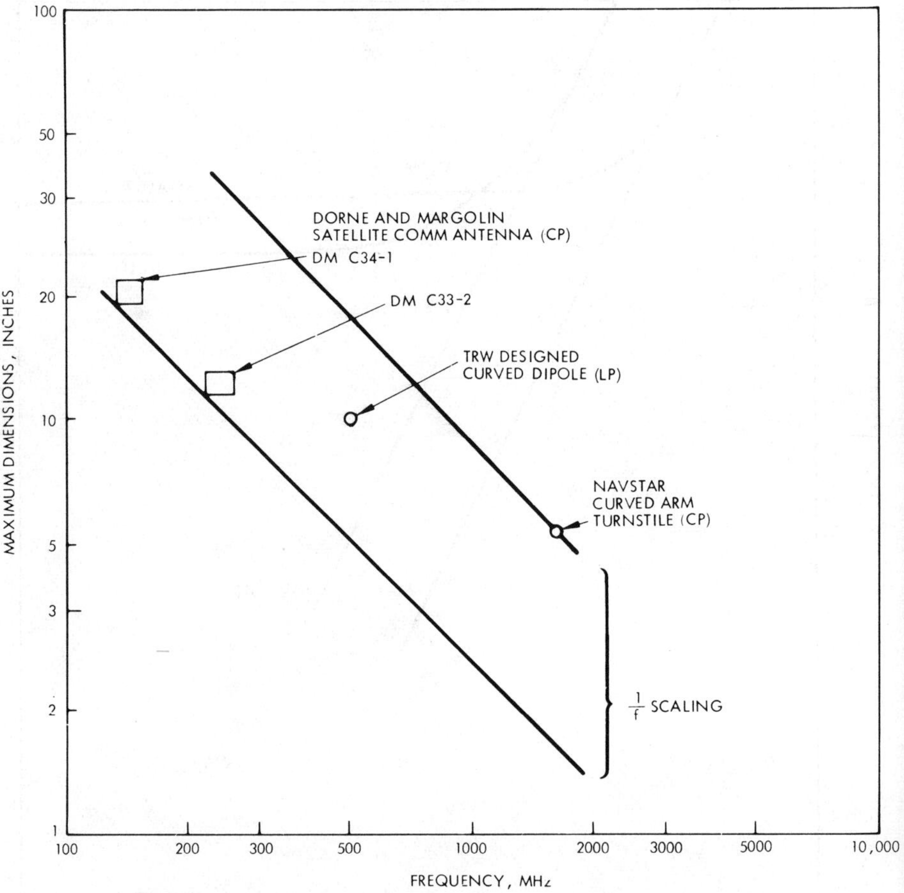

100
50
30
20
10
5
3
2
1
MAXIMUM DIMENSIONS, INCHES
DORNE AND MARGOLIN
SATELLITE COMM ANTENNA (CP)
DM C34-1
DM C33-2
TRW DESIGNED
CURVED DIPOLE (LP)
NAVSTAR
CURVED ARM
TURNSTILE (CP)
1/f SCALING
100
200
300
500
1000
2000
3000
5000
10,000
FREQUENCY, MHz

Fig. 5 Typical maximum dimensions of upper hemisphere coverage aircraft antennas.

ANTENNA
(OPTIONAL)
LIT
STATUS
LIT

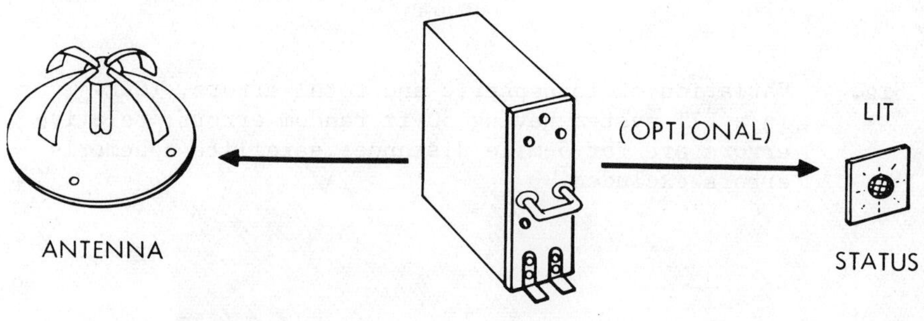

Fig. 6 Location/identification transmitter (LIT) system at 1600 MHz.

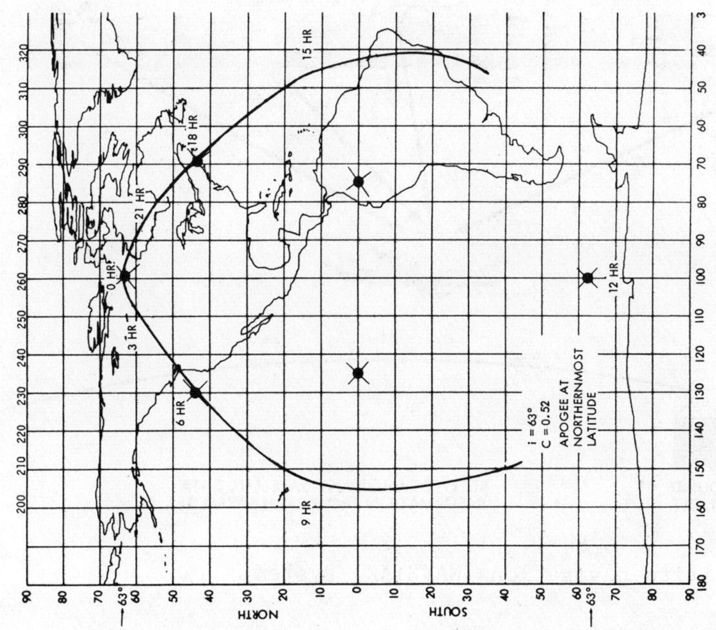

Fig. 8 High inclination constellation.

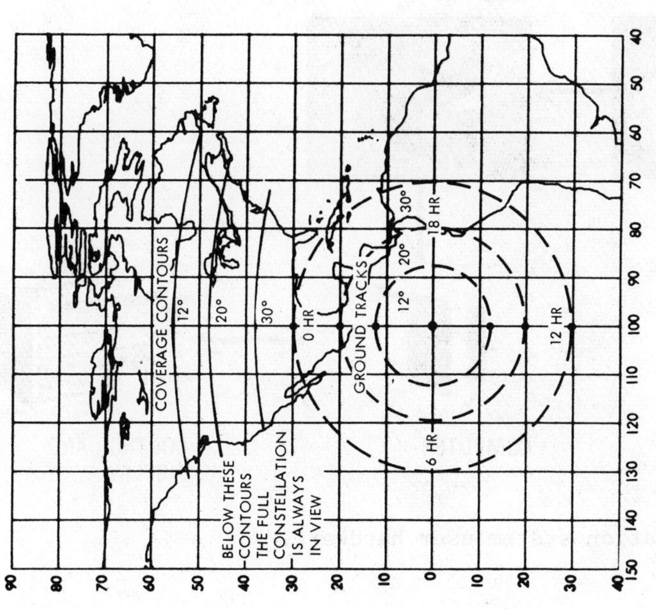

Fig. 7 Orbit ground tracks and coverage contours.

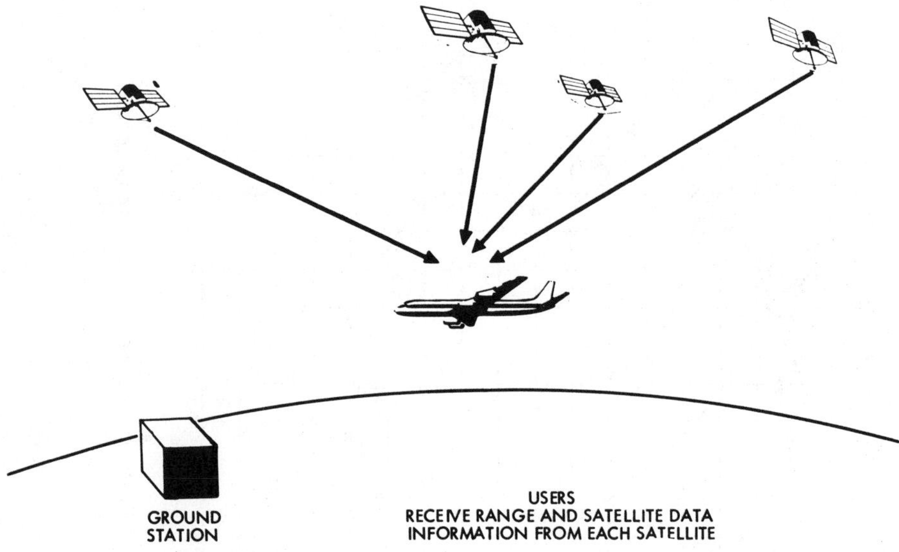

Fig. 9 Satellite navigation (NAVSTAR) system.

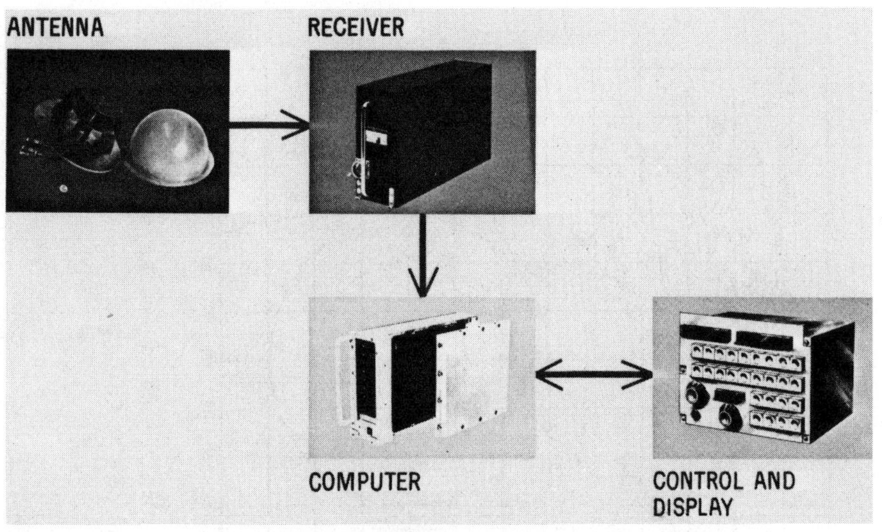

Fig. 10 Navigation system user hardware.

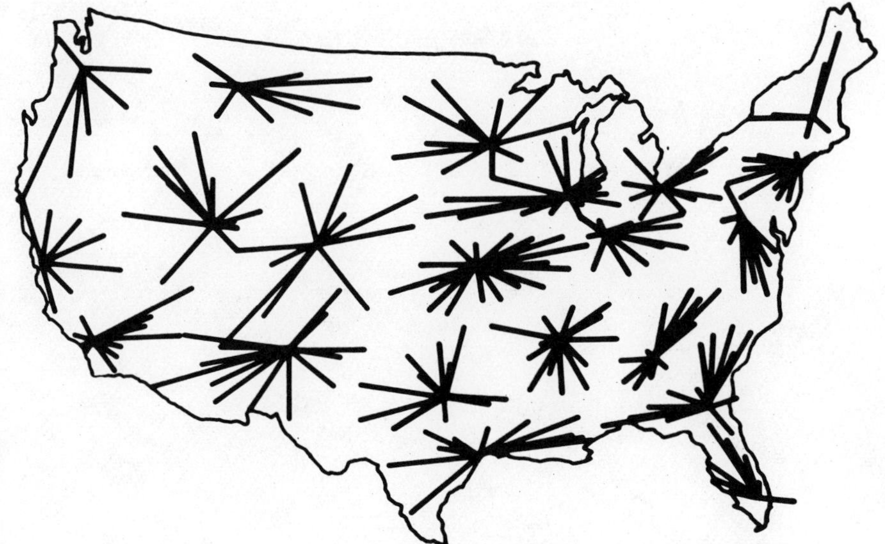

Fig. 11 Configuration of air-ground communication network.

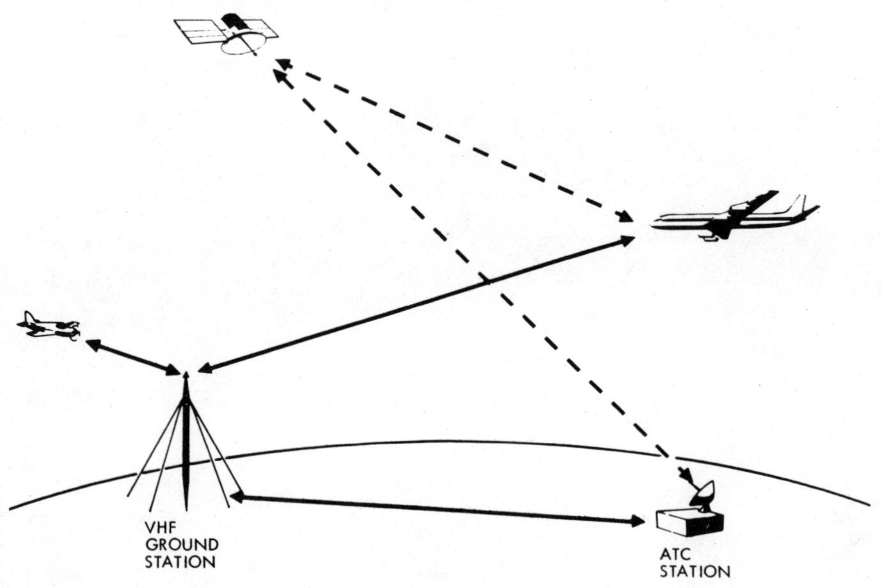

Fig. 12 Satellite communications network.

EXPERIMENTAL EVALUATION OF VHF FOR POSITION FIXING BY SATELLITE

Roy E. Anderson*

General Electric Company, Schenectady, N. Y.

Abstract

The VHF transponders of the ATS-1 and ATS-3 geostationary satellites were used in ranging and position fixing experiments. An interrogation signal was transmitted from a ground terminal to ATS-3, which relayed it to the vehicle transponders. The vehicle that was addressed repeated the signal, and its response was relayed back through both satellites to the ground terminal, where propagation times were measured; lines-of-position and fixes were computed. The 0.43-sec. "tone-code" ranging signal contained a single audio tone frequency. Ambiguity was resolved and user craft identified by a simple digital code. Five vehicles were used in the tests: two aircraft, a ship, an oceanographic buoy, and a truck. Ionospheric and multipath effects were studied. It is concluded that a VHF system could have an accuracy of ± 1 nmi. for ships and aircraft if calibration transponders are used to monitor the ionosphere.

Presented as Paper 70-489 at the AIAA 3rd Communications Satellite Systems Conference, Los Angeles, Calif., April 6-8, 1970. Many persons in several organizations contributed with skill and enthusiasm to the success of the experiment. Major contributors included the following: NASA (E. Ehrlich, D. Fordyce, G. Melton); FAA (O. DeZoute, F. Jefferson); US Coast Guard (M. Johnson, R. Basteck, Officers and crew of Coast Guard Cutter Valiant, WMEC-621); General Electric Co. (R. C. Sisson, Responder Circuit Design; R. W. Garrett, Ground Terminal Operations; J. R. Lewis, Ground Terminal Operations, Digital Recording; R. C. Rustay, Performance Analyses, Computer Program Design; L. Cheney, Sea Robin Project Coordinator; G. H. Millman, Consultant on Ionospheric Propagation; L. Loeber, Secretarial Duties). Navy support thru ONR on N00014-68-C0467.

*Consulting Engineer, Research and Development Center.

Introduction

The Applications Technology Satellites, ATS-1 and ATS-3, of the National Aeronautics and Space Administration (NASA) were used in a series of tests to evaluate VHF for locating vehicles by range measurements from satellites. NASA supported the experiment under contract NAS5-11634. There has been lack of agreement concerning the usefulness of VHF for position determination. A major objective of the ranging and position fixing experiments was to provide definitive data for the resolution of this uncertainty. If it is shown that the accuracy is sufficient to meet current needs in applications, such as air and marine traffic control, VHF is attractive because a first generation operational system can be implemented with adaptations of available equipment.

The VHF aircraft band, 118-136 MHz, is approximately the lowest frequency that could be considered for range measurements from satellites because the ionosphere introduces propagation delays that change with time and are not accurately predictable. In addition, the signal may arrive at the antenna over more than one path, the direct path from the satellite to the craft, and reflection by the ground or the sea. The multipath effect can introduce ranging errors, especially for aircraft. The use of higher frequencies such as L-band, 1540-1660 MHz, would virtually eliminate the atmospheric propagation delay uncertainties, and L-band antenna designs may offer better discrimination against multipath. L-band appears to offer potentially higher accuracy, but requires further technical development, and may be more costly to implement.

The parameter values selected for the VHF experiment were restricted by practical considerations; since it is recognized that these must be taken into account in any application of the techniques that were tested. The bandwidths were limited to those of VHF mobile communications, 3.6-kHz audio bandwidth and 15-kHz radio frequency bandwidth. The modulation selected was entirely compatible with voice and digital communications; so that existing or planned user equipment for VHF satellite communications could be used with only the attachment of an additional electronic circuit. Satellites suitable for VHF mobile communications needed no modification to adapt them for range measurements. The time necessary for a range measurement could be interleaved with the communications without adding significantly to the time occupancy of the communications channel.

Within the restrictions imposed by propagation and the practical limits selected for the tests, the results indicate that an operational surveillance system could have an accuracy better than ± 1 nmi for ships and approximately ± 1 nmi for aircraft, one sigma.

Tone-Code Ranging Experiment

Range measurements were made with a simple "tone-code" technique that proved to be efficient in the use of satellite energy and compatible with presently used mobile communications equipment and bandwidth allocations.

Five vehicles were used in the tests: a Coast Guard Cutter in the Gulf of Mexico; two aircraft, a DC-6B and a KC-135 of the Federal Aviation Administration; a buoy moored in deep water off Bermuda; and a panel truck in upstate New York. Each was equipped with conventional mobile communications transmitters, receivers, and antennas, and each had a 6-in. by 8-in. by 10-in., 6-lb experimental tone-code "responder" unit attached between the receiver and transmitter. The combined receiver, transmitter, and responder is termed a "transponder."

When a vehicle was to be located, a ground station transmitted a 0.43-sec tone-code signal to one of the satellites, the "interrogating satellite," usually ATS-3. The signal consisted of a 2.4414-kHz tone burst, followed by the individual user address formed by suppressing an audio cycle for a digital "zero" and transmitting an audio cycle for a digital "one." The tone-code signal was frequency modulated on a 149.22-MHz carrier with a narrow deviation so that the rf bandwidth was within the 15-kHz bandwidth of the mobile receivers.

The satellite repeated the signal on 135.6 MHz. All of the activated vehicle equipments received the signal, and each matched the phase of a locally generated audio tone to the received tone phase. The one vehicle that was addressed responded with a short burst of its properly phased locally generated tone, followed by its address code; introducing a very precisely known time delay between reception and retransmission of the code. The vehicle response on 149.22 MHz was through a broad beamwidth antenna. If both satellites were in range of the vehicle, they both repeated it on 135.6 MHz.

The ground station received the returns from the two satellites separately with narrow beamwidth antennas. It measured the time interval from its initial transmission of the signal to the first return from the interrogating satellite and to the two returns from the satellites as they were relayed back from the user. From these measurements the ranges from the two known positions of the satellites to the vehicle were determined. These ranges, together with vehicle altitude and corrections for ionospheric delay, were used to compute the vehicle location. When only one satellite was in range, a line of position was computed and a fix defined as the crossing of the line with latitude or longitude of the vehicle, determined by other means.

The time required for the interrogation and response was about 1.3 sec except when a data transmission was included with the user response or, for the aircraft, the equipment required a longer time to switch from receive to transmit. The usual interrogation rate was once every 3 sec, although a once-per-two-second rate was also used.

The tone-code ranging technique has the following characteristics: 1) useful accuracy can be achieved within the modulation and radio-frequency bandwidths of present-day mobile communications; 2) the technique can be used with wide bandwidth for high accuracy; 3) it requires only one channel for range measurement, receiving and transmitting in the simplex mode if desired without need for an antenna diplexer; 4) the time required for a range measurement is a fraction of a second so that it can time-share a communication channel with little additional time usage of the channel; 5) it can be implemented by the addition of an inexpensive, solid-state responder unit attached to a communication receiver-transmitter; 6) it can, but need not, employ digital or digitized voice transmissions to provide synchronizing of the user responder, thereby further increasing the efficiency of channel usage; 7) there are no "lane" ambiguities in the range measurements; and 8) user identification is simple and is confirmed in the return signal.

Factors Affecting Accuracy

Geometrical Effects

Geometrical dilution of position affects the accuracy of position fix determinations independently of the radio frequency used in making the measurements. These geometrical effects are listed below, but not discussed, since they are not the subject of this paper and have been presented elsewhere.[1]

1. A ranging measurement error is multiplied many times to produce a large line-of-position error in the region close to the subsatellite point. The magnitude of the line-of-position error decreases rapidly and becomes equal to the range measurement error for points on the Earth where the satellite is on the horizon.

2. Uncertainty in satellite altitude produces an effect similar to a range measurement error.

3. A lateral position uncertainty for the satellite produces an error corresponding to the error in the subsatellite point, if an Earth-centered coordinate system is used for fix determination.

4. A ranging measurement from a satellite produces a circle of position on the Earth or on an Earth-centered sphere having a radius equal to the Earth's radius plus the aircraft altitude. When a fix determination is made by range measurements from two satellites, the best accuracy is achieved when the circles of position intersect at a right angle, but the fix accuracy becomes poorer as the angle becomes smaller. When the circles of position are tangent, only a line of position can be determined. When pairs of geostationary satellites are used, position fixes are inaccurate near the equator, although good longitude determinations are possible.

5. Aircraft altitude uncertainties introduce a position fix uncertainty that is large if the aircraft is under the satellite, but decreases to zero as the elevation angle to the satellite approaches zero.

Atmospheric Propagation Effects

The most frequently cited objection to the use of VHF for position fixing with satellites is the uncertainty introduced into the range measurements by the ionosphere. Free electrons in the ionosphere cause its refractive index to be different from that of free space; so that the propagation velocity of a radio wave through the ionosphere is reduced in proportion to the number of electrons along the ray path. The ionosphere also bends the ray so that the actual propagation path length is increased. In addition, the magnetic field and the ionosphere interact to produce a change in the polarization of the wave, an effect called Faraday rotation. Lateral variations in electron density and movement of these electron clouds produce widely varying signal strengths at a receiver, an effect called amplitude scintillation. All of these factors are experienced at VHF frequencies, but the effects decrease as $1/f^2$ so that higher frequencies would be less subject to communication signal strength variations and large ranging errors resulting from these ionospheric effects. An optimum frequency for radio determination using satellites would be approximately 400-700 MHz, where the problems of ionospheric propagation are less severe than at VHF, and the problems of user antenna design are less severe than at L-band.

During undisturbed daytime and nighttime conditions, the one-way range bias resulting from tropospheric and ionospheric retardation is as shown in Fig. 1.[2] Tropospheric effects are independent of frequency. It is expected that corrections based on predictions of electron content in the ionosphere may reduce the bias errors of the ionosphere to within 20-40% of the total bias.

If calibration stations were used in conjunction with range determination measurements, it is probable that corrections to less than 10% of the total range bias might be achieved in the region surrounding the station,[3,4] assuming that the distance from the satellite to the station was accurately known. The calibration stations could be equipped with repeaters like those used in mobile craft. The known range from the satellite would be subtracted for a calibration range measurement to derive an accurate range correction.

Unpredictable solar disturbances can cause changes in the ionosphere that can increase range bias by more than 50% over the values shown in Fig. 1 at high latitudes and cause even larger changes in tropical regions. If range measurements from two satellites were used for fix determination without an applied correction, the effect of the disturbance would be an apparent systematic displacement of all the craft in a limited area. The relative position determinations of these craft would not be significantly affected.

More data are needed on the size of the area in which the information from a calibration station can be used, and therefore the number of calibration stations required in the area to be served. However, an estimate can be made on the basis of an extreme case reported by Mendillo, et al.[5], (see Fig. 2). In that instance, sunset occurred at the peak of the disturbance, and the total electron content dropped by a factor of 10 in 1 hour. It would have caused a change in total range bias from approximately 2700 to 270 meters for an elevation of 20°. Assuming that the principal cause of the decay is related to sunset, the east-west gradient of electron density and hence, range bias would have been over a distance approximating the distance the terminator moved over the Earth in 1 hour, or approximately 1000 km. Calibration stations at 1000-km intervals probably would have been sufficient to determine the range bias correction during that event.

Sea Reflection Multipath

Range error due to sea reflection multipath is a function of the relative amplitudes of the reflected and direct signals, the delay of the reflected signal behind the direct signal, and the radio frequency phase difference between the direct and reflected signals. As an aircraft flies through a distance such that the radio-frequency path length difference of the direct and reflected signals changes by one wavelength at the radio frequency, the one-way range error resulting from specular sea reflection will vary in a manner typified by one of the curves in Fig. 3. The aircraft may fly a considerable distance in straight and level flight to change the path length difference by one wavelength, but the aircraft need only move vertically a few feet to change the path length difference by one wavelength at VHF.

The range error probability distribution is depicted in Fig. 4. The solid line curve was calculated from the computed results of the 0.1 amplitude curve of Fig. 3. The dashed line curves have limits that were computed, but their exact shape is approximated. The area under each curve represents unity probability. Although it was not proven by rigorous mathematical analysis, examination of the computed curves suggests that the probability of a single range measurement being longer than the true value is equal to the probability that it is shorter than the true value. The magnitudes of the errors in the long direction can be larger than the magnitudes of the errors in the short direction, but the probability of a large error in the long direction is low. The average error for a large number of range measurements tends towards zero in the presence of specular sea reflection.

The down-link and up-link frequencies for the ATS ranging and position fixing experiments were different, being 135.6 on the down-link and 149.22 on the up-link. Sea reflection multipath can affect both links. Because of the very large number of rf wavelengths between the satellite and the aircraft, the radio frequency phase differences on the two paths are dependent, and therefore the total effect of sea reflections on the two paths must be determined by convolving the range error probability distribution for the one-way path (see Fig. 4). The result of the convolution is depicted in Fig. 5 for the case of an amplitude ratio equal to 0.1. The convolution extends the limits to twice the limits of the one-way ranging curves with equal probabilities of errors in the long or short range directions. The maximum error expected in the short direction is less than the maximum error in the long direction. Maximum errors expected for the 0.25 amplitude ratio dashed line of Fig. 4 would be approximately -10 and +16 μsec, and, for the 0.5 amplitude ratio, approximately -16.5 and +36 μsec.

Experimental variations in range measurement are due to many factors, including signal-to-noise ratio, equipment time delay variations as a function of signal amplitude, and diffuse multipath effects as well as the specular multipath effects that were considered in the foregoing analysis. Some of the data obtained on over-water flights were examined to see if the effects could be observed. Time delay was plotted as a function of time for some of the over-water flights. In some cases, a cyclic pattern of range variation about the mean suggested the shape of the curve shown in Fig. 3. An example is the plot of a portion of the June 13, DC-6B flight shown in Fig. 6. The DC-6B craft was at 21,000 ft. over Lake Michigan using the horizon mode of a Dorne & Margolin Satcom antenna. The Satcom antenna is circularly polarized, essentially omnidirectional in the azimuth plane. The horizon mode is useful from approximately 10° to 40° elevation; the zenith mode is useful above 40° elevation. Maximum gain for linearly polarized signals is approximately +3 dB. A portion of these data, extending over a period longer than the 2.5-min period of Fig. 6 was plotted as a histogram and is presented in the lower portion of Fig. 7. Although it cannot be concluded definitely that the envelope shape of the histogram is the result of sea reflection multipath, it is interesting to compare it with Fig. 5. The corresponding histogram for the responses through ATS-1 is also plotted in Fig. 7.

Range measurements from ATS-3 to the KC-135 aircraft over the North Atlantic at 39,000 ft. are plotted as a histogram in Fig. 8. The envelope of the histogram suggests the shape of a distribution curve for two-way ranging that would result from convolving with itself one of the dashed line curves of Fig. 4. The limits of the experimental data are -21 and +44 μsec , with comparatively few measurements near the largest, longer range values. The aircraft was using a conventional VHF blade antenna that is sensitive to sea reflections.

Multipath errors are expected to increase with the altitude of the aircraft if the elevation angle to the satellite does not change. During the June 6 flight off the coast of New Jersey, the aircraft flew at 20,000 ft. and also at 5000 ft. Standard deviations of the range measurements were 2.1 μsec. at 5000 ft and 3 μsec at 20,000 ft.

Equipment Time Delay Variations

Equipment design, performance, and usage influence the precision and accuracy of ranging measurements. The experiment provided an opportunity to evaluate limitations imposed by the equipment.

Bandwidth of the ranging signal influences the measurement precision. Radio range measurements can only be made by measuring propagation time, and that in turn by measuring the interval between a transmitted and received waveform. Although the modulating wave may take many forms, the highest range resolution that can be achieved is set by the ability to measure the phase of the highest frequency component in the waveform. Phase measurement precision is relatively independent of frequency. The higher the frequency, the better the timing resolution. A phase measurement precision of the order of one degree is usually achievable without refined techniques.

Measurement precision is affected by noise, for noise causes a "jitter" of the signal phase. If the noise is random, the error may be reduced by averaging a number of measurements. As the signal-to-noise ratio becomes poorer, the amount of jitter increases, and the measurement precision decreases even if averaging is used. It is important in a ranging measurement that the precision be within acceptable limits at the poorest signal-to-noise ratio that is experienced. In the experiment, this was insured by the use of frequency modulation that yielded better signal-to-noise ratio of the detected signal than that of radio frequency signal, and by post-detection filtering. An acceptable phase measurement precision was possible whenever the radio frequency signal was strong enough to be detected.

Accuracy is affected by the time delay variations of the equipment. The absolute value of the time delay is not important. If it does not change, it is subtracted from the measured interval in calculating the range. However, if the delay changes in an unknown way, it introduces an error that cannot be compensated. Electronic circuits are subject to changes in delay time, with the delay variations tending to be greater in narrow-bandwidth circuits than in wide-bandwidth circuits. Delay variations can occur in receivers as a function of signal level. Bandpass filters such as receiver IF stages may have time delay changes if the circuits are slightly detuned. Signals from different users operating in the same channel may experience difference time delays if their carrier frequencies are not exactly the same. The effects may be minimized by the use of temperature-stable components and by careful alignment to assure a linear phase characteristic across a bandwidth that includes all significant frequency components of the signals. FM limiters may introduce time delay variations if they do not limit symmetrically about the IF zero crossings, and discriminators may introduce changing time delays if they are improperly aligned.

The magnitudes of equipment time delay effects are a function of bandwidth and are independent of radio frequency. Wider bandwidth allocations at L-band than at VHF may be important if full advantage is to be realized from the better propagation characteristics.

All of the factors influencing precision and accuracy, including equipment delay characteristics and signal-to-noise ratio effects, were measured for the various equipments used in the experiment.

Each electronic circuit contributes a time delay to the interval between the initiation of an interrogation and the final readout of the measurement at the ground station. The sum of these delays must be known precisely and subtracted from the total delay to yield the propagation interval. The range uncertainty introduced by uncertainty in equipment delay is the same as that for propagation time uncertainty, or approximately ± 500 ft for 1.0 μsec. Although the equipment delays may total several hundred μsec, they must be known to within limits appropriate to the accuracy required of the measurements.

For a statistical estimate of system performance the delay uncertainties in ground equipment, satellite, and user equipment may be taken as the root-sum-square of their individual uncertainties, since they are independent. The contribution of the ground station and the satellite equipments can be maintained at a low value because they are in stable environments, and because they are available to the ground terminal for frequent measurement and compensation or calibration.

An operational system would include many user equipments subject to a wide variety of environmental conditions and maintenance skills. It may be impractical to expect that each user equipment be preset and maintained to have a predetermined time delay within the limits assigned for equipment time delay in the error budget of the system. If this is true, a better procedure may be for the ground terminal to determine each user's time delay when the user is at a known location. The delay may then be stored with the user's address in the computer memory. Computer program designers advise that automatic calibration in this manner is simple and does not tax the computer memory.

Narrow-bandwidth circuits are more apt to contribute time delay uncertainty than wider bandwidth circuits, because a phase error caused by a nonlinear phase characteristic in a filter represents a larger time error at a low frequency than the same phase error at a higher frequency. Experience in designing equipment for the experiment indicates that filters employing active components are subject to larger time delay uncertainties than filters employing only passive components. Digital switching and logic circuits can be designed to keep their time delay variations to a negligible value.

Tone-code ranging has a distinct advantage because it does not require an antenna diplexer. In a separate program, a diplexer for reception on 157.86 MHz and transmission on 154.89 MHz was tested in a vehicle on a cold winter day and found to have intolerable time delay variations, depending on its temperature. A diplexer for use with a ranging technique that requires simultaneous reception and transmission by the user equipment would have to be carefully designed to have a constant time delay over a wide temperature range. A diplexer co-located with an aircraft or ship antenna would be subject to large temperature changes.

For any active ranging technique, the user receiver must be designed for minimum time delay variations. Receivers designed for communications may have time delay variations with tuning and with signal amplitude. Tuning variations can be minimized by the use of crystal control and proper alignment for a linear vs. frequency characteristic in the IF and other bandwidth limiting circuits. Carrier phase locking is helpful if the received signal frequency is not accurate.

During the experiment, no significant changes in time delay with tuning were observed with any of the crystal-controlled receivers. Time delay variations with signal amplitude were observed. One Bendix RTA41B aircraft receiver was measured to have approximately 1.5 μsec. total variation in delay over its dynamic range. It is understood that an objective in the design of the receiver was to minimize time delay variations. A GE ER-52-A receiver, designed for mobile communications without consideration to time delay variations, was found to have a total variation in time delay of approximately 7 μsec. A different limiter was substituted for the original and the total variation reduced to about 1 μsec., as shown in Fig. 9.

Signal-to-Noise

Noise added to the demodulated tone causes jitter of the zero crossings that are used for the timing measurement. Probable phase error on one cycle may be estimated by a vector addition of the signal and noise phasers. Assuming that the noise has a gaussian distribution, which is the usual case when receiving signals from the satellite, the one-sigma probability phase error is the angle in radians between the signal vector and signal-plus noise resultant when the signal and noise vectors are at right angles, as follows: Signal-to-noise ratio, dB = 20 log E_s/E_n, where E_n = rms noise voltage, E_s = signal voltage, and $\emptyset_\epsilon$ = 1-sigma phase error in radians. The phase measurement precision is improved by the square root of the number of cycles averaged.

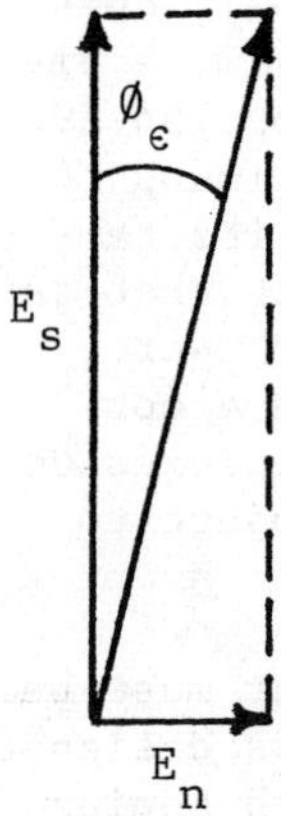

For the narrow-bandwidth frequency modulation used in the experiment, the minimum signal-to-noise ratio for detection is 5 dB in the 15-kHz bandwidth of the receiver. The equivalent signal-to-noise ratio out of the 4-kHz bandwidth detector is 10 dB. In a laboratory test the standard deviation was determined to be 1.283 μsec. All of the experimental data confirm that the standard deviation stays within reasonable limits at the lowest signal levels for which the signal can be detected. The laboratory test shows a standard deviation of 0.4 μsec when the signal-to-noise exceeded 20 dB. Some experimental data provided even better results than the laboratory test, as evidenced by Fig. 10, in which range measurements from the GE Observatory to the satellite had a standard deviation of 0.31 μsec, and for range measurements through the satellite to a user equipment the standard deviation was 0.43 μsec (see Fig. 11).

It is important that the time delay through the user equipment remain constant within specified limits over a period of time between calibrations. Tests with the Sea Robin equipment indicate that such a requirement can be met. Latitude determinations on the buoy were made while the buoy was moored in a harbor in March and at sea between April 15 and 25. The buoy was then removed from the sea and later returned to the deep sea mooring where further latitude determinations were made on May 9 and 10. No adjustments were made to the ranging transponder during that time. The same equipment calibration was assumed for all latitude determinations, in the harbor and for the two periods at sea. There was no discernible effect due to change in equipment time delay over that period.

Results of Ranging from Satellites to Five Vehicles

Aircraft Tests

The DC-6 aircraft of the Federal Aviation Administration was flown from Omaha to Detroit via Minneapolis on June 13, 1969. A major objective of the June 13 flight was a test of the two-satellite position fixing performance under good signal conditions over land and water. A portion of the flight path is shown on the map, Fig. 12. The numbered positions refer to the satellite and VORTAC position fixes, shown in Table 1.

Interrogations were made at 3-sec intervals through ATS-3, and responses were returned from the aircraft through both ATS-1 and ATS-3. The horizon mode of the Satcom antenna provided the best performance, the elevation angle to ATS-1 being 9° and ATS-3 beint 33°. Many hundreds of two-satellite ranging measurements were made during the flight. From these, ten were arbitrarily selected for comparison with VORTAC positions, choosing ranging measurements made within 3 sec of a VORTAC fix.

Table 1 Comparison of satellite and VORTAC position fixes DC-6B aircraft, Minneapolis to Detroit, June 13, 1969

	Time-GMT	Latitude	Longitude
Satellite	13 21 54	44°59'08"	91°15'33"
VORTAC	13 21 51	44°57'	91°16'
Difference		+02'	0
Satellite	13 25 00	44°54'19"	90°56'21"
VORTAC	13 25 00	44°55'	90°55'
Difference		-01'	+01'
Satellite	13 30 45	44°48'51"	90°17'13"
VORTAC	13 30 45	44°50'	90°15'
Difference		-01'	+02'
Satellite	13 43 21	44°41'47"	88°51'29"
VORTAC	13 43 24	44°39'	88°53'
Difference		+03'	-02'
Satellite	13 57 57	44°24'28"	87°18'12"
VORTAC	13 57 57	44°22'	87°19'
Difference		+02'	-01'
Satellite	14 00 03	44°20'41"	87°06'14"
VORTAC	14 00 00	44°19'	87°04'
Difference		+02'	+02'
Satellite	14 06 00	44°12'12"	86°30'53"
VORTAC	14 06 00	44°12'	86°31'
Difference		0	0
Satellite	14 10 30	44°07'31"	86°04'15"
VORTAC	14 10 30	44°07'	86°04'
Difference		+01'	0
Satellite	14 16 06	43°58'07"	85°31'55"
VORTAC	14 16 06	43°59'	85°31'
Difference		-01'	+01'
Satellite	14 20 36	43°50'33"	85°06'42"
VORTAC	14 20 33	43°52'	85°07'
Difference		-01'	0

The first satellite position fix was recomputed several times in order to determine the correct equipment delay biases. Once these had been determined, the same biases were used for all of the other position determinations. This step was necessary as changes had been made to the aircraft equipment and to the Observatory equipment between the time the aircraft transceiver was integrated with the responder at Schenectady and installed on the aircraft. In all cases throughout the experiment it was found easier to calibrate equipment delay biases with the user equipment at a remote location rather than by measurement in the laboratory. Once the biases were determined for the June 13 flight, the same values were used for the later processing of data from the June 12 flight. There was no significant difference in the equipment delay biases for the two days.

With the bias corrections properly inserted, the computations for the arbitrarily selected ten sample measurements of the June 13 flight were made and compared with the VORTAC position determinations as provided by the FAA. The satellite and VORTAC position determinations are in such close agreement that they could not be separately plotted to the scale of the map in Fig. 12. They are compared in Table 1. The VORTAC positions were determined to the nearest minute. One minute represents 1 nmi in latitude and approximately 0.7 nmi in longitude. All of the arbitrarily selected satellite position fixes agree with the VORTAC positions within ± 3 min of latitude and ± 2 min of longitude.

Short-term variations in satellite position fixes are indicated in Fig. 13, a plot of ten fixes made within 45 sec. They are numbered in time sequence. VORTAC positions for times before and after the ten satellite position fix determinations are plotted as the larger circles on the chart. The data are also presented in Table 2.

Precision of the fixes was plotted relative to the mean range measurements for the period between 1354 and 1402 GMT, while the aircraft was making a land-to-water transition. It shows that only 4 of the 74 dual responses would produce fixes in error by more than 1 nmi (see Fig. 14).

The plot was made as follows: 1) A best fit curve was computed for the range measurements from each satellite during the eight minute period. 2) The displacement of a range measurement from the best fit curve was determined and its projection on the Earth for the elevation to the satellite was calculated. 3) A line of position was plotted at right angles to the satellite azimuth direction, and displaced from the center reference point by the distance determined in step 2. 4) A second line o position for the same interrogation was plotted from the range

Table 2 Comparison of satellite and VORTAC position fixes, short sequence, DC-6B aircraft near Minneapolis, Minn., 6/13/69

Time-GMT	Latitude	Longitude
Satellite position fixes		
13 22 18	44°58'40"	91°12'35"
13 22 24	44°58'26"	91°12'48"
13 22 27	44°58'50"	91°11'06"
13 22 30	44°58'28"	91°11'43"
13 22 39	44°58'21"	91°10'03"
13 22 42	44°58'09"	91°10'41"
13 22 45	44°58'50"	91°09'31"
13 22 51	44°58'14"	91°08'36"
13 22 57	44°58'19"	91°08'25"
13 23 03	44°58'11"	91°07'24"
VORTAC position fixes		
Time-GMT	Latitude	Longitude
13 21 51	44°57' N	91°16' W
13 24 24	44°55' N	91°00' W

measurement of the other satellite. The crossing of the two lines of position determines individual fix error in magnitude and direction relative to the mean of all the fixes during the time period.

Precision plots show the deviation of individual fix determinations from a mean value determined from best-fit curves computed from the range measurements from each satellite. Factors not included in the fix position plots are bias errors due to ionospheric propagation delay, satellite tracking uncertainty, aircraft altitude uncertainty, and equipment time delay bias. With the possible exception of aircraft altitude uncertainty, these bias changes have periods of hours, days or longer. All of the factors that affect the short term variations in accuracy are included in the precision plots. These are sea reflection interference multipath, changing signal-to-noise ratio, changing signal level and its effect on receiver time delay, as well as short term fluctuations in ionospheric time delay.

The June 12 flight from Joliet to Omaha via Minneapolis was a test of two-satellite position fixing performance over land. It was the first test made with an aircraft in flight where the elevation angles from both satellites to the aircraft were high enough to correspond with the minimum elevation angles that would be used in an operational system; although the path from ATS-1 to the GE Observatory, which influences the accuracy of the system, was much below the elevation angle anticipated for an operational system. The signal-to-noise ratio at the aircraft receiver was poor because the aircraft was flying in the vicinity of thunder storms. As a result, the link performance was worse than that of the following day when the aircraft was in the same geographical area. All three antenna modes were tested: the horizon and zenith modes of the Satcom antenna, and the VHF blade. Interrogations were made through ATS-1 as well as through ATS-3. Although the link performance was marginal throughout most of the tests, the accuracy of the position fix measurements made was not significantly different from the results obtained under better signal conditions the following day.

On June 20, a flight test was made to evaluate the precision of ranging measurements through ATS-3 to a high-performance jet aircraft on the North Atlantic principal routes. The KC-135 aircraft flew at 39,000 ft. The blade antenna was used throughout the test so that the effects of multipath combined with ranging fluctuations from low signal-to-noise and receiver delay characteristics to produce scatter in the readings. Standard deviation of the sub-sets of measurements varied from approximately 8 μsec. to 11 μsec. A histogram of all the range measurements is plotted in Fig. 8. The envelope of the histogram suggests the shape of range error probability distribution that was obtained by analysis and shown earlier in Fig. 3. The histogram shows a maximum deviation range measurement in the long direction, which is approximately twice that of the largest deviation in the short direction, as expected from the analysis when the multipath delay is long and the amplitude of the reflected signal is relatively large. The lower scale shows the approximate cross-track error if the range measurements are projected on the Earth for a satellite having an elevation angle of 25°. It indicates the precision of cross-track monitoring that would be achieved for a satellite at midocean longitude while monitoring cross-track positions of aircraft flying at maximum altitudes on North Atlantic routes with signal conditions like those in the experiment. It is expected that an operational system would have stronger signals, antennas with better discrimination against multipath than the VHF blade, and better ground station receiver characteristics than were used in the experiment. Nevertheless, the experimental result, as it stands appears to have sufficient precision for useful cross-track surveillance.

Ship Tests

The Coast Guard Cutter Valiant, WMEC-621, based at Galveston, Texas, is a 210-ft ship like the one shown in Fig. 15. It was equipped with a ranging transponder, also shown in Fig. 15. The lower unit is a General Electric type DM76LAS mobile radio base station. Within the base station unit, the tone-code responder was connected between the receiver and transmitter, together with its solid-state power supply. A Parks electronic type 144-1P preamplifier was also mounted in the unit and connected between the antenna and receiver. The 35-W output of the mobile radio transmitter was applied to a Gonset Model 903 Mark II 300-W. power amplifier, shown above the base station unit. The oscilloscope was used for display of the signal, and it is not a part of the transponder. The equipment was mounted on the bridge of the ship.

The antenna shown in Fig. 15 was mounted on the flying bridge. Originally designed for the NASA OPLE project, the antenna consists of a pair of crossed dipoles connected for circular polarization. Its gain is 3 dB toward the zenith. It has 0 dB for circular polarization at 45° elevation. Variation in azimuth gain is approximately 2 dB. When used with the linearly polarized antenna on the satellite, the antenna gain is effectively reduced by 3 dB, so that its zenith gain is approximately 0 dB; its gain at 45° is -3 dB. It is even lower at the elevation angles to the satellites, which were 29° for ATS-3 and 17° for ATS-1. Fig. 15 shows the location of the antenna on the flying bridge relative to other structures of the ship. For certain headings of the ship, the mast shielded the antenna from one satellite or the other.

Good voice communications were relayed through the ATS-3 satellite between the ship and the GE Observatory. The equipment was tested with the ship at its berth with a southerly heading on June 27, 1969. Interrogations through ATS-3 had a standard deviation as small as 0.7 μsec.

On July 1, 1969, the Valiant sailed straight-line courses at various headings and three complete circles, each approximately 1 mile in diameter. Except during short periods of voice communication, the ship was interrogated through one of the satellites at 3-sec intervals. Responses from the ship were relayed back through both satellites to the Observatory.

Figs. 16 and 17 show the ranging time intervals for the satellites ATS-1 and ATS-3 as the ship sailed a circle. The number of returns as well as the magnitude of the variations of the time intervals clearly shows the effect of the mast and perhaps other ship structures on the antenna pattern. The performance changed with ship heading, so the data were examined separately for short time intervals during the circle.

A plot of fix precision was made for each time interval in the manner described previously. Each small circle is a measure of fix precision, including the effects of instrument measurement resolution, time delay variations in the instrumentation with signal amplitude or detuning, the effects of noise on the signal, and geometrical dilution of position. It does not include bias errors resulting from the ionosphere, error in the estimates of the satellites' positions, or in the estimate of equipment time delay.

During the first period from 0303 to 031418 GMT (Fig. 18), the signal paths were relatively good. Only one of the 26 fix estimates falls outside of a 1-nmi. radius circle. The period from 030420 to 0307 GMT had generally good signal levels, but there are several points which are displaced farther from the best-fit curve than is typical of the previous or following time periods. The reason has not been definitely assigned, but it is most probable that it represents a known characteristic of the receiving equipment. The receivers exhibited a change in time delay of 5-7 μsec with changes in received signal level. A comparison of the range measurements for ATS-1 and ATS-3, as shown in Figs. 16 and 17, reveals that the displacement of the range measurements is highly correlated in the returns from the two satellites, indicating that the variance in time delay occurred in the receiver on the ship as it received the interrogation signal from ATS-3. Because the error causes an advance or delay in the transmission of the response from the ship, it adds or subtracts an equal delay to the returns from the two satellites. The range difference of the measurements remains unchanged and therefore, the fix errors tend to lie along a hyperbolic line of position. The effect is clearly evident in Fig. 18 where the position fixes are concentrated along a generally north-south line.

Buoy Tests

The Sea Robin buoy was developed and constructed by the General Electric Co. to test key technologies applicable to remote, unattended data collection buoys for sensing and relaying oceanographic and meteorological data. Sea Robin is a modified spar buoy, approximately 4 ft in diameter and 15 ft long, with stabilizing means designed for mooring in the deep ocean or for free-floating in all sea states.

During the period, March through May 1969, the buoy was teste ashore, in a harbor, and at a deep sea mooring near Bermuda, at 32°10'00" N, 64°55'30" W, in a joint Navy-GE experiment. The experiment verified the sea-worthiness of the buoy when moored where the ocean depth was $>$4000 ft., using a 7000-ft line, under a variety of weather and sea states.

For the data transmission, the buoy sensed 19 housekeeping, weather, and oceanographic conditions. Each data transmission totaled 336 bits. Data were transmitted directly by telemetry to a van on shore, which also commanded functions aboard the buoy and monitored its position by radar and telescope.

Data transmission through the satellite was accomplished by the use of the transmitter-receiver-responder unit used in the VHF ranging experiment. A data transmission followed each range interrogation. Two data rates were tested, 2.4414 kbits/sec. and 305 bits/sec. At the higher rate, a "one" bit was formed by transmitting an audio cycle, a "zero" by suppressing a cycle. At the lower rate, eight audio cycles were transmitted for a one, and eight suppressed for a zero. A complete response from the buoy consisted of the transmission of 1024 cycles at 2.4414 kHz followed by the address code, consisting of 30 bits at the 2.4414 rate, and the address code followed by a 1.25-sec data transmission to accommodate slightly more than one complete data frame at the low rate. The radio frequency energy transmitted from the buoy was approximately 50 W-sec. for the ranging signal (120 W for 0.43 sec) and 150 W-sec (120 W for 1.25 sec)c for the data transmission.

The transponder aboard the Sea Robin consisted of a 35-W, solid-state FM mobile radio transmitter-receiver with a 120-W solid-state amplifier, and the first experimental responder unit for phase matching, address correlating, clocking for the digital data readout, and switching of the receiver and transmitter. Despite the early state of the responder development, it performed reliable throughout the test at the mooring.

The satellite antenna is linearly polarized. Horizontal and vertical linear, and circularly polarization were available at the GE Observatory. Sea Robin had two linearly polarized antennas mounted at right angles with a switching arrangement to permit transmission and reception on either, or reception on one and transmission on the other. These antenna arrangements were designed so that information could be obtained about the difference in Faraday rotation between the up-link frequency of 149.22 MHz and the down-link frequency of 135.6 MHz. Separate receive and transmit polarization angles were not available at the Observatory during the test period. The capability was added later. A difference of approximately 90° in the Faraday rotation of polarization at the two frequencies was observed frequently, making it desirable to receive on one linear polarization and transmit on the other. A switching sequence was deliberately introduced at the buoy to observe the effect, resulting in an anticipated failure to respond to some interrogations.

While the buoy was at its deep sea mooring, between April 14, 1969 and April 25, 1969, it was interrogated at 3-sec intervals during 3-min periods at various times of the day and night. A latitude determination was made for each range measurement. They are plotted as histograms for each interrogation period in Fig. 19. A total of 759 latitude determinations were made. All but one of them were within 1.8 nmi N and 2.8 nmi S of the mooring latitude. The remaining determination was 3.4 nmi N of the mooring latitude. However, because of the 7000-ft length of the mooring line, the buoy was free to move with respect to the mooring latitude. The median value of the fix determinations for each interrogation period are shown as the arrowheads by the histograms in Fig. 19.

The averages of the median values for each day are plotted as the dots within the triangles on Fig. 20. The wind direction and velocity as furnished by the Navy for the area are shown by the vectors near the top of Fig. 20. It is expected that the buoy, which has a small cross-section to the wind and much drag at the mooring line, responded slowly to the changes in wind direction. The dashed line suggests the actual positions of the buoy as determined from the satellite measurements. They appear to correlate with the wind directions. The average value for April 15 is displaced south from the other average values. It is interesting to note that the electron density of the ionosphere measured on that day at Hamilton Hill, Mass. (private communication from J. Klobuchar, AFCRL) was lower than the monthly average. If it is assumed that the dashed curve of Fig. 20 represents the true latitude of the buoy, the position determinations replotted with respect to the buoy position provide an estimate of the accuracy that was actually achieved, as shown in Fig. 21.

The median values for each interrogation period are plotted vs. time of day in Fig. 22. The diurnal variation is evident and the dashed line suggests its phase and amplitude. An ionosphere model was included in the computer program to correct for the diurnal change in radio propagation path delay. Figure 22 shows that the latitude determinations in the daytime hours are south of the nighttime determinations, as would be the case if the ionosphere model estimated a diurnal change in path delay that is larger than the actual change. However, an examination of the magnitude of the change shows it to be larger than the total ionosphere correction inserted by the model. Some other diurnal effect must have affected the latitude determinations.

The satellite position is the only other factor with a diurnal variation that we can identify. The satellite orbit is slightly inclined to the Earth's equatorial plane, so that the subsatellite point traces a figure-eight pattern on the Earth, moving north and south of the equator, with a 24-hr period. On April 14, the maximum northward excursion of the satellite was 0.313° at approximately 1430 GMT, and the maximum south excursion was also 0.313° at approximately 0230 GMT. By April 25, the motion had changed slightly, to maximum north and south excursions of 0.286° at 1400 GMT and 0200 GMT, respectively. It is interesting to note that the extremes of the latitude displacements occur at corresponding times.

If the ionospheric correction was of the right magnitude, there is a residual diurnal variation of approximately ±3000 ft, representing an imprecision of approximately $\pm 0.01^{\circ}$ at the extremes of the satellite positions. The imprecision could be the result of computer round-off in our linear interpolation between the half-hourly satellite position predictions of NASA, inaccuracy in the predictions, or a combination of such factors.

If it is assumed that an operational system can correct for the diurnal effects shown in Fig. 22, an estimate of the line-of-position accuracy that would be achieved in an operational system is obtained by shifting the data points of Fig. 36 to correct for the diurnal effects as estimated from the dashed curve of Fig. 22. The resulting histogram, Fig. 23, has a standard deviation of approximately ± 0.5 nmi, 1 sigma, and is believed to represent the accuracy that can be achieved in an operational system, even if the equipment, techniques, and signal conditions are no better than they were in the April 14-25 Sea Robin tests.

Van Tests

A Ford Econoline van was equipped with a GE mobile radio, as used in taxicabs and police cars. A responder unit was connected between the receiver and the transmitter. A Parks Electronics Laboratories preamplifier model 144-1P preceeded the receiver. The transmitter output power was 80 W. Separate dipole antennas were provided for the receiver and the transmitter.

On March 27, the van was driven from the village of Manny Corners, N.Y., northward through the village of Hagaman, N.Y., to Route 29, then eastward on Route 29.

Range measurements were made through ATS-3 during the 1-hr test period. The dipole antennas were separately oriented for best signals.

In previous mobile radio ranging tests not using satellites, it was observed that reflected signals can arrive at the receiving station antenna in any phase relative to the direct signal, and severely affect the received signal level. A vehicle position change of only a few feet can result in a large change in signal amplitude at the receiver.

During the first portion of the March 27 satellite ranging experiment, the van was moved in small increments: 1 ft, then 2 ft, then 4 ft, etc., to determine if a similar effect occurred between the vehicle and the satellite. The amplitude change was observed. No effects on range measurements were observed that could not be attributed to the separately measured change in receiver time delay with signal amplitude. The receiver was an unmodified production unit with a time delay change of approximately 7 μsec. with signal amplitude change.

Fig. 24 is a plot showing every range measurement made during the March 27 test. The shape of the data curve matches the changing range from the satellite to the vehicle. Events during the test are marked on the plot. There is a change in measured range, noticeable during the first part of the test, greater than the movement of the vehicle. The change in range is due to satellite motion. Gaps in the data occurred because the range measurements were stopped when voice communications were used to coordinate the experiment. Scatter of the data points includes the variation in the time delays through the van and GE Observatory receivers.

Some of the range measurements were used to compute the intersection of the line of position with the road. The computation included an estimate of the equipment and ionospheric time delays. A first estimate for the delays resulted in a bias error to the south of approximately 1.5 mi. A new estimate was made to place the first location at Manny Corners, then the same equipment and ionospheric time delays were used for all the rest of the computations. Fig. 25 shows the route which was traced from a topographic map. The actual positions of the vehicle are plotted as dots, and the positions as determined from the satellite range measurements are plotted as short horizontal line segments. The satellite was nearly due south of the area at the time.

Conclusions

The test results suggest that an accuracy better than ±1 nmi. 1 sigma for ships and approximately ±1 nmi, 1 sigma for aircraft can be achieved at VHF. To achieve that accuracy, it will be necessary to employ automatic, unmanned calibration transponders at fixed, known locations with approximately 600 mile spacing, and interrogate each one a few times/hour to determine range measurement corrections. The calibration transponders are needed especially during solar disturbances when unpredictable changes in ionospheric propagation delay can occur within a short time. They also serve to correct for small uncertainties in satellite position. It is recommended that calibration of vehicle equipment time delay be accomplished at the ground terminal by interrogating each craft when it is at some known location. The time delay calibration is then stored in the computer with the vehicle address. It will be necessary to employ aircraft antennas that discriminate against sea reflections, so that the reflected signal is more than ~ 10 dB below the direct signal. The use of circular polarization for the satellite and aircraft antennas is recommended.

Position fixing is feasible by simultaneous range measurements from two geostationary satellites operating on the same frequency; it was demonstrated by interrogating any one of several user craft through one satellite and receiving a response from the addressed user relayed through both satellites.

The equipment limitation on ranging resolution was approximately ±200 ft (0.4 μsec timing resolution) when the bandwidth limits were like those presently used for aircraft mobile communications; i.e. 4 kHz for data baseband, and 15 kHz for the radio-frequency spectrum.

Ranging precision for surface craft, including VHF transmission links, was better than ± 500 ft (1 μsec); one sigma for IF signal-to-noise ratios at the 5-dB FM signal detection threshold; and better than ± 5000 ft for an aircraft at 39,000 ft over the North Atlantic using a VHF blade antenna and experiencing sea reflection multipath.

Position fix precision was better than 1 nmi, one sigma for a ship at sea off Galveston, Texas when the antenna was not shielded from a satellite by the mast or other structures. When the antenna was shielded, some interrogations did not result in responses. Fix precision was reduced to approximately 1 nmi, one sigma, and in that particular test the errors were along a hyperbolic line of position.

Position fix precision was approximately 1 nmi one sigma for aircraft, including fixes made over water when using a Dorne and Margolin Satcom antenna in the azimuth mode.

Line of position accuracy for a buoy moored in deep water at middle latitudes was better than ± 1 nmi one sigma using an unattended transponder over approximately a 2-wk period, including the effects of diurnal and other ionospheric delay changes, pitch and roll of the buoy, and signal-to-noise ratio variation down to the detection threshold.

Lines of position for aircraft at jet speeds and altitudes over the North Atlantic had a standard deviation of approximately ± 1 nmi when using a VHF blade antenna. Individual measurement displacements as large as 2 miles toward the subsatellite point and 4 miles away from the subsatellite point were attributed to multipath.

Fixes determined by two-satellite ranging at VHF agreed with VORTAC fixes within ± 3 miles over the midwestern United States. At best, the VORTAC fixes are accurate to ± 1 nmi.

Ranging and position fixing are compatible with mobile voice and data communications. No antenna diplexer is required for tone-code ranging, and simplex operation is feasible. No modifications are necessary to satellites designed for relaying mobile communications.

Ranging and position fixing measurements for locating a transponder can be made from a ground terminal through geostationary satellites within 1 sec of time, with ranging signal transmissions shorter than 1/2 sec.

Ranging and position fixing can be performed with currently available mobile radio transmitters and receivers with the addition of an inexpensive solid-state circuit, although some receiver-transmitters, as supplied commercially, may have time delay variations with signal level or tuning that introduce ranging errors as large as several thousand feet. The time delay variations can be corrected by circuit modifications such as the substitution of a different limiter.

Narrow-bandwidth VHF user equipment in an unattended buoy moored in deep ocean retained its time delay calibration without discernible change for several weeks.

References

[1]Anderson, R. E., "A Navigation System Using Range Measurements from Satellites with Cooperating Ground Stations," Jr. of the Inst. of Navigation, Vol. 11, No. 3, Autumn 1964, p. 315.

[2]Millman, G. H., "A Survey of Tropospheric, Ionospheric, and Extraterrestrial Effects on Radio Propagation Between the Earth and Space Vehicles," Rept. TIS R66EMH1, 1966, GE Co., Syracuse, N. Y.

[3]Klobuchar, J., "Ionospheric Propagation Delays at VHF," paper presented at the Symposium on the Application of Atmospheric Studies to Satellite Transmissions, Boston, Mass., sponsored by USAF-AFCRL, Bedford, Mass., 1969.

[4]DaRosa, A. V., "Propagation Errors in VHF Satellite-to-Aircraft Ranging," IEEE Transactions on Antennas, Vol. AP17, No. 5, September, 1969, p. 628.

[5]Mendillo, M., Papagiannis, M. D., and Klobuchar, J., "Mid-latitude Ionosphere Variations During Magnetic Storms," paper presented at the Symposium on the Application of Atmospheric Studies to Satellite Transmissions, Boston, Mass., sponsored by USAF-AFCRL, Bedford, Mass., 1969.

[6]Anderson, R. E., "Interim Report - VHF Ranging and Position Fixing Experiment Using ATS Satellites," Rept. S-70-1003, Jan. 1970, General Electric Co., Schenectady, N. Y.

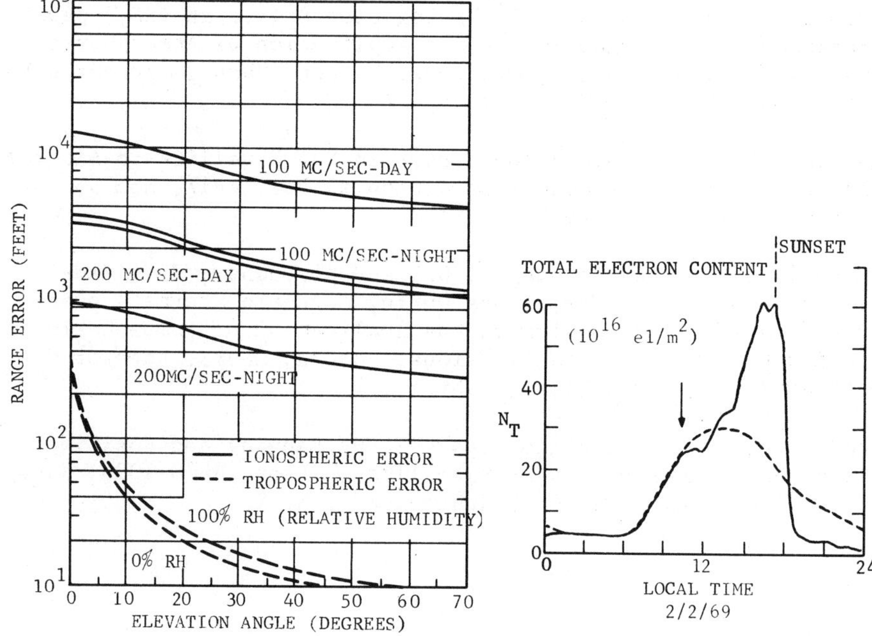

Fig. 1 One-way range bias resulting from tropospheric and ionospheric retardation.

Fig. 2 Electron content resulting from magnetic storm.

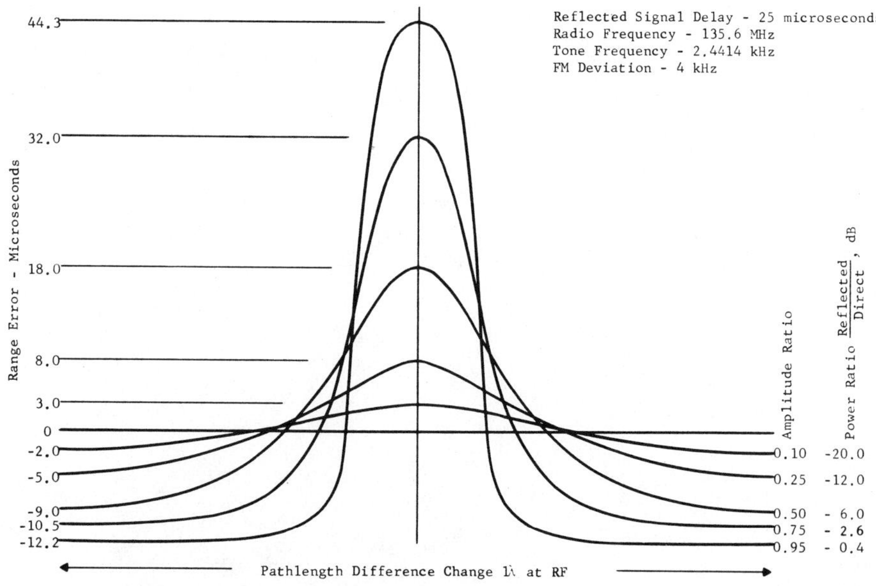

Fig. 3 Range error resulting from specular sea reflection (one-way).

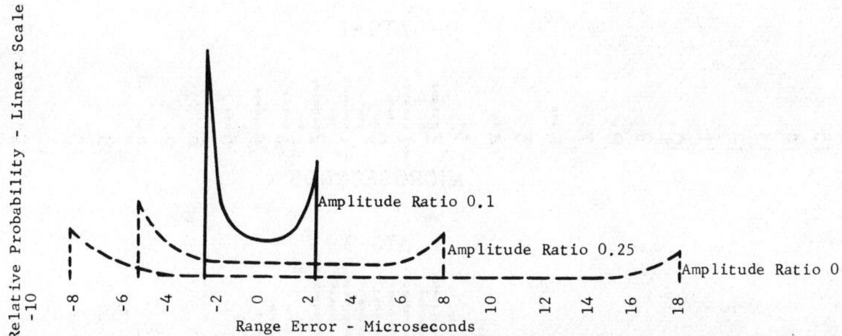

Fig. 4 Range error probability distribution delay of reflected signal; 25μsec ; one-way ranging.

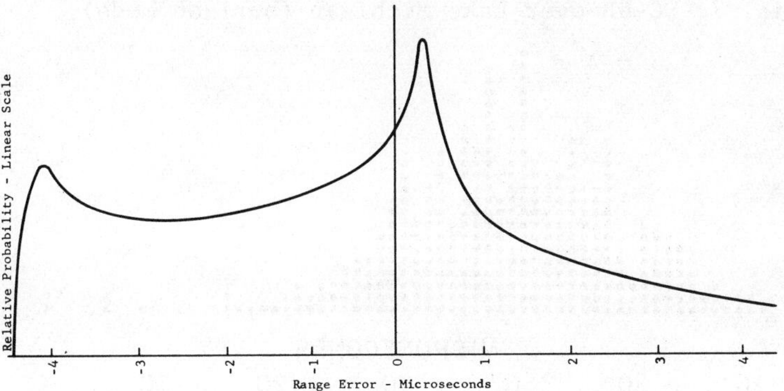

Fig. 5 Range error probability distribution delay of reflected signal; 25 μsec ; two-way ranging.

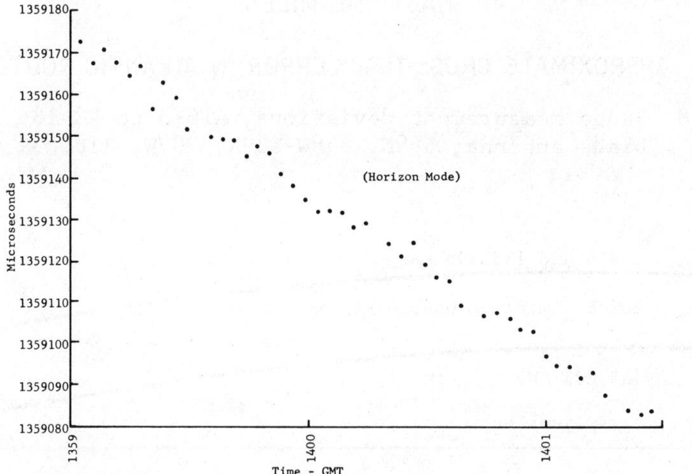

Fig. 6 Range measurements from ATS-3 to a DC-6B aircraft at 21,000 ft over Lake Michigan.

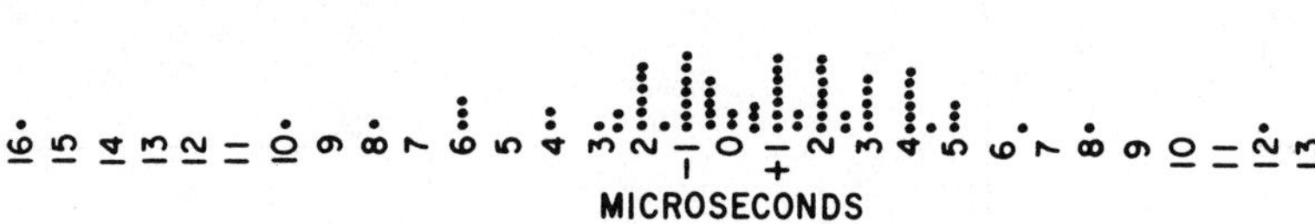

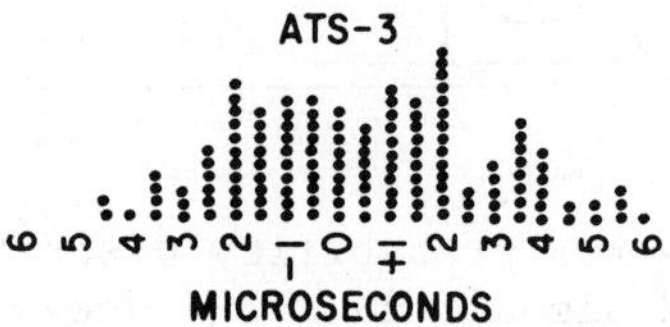

Fig. 7 DC-6B over Lake Michigan (horizon mode).

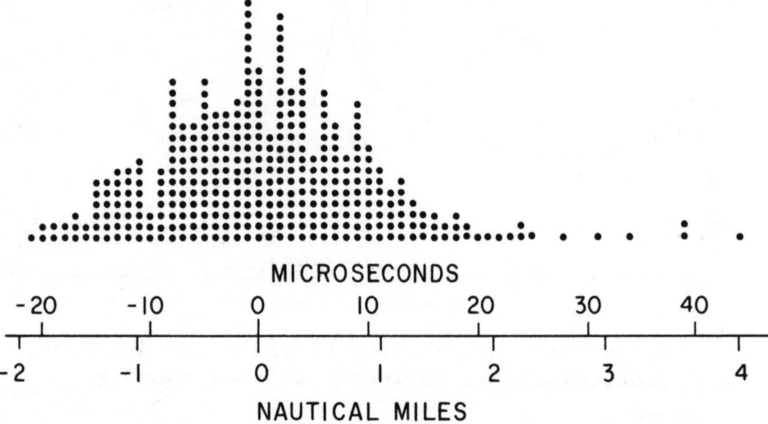

Fig. 8 Range measurement deviations, ATS-3 to KC-135 aircraft, blade antenna; 56°N, 55°W-63°N, 30°W; Alt. 39,000 ft.

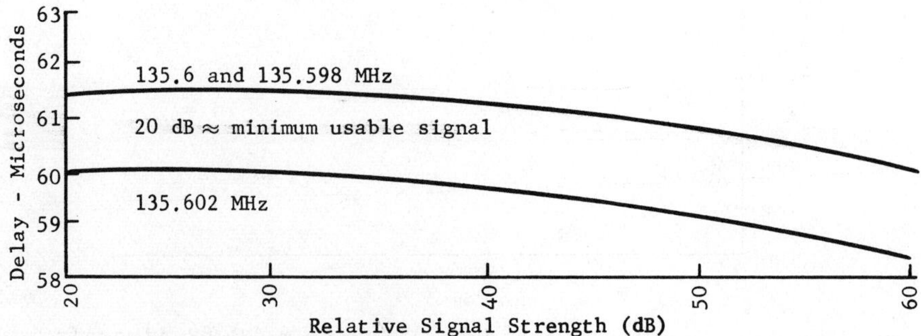

Fig. 9 GE-type ER-52-A monitor receiver; original limiter replaced by special tuned limiter.

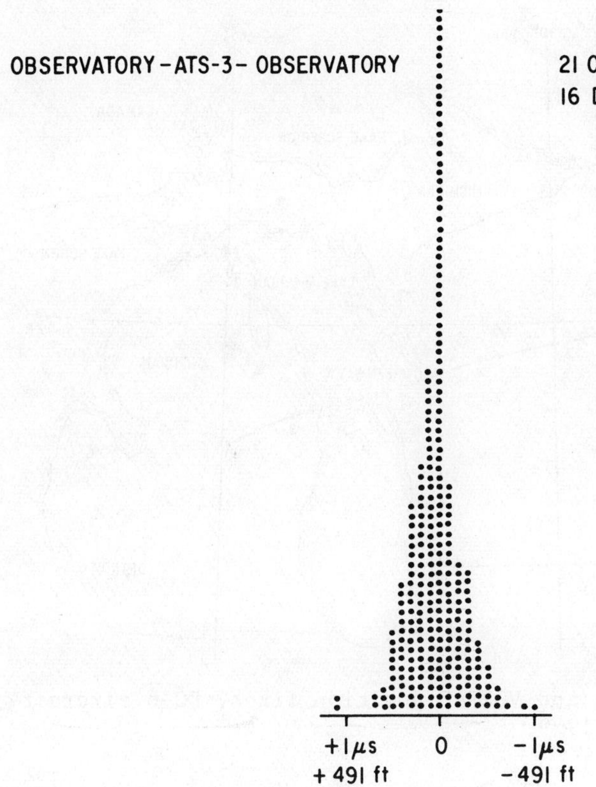

Fig. 10 Range measurement precision.

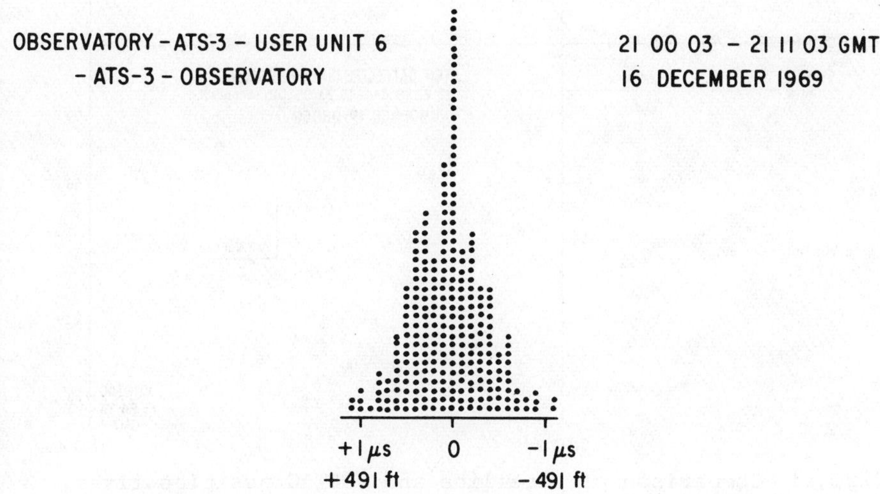

Fig. 11 Range measurement precision.

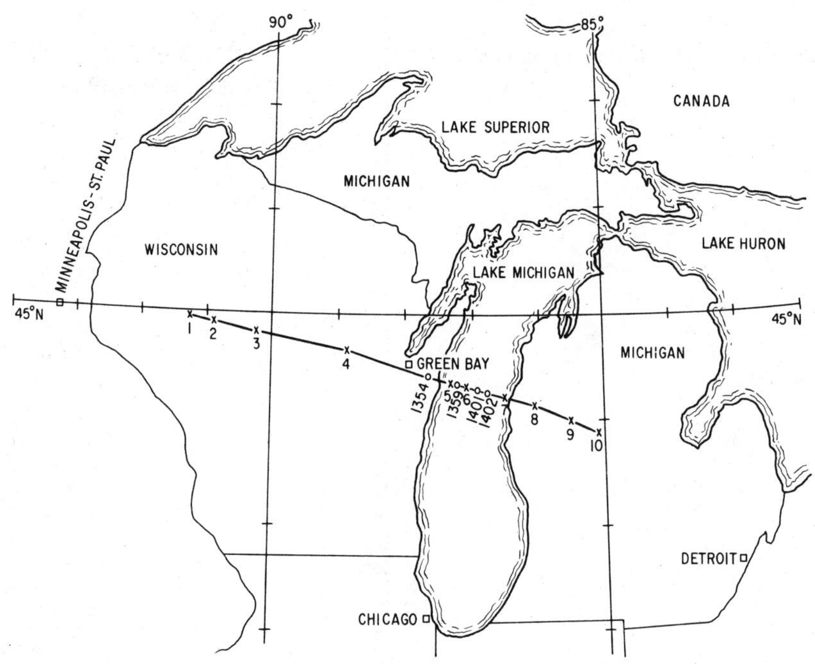

Fig. 12 Satellite and VORTAC position fixes, DC-6 aircraft, June 13, 1969.

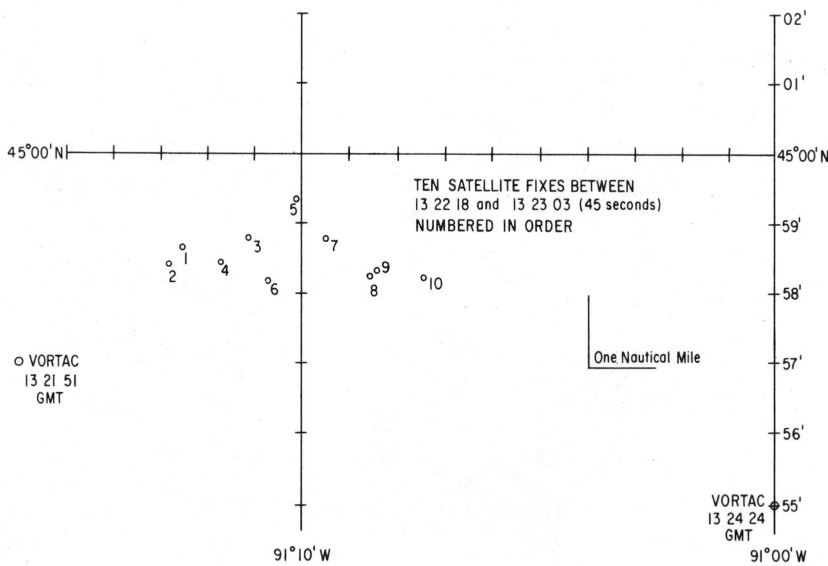

Fig. 13 Comparison of satellite and VORTAC position fixes, short sequence; DC-6 aircraft near Minneapolis, Minn., June 13, 1969.

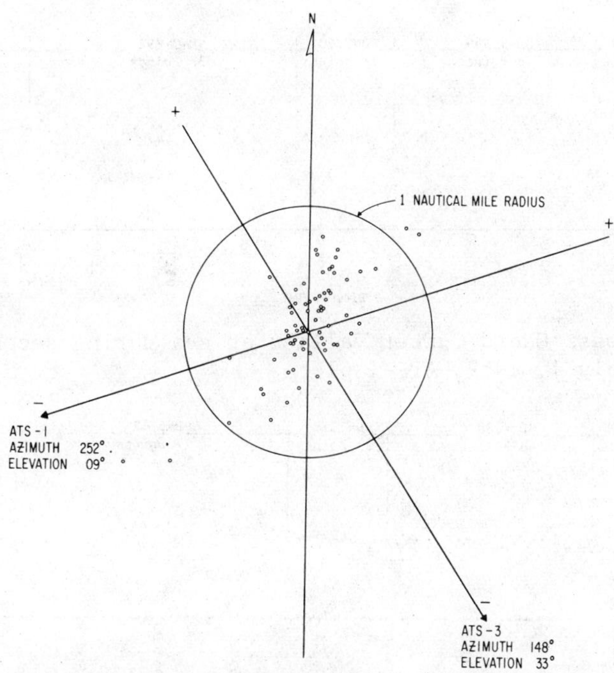

Fig. 14 Fix error distribution.

Fig. 15 Sister ship of Valiant, equipment used in ship tests.

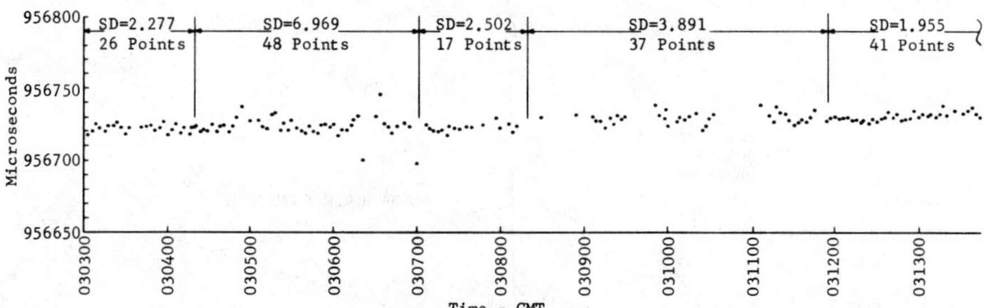

Fig. 16 Coast Guard Cutter Valiant at sea during second turn, July 1, 1969, ATS-1 plot.

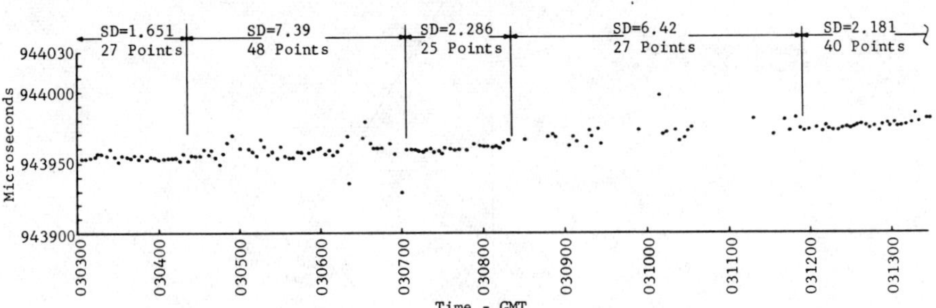

Fig. 17 Coast Guard Cutter Valiant at sea during second turn, July 1, 1969, ATS-3 plot.

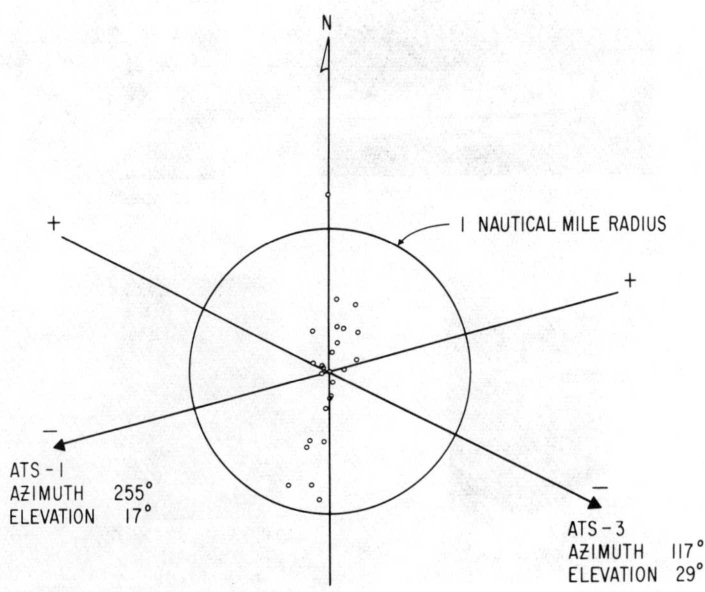

Fig. 18 Fix error distribution.

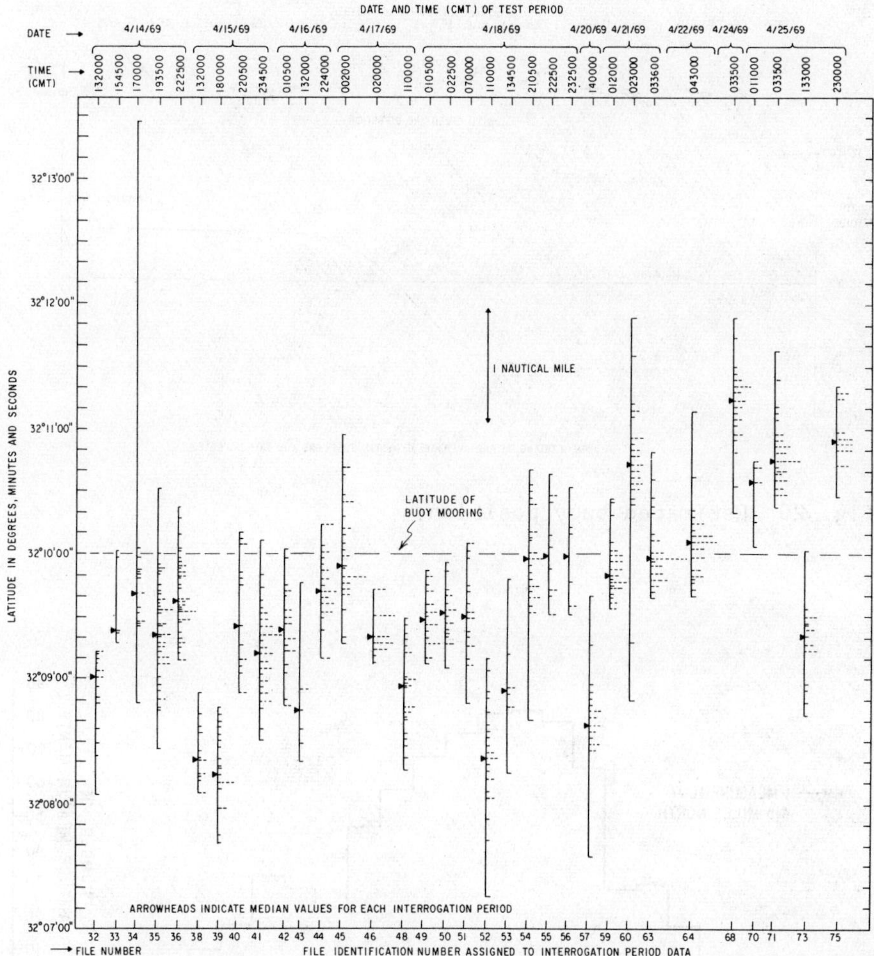

Fig. 19 Latitude determinations.

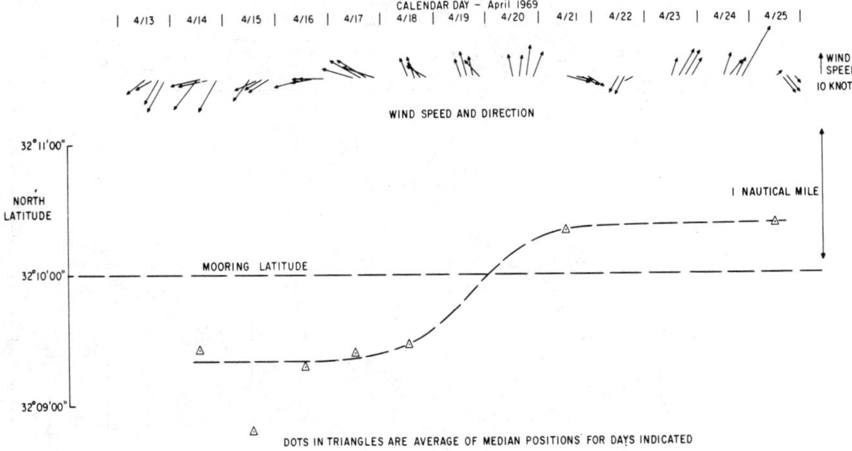

Fig. 20 Estimated buoy position.

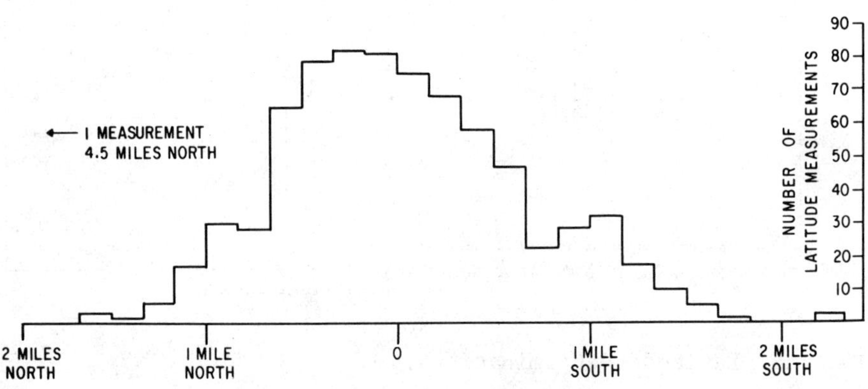

Fig. 21 Distance from estimated buoy position (estimate of accuracy achieved).

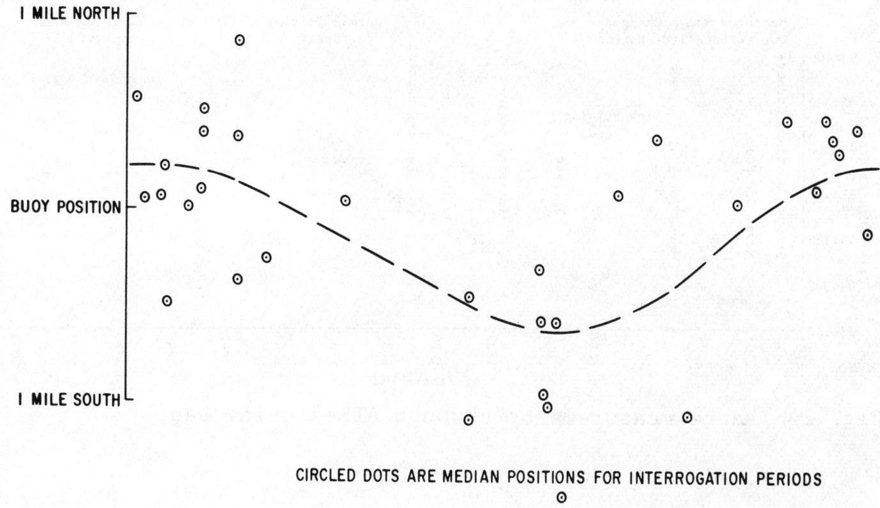

Fig. 22 Diurnal variation of latitude measurement error.

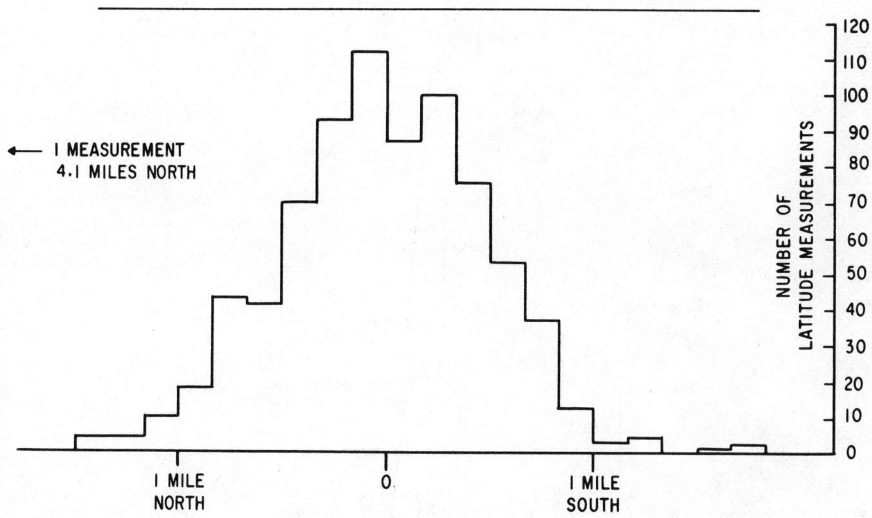

Fig. 23 Estimate of residual errors after correction for ionosphere and satellite position uncertainty (estimated accuracy if calibration station had been used).

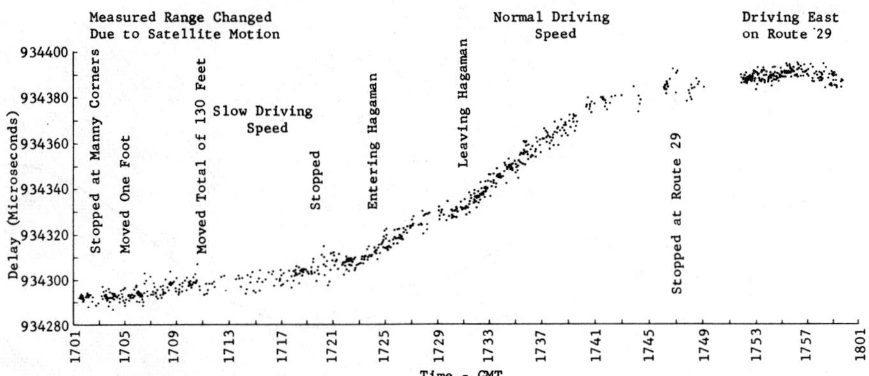

Fig. 24 Range measurements through ATS-3 using van.

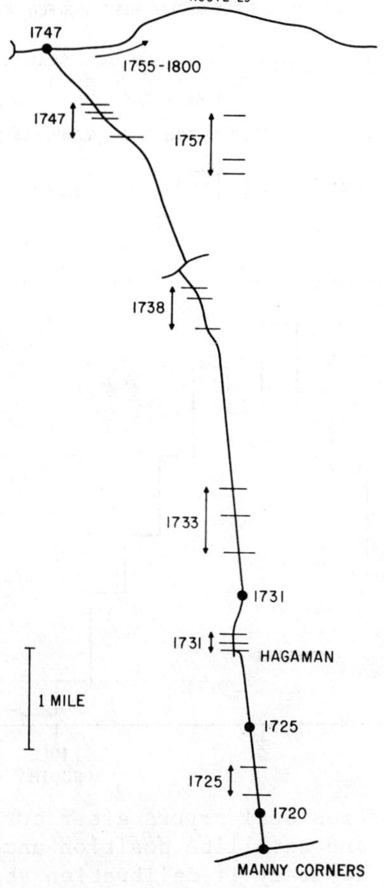

Fig. 25 Route of March 27, 1969 test.

VI. EARTH RESOURCES SATELLITE COMMUNICATION

COMMUNICATION SYSTEM CONFIGURATION FOR THE EARTH RESOURCES SATELLITE

G. R. Welti* and S. H. Durrani†

Communications Satellite Corporation, Clarksburg, Md.

Abstract

The communications requirements of the Earth resources disciplines and other programs for the 1970's are estimated. Several methods of providing a continuous real-time wideband communication link between the Earth resources satellite and a central ground station using relay satellites are considered. The feasibility of using commercial communication satellites is examined. It is found that INTELSAT IV (to be launched in 1971) can provide data links with mission spacecraft and data collection platforms, but that a modified INTELSAT IV would be needed for a television link. Other systems examined employ dedicated medium-altitude satellites and shared synchronous-altitude commercial satellites. An example of a shared system is presented, which is capable of serving Earth resources satellites, in situ data collection platforms, and manned space stations, as well as providing communications services to other users.

I. Introduction

A large number of Earth resources programs‡ have been proposed in recent years by both government and industry. The first satellite in the Earth Resources Technology Satellite

---

Presented as Paper 70-326 at the AIAA Earth Resources Observations and Information Systems Meeting, Annapolis, Md., March 2-4, 1970, and also at the AIAA 3rd Communications Satellite Systems Conference, Los Angeles, Calif., April 6-8, 1970. This paper is based upon work performed at COMSAT Laboratories under corporate sponsorship.

*Manager, Systems Analysis Lab, COMSAT Laboratories.

†Manager, Systems Analysis Branch I, COMSAT Laboratories.

‡The terms "Earth resources program" and "Earth resources satellite" (ERS) are used in this paper in the generic sense and do not indicate a specific system being proposed or implemented.

(ERTS) Program is expected to be launched in 1973, and other satellites will follow to meet the expanding needs of several Earth resources disciplines.

In its simplest form, the Earth resources satellite (ERS) would have a communication system depending entirely on direct links between the satellite and designated Earth stations (ESs); the information collected by an ERS would be stored on board until it can be dumped to the ES. However, to derive maximum benefit from the data gathered by the ERS, it is necessary to maintain a continuous real-time communication link between the satellite and a central ES. Such a link can be provided with the help of relay satellites (RSs).

Several configurations of the communication system are possible, using RSs which are either dedicated exclusively to the ERS program or are shared with other users. In particular, commercial communications satellites may be modified to operate in conjunction with the ERS while simultaneously serving other users. The feasibility and relative merits of the various configurations depend on the communication requirements and related operational constraints, which are discussed next.

II. Communications Requirements

A comprehensive ERS program would require the collection of data relating to several disciplines. The National Academy of Sciences report[1] identifies the needs of these disciplines in five groups: forestry, agriculture and geography (FAG); geology; hydrology; meteorology; and oceanography. The required data can be collected by either of two approaches. The first approach uses five separate systems (one for each discipline) and requires six satellites in low-altitude near-polar orbits and six satellites in geostationary orbits. The second approach uses a common set of satellites to meet the common needs of the several disciplines and is more economical. It requires a total of four low-altitude and four geostationary satellites, as listed in Table 1. Both types of satellites carry remote sensors; in addition, the geostationary satellites receive data from in situ sensors (located at the ground, in balloons or in buoys) and relay it to an Earth station.

Quantitative estimates of the communication requirements of the ERS program and several other potential users of a Relay Satellite System have been examined in many studies.[2-8] A summary of the estimated requirements is given in Table 2. Note that (except for manned terminals) the data rates from the terminal to Earth are generally much larger than those from Earth to the terminal, since the latter is simply a command link.

Table 1 Requirements of Earth resources disciplines
Common approach

Discipline	No.of sat.	Satellite orbit	Function
Meteorology	1	Polar	IR mapper, IR sounder
FAG,[a] geology, hydrology, oceanography	3	Polar sun sync	Color photography, IR mapper, Scatterometer, Spectrophotometer, Radiometer, etc.
All 5 disciplines	4	Geostationary	Data collection from balloons, buoys, gauges, etc.; also, spin-scan color camera, IR imager and sounder

[a]FAG = Forestry, Agriculture, Geography

Table 2 Communications requirements in late 1970's

User	Peak requirements	
	From terminal	To terminal
Earth resources disciplines		
E.R. satellite	Video/FM or 100 Mbit/sec; 150 kbit/sec	1 kbit/sec
In situ sensors	25 kbit/sec	1.2 kbit/sec
Other users		
Explorer type satellite	20 kbit/sec	1 kbit/sec
Astronomical satellite (wideband sensors)	100 Mbit/sec	1 kbit/sec
Space station	1 color TV channel; 120 kbit/sec; 10 voice channels	1 color TV channel; 120 kbit/sec; 10 voice channels
NASA stations	5 Mbit/sec	60 kbit/sec
Small users	30 kbit/sec	10 kbit/sec

III. System Considerations

General

In the early phases of the ERS program, the ERS sensor data will be delivered to a central ES in the United States on an intermittent basis by first dumping the data to a few NASA or ESSA stations and then using radio (or flying the tapes) to transmit the data to the U.S. Another mode of operation is to interconnect Earth stations via commercial communications satellites. This mode is currently employed to meet NASA's needs for communication with orbiting spacecraft via overseas stations and INTELSAT II or III satellites. In this mode, however, the ERS does not have direct access to the relay satellite, nor is there continuous contact between the central ES and the ERS.

A communications system capable of relaying data from an Earth resources satellite via a medium-altitude or synchronous-altitude relay satellite to the ground requires considerably more sophistication and power than a communications system which links the Earth resources satellite directly with the ground. The reasons are illustrated in Table 3, which contrasts the two approaches. The table shows that the transmitter in the Earth resources satellite must send signals over considerably longer distances to receivers with smaller antennas and higher system temperatures when the relay approach is used than when the direct approach is used. If we assume that for either approach, the transmitter power is limited to 25 w, the required antenna gain for the relay approach is 46 db against 8 db for the direct approach. The result is that, for satellite relay, the Earth resources satellite will need an antenna diameter of 100λ, producing a beamwidth of approximately 1°, in contrast with the 1-λ antenna having 90° beamwidth that can be used for direct-to-ground transmissions. If we further assume that the antenna diameter is limited to two meters, the minimum frequency for the relay approach is 15 GHz as opposed to 150 MHz in the direct approach.

Offsetting all the disadvantages of the relay approach enumerated above, we have the following advantages: the relay approach provides continuous instead of intermittent system availability; it provides global instead of local coverage; the service is real-time instead of delayed; and on board video recording is eliminated.

Table 3 Direct vs relay satellite

Parameter	Direct	Relay
Distance to receiver	2000 km	40,000 km
Receiver antenna diameter	10 m	2.5 m
Receiver system temperature	300°K	2400°K
Antenna gain for 25-w transmitter	8 db	46 db
Transmitter antenna diameter	λ	100λ
Transmitter beamwidth	90°	1°
Minimum frequency for 2-m antenna	150 MHz	15 GHz
System availability	Intermittent	Continuous
System coverage	Local	Global
Type of service	Delayed	Real-time
Onboard storage	Needed	Not needed

The Feasibility of Using INTELSAT IV

It is instructive to examine the feasibility of using ERSs specifically designed for operation with the INTELSAT IV series, which will be in service from 1971 until the late 1970's. The INTELSAT IV technical characteristics are given in Table 4. The satellite operates in C band (6-GHz receive and 4-GHz transmit) with either global-beam or spot-beam antennas. It has 12 transponders, each with a 36-MHz bandwidth and 6.3 w of saturated rf power.

The standard Earth stations for the INTELSAT system have parameters shown in Table 5.

For the proposed operation, the ERS transmits in the 6-GHz band to the INTELSAT IV, which retransmits the signal to Earth in the 4-GHz band; in the opposite direction the transmissions are in the 6-GHz band from Earth to RS and in 4-GHz band from RS to ERS. Thus, the INTELSAT IV would be used without any modification to serve as a relay satellite for the ERS, while simultaneously carrying other commercial traffic.

Since the INTELSAT satellites are designed for operation with large Earth stations, their use in conjunction with a small terminal such as the ERS entails a heavy penalty. Thus assume the ERS has the parameters listed in Table 6. Then the ERS has a transmit e.i.r.p. 57 db less than an INTELSAT Earth station, and a receive G/T which is 34.7 db less than that of the ES, while the distances of these two terminals from the INTELSAT IV are essentially the same. Moreover, the total gain in the amplifying stages of the INTELSAT IV transponder

Table 4 INTELSAT IV parameters

Diameter	8.9 ft
Height	18 ft
Weight	1200 lb
d.c. power	500-600 w
Number of transponders	
Global only	4
Spot or global	8
Frequency bands	
Transmit	4 GHz
Receive	6 GHz
Half-power beamwidths	
Global	17°
Spot	4.5°
Beam-pointing accuracy (3σ)	±0.35°
Usable bandwidth per transponder	36 MHz
TWT rf power per transponder	6.3 w (8 dbw)
Antenna gain of global beam, beam center	19.5 db[a]
e.i.r.p. available per transponder with global beam	
Beam center	25.5 dbw[a]
Beam edge	22.5 dbw[a]
9.8° off-axis	21.0 dbw
Receive G/T	
Beam edge	-17.1 db/°K
9.8° off-axis	-18.6 db/°K
Minimum on-axis flux density needed at transponder to obtain saturated rf output power with maximum amplifier gain	-73 dbw/m^2

[a]The gain and e.i.r.p. are increased by 11.5 db if a spot beam is used.

Table 5 Standard INTELSAT Earth station parameters

rf power	5 kw at 6 GHz
Antenna size	97 ft diameter
Antenna gain, G	63 db at 6 GHz
Receive system noise temperature, T	77°K
Receive G/T	40.7 db/°K at 4 GHz
Transmit e.i.r.p.	100 dbw

Table 6 Earth resources satellite parameters

rf power	25 w
Antenna	2 m
Receiver	Tunnel diode amplifier or transistor (uncooled)
Receive G/T	6 db/°K at 4 GHz
Transmit e.i.r.p.	53 dbw

is limited to a value which is appropriate to INTELSAT's current missions, but inadequate for optimizing communications between an ERS and an INTELSAT Earth station. Transmission of a good quality TV picture requires a total carrier-to-noise-density ratio of C/N_o = 90 db-Hz for the two cascaded links, ERS-RS and RS-ES. With the parameters shown in Tables 4 and 5, this requires an ERS e.i.r.p. on the order of 81 dbw at 6 GHz; Table 6 shows that an e.i.r.p. of only 53 dbw is available. It follows that TV cannot be transmitted from ERS to ground via an INTELSAT IV.

The difference between required and available e.i.r.p. at the ERS is so large that slight modifications of INTELSAT IV cannot solve the problem. Major modifications are needed to permit operation with a much larger antenna and/or at a much higher frequency. Such modifications may be available in some future INTELSAT series.

Although an unmodified INTELSAT IV is incapable of providing a TV link between an ERS and the ground, analysis shows that it can easily support high-quality data links operating at 12,000 bit/sec in each direction. Together, these channels would occupy about 3 MHz in one INTELSAT IV transponder and would require about 1 w of rf power. The bandwidth requirement is imposed by the fact that the 25-w carrier, a portion of which is radiated by the ERS through its antenna sidelobes, must be spread over this amount of spectrum to comply with CCIR restrictions on flux densities radiated by satellites. The 1-w carrier from the INTELSAT IV would permit the ERS antenna to track the INTELSAT IV as well as to receive command transmissions. The 12,000-bit/sec telemetry from the ERS would be relayed to the ES using an INTELSAT IV carrier power less than 1 mw.

The ERS transmitting frequency must be chosen to avoid uplink interference from other terrestrial radio transmitters. The regulations allow such transmitters to radiate up to 47 dbw e.i.r.p. in the direction of the synchronous orbit. Since this is only 6 db below the assumed ERS e.i.r.p., the ERS could not operate on frequencies employed at this power level by terrestrial radio relay transmitters aimed at the synchronous relay satellite.

The command receiver in the ERS will occasionally be subjected to severe interference from terrestrial radio relay transmitters. This type of interference will occur when the ERS passes through the main beam of a radio relay transmitter operating on the command frequency and will be received through sidelobes of the ERS antenna. Care must be taken to render such interference harmless by suitable encoding of command transmissions.

Telecommunication with in situ data collection platforms (DCP) via an INTELSAT IV relay satellite is also feasible, provided a suitable frequency channel is used and the platform has an e.i.r.p. of at least 24 dbw in the 4-GHz band. With a 1-w carrier and less than 100 kHz of bandwidth in the INTELSAT IV, a duplex channel operating at 10 bit/sec can be supported. In order to comply with the CCIR flux-density limitation, the 10-bit/sec command link to the DCP must be spread over a spectral bandwidth of at least 25 kHz. This corresponds to a spectral spreading ratio of 2800 and entails an undesirable degree of complexity in the DCP. However, the whole telecommand problem can be circumvented by the use of unregimented transmit-only DCPs.

New Types of RS Systems

A new system using relay satellites may deploy the RSs in medium-altitude or geostationary-altitude orbits. The former altitude has the following advantages over the latter: a) higher payload capability for a given booster, b) smaller path losses on both the RS-ERS and RS-ES links, and c) fewer handovers between RSs on the ERS-RS link (26 handovers per day for 8600 statute-mile altitude vs 30 for geostationary latitude).

The medium-altitude system has the following disadvantages as compared to the geostationary system: a) The ES and other terrestrial users need complex tracking antennas to track the RS. It is difficult for the in situ sensor platforms to meet this requirement. b) The proximity to Earth makes the RS more liable to interference from terrestrial sources. Note also that according to CCIR recommendations, these sources may radiate only 47 dbw towards the geostationary orbit, but may radiate up to 55 dbw in other directions. The increased interference at the RS makes it necessary to provide a bigger e.i.r.p. in the ERS and the ES. c) The medium-altitude system cannot share a frequency band with existing geostationary systems since the line of sight between ES and RS (and also ERS and RS) would occasionally intersect the geostationary orbit and cause interference. The disadvantages of the medium-altitude system outweigh the advantages; therefore, this system is not considered further.

A simple system using two geostationary RSs is shown in Fig. 1. The two RSs are 146° apart with a ± 3° station-keeping tolerance. For the maximum separation of 152°, both of them see the same ES with a 5° elevation angle. For the minimum separation of 140°, full coverage of the ERS orbit is obtained for orbit heights of 440 statute miles or more, as seen from the curve of Fig. 2. The system suffers from the drawback that there is a coverage gap on the surface of the Earth which is about 60° wide at the equator; the RSs cannot collect data from in situ sensors located there (or at the polar caps). A spare RS is located midway between the others.

Another system using three geostationary RSs roughly 120° apart is shown in Fig. 3. Two ESs are used, one in the U.S. and the other in the Atlantic (or Pacific) regions. One of the RSs is used to connect the overseas ES with the United States ES. A spare RS is collocated with this RS in the same ES antenna beam.

Several other system configurations are possible with two, three or four RSs in geostationary orbit, but the two systems just described are taken as typical.

The RSs may be expected to have a 5- to 7-yr lifetime. Thus, spares are not needed if the ERS program requires service for five years or less.

Interference Constraints

The relay satellite system must be coordinated in frequency and space with existing terrestrial and satellite systems. The problem is complex if the RS system is "dedicated" exclusively to the ERS program, and operates in a shared frequency band between 1 and 10 GHz. In such a case it is necessary to consider possible interferences to and from INTELSAT, TACSAT, and other satellite systems, as well as to and from Earth stations and terrestrial radio relay links.

The situation is simpler if the RS system is "shared" with other users so that the same RS and ES serve both the ERS program as well as other (commercial) users. Then the interference environment in the 4- and 6-GHz bands is that shown in Fig. 4.

Interference A is caused by terrestrial radio relay links at the ERS receiver. Terrestrial radio relay transmitters are permitted to radiate as much as 55 dbw of e.i.r.p. into space. This necessitates the provision of high e.i.r.p. in the RS. High RS e.i.r.p. illuminates the Earth with a flux density which must not exceed maximum levels permitted by CCIR (see interference B); therefore, it is necessary to spread the RS signal over a suitably wide bandwidth. For grazing incidence, the flux density of satellite signals must not exceed -152 dbw/m^2 per 4 kHz of bandwidth.

Interference C is caused by terrestrial radio relay transmission entering the RS receiver. It results in the requirement of a high e.i.r.p. in the ERS.

Interference D is caused by radiation from the ERS transmitter through the sidelobes of the ERS antenna into the main lobe of a terrestrial radio relay receiver. This interference must be held to an acceptably low level by using sufficient bandwidth to comply with existing flux-density limitations. Some easing of this problem can be derived by careful design of the ERS antenna to reduce sidelobe radiation.

The quantitative effects of these four interferences depend on the system geometry, relative antenna directivities, locations and strengths of the interference sources, and the exact frequencies involved. Several other interference entries are possible,[9] but they are negligible compared to those discussed above.

IV. Example of a Shared Relay Satellite System

Figure 5 shows the signal flow for a conceptual multifunctional relay satellite. In this concept, three frequency bands are employed. L band is employed for communication to and from in situ sensors as well as low-data rate communications to and from ships, aircraft, and low-orbiting satellites. The conventional C-band frequencies of 4 and 6 GHz are used to link the satellite with a standard INTELSAT Earth terminal. The satellite also receives imagery transmissions from Earth resources satellites and/or manned space stations at 15 GHz. To facilitate tracking at the Earth resources satellite, the relay satellite transmits a beacon at 4 GHz in the direction of the Earth resources satellite.

In this concept, the relay satellite employs four antennas. Two of these are conventional C-band antennas with global coverage, one for reception, the other for transmission. The L-band antenna is a phased array with multiple switched beams. This array is capable of both transmission and reception. A dot-beam antenna which is steerable over the Earth's face is also employed. This antenna transmits at 4 GHz and receives at 15 GHz. More detail of this system is shown in Fig. 6, which includes some of the principal link parameters. The system can relay 100 Mbit/sec digital TV from the ERS, and 2400 bit/sec data to and from DCPs. The DCPs need 10 w rf power with a 0 db gain antenna.

A block diagram of the relay satellite is shown in Fig. 7 and an illustration of a possible spacecraft configuration appears in Fig. 8. On the right of Fig. 8, the spacecraft is shown in the stowed position with the solar panels retracted; the phased array is folded around the box structure, and the umbrella type paraboloidal reflector is furled. This configuration would use two mutually perpendicular solar panels extended by masts pointing north and south. The panels would be nonrotating, and both sides would be active. The required d.c. power for the satellite would be 1-2 kw.

Table 7 shows a weight breakdown for the relay satellite. The weights listed include prorated shares of the power supply weights. Such a satellite could be launched with an Atlas-Centaur launch vehicle. The table shows that one third of the communication payload is utilized by the Earth resources application, while the other two thirds is available for other applications such as international telephony, maritime mobile service, etc.

Table 7 Relay satellite weights breakdown

System	No. on board	No. for ERS	Weight on board	Weight for ERS
L-band transponders	9	1	180 kg	20 kg
C-band transponders	6	2	135 kg	45 kg
K-band receiver	1	1	5 kg	5 kg
Dot-beam antenna	1	1	55 kg	55 kg
Total communications payload			375 kg	125 kg
Housekeeping system			125 kg	
Total satellite			500 kg	

It is not unreasonable to base a cost-sharing formula on the weight fraction of the relay satellite. Assuming a rather expensive R & D program leading to the launching of four satellites with a 75% probability of launch success, the estimated annual cost for such a service might be on the order of $20 million/yr.

Table 8 shows the estimated power and weight for the equipment which must be placed on board the Earth resources satellite. The equipment would weigh approximately 50 kg and consume 150 w of d.c. power. A suitable location would have to be found to deploy a two-meter diameter paraboloidal antenna capable of pointing in any direction away from the Earth. On the Earth resources satellite, the weight increase required to accommodate the transmitting equipment would be partially compensated for by the elimination of video tape recorders.

We conclude that communication via a shared commercial satellite relay can fulfill the data rate requirements of an Earth resources satellite and in situ data collection platforms at a reasonable cost and thereby provide the advantages of continuous real-time global communications.

Table 8 ERS power and weight

d.c. power	150 w
Antenna diameter	2 m
Communication system weight (transmit & receive equipment, antenna assembly, pointing subsystem)	50 kg

Appendix. Supporting Calculations

Nomenclature

C/I = carrier-to-interference ratio

C/N_o = ratio of received carrier-power to noise-power spectral density (Hz)

d_{ir} = distance from interference source to receiver (m)

d_{ti} = distance from transmitter to receiver of interference (m)

d_{tr} = distance from transmitter to receiver (m)

E/N_o = ratio of received energy per bit to noise power spectral density

F = transponder gain correction factor

G_{rt} = receiving antenna gain in direction of transmitter

G_t = maximum transmitting antenna gain

G_{tr} = transmitting antenna gain in direction of receiver

k = Boltzmann's constant ($= 10^{-22.86}$ w/Hz°K)

K/K_{max} = transponder gain relative to maximum transponder gain

L = free-space path loss

P = radiated carrier power (w)

P_i = equivalent isotropically radiated interference power in the direction of the receiver (w)

P_{sat} = saturated transponder output power (w)

R = data rate (bits/second)

S/N_o = ratio of received signal-power to noise-power spectral density (Hz)

T = receiving system noise temperature (°K)

W_{min} = minimum spectral bandwidth satisfying CCIR flux-density limitations (Hz)

Φ = received power flux density (w/m^2)

Φ_{max} = maximum power flux spectral density allowed by CCIR regulations ($w/m^2 Hz$)

Φ_n = power flux density received from direction of maximum gain which is required to saturate transponder when transponder gain is set to maximum value (w/m^2)

λ = wavelength (m)

DCP = data collection platform

ERS = Earth resources satellite

ES = Earth station

INT = INTELSAT IV

RS = relay satellite

TRR = terrestrial radio relay

Formulas

For each link the receive carrier-to-noise power density ratio is given by the relation

$$C/N_o = G_{tr}G_{rt}/LkT$$

where

$$L = (4\pi d/\lambda)^2$$

The flux density at the receiver is

$$\Phi = PG_{tr}/4\pi d^2$$

and the transponder output carrier power is

$$P = \Phi G_{rf} KFP_{sat}/\Phi_n G_r K_{max}$$

The value of the gain correction factor F is in the range between 0.25 and 2.5, and is determined by the statistical properties of all the carriers (including interfering ones) in the transponder and by the nonlinear characteristics of the transponder. Therefore, F can be accurately determined only from a complete description of these properties and characteristics. For the purpose of the present analysis estimated values of F are used.

The signal-to-noise power density ratio for a channel composed of an up-link to the RS, a nonlinear transponder, and a down-link from the RS is determined from the two link carrier-to-noise power density ratios by the following approximation:

$$1/(S/N_o) = 2/(C/N_o)_{up} + 1/(C/N_o)_{down}$$

In a precise analysis, the interference resulting from intermodulation distortion in the relay satellite transponder, which is determined by the carrier arrangement and backoff used, must be computed for each specific arrangement. These computations are not in the scope of the present study, which assumes that intermodulation products can be limited to sufficiently low levels to make them equal to the up-link noise.

The formula for E/N_o is

$$E/N_o = (S/N_o)/R$$

The carrier-to-interference ratio in the down-link to the ERS (type A interference) is approximately

$$C/I = PG_{tr}G_{rt}d_{ir}^2/P_i d_{tr}^2$$

The exact value depends on the receiving antenna pattern and the angle of arrival of the interference, which is variable. The above formula assumes that the receiving antenna gain in the direction of the interference equals unity.

The carrier-to-interference ratio in the up-link (type C interference) is

$$C/I = PG_{tr}/P_i$$

The maximum values of P_i permitted by the CCIR are assumed. These values are 50 kw in the direction of the geostationary arc and 320 kw in other directions.

To comply with CCIR limitations which guard against interference from the satellite, spectral band spreading is required over a minimum bandwidth of

$$W_{min} = PG_t/4\pi d_{ti}^2 \cdot \Phi_{max}$$

The value for d_{ti} for type B interference is taken as $4x10^7$ m. The value of Φ_{max} depends on satellite elevation angle. For simplicity, we apply the minimum value of Φ_{max}, which equals $10^{-18.8}$ w/m^2Hz. For type D interference the following approximation is used:

$$W_{min} = P/4\pi d_{ti}^2 \cdot \Phi_{max}$$

This approximation again assumes unity gain for the ERS antenna in the direction of the interference receiver. For d_{ti} we use $2x10^6$ m in this case. This choice is a compromise between the satellite altitude and the satellite-to-horizon distance which takes into account a reasonable antenna sidelobe envelope.

System Calculations

It is possible to determine the system parameters of interest by using the formulas given previously. The calculations are performed conveniently in the form of tables.

Table 9 shows the C/N_o performance of the up-link for assumed values of transmitter power and antenna gain. The resulting flux density at the transponder produces a power output P, which results in the indicated C/N_o performance of the down-link. Combining the C/N_o values for the up- and down-links, one obtains the over-all signal to noise density ratio S/N_o. Assuming certain values for E/N_o (corresponding to the desired probability of error), this leads to the data rate that can be supported by the over-all channel.

Table 10 shows similar calculations when the INTELSAT is replaced by a conceptual relay satellite.

Tables 11-14 show interference calculations, i.e., the C/I resulting from type A and type C interference, and the minimum band spread needed to comply with CCIR flux-density constraints for type B and type D interference.

Tables 15 and 16 summarize the system performance using an unmodified INTELSAT IV or a conceptual relay satellite, respectively.

Table 9 Links using unmodified INTELSAT IV

Channel						
Channel	From		ES	ERS	ES	DCP
	Via		INT	INT	INT	INT
	To		ERS	ES	DCP	ES
Up-link	P	(dbw)	27	14	27	14
	G_{tr}	(db)	63	39	63	10
	G_{rt}/T	(db/°K)	- 17.1	- 18.6	- 17.1	- 17.1
	1/L	(db)	-199.5	-199.5	-199.5	-199.5
	1/k	(db°K-Hz/w)	228.6	228.6	228.6	228.6
	$(C/N_o)_{up}$	(dbHz)	102.0	63.5	102.0	36.0
Transponder	PG_{tr}	(dbw)	90	53	90	24
	$1/4\pi d^2$	(db/m²)	-163	-163	-163	-163
	G_{rt}/G_r	(db)	- 3	- 4.5	- 3	- 3
	$1/\Phi_n$	(dbm²/w)	73	73	73	73
	KF/K_{max}	(db)	- 4	- 2	- 4	- 2
	P_{sat}	(dbw)	8	8	8	8
	P	(dbw)	1	- 35.5	1	- 63
Down-link	P	(dbw)	1	- 35.5	1	- 63
	G_{tr}	(db)	15	16.5	15	16.5
	G_{rt}/T	(db/°K)	6	40.7	- 24.8	40.7
	1/L	(db)	-196	-196	-196	-196
	1/k	(db°K-Hz/w)	228.6	228.6	228.6	228.6
	$(C/N_o)_{down}$	(dbHz)	54.6	54.1	23.8	26.8
Over-all channel	$(C/N_o)_{up}$	(dbHz)	102.0	63.5	102.0	36.0
	$(C/N_o)_{down}$	(dbHz)	54.6	54.1	23.8	26.8
	S/N_o	(dbHz)	54.6	53.1	23.8	25.8
	E/N_o	(db)	13.8	12.3	13.8	15.8
	R	(dbbit/sec)	40.8	40.8	10	10

Table 10 Links using conceptual RS

			ES	ERS	ES	DCP
Channel	From		ES	ERS	ES	DCP
	Via		RS	RS	RS	RS
	To		ERS	ES	DCP	ES
Up-link	P	(dbw)	30	14	14	10
	G_{tr}	(db)	63	47.5	63	0
	G_{rt}/T	(db/°K)	- 17.1	15.7	- 17.1	- 3.5
	1/L	(db)	-199.5	-208.0	-199.5	-188.5
	1/k	(db°K-Hz/w)	228.6	228.6	228.6	228.6
	$(C/N_o)_{up}$	(dbHz)	105.0	97.8	91.0	46.6
Transponder	PG_{tr}	(dbw)	93	61.5	77	10
	$1/4\pi d^2$	(db/m^2)	-163	-163	-163	-163
	G_{rt}/G_r	(db)	- 3	0	- 3	0
	$1/\Phi_n$	(dbm^2/w)	73	92.5	89	96
	KF/K_{max}	(db)	0	- 4	0	- 4
	P_{sat}	(dbw)	8	16.5	7	16.5
	P	(dbw)	8	3.5	7	- 44.5
Down-link	P	(dbw)	8	3.5	7	- 44.5
	G_{tr}	(db)	37.9	27	28	27
	G_{rt}/T	(db/°K)	6	40.7	- 28.5	40.7
	1/L	(db)	-196	-196	-187.5	-196
	1/k	(db°K-Hz/w)	228.6	228.6	228.6	228.6
	$(C/N_o)_{down}$	(dbHz)	84.5	93.8	47.6	55.8
Over-all channel	$(C/N_o)_{up}$	(dbHz)	105.0	97.8	91.0	46.6
	$(C/N_o)_{down}$	(dbHz)	84.5	93.8	47.6	55.8
	S/N_o	(dbHz)	84.5	92.3	47.6	45.9
	E/N_o	(db)	14.5	12.3	13.8	12.1
	R	(dbbit/sec)	70	80	33.8	33.8

Table 11 Type A interference

Link		INT to ERS	RS to ERS
Interference		TRR to ERS	TRR to ERS
PG_{tr}	(dbw)	16	45.9
G_{rt}	(db)	36	36
$(d_{ir}/d_{tr})^2$	(db)	-26	-26
$1/P_i$	(db/w)	-55	-55
C/I	(db)	-29	0.9

Table 12 Band spreading to avoid type B interference

Link		INT to ERS	INT to DCP	RS to ERS	RS to DCP
Interference		INT to TRR	INT to TRR	RS to TRR	RS to TRR
P	(dbw)	1	1	8	7
G_t	(db)	19.5	19.5	37.9	30
$1/4\pi d_{ti}^2$	(db/m^2)	-163	-163	-163	-163
$1/\Phi_{max}$	(dbm^2Hz/w)	188	188	188	188
W_{min}	(dbHz)	44.5	44.5	70.9	62

Table 13 Type C interference

Link		ERS to INT	DCP to INT	ERS to RS	DCP to RS
Interference		TRR to INT	TRR to INT	TRR to RS	TRR to RS
PG_{tr}	(dbw)	53	24	61.5	10
$1/P_i$	(db/w)	- 47	- 47	- 47	- 47
C/I	(db)	6	- 23	14.5	- 37

Table 14 Band spreading to avoid type D interference

Link		ERS to INT	ERS to RS
Interference		ERS to TRR	ERS to TRR
P	(dbw)	14	14
$1/4\pi d_{ti}^2$	(db/m^2)	-137	-137
$1/\Phi_{max}$	(dbm^2Hz/w)	188	188
W_{min}	(dbHz)	65	65

Table 15 System summary using unmodified INTELSAT IV

Channel	To ERS	From ERS	To DCP	From DCP
Frequency	4 GHz	6 GHz	4 GHz	6 GHz
ERS or DCP power	...	25 w	...	25 w
ERS or DCP antenna	2 m	2 m	10 db	10 db
Data rate	12 kbit/sec	12 kbit/sec	10 bit/sec	10 bit/sec
Minimum bandwidth[a]	28 kHz	3 MHz	28 kHz	...
C/I	-29 db	6 db	...	-23 db
Remarks	Occasional strong type A interference	Mild type C interference at some frequencies and orbital longitudes; 250:1 band spreading required; TV not feasible	2800:1 band spreading required	Strong type C interference at some frequencies and orbital longitudes

[a]Subject to CCIR constraint on Earth surface flux density.

Table 16 System summary using conceptual RS

Channel	To ERS	From ERS	To DCP	From DCP
Frequency	4 GHz	15 GHz	1.5 GHz	1.6 GHz
ERS or DCP power	...	25 w	...	10 w
ERS or DCP antenna	2 m	2 m	0 db	0 db
Data rate	10 Mbit/sec	100 Mbit/sec	2400 bit/sec	2400 bit/sec
Minimum bandwidth[a]	12.3 MHz	3 MHz	1.6 MHz[b]	...
C/I	0.9 db	14.5 db	...	-37 db
Remarks	Occasional type A interference	Essentially interference-free; TV feasible	[b]670:1 band spreading required, if flux limit applied	Frequency coordination with other L-band users necessary to avoid type C interference

[a]Subject to CCIR constraint on Earth surface flux density.

References

[1]Reports of Panels 1-5 and Summary Report, National Academy of Sciences, Summer Study on Useful Applications of Earth Oriented Satellites, Contract NSR 09-012-909, 1967, NASA.

[2]Dorfman, S., "Collection of Data from in-situ Sensors via Satellite," Aug. 1969, Session 12, Western Electronic Show and Convention, San Francisco, Calif.

[3]Heffernan, P. and Gilchriest, C., "A Multiple-Access Satellite Relay System for Low Data Rate Users," Aug. 1969, Session 12, Western Electronic Show and Convention, San Francisco, Calif.

[4]Barritt, P. and Andrus, A. G., "Tracking and Data Relay Satellite Systems," Session V, 1969 International Telemetry Converence, Washington, D.C.

[5]Stampfl, R. and Jones, A., "Tracking and Data Relay," Preprint X-701-69-420, Oct. 1969, NASA-Goddard Space Flight Center; also, IEEE Transactions on Aerospace and Electronic Systems, Vol. AES-6, No. 3, May 1970.

[6]Durrani, S. H., Hanell, S. H., and Welti, G. R., "Data Relay Service Using Satellites," TM CL-24-69, April 1969, COMSAT Laboratories, Clarksburg, Md.

[7]Welti, G. R., "System Concepts for Global Data Relay via Satellite," 1969 IEEE International Convention Digest, Institute of Electrical and Electronics Engineers, March 1969.

[8]Welti, G. R., "Satellite Communications for Small Users," AIAA Paper 69-1073, Anaheim, Calif., 1969.

[9]Durrani, S. H., "Relay Satellite System for the Space Station Program," Tech. Report CL-TR-1970-1, March 1970, prepared for McDonnell Douglas Astronautics Co. by COMSAT Corp., Clarksburg, Md.

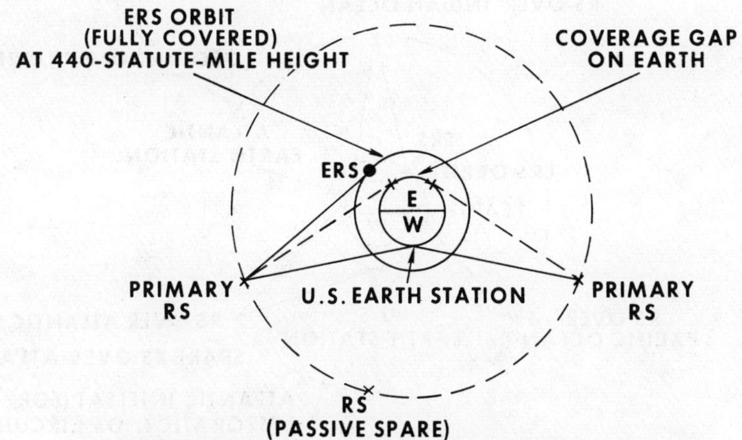

Fig. 1 System using two primary geostationary relay satellites and a single Earth station.

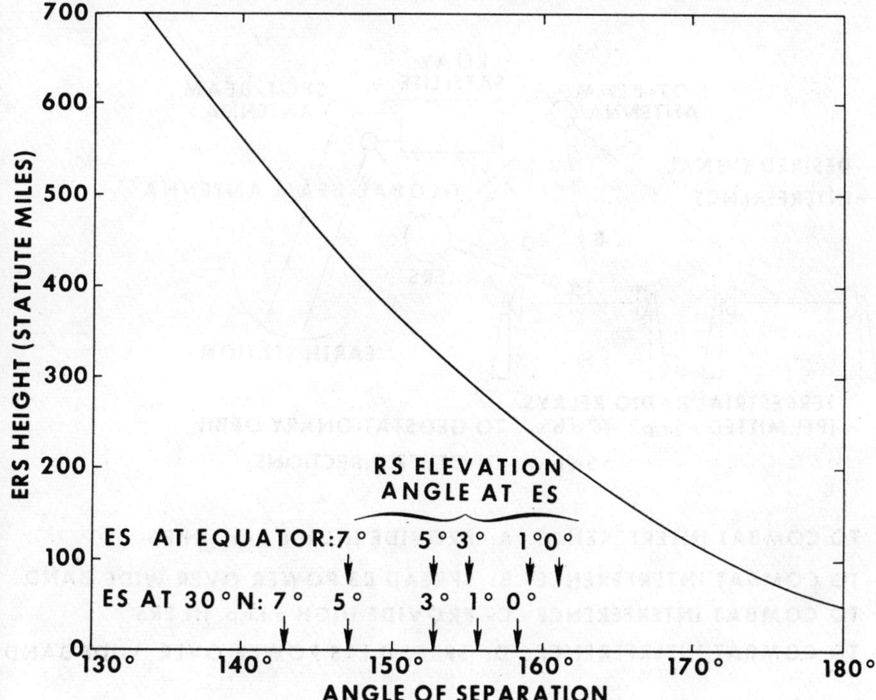

Fig. 2 Angle of separation between two geostationary relay satellites for 100% coverage of ERS orbit.

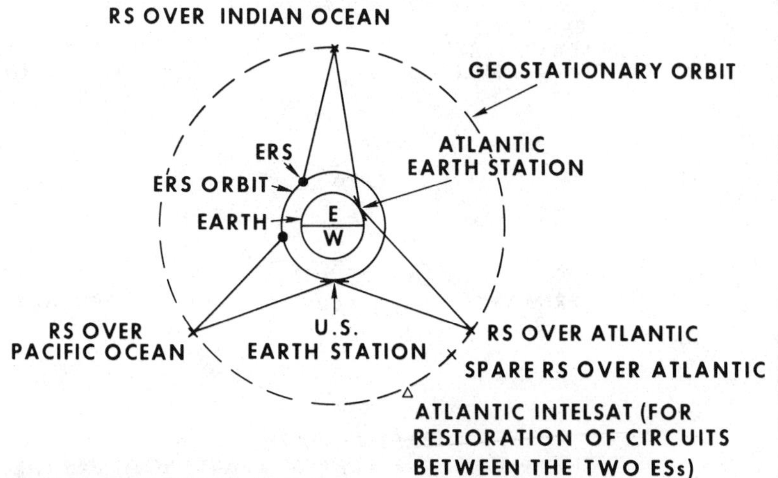

Fig. 3 System using three primary geostationary relay satellites and two Earth stations.

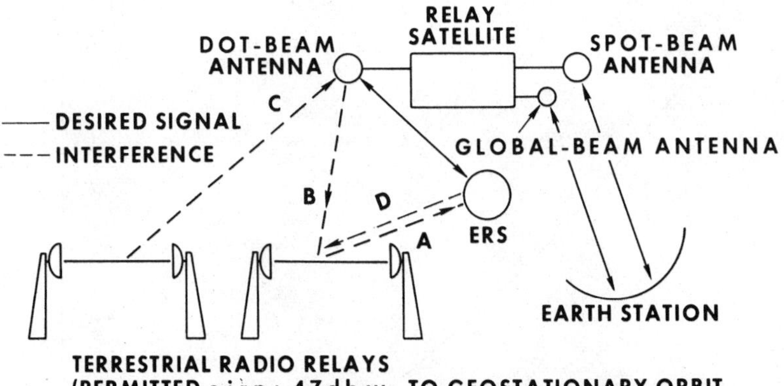

Fig. 4 Interference considerations.

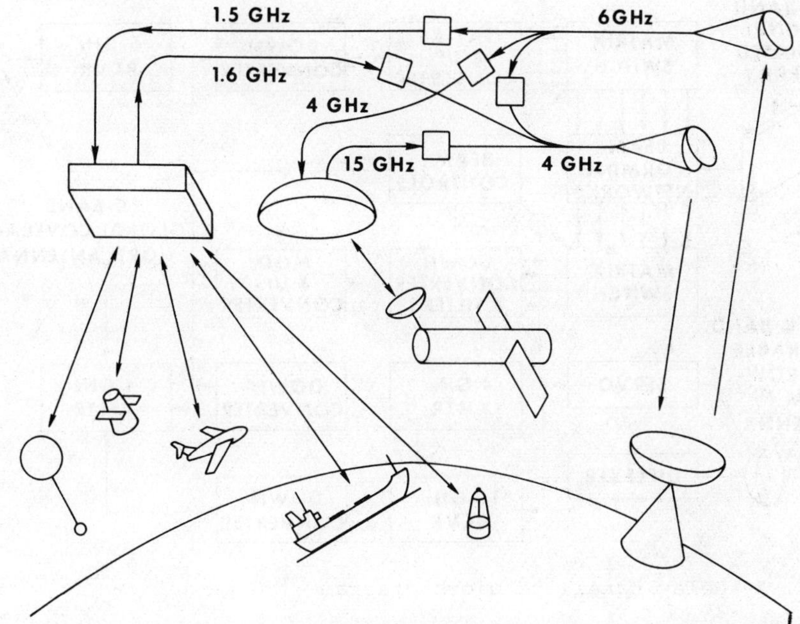

Fig. 5 Signal flow for multifunctional relay satellite.

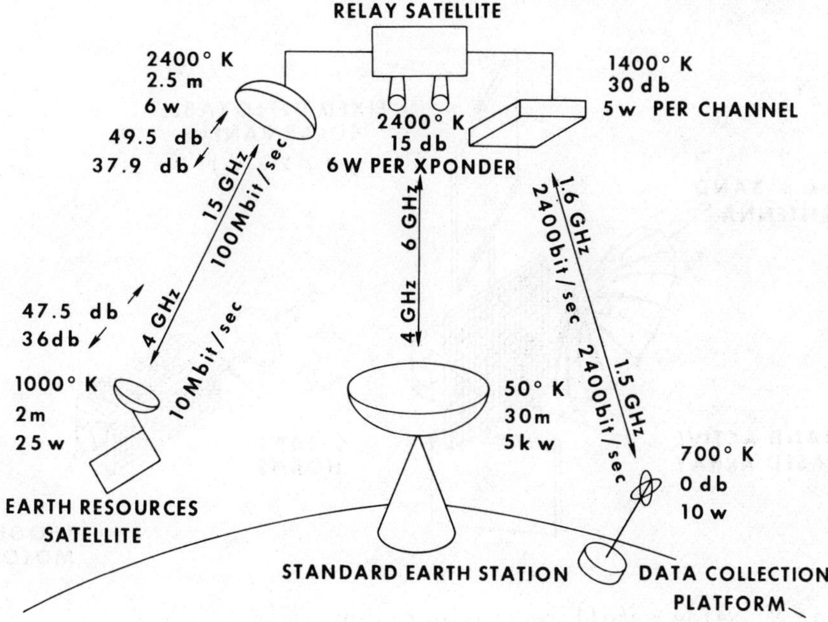

Fig. 6 Link parameters.

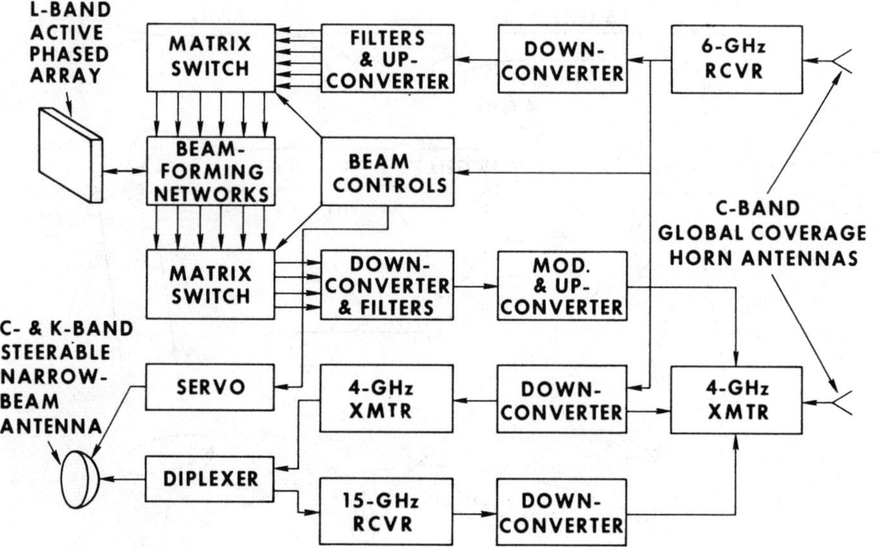

Fig. 7 Relay satellite block diagram.

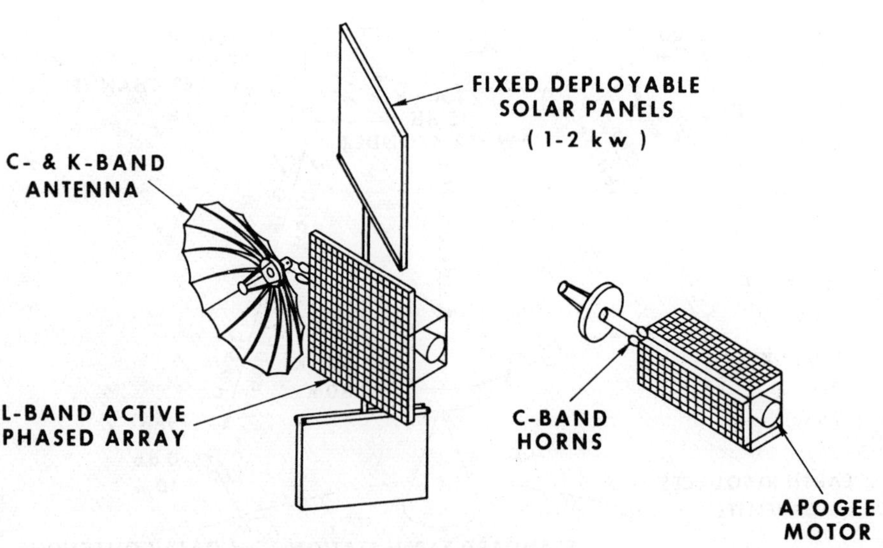

Fig. 8 Relay satellite configuration.

EARTH RESOURCES DATA COLLECTION BY COMMUNICATION SATELLITE: PROBLEM, SCOPE, AND SOME POTENTIAL SOLUTIONS

G. M. Northrop*

The Center for the Environment and Man, Inc., Hartford, Conn.

Abstract

Earth resources data are anticipated to come primarily from two general sources: 1)thousands of widely distributed data collection platforms on or near the surface of the earth, and 2)a small number of Earth-orbiting satellites, usually in low-altitude orbit. The sensor platforms will provide data of use in meteorology, oceanography, hydrology, agriculture, forestry, and geology. The Earth-based sensor platforms may be fixed or mobile on land and in water, moored in the ocean, or freely drifting in the atmosphere or ocean. The Earth-orbiting satellites will likely be capable of providing continuous, wideband data from low altitude and geostationary orbit. This paper outlines some facets of the expected diverse data collection requirements for Earth resources data and some potential solutions for meeting these requirements within the scope of the communication satellite technology of the early 1970s.

1. Introduction

The decade of the seventies will see major advances in the use of satellite communications. One very important area where a new field of endeavor will develop is in the collection and relaying of Earth resources data. These data may be acquired remotely from satellite platforms or directly by in situ data

---

Presented as Paper 70-327 at the AIAA Earth Resources Observations and Information Systems Meeting, Annapolis, Md., March 2-4, 1970, and also at the AIAA 3rd Communications Satellite Systems Conference, Los Angeles, Calif., April 6-8, 1970. This paper is based on the report of the Points-to-Point Communication Panel of the NAS/NRC Summer Study on Space Applications performed for NASA under Contract NSR 09-012-909 (1967). Richard Marsten was Chairman of the Panel. S.D. Dorfman, A.R. Garfinkel, A. Valakis, and G.M. Northrop were Panel Members.

*Director, Environmental Monitoring Systems.

collection platforms. In either event, data relay by satellites to regional ground data handling centers appears to be both feasible and cost effective when compared with alternative means.

The perishability of Earth resources data hardly can be measured by 1970 standards. Much environmental information now being collected is slow in being returned for processing and, hence, is used primarily for research or in a climatological sense. Around the world there is need for more information available synoptically in the form of weather reports, stream and tide levels, Earth moisture content, earthquake and volcanic activity and oceanographic characteristics. Availability of such synoptic data in increased volume over broader areas, or, as appropriate, in greater density doubtless will lead to improved and new forecasting techniques, warning systems, and real-time uses of Earth resources information.

Certainly, many years of work with limited capability experimental Earth resources data-collection-and-relay satellites and much research involving data analysis and data product use must be accomplished before a truly satisfactory (and internationally acceptable) operational system evolves for Earth resources data collection, processing, and dissemination. It is likely, however, that by the end of the decade of the seventies, a system will be operational, comprising remote and in situ data collection from satellites in low altitude orbit and geostationary orbit. It is also possible that such a system will include the use of men operating in orbital space stations. (This is not to suggest that an Earth resources program would hinge on the availability of man in space. Indeed, this author believes that, once space stations have been developed for other purposes, then man will find his place in the space element of the Earth resources system. This is because of his capability to repair equipment, optimize the use of time and bandwidth, and flexibly discriminate desired information from unwanted signals and noise and also because the other reasons for establishing man in space are quite compatible with the aims of an Earth resources program.)

In the United States, interest in Earth resources data is shared by the Departments of Commerce, Agriculture, and Interior. Similar interests exist in all the major nations of the world. In fact, at least 14,000 Earth-based data collection platforms (DCPs) existed in 1970. With the addition of Earth-to-satellite communications, these DCPs could form a substantial basis for synoptic reporting systems for weather and hydrology. (The World Weather Watch presently makes use of land communication networks to channel synoptically collected data into the three world meteorological centers in Washington,D.C. Moscow, and Melbourne.) Other nations of the world can be expected to show similar interest in other environmental monitoring systems that aid their agriculture and commerce and to provide warnings of natural disasters, such as earthquakes, volcanic eruptions, or tsunamis.

The National Academy of Sciences Summer Study on the Useful Applications of Earth Oriented Satellites[1] undertook to address in detail the many ways that satellites can be of use in advancing forestry, agriculture, hydrology, meteorology, oceanography, geology, geography, and navigation and traffic control. Three basic types of communication satellite were considered during the summer study: broadcast, point-to-point, and points-to-point. The points-to-point communication satellite forms the basis for this paper, for it is most directly associated with Earth resources data collection and relay.[2,3]

2. Data Collection Requirements

Most readers are familiar with the statement that the world's oceans occupy approximately 70% of the Earth's area of 196 million sq miles. Some other characteristics of our Earth are less generally known, however. For example, approximately two-thirds of the Earth's land mass is found in the northern hemisphere, and the northern hemisphere also contains approximately 90% of the world's population (i.e., all but the nearly 400 million people in Australia and New Zealand and the southern hemisphere sections of Africa, South America, and Indonesia). Barren Antarctica occupies nearly one-third of the total land mass in the southern hemisphere. Furthermore, virtually all of mankind is found between 60°N and 40°S latitudes. With the exception of certain meteorological, oceanographic, and geological considerations, these bounds tend to circumscribe and guide what likely will be the bulk of an Earth resources program in the decade of the seventies.

These brief comments are intended to suggest that the Earth resources program will be concerned primarily with the northern hemisphere, simply because it is where most of the world's 4100 meteorological land stations are found and where most of the ship and air traffic occurs. Hydrologic gauges and agriculture and forestry ground sensors are found predominantly in the more advanced nations in the northern hemisphere—e.g., in the U. S. there are more than 9000 operating hydrologic gauges.

In 1968 the Points-to-point Communication Panel of the NAS Summer Study on Space Applications solicited data needs from other panels of the study and from members of government departments and agencies. The results of this effort have been condensed in Table 1, which highlights types, distribution, and numbers of data collection platforms. Estimates are given for message length and required reporting interval. Of course, Table 2 represents only broad estimates of potential data requirements for the 1975 period, with primary emphasis on Earth resources data of interest to U.S. departments and agencies. Actual U.S. Government Earth resources requirements can stem only from an official government program.

Consideration of Table 1 leads to four important points.

1) If this 1975 data collection system is implemented, approximately 6 million bits of data would be collected during each 6-hr synoptic interval. Most of it would be collected within an hour or two of the synoptic reporting time.

2) Of the approximately 26,000 data collection platforms (DCPs) identified for the 1975 system, about 14,000 are in existence today. In 1975, over 16,000 of the 26,000 DCPs would be on, over, or near the shores of North America.

3) Four general types of data collection from in situ DCPs are evident: a) synoptic weather and oceanographic data, which most likely will be acquired within as short a time interval as possible following GMT 00:00, 06:00, 12:00, and 18:00; b) synoptic hydrologic, agricultural, and forestry data, most of which likely will be acquired just prior to local sunrise or local 8:00 AM, thus resulting in a moving data acquisition "time window" as the terminator moves around the earth; c) emergency data associated with volcanic and seismic activity, which can be collected routinely and forwarded periodically, except when activity exceeds a preset threshold; and d) data from remote users, which is most easily characterized by the need to transfer research data collected by an oceanographic vessel to a shore-based computer center for processing and preliminary analysis, and, possibly, for near-real-time control of selected data collection activities aboard ship.

4) Only the horizontal sounding balloons (HSBs) and free-drifting buoys require position fixing via the data collection and relay satellite (DCRS).

The first of these points—the need to collect and relay 6 million bits each 6 hrs—is of considerable importance to this part of an Earth resources system. As will be shown in a subsequent section, this volume of data can be accommodated

Table 1 Initial traffic estimates, neglecting position-location requirements

Platform type	Distribution	Position fix required	No. of platforms	Message size (bits)	Reporting interval (hr)	Total quantity of information (kilobits)/reporting interval	Allowable reporting time (hr)/reporting interval	Minimum allowable data rate (bps)	Comments		
									Future system	% of system existing	Existing platforms
1. Balloons	Global	Yes	4500	100	6	450	1(twice in 3 hr)	125	Yes		
2. Land weather stations with U/A[a]	Global	No	500	1400	6	700	1.5	130		100	500
3. Land weather stations without U/A[a]	Global	No	3600	225	6	810	1.5	152		100	3600
4. Ships-of-opportunity with U/A[a]	Global	No	60	1400	6	84	1.5	15.6		50	30
5. Ships-of-opportunity without U/A[a]	Global	No	800	275	6	220	1.5	40.8		100	800
6. Ocean-station vessels	Global	No	25	1400	6	35	1.5	6.5		40	10
7. Buoys (moored)	Global	No	500	2000-3000	6	1000-1500	1.5	278	Yes		
8. Buoys (drifting)	Global	Yes	200	500	6	100	1.5	18.5	Yes		
9. Buoys (manned)	Global	No	5	3000-10,000	6	15-500	1.5	93.0		40	2
10. Aircraft-of-opportunity	Global	No	200	500	6	100	1.5	18.5		20	40
11. Hydrologic (A)	U.S.	No	7300	200	24	1460	1.5	270		100	7300
12. Hydrologic (B)	U.S.	No	1100	1000	7 days	1100	1.5	204		100	1100
13. Hydrologic (C)	U.S.	No	600	2000	30 days	1200	1.5	222		100	600
14. Offshore platforms (ocean, estuaries, and lakes)	U.S.	No	500	500	6	250	1.5	46.3	Yes		
15. Seismic	Global	No	240	350	6	84	1	23.4		12	30
16. Volcanic	Global	No	50	1800	6	90	1	25.0		30	15
17. Agriculture	U.S., Canada	No	4000	100	6	400	1	114		5	200
18. Forestry	U.S., Canada	No	2000	100	6	200	1	55.5		4	75

[a] U/A=upper-atmosphere capability.

readily with a 50 kHz channel broken down into subchannels. Antenna size, transmitter power, frequency allocation, and satellite weight all appear to be feasible. Furthermore, by taking advantage of times when certain channels have low use, system growth by several factors might be accommodated.

The second point—number of data collection platforms in the future—is open to considerable interpretation. For example, the Summer Study Panel on Hydrology has outlined an evolutionary system of 10,000 sensor platforms initially, with growth to 25,000 and then 50,000 platforms. Not all of these platforms need report once a day, but the ability to respond to an interrogation at any time could be a standard feature that would enhance greatly the utility of a hydrologic data collection system without seriously compromising the cost of a station capable of responding to satellite interrogation. Furthermore, the estimate that more than 60% of the 26,000 DCPs in 1975 would be on, over, or near the North American continent makes it clear that a single DCRS in geostationary orbit in the vicinity of 90°–105°W would provide almost full coverage of the U.S., Canada, and the Central American and South American countries, as shown in Fig. 1.

It would take 4 geostationary DCRSs to give adequate global coverage, and it is not yet clear when the U.S. will get sufficient international cooperation to undertake the implementation of a full-scale DCRS system. Therefore, it is important to note that a western hemisphere Earth resources system could be established with a single geostationary satellite (and upgrading the existing and planned in situ sensors with a ground-to-satellite telemetry capability).

The foregoing comments and some further points concerning system data collection are summarized in Table 2. For example, the table indicates that more than one-third of the "new" DCPs would be HSBs. The number of HSBs that may be operational at any given time in the late seventies is open to conjecture. The value of 4500 is based on the assumption that there will be balloons at 3 levels above 500 mb and that balloons will be equally dense in the northern and southern hemispheres. More definitive answers to this question will have to await results from the Global Atmospheric Research Program (GARP) experiments in the mid-seventies, as well as the normal feedback that occurs with an evolutionary growth program.

The unmanned data buoy program is now underway in earnest, following approval in Fiscal Year 70 for the first major budget for the U.S. Coast Guard National Data Buoy Development Project.

Table 2 Summary of sensor platform types and numbers

Type	On, over, near North America	Planned for 1975	Total Number	Assumptions/comments
Land weather stations	1,000	...	4,100	Part of World Weather Watch.
HSBs	200	4,500	4,500	Primarily over oceans.
Buoys	200	700	700	Up to 350 in Coastal North American waters.
Ships	200	30	850	Most in normal shipping lanes.
A/C of Opportunity	50	160	200	Most in normal air traffic lanes.
Offshore platforms	100	500	500	
Hydrologic	9,000	...	9,000	Many sites part of ESSA weather broadcast networks.
Seismic and volcano	50	245	290	8 zones of 30 siesmic sensor clusters each. Five clusters of volcanic sensors.
Agriculture	3,500	3,800	4,000	3/4 in U.S.; most others in Canada.
Forestry	2,000	1,925	2,000	3/4 in U.S.; most others in Canada.
Totals	16,300	11,860	26,140	

However, whether the development efforts result in large numbers of small—possibly free-drifting—buoys or lesser numbers of large, moored high-capability buoys remains to be seen. It may be expected that other nations will implement data buoy systems of their own.

The approximately 6000 agriculture and forestry sensors to be installed throughout North America by 1975 represent a major phase of the Earth resources program, for this will be a new system, rather than an improvement, as is the case with the weather and hydrologic networks. To a great extent, the same is true for volcanic and seismic warning networks. Although the feasibility of such monitoring and warning networks already has been clearly established, a centrally controlled system of sensors and central data processing and warning dissemination remains in the future.

The four types of data collection from in situ sensors give rise to numerous features that will be of concern to a ground command-and-data-acquisition (CDA) facility. Some of these features are discussed in the remainder of this section.

Assuming that a 50-kHz rf bandwidth is available, Fig. 2 shows a typical example of how frequency and time division multiplexing might be accomplished over a 6 hr reporting interval. In this example, the 250 bit/sec channel is the basic "working block," with a 200-bit/sec information rate assumed feasible (i.e., 50 bps is allotted for the guard band). Within this framework, Table 3 indicates the time that would be required to accommodate from a single geostationary satellite the DCPs that might be expected to be within view. Figure 3 indicates one possible arrangement for accomplishing the synoptic collection of weather data.

Table 4 gives preferred data collection times for certain land-based DCPs. The requirement for data collection before local sunrise creates the need for a data collection "time-window" that varies seasonally, as shown in Fig. 4, for U.S., Canada, and the Caribbean. Figure 5 shows one way that hydrologic, agriculture, and forestry data could be collected around the 1-hr synoptic reporting interval used for weather and oceanographic data collection.

The fourth and last point concerns the location of horizontal sounding balloons and drifting data buoys. In the Summer Study Points-to-Point Communication Panel Report it was shown that use of Omega position locating (OPL) would require approximately 1.5 min/200-bps channel to determine the location of an HSB or drifting buoy. Then an assignment of 100 channels (200 bps each) to locate 1200 HSBs that might be within view of one DCRS results in 18 min required to locate all the balloons once. Thus, within a 1-hr synoptic reporting interval all the balloons can be located twice, with a 42-min wind-integration period between location times. Temperature, humidity, and pressure data would occupy about 1 sec and could be collected at the time of transponding location signals. Unless the short-term integrated effect of currents and surface winds is desired, the drifting data buoys would have to be located only once each synoptic reporting interval.

The 100 balloon location channels could be unoccupied 5 out of each 6 hrs. This data collection and relaying capability (20,000 bps) could be used for collecting other data, or, possibly, for dissemination of environmental data products, such as facsimile pictures.

Table 3 Data-collection times for a 200-bps channel

Platform type	Information/bits/message	Secs[a]/message	Expected no.of platforms	Data collection time[a] sec	Data collection time[a] hr (approx)
Superpressure balloons	100	0.5[b]	1200	600	1/6
Land weather stations (U/A)[d]	1400	7.5[c]	500	3750	1-1/24
Land weather stations (No U/A)	225	1.5	1000	1500	1/2
Ships-of-opportunity (U/A)	1400	7.5	40	300	1/12
Ships-of-opportunity (No U/A)	275	1.5	400	600	1/6
Ocean station vessels	1400	7.5	11	82	
Buoys (moored)	2000	10.5	350	3670	1
Buoys (drifting)	500	3.0	100	300	1/12
Buoys (manned)	3000	15.5	5	80	
Aircraft-of-opportunity	500	3.0	140	420	1/9
Hydrologic (Class A)	200	1.5	7300	10,950	3
Hydrologic (Class B)	1000	5.5	1100	6060	1-2/3
Hydrologic (Class C)	2000	10.5	600	6300	1-3/4
Offshore platforms	500	3.0	300	900	1/4
Seismic	240	1.5	100	150	1/24
Volcano	50	0.5	30	15	
Agriculture	100	0.75	2500	1870	1/2
Forestry	100	0.75	1300	970	1/4

a For a 200 bps channel.
b No acquisition time included.
c Acquisition time of 1/4 to 1/2 sec assumed for all platforms except balloons.
d Upper-atmosphere capability.

3. Alternatives in System Design

For an Earth resources system, the principal system design alternatives involve, but are not limited to, the following:

a) All data observed remotely by low-altitude satellites and relayed to ground data handling centers by geostationary satellites.

Table 4 Data-collection times

Platform Type	Data-collection time
Hydrologic (Class A)	At least once/day, preferably before 8:00 AM local time.
Hydrologic (Class B)	At least once/week.
Hydrologic (Class C)	At least once/month.
Seismic	Preferably 4 times/day.
Volcano	Preferably 4 times/day.
Agriculture	At least once/day, preferably before local sunrise.
Forestry	At least once/day, preferably during local midafternoon.

b) Data observed primarily by in situ sensors and collected and relayed by satellite to ground data handling centers.

c) Data observed primarily by in situ sensors and collected and relayed by conventional HF and landline networks to ground data handling centers.

d) All data collected by satellite and relayed in real time to regional sites.

e) Data collected and stored by satellite and delay-relayed on command.

f) Limited numbers of regional sites with acquisition to satellite-relayed data.

g) Essentially limited access to satellite-relayed data on a regional basis.

h) Alternatives in satellite altitude and inclination.

i) Alternatives in antenna design, operating frequency, transmitter power, modulation technique, etc.

j) Alternatives in position location techniques, e.g., range-range (IRLS), range-rate (EOLE) Omega, etc.

Although it is beyond the scope of this paper to consider in detail all the alternatives given above (and the list is not exhaustive), there are several summary comments that can be made.

Remote Data Collection by Satellite

To collect Earth resources data remotely by satellite in the seventies will require low-altitude orbits. When global coverage and synoptic reporting are among the requirements, sun synchronous circular polar orbits of about 350 to 820 naut mile altitude often are used, with inclinations of about 98° to 102°, respectively. (The indicated relationship of 0.85° increase in inclination per 100 naut-mile increase in altitude is a good approximation here.) The corresponding periods are of the order of 98 to 114 min. With these orbits, the sub-satellite point retraces the same points on the ground. Satellites observing meteorological parameters can be operated at the higher end of the altitude range, while satellites observing agriculture, forestry, and hydrologic parameters normally would be at lower altitudes. At least two meteorological satellites and three ERS satellites would be required for global synoptic reporting. Two wideband (perhaps 90-MHz to 1.0-GHz) data-collection-and-relay satellites in geostationary orbit would be required to keep these 5 low-altitude satellites and some low-altitude manned orbital missions in real-time contact with the three ground data handling centers that would be needed. Such a system of unmanned satellites and manned ground stations has been grossly estimated to cost $310 million, covering 3 yr of R & D and 4 yr of prototype operations, and including $149 million for maintaining three Earth resources satellites in orbit.

Data Collection from in Situ Sensors

In previous discussion, the point has been made that all of the 26,000 DCPs identified as representative of a 1975 system of in situ sensors could be interrogated by a group of 4 geostationary data-collection-and-relay satellites each having an rf bandwidth of 50 kHz. Interrogation could take place within the framework of frequency and time division multiplexing, as noted in Fig. 2. Actually, this rather nominal bandwidth requirement provides considerable unused capacity between synoptic reporting intervals. And, because 16,000 of the 26,000 DCPs could be covered by a single DCRS (Fig. 1), the data collection demands on the other three DCRs would be correspondingly less. However, if there are four geostationary DCRSs, then coverage overlap over the U.S. would minimize ground data relaying to the ground data handling center in the U.S., for only two ground relay stations would be needed: one in the U. S. and one in India, Pakistan or Burma.

Data collection satellites such as these could be operated in the VHF portion of the frequency spectrum: possibly in the 136-138 MHz band allocated to space telemetry. Once the downlink frequency is selected, it is desirable to choose an uplink frequency sufficiently close to allow shared use of the antenna; 149 MHz is selected for the uplink frequency. A parabolic antenna, operating at these frequencies and providing half-power points at ± 10° to allow Earth coverage, would be 25 ft in diameter and would provide 18-db gain. A folded parabolic dish of this size would stow in 13 ft and weigh less than 40 lb. As an alternative, a 5-element helix array might be used with reasonably comparable weight and stowed dimensions. The weight of these satellites would be of the order of 350 lb and launch into geostationary orbit could be accomplished by the Thor-Delta launching rocket and a 353-lb SVM-2 apogee motor.

The cost of a DCRS system as described has been grossly estimated to be of the order of $90 million, covering 3 years of R & D and 4 years of prototype operations.

Use of Conventional HF and Land Lines

Gross cost estimates have been made for a conventional HF/land line system to communicate with the 26,000 DCPs estimated for 1975. The 4-year operating costs were estimated to lie in the range of $1.56 billion to $2.12 billion. While these costs are at best only indicators, their composition is of some interest. For example, it is assumed that nearly 7000 DCPs use HF: the 4500 HSBs, the 860 ships, the 700 buoys, etc. Over 19,000 DCPs use conventional land lines: the 9000 hydrologic sensors, the 4100 land meteorological stations, the 6000 agriculture and forestry DCPs, etc. However, the costs are in opposite proportion: the global HF system costs $1.43 - $1.9 billion for 4 years, but the conventional land lines cost only $129 - $209 million. Obviously, collecting Earth resources data from over the oceans by conventional HF is very expensive. If the balloons were eliminated, the cost might come down considerably, because the other ocean platforms can provide higher power output and, hence can expect to communicate at greater HF range. There are many possible trade-offs; it is of greatest interest to note that the four-geostationary-DCRS system costs less than the land line segment alone for the conventional communication system. Furthermore, the four-DCRS system provides global coverage and considerable inherent growth potential.

Direct Regional Readout

This is a highly hypothetical concept, considered here simply to give a gross cost comparison. Such a system might comprise 3 near-polar orbiting, sun-synchronous, Earth resources sensing, 500-naut-mile altitude satellites and 120 regional surface stations, 84 of which would be ships (the USSR and China are not covered). Ground stations have been estimated to cost only $0.8 million, but ship stations would be $5.4 million. Three years of R & D followed by 4 years of prototype operations were estimated to cost $1.13 billion, of which $790 million is attributable to implementation and operation of the 84 ship stations. Again, collection of Earth resources data over the oceans is seen to be expensive. Cost of this system is about 13 times that of the four-DCRS system.

Direct Data Collection/Delayed Readout

A variation on the theme, immediately above, involving the same satellites, but including development and acquisition of satellite-borne tape recorders, would make it possible to acquire the desired data (with limited delay) at only three ground stations, each costing $4 million to implement and $2 million/yr to operate. Gross cost estimates for 3 years of R & D and 4 years of prototype operations is $193 million. This is essentially twice the cost of the four-geostationary-DCRS system for collecting data from in situ DCPs.

Limited Numbers of Regional Sites/Direct Data Collection

If the 84 ship stations are eliminated from the direct regional readout system described above, the cost estimate under equivalent conditions is $237 million. However, this limited coverage system still costs more than the direct data collection, delayed readout system. Its cost exceeds that of the four-DCRS system by a factor of about 2.6.

Limited Access by Regional Sites

Considerable international cooperation may have to be obtained, if global Earth resources data from low altitude remote sensing satellites capable of 100 ft (or less) resolution are to be freely exchanged and used openly. (Such data may be obtained surreptitiously and used covertly, but that is outside the scope of this discussion.) A limited amount of international cooperation might be achieved, if a) the satellite is designed to afford a narrow angle transmitting beam, steerable to the proper regional site; or b) the transmitted video is digitized and encrypted, with separate encryption keys for each user; or c) ground data handling centers are staffed with "bonded" representatives from all participating countries. The first of these techniques might be accomplished with a satellite antenna using a motor-driven parabolic dish or an electrically steered retrodirective array. Operating in the gigaHertz range (to get the required wide bandwidth), the dish would be only 1 to 3 ft in diameter or the retrodirective array might be of the order of 1 to 16 sq ft in size. Encryption of video requires digitization of the analog signal, thus requiring at least a tripling of bandwidth and some reduction of signal-to-noise ratio because of elimination of the gain possible from FM improvement. The use of "bonded" personnel in ground data-handling centers is not a technical solution to this problem of Earth resources data privacy. Using "bonded" personnel has the advantage that regional sites would not be required, but each participating nation would get only its own data and any data previously agreed upon to be shared. The present global sharing of weather data and data products through World Weather Watch gives an indication that such an approach is feasible.

Position Location Techniques

The position location of horizontal sounding balloons provides a major technical obstacle to be overcome in the development of a DCRS for use with in situ DCPs. If, indeed, a population of 4500 to 10,000 HSBs becomes reality in the late seventies, then a major portion of the narrow band DCRS channel capacity will have to be devoted to balloon interrogation and location. At present, three approaches to this problem are under investigation: one is based on the U.S. Navy's global Omega position location system of (ultimately) eight VLF transmitters. NASA has been carrying out the Omega position location experiment (OPLE) since 1967, and has succeeded in operating within a 2500 Hz bandwidth using the ATS-3 satellite. DCP position fix has been obtained within 3 minutes. The Points-to-Point Communication Panel report suggests that if all Omega ground stations operate at full-rated transmitter power of 10 kw, then the improved location-data signal-to-noise ratio might increase by as much as 3 db, thus reducing integration time to as little as 1.33 mins. Furthermore, by using an appropriate mixing process, it is suggested that only 250 Hz bandwidth is required. Then 100 such channels could locate 4500 balloons within 1 hr. Adding 20% for guard bands, the required bandwidth would be about 30 KHz.

Another position location scheme is based on interrogation, response, and time counting for range determination (IRLS). Like OPLE, IRLS requires two ranging fixes between a mobile DCP and known references. In IRLS, triangulation from known satellite ephimerides to the balloon is used, thus essentially prohibiting the use of IRLS with any but relatively low-altitude satellites because of the very long time required between time fixes when the interrogating satellite is more than a few thousand miles above Earth.

The third technique is the French EOLE system, which requires only 16 sec of integration time when used with a 700 naut-mile satellite. This short time is achieved in part by maintaining a constant satellite beacon signal. As with IRLS, the integration time required for operation with a geostationary satellite is projibitively long. A summary of some of these comments is given in Table 5.

4. Characteristics of a Representative System

To show the feasibility of providing global Earth resources data collection from in situ sensors, the Points-to-Point Communication Panel prepared the characteristics of a representative system comprising four geostationary DCRSs equally spaced about the equator, and two command and data acquisition stations located in overlap zones in the U.S. and India or Pakistan or Burma. The location of satellites and overlap zones is illustrated in Fig. 6. In its report, the Panel stressed that the system presented was not an optimum one. Much more detailed analysis must be performed and the uncertainties relative to traffic loads reduced before it will be possible to approach optimum system design.

The Panel considered, among others, the following topics: channel utilization, transponder design (frequency selection, antenna selection, and link calculations), satellite design, and command and data-acquisition stations. The reader may recognize that many of these system characteristics have been mentioned in previous sections of this paper. Some of that information will be brought into better focus here.

Channel Utilization

The requirement to collect in situ data at synoptic intervals and, possibly, to locate thousands of mobile DCPs using improved OPL, led to the selection of a 50-kHz rf bandwidth. This 50 kHz is divided into twenty 2500-Hz channels, and each of these channels is further divided into 10 subchannels of

Table 5 Data-collecting techniques

Satellite	No. of satellites required and coverage	Location	Advantages	Disadvantages
1. Low-altitude polar orbiter	1: daily 4: 6 hr 6: synoptic [a]	Range-range (IRLS) or range-rate (EOLE) [b]	Relatively low DCP power [c] Obtains polar data Can share function, i.e., imaging Ranging geometry simpler	Difficult to obtain a synoptic coverage Storage with irregular readout or geostationary relay required Does not obtain continuous coverage Requires deployment of Omega
2. Geostationary	4, equatorial, synoptic [d]	Omega	Synoptic coverage Simpler data logistics particularly for fixed platform High-gain antenna may be used.	Requires deployment of Omega system for location Poor polar coverage Omega requires relatively wide band widths, long integration time
3. Geostationary	6 to 8 equatorial, synoptic [e]	Range-range using cooperative satellites	Same as above No need for Omega Allows shorter integration time	Poor polar coverage Requires 2 to 4 more satellites
4. Geostationary and polar orbiters	3+2 [f]	Omega range-range	Obtains global and synoptic coverage (but not totally continuous)	Same as above for Omega Requires two extra satellites
5. Geosynchronous inclined	4 inclined at 48° [g]	Omega	Obtains global coverage	Synoptic-interval problems—only one coverage per day per satellite for poles—no wind vector Absentee problem

a One satellite can report an area once per day; two can report an area every 6 hours; six can provide synoptic coverage—i.e., worldwide coverage within the synoptic reporting interval.

b Interrogation Recording and Location Subsystem (IRLS) requires two range measurements during one orbital pass to obtain trilateration data to fix balloon positions: EOLE measures range and range-rate simultaneously.

c Only 1/18 of the power required for geostationary satellite communication, but fixed DCP may use high-gain antenna, reducing the importance of this advantage.

d Four satellites in 24-hour equatorial orbit give nearly full coverage between 60° N and 60° S latitudes for elevation angles greater than 15°.

e Six to eight satellites required for full location and coverage of balloons between 60° N and 60° S latitudes for elevation angles greater than 15°

f Three geostationary satellites and two low-altitude satellites in highly elliptic polar orbits, with orbiting periods ranging from 6 to 24 hours.

g Four satellites in 24-hour orbits inclined at 48°. Each satellite sees each pole once per day.

250 Hz each. Each basic subchannel is assumed to have an information capacity of 200 bps. For mobile platform location, one 250-Hz channel will require 1.5 mins of integration time. The DCP will have a phase-lock loop with 10^{-7} frequency stability.

The allocation of the twenty 2500-Hz channels has been shown in Fig. 2. In Fig. 2, the first channel is used to interrogate DCPs. Five subchannels, to be used with five satellites, are allowed. The control and data acquisition station (DCA) operates through the satellite to interrogate DCPs. Agriculture and forestry data are handled in Channel 2, which operates continuously, as does Channel 1. A single 250-Hz subchannel could accommodate the 1975 traffic. The other 9 subchannels provide for growth and isolation.

All seismic and volcanic sensors can be accommodated in Channel 3, with considerable room for growth. Channel 4 is for hydrology, and, since this is collected on a moving "time window" basis, it operates continuously. The next two channels are for alarms resulting from extraordinary volcanic, seismic, or other natural phenomena activity. Disaster warnings always can be transmitted on these "hot-lines." Channel 7 is for administrative traffic in either direction between CDA and DCP. Channel 8 is a spare.

Fixed station meteorological and oceanographic data are collected synoptically through Channels 9 and 10. Mobile DCPs (principally balloons) are located and data collected using Channels 11 through 20. Thus, it is clear that essentially one-half the capacity of this system is concentrated on locating horizontal sounding balloons. Of course, the duty cycle for these channels is quite low. (In fact, with estimated 1975 traffic, 75% of the 50-kHz capacity is left for future growth.)

Transponder Design

As noted earlier, the low bandwidth requirement (50 kHz) and availability of spectrum in the VHF range for space telemetry suggests use of 137 MHz. Then 149 MHz is an appropriate down frequency, since it has been authorized for space telecommand.

The characteristics of a furled parabolic dish satellite antenna of 25-ft diam and a compressed helix antenna have been noted earlier. The dish would weigh 40 lb and the helix about 25 lb. They would provide 18 db and 15 db gain, respectively.

Because this data collection satellite system need handle only low data rates, low transmitter powers in the range of 1.0 to 20 w can be used. Link calculations are presented in a comparative format in Table 6. This table indicates that an Earth coverage satellite antenna is used with 18-db maximum gain and 15-db gain at the Earth's horizon. DCPs are assumed to have omni-directional antennas, although this obviously need not always be the case. Circular polarization is assumed and the 1.2 db polarization loss is for the pessimistic case where the line-of-sight from DCP to satellite is at 15° elevation angle. Table 6 indicates that sufficient carrier-to-noise ratio can be obtained in all links.

Satellite Design

The four geostationary satellites of Fig. 6 are shown stationed at 50°W, 140°W, and 120°E longitude. Even when 15° elevation angle is assumed to limit operation, it is clear that data can be collected between 60°N and 60°S latitude. It has been noted previously that this embraces the vast majority of mankind and the habitable regions of the Earth.

The geostationary satellite described herein would be relatively simple and would weigh about 350 lb. The suggested 20-w transponder is essentially state-of-the-art. The antenna would require development. This satellite would be inserted into geostationary orbit with a DSV-2N (TE-364-3) Thor-Delta, with an SVM-2 apogee motor. Launch cost would be about $5 million; spacecraft R & D to achieve a 5-yr lifetime would be about $7 million. Thus, the spacecraft portion of this system would cost on the order of $50 million for development and 5 yr of operations.

A minimum of two command-and-data-acquisition stations would be required, operating in the overlap zones over the U.S. and Asia. Data exchange between CDAs could take place by time-division multiplexing through the DCRS themselves, but this poses an added complexity. Conventional commercial communication satellited may offer the simplest method of data exchange, although other methods are possible. Two CDAs could be acquired and operated 4 yr for about $27 million, with cost for systems management included. Total system cost for development and 4-yr prototype operations would be about $80 million. To be on the conservative side, elsewhere in this paper the system cost has been taken to be $90 million.

These data collection and relay satellite system representative characteristics are summarized in Table 7.

Table 6 Link calculations

Parameter	Platform to satellite (DCP response)	Satellite to platform (interrogation)	CDA[a] to satellite (interrogation)	Satellite to CDA (DCP response)
Link frequency	Up 149 MHz	Down 137 MHz	Up 149MHz	Down 137 MHz
Radiated power:	7.0 dbw	0.0 dbw		
(5 w nominal)			7.0 dbw	
(20 w nominal)				13.0 dbw
Antenna gain:				
(platform: 15°elec. angle)	0.0 db			
(satellite antenna: min. gain)	15.0 db	15.0 db	15.0 db	15.0 db
(ground station)		0.0 db(HSB)[b]	14.0 db	22.0 db
Polarization loss	1.2 db	1.2 db	1.2 db	1.2 db
Path loss	167.9 db	107.0 db	107.9 db	167.0 db
Cable loss	1.3 db	1.0 db	3.0 db	3.0 db
Receiver noise	198.0 dbw/Hz	197.1 db2/Hz	198.0 dbw/Hz	194.8 dbw/Hz
Bandwidth:				
(250 Hz)	24.0 db			
(2500 Hz: interr. lock-on)		34.0 db	34.0 db	
(500 Hz: signal)		27.0 db		
(50 kHz)				57.0 db
Carrier-to-noise ratio:				
(in 250 Hz)	26.0 db			
(in 2500 Hz)	16.0 db		27.9 db	
(lock-on; 2500 Hz)		8.9 db		
(signal: 500 Hz)		15.9 db		
(in 50 KHz)				12.6 db

a CDA: Command-and-data-acquisition station.
b HSB: Horizontal sounding balloon.

Table 7 A representative data collection satellite system for 1975

Minimum DCP	5 w; 0 db (15°)
Four geostationary DCSs	
Balloon coverage	60° latitude
Others	70° latitude
Two control and data acquisition (CDA) stations	
VHF channelized transponder - 50 Hz	
• 149 MHz up, 137 MHz down	
• 20 channels, 10 subchannels each; 2.5 kHz	
• 20 w transmitter	
Earth coverage, 18 dB gain, VHF antenna	
Weight in orbit	350 lb
Power	105 w
Launch vehicle	Thor delta
Estimated 5-yr cost	
• Satellites (four orbited plus spare)	$25 million
• Launch vehicle (5 at $5 million)	$25 million
• CDA plus maintenance	$25 million
Subtotal	$75 million
20% additional for contingencies	15 million
R & D and 4-yr operations (Total)	$90 million

5. Conclusions

In the decade of the seventies it may be anticipated that efforts will be made to collect synoptically Earth resources data from thousands to tens of thousands of in situ data collection platforms. Collection by satellite appears to afford economies of scale, as well as ample growth potential. Total required rf bandwidth appears to be modest (of the order of 50 kHz), and operation in the VHF portion of the spectrum looks feasible. Of the required rf bandwidth, a sizable portion (perhaps 50%) may be required to locate mobile data collection platforms, such as horizontal sounding balloons and free-drifting data buoys. Planned experiments and development efforts in the early seventies should give more insight into this aspect of the system. More detailed system design is warranted but it should be preceded by formal governmental efforts to establish user requirements and the anticipated evolutionary growth of traffic loading. Tradeoff studies should be performed to determine what system configuration must effectively satisfies U.S. and/or international Earth resources data needs. The development and 4 yr of operations of a global Earth resources data collection and relay satellite system may cost under $100 million.

References†

1 "Useful Applications of Earth-Oriented Satellites. Summaries of Panel Reports," Summer Study on Space Applications, Division of Engineering, National Research Council, Washington, D. C., 1969.

2 "Useful Applications of Earth-Oriented Satellites, Points-to-Point Communication," prepared by Panel 7 of the Summer Study on Space Applications, Division of Engineering, National Research Council, Washington, D. C., 1969.

3 Dorfman, S. D., "Collection of Data from in situ Sensors via Satellite," Western Electronic Show and Convention, August 19-22, 1969.

† A comprehensive bibliography is given in Appendix B, Ref. 2.

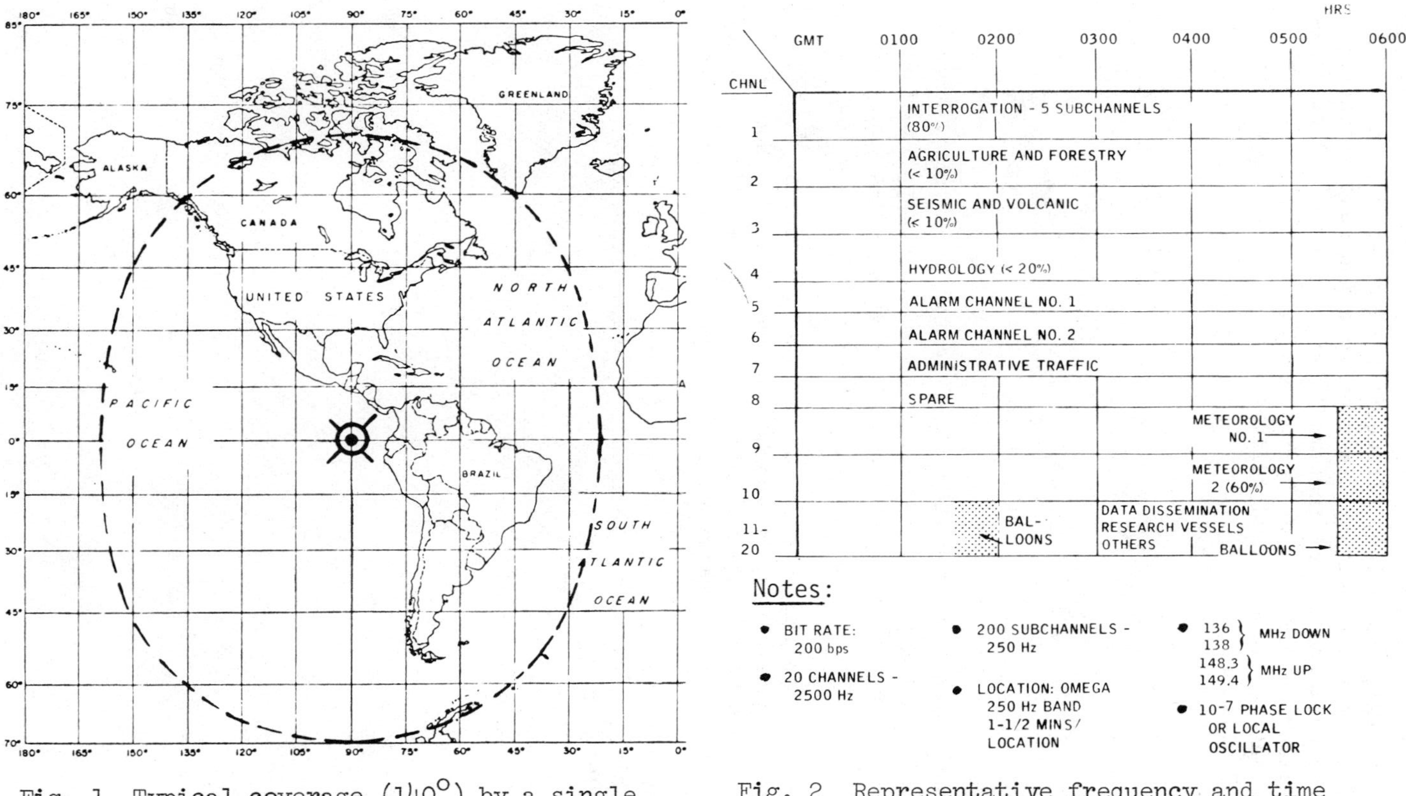

Fig. 1 Typical coverage (140°) by a single data-collection-and-relay satellite.

Fig. 2 Representative frequency and time division multiplexing.

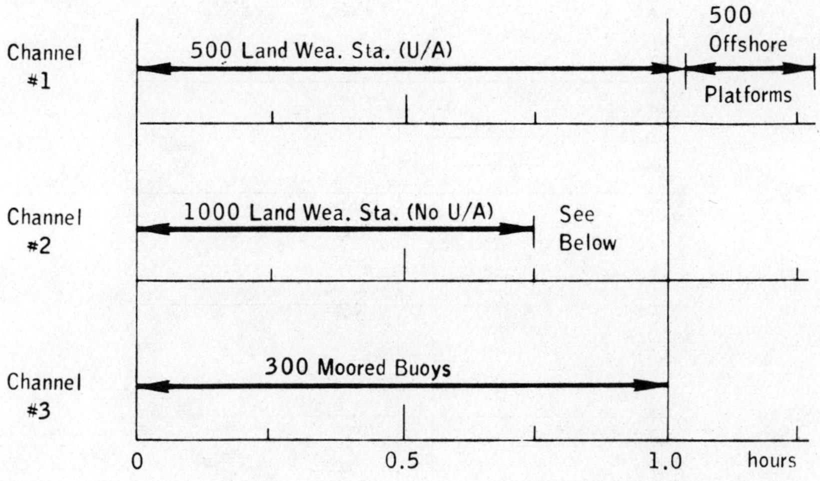

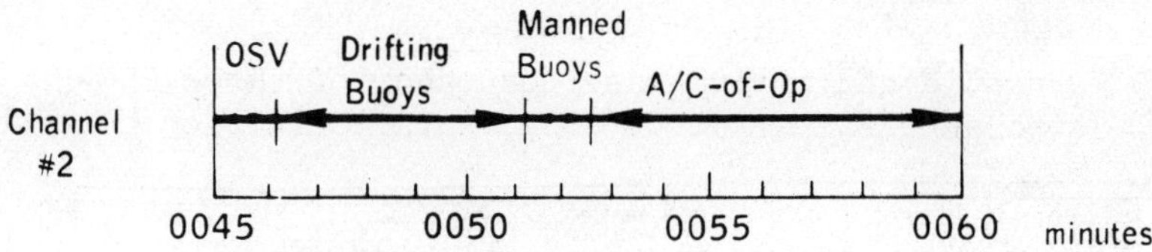

Fig. 3 Data-collection schedule.

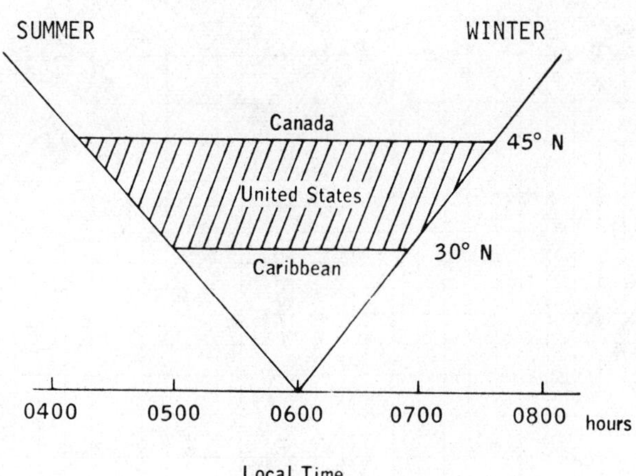

Fig. 4 Seasonal variation of local sunrise.

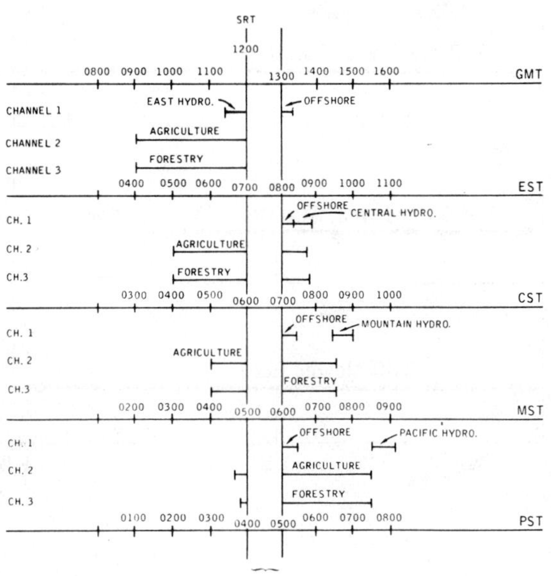

Fig. 5 Synoptic and moving-base occupancy of three 250-Hz subchannels for 26,000 small data collection platforms.

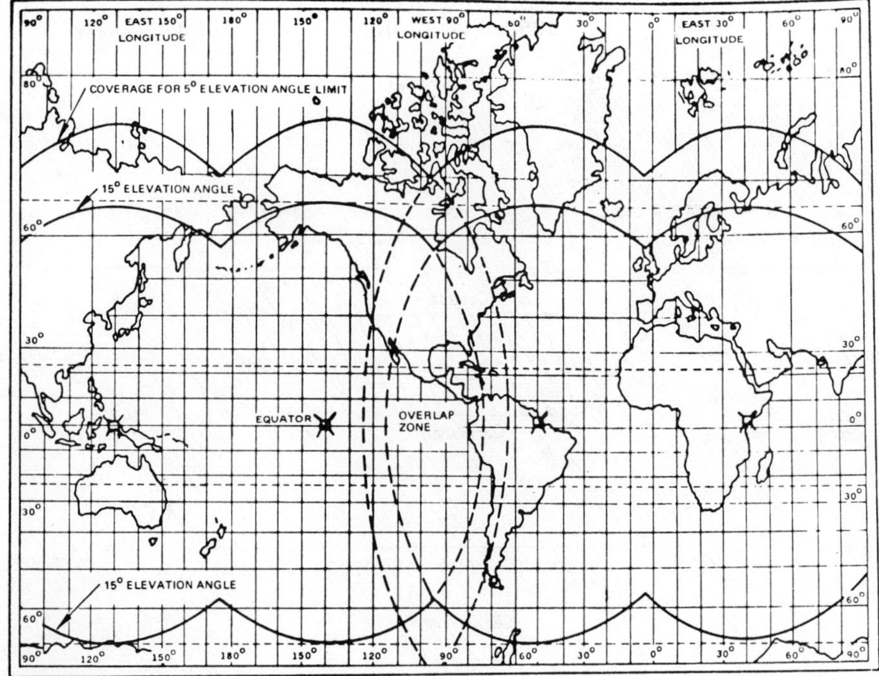

Fig. 6 Reference system coverage.

VII. DEFENSE SYSTEMS

IDCSP/A Satellite: Concept and Performance

Douglas G. Dwyre*

Philco-Ford Corporation, Palo Alto, Calif.

Abstract

Established as a program for allied participation in the Defense Satellite Communication Program, the Initial Defense Communication Satellite Program/Augmentation (IDCSP/A) services vital links in the United Kingdom (U.K.) Defense Communication Network that is part of the Skynet Program. The communications transponder is a modification of the standard IDCSP transponder that is used in the network of United States (U.S.) satellites supporting the Defense Satellite Communications System. Interoperation between the IDCSP and IDCSP/A systems is assured by using the IDCSP 20-MHz channel. However, additional flexibility is available for special U.K. nets in a separate 2-MHz channel. The spacecraft integrates a mechanically despun antenna, self-contained orientation control, 375-400 MHz telemetry/tracking/command, and a hydrazine monopropellant secondary propulsion system. With an initial launch on November 21, 1969, initial orbit data are available for comparison with design specifications.

Introduction

On November 21, 1969, the first IDCSP/A satellite (Fig. 1), was launched successfully into a synchronous transfer orbit. This accomplishment established a new dimension in military satellite communications. Progressing from early Courier experiments, through the system evaluation of the Initial Defense Communication Satellite Program (IDCSP), the Department of Defense (DOD) established a satellite network for global communications. Beginning in the summer of 1967, this network provided operational circuits for the Defense Communication System.

Presented as Paper 70-492 at the AIAA 3rd Communications Satellite Systems Conference, Los Angeles, Calif., April 6-8, 1970.

*Manager, Advanced Systems Department, Western Development Laboratories.

The DOD offered to share IDCSP circuits for R&D. The U.K., acting on a bilateral Memorandum of Understanding with the U.S., accepted this U.S. offer of participation in the Defense Satellite Communications System.

To meet U.K. operational requirements, the Memorandum of Understanding (dated September 19, 1966) provided for procurement and launch of the IDCSP/A (designated SKYNET) satellites by the U.S. on behalf of Her Majesty's Government, England. The U.S. Air Force Space and Missile Systems Organization (SAMSO) procured and supervised satellite manufacture and launch for the U.K. Ministry of Technology in response to U.K. Ministry of Defense requirements.

Two major features characterized U.K. requirements as distinct from the IDCSP: 1) increased satellite circuit capacity over a restricted coverage area, and 2) a dual channel repeater for netting incompatible Earth terminal networks.

U.K. communications coverage requirements can be satisfied by a geostationary satellite. The success of the Syncom and Intelsat satellites demonstrated the feasibility and reliability of geostationary satellites. The IDCSP/A satellite, developed by Philco-Ford, is based upon communications technology, which was developed in the IDCSP, and stationkeeping techniques, which were demonstrated in NASA/Intelsat programs.

Mission Profile

The spacecraft mission has four distinct, sequential phases: 1) boost lift-off through Stage III burnout, 2) transfer orbit through apogee boost burnout, 3) station acquisition and drift from apogee boost-to-station, and 4) stationkeeping and final orbit maintenance. The geometry of the mission profile is shown in Fig. 2.

Each phase has particular requirements and constraints which are generally interdependent. This paper contains a summary of these parameters plus an optimized set of mission profiles.

System requirements specify that, during a 5-yr lifetime, the final orbit shall be maintained within a $\pm 3^{o}$ inclination and shall also be controlled in longitude to within $\pm 3^{o}$.

Launch Vehicle and Apogee Boost Motor

The Thor-Delta rocket (DSV-3L/TE-364-3) launched the spacecraft from the Eastern Test Range. The first stage booster was a combination of the Long Tank Thor, and three Thiokol TX-354-5 Castor II solid propellant strap-on motors. The Improved Delta was the second stage, and a TE-364-3 solid propellant motor provided the boost for the second stage.

The Apogee Boost Motor (ABM) was an extended Thiokol TE-M-479, which originally qualified for the Radio Astronomy Explorer spacecraft. The ABM weighed 275 lb (247 lb of propellant) with a specific impulse (I_{sp}) of 290 sec which provided a velocity increment of 5810 ft/s at final apogee. This was sufficient to remove most of the transfer orbit inclination and to circularize the orbit.

Errors and Constraints

The errors and constraints influencing the mission profile can be roughly categorized as follows: 1) boost errors, 2) lunisolar effects, 3) launch trajectory inclination, 4) thermal/power constraints, 5) tracking coverage, and 6) tracking errors.

Boost errors include velocity and position dispersion at transfer orbit injection plus the following errors contributed by off-nominal performance in the apogee motor burn: 1) attitude orientation, 2) variations in specific impulse, 3) thrust vector misalignment, and 4) propellant loading tolerances. The errors in synchronous orbit are divided into "in-plane" components, which require correction by the Secondary Propulsion System (SPS), and "out-of-plane" components, which contribute to inclination error. To some extent, off-nominal booster performance can be compensated by apogee burn vector alignment and/or rotating the line of nodes of the drift orbit with final orbit inclination constraints.

Lunar and solar gravitation combine to gradually increase the inclination of a nominally geostationary orbit. Because communications coverage requirements permitted a tolerance of $\pm 3^{\circ}$ in final orbit inclination, the initial orbit inclination was biased to a nominal $-2.5 \pm 0.5^{\circ}$. This inclination will increase to nearly $+3^{\circ}$ within the 5-yr orbit lifetime.

To limit orbit plane inclination to less than 3° for a 5-yr period, the right ascension of the transfer orbit ascending node must be 281 ±16° (when worst case combinations of nodal and initial inclination errors are assumed). Because of launching safety considerations, the minimum launch trajectory inclination was limited to 26°.

Thermal/power constraints are placed upon the mission profile for the following reasons: 1) to prevent excessive cooling of critical components during the five periods of the transfer orbit and 2) to ensure sufficient illumination of the solar array (this prevents excessive discharge of the batteries). There are two kinds of constraint: 1) the sun angle (between the satellite-sun line and a plane normal to the satellite spin axis) must be less than 30°, and 2) occultation of the spacecraft must not exceed 60 min. The combination of these constraints results in the launch window restrictions illustrated in Fig. 3. These launch windows are centered about noon and midnight, GMT.

Tracking coverage during launch, ascent, injection, and station acquisition was provided by the Satellite Control Facility (SCF) stations. Turn-over to the U.K. Control Station was accomplished after station acquisition and complete satellite checkout. Adequate tracking must be provided prior to and following all critical events, particularly for a sufficient period prior to ABM burn.

By allowing several revolutions in the transfer orbit, for the acquisition and smoothing of tracking data, tracking errors are minimized. Figure 4 shows the improvement in position uncertainty during approximately seven periods of transfer orbit, using azimuth, elevation, and range-rate data. The period for the transfer orbit is approximately 10-1/2 hr.

Boost Phase

To permit launching any time during the year and to minimize usage of the Secondary Propulsion Subsystem (SPS) (while considering launch vehicle and ABM error contributions), two trajectories were selected for the boost phase (see Blocks A & B below). The launch date determines which trajectory will be used. Summary characteristics of each trajectory and the resulting transfer orbit injection conditions are shown in Table 1.

Transfer Orbit

Transfer orbit parameters (apogee altitude, ascending node, and inclination) are determined by the requirements to permit launch during any day of the year, and to minimize consumption

Table 1 Boost phase trajectories

	Block A	**Block B**
	Block A	**Block B**
Launch period	**Nov-Feb; May-Aug**	**Feb-May; Aug-Nov**
Apogee altitude	**Synchronous +400 naut mile**	**Synchronous + 800 naut mile**
Perigee altitude	**150 naut mile**	**150 naut mile**
Inclination, transfer orbit	**28°**	**26°**
Latitude of perigee	**0°**	**0°**
Ascending node	**$281^{\circ} \pm 16^{\circ}$**	**$281^{\circ} \pm 20^{\circ}$**
Inclination, final orbit	**2.5°**	**2.75°**

of SPS fuel. After separation from the third stage, the satellite is rotated to maintain nearly normal illumination of the solar array during the several periods of transfer orbit (usually 4 1/2). Approaching final apogee, the satellite is commanded into an attitude which will properly align the thrust vector of the apogee burn. During transfer orbit, attitude information is derived from telemetered Earth and sun sensor data. Position and velocity information derives from tracking data received at the several SCF stations (Fig. 4).

Station Acquisition

During the period of drift to the desired longitude, the primary consideration is to minimize SPS fuel consumption within the following constraints: 1) the characteristics of the initial drift orbit following apogee burn, 2) the initial longitude difference between the station and the apogee burn, and 3) the time permitted for station acquisition (3 weeks max). The initial drift orbit results in an easterly or westerly drift to station.

Orbit corrections are made with SPS burns at apogees and perigees of the drift orbit after tracking information has been accumulated for several days and before fine corrections are made (approximately 5 ft/sec are required).

Stationkeeping

A two-correction program is used to maintain station within a $\pm 3^{\circ}$ longitude limit. Figure 5 illustrates satellite motion (radius and longitude) relative to the station which is located at the origin. Satellite drift, starting at point "0" (intentionally biased east of station and at super-synchronous radius), takes the satellite west of station, through synchronous altitude and back to a longitude east of station at a subsynchronous radius. At point "1" a velocity correction is made to raise the apogee. At point "2" another velocity increment is added to circularize the orbit. The drift cycle starts once again.

Spacecraft

Following are seven subsystems for the IDCSP/A spacecraft:

1) Communications - a mechanically despun antenna and redundant X-band transponders;

2) Telemetry, Tracking, and Command (TTC) - a UHF array providing near-spherical antenna coverage, redundant UHF transponders, and command/telemetry processing equipment;

3) Electrical Power - a cylindrical array of silicon solar cells capable of providing 97 w of prime power throughout 5 yr of orbit life, a power control unit (PCU), and two redundant 6 ah nickel-cadmium batteries;

4) Attitude and Orbit Control (AOCS) - two Earth-horizon sensors, sun-pulse sensors, two sun-angle sensors, and electronics for processing sensor signals;

5) Secondary Propulsion Subsystem (SPS) - monopropellant hydrazine tanks, manifold fueling, and redundant radial and axial thrust chamber assemblies (TCA);

6) Spacecraft Structure and Harness assemblies - includes thermal radiation shields; and

7) Solid propellant Apogee Boost Motor.

The mechanical layout of the spacecraft is shown in Fig. 6. By using a V-band clamp, the spacecraft is attached to a standard payload adapter on the TE-364-3 rocket stage. A preseparation spin rate of 90 r/min ±10% is imparted by the DSV-3L spin table. Electronic equipment is mounted on eight vertical panels. Major axial structural loads, occurring during boost, are transferred through the truss structure to the central cylinder.

The ABM attaches directly to a mounting ring on the central cylinder. The eight solar panels are supported by a framework of laminated fiberglass. A summary weight and power budget is presented in Table 2.

Table 2 Weight and Power Summary

Component		Weight lb	Power w
Attitude and orbit control		12.76	6
Communications:	transponder	44.94	30
	antenna	4.76	7
Electrical power		54.09	5
Reaction control equipment (incl. fuel)		32.88	
Structure		34.77	
Telemetry, tracking and command		32.14	16
Mechanical integration		6.55	
Electrical integration		14.78	
Thermal control		13.54	
ABM:	case	24.61	
	consumable inerts	3.30	
	propellant	247.00	
Ballast and balance		9.88	
Total		535.00	64

On-oribt weight: 280 lb
Minimum MOI ratio: at launch, 1,306
on-orbit 1,471

Communications Subsystem

The communications subsystem receives, translates in frequency, amplifies, and retransmits X-band signals. Two channels, 20 and 2 MHz bandwidth, are provided for netting communications, with output power equally divided between channels.

Communications Transponder

The communications transponder block diagram (Fig. 7) indicates a single thread path. The complete equipment redundancy, and cross-strapping which is employed to achieve reliability, is not shown in this figure. Selection of either set of communications equipment, operating with either traveling wave tube amplifier (TWTA), is accomplished by ground command.

The received signal is 1) isolated by polarization diversity in the orthomode transducer, 2) split into separate channels for amplification and hard limiting, 3) recombined and up-converted to output frequency, 4) amplified to output power in the TWTA, and 5) introduced into the motor drive assembly through the orthomode transducer. A signature tone timing signal is frequency modulated on the beacon carrier and introduced into the communication band in the channel summer. The RF power budget for the communications subsystem, for nominal (beam-center) and worst case (9.5° off beam-center) conditions, is presented in Table 3.

Communications Antenna

The communications antenna (Fig. 8) consists of the RF assembly and the motor drive assembly (MDA). The Earth sensor assembly and control unit are included in the AOCS.

1) RF Assembly: radiating aperture (horn, lens, reflector), orthomode transducer, polarizer, and RF rotary coupler, and

2) Motor Drive Assembly: motor, lubricated bearings, resolver, angular rate, and position pickoffs.

Radio frequency energy is circularly polarized, collimated into a plane wave front and focused upon the flat-plate reflector. The beam axis, reflected through a 90° angle, is continually directed toward the subsatellite point by the

Table 3 RF power budget

	Nominal [a]	Worst-case [b]
TWT power output (dbm)	**36.0**	**35.4**
Losses: TWT level set (db)	**−0.3**	**−0.5**
Ferrite switch (db)	**−0.2**	**−0.2**
Filter/detector/coupler (db)	**−0.3**	**−0.3**
Line loss (db)	**−0.3**	**−0.3**
2 channel power split (db)	**−3.0**	**−3.0**
0.5 db power unbalance (db)	**−0.2**	**−0.3**
Cross product loss (db)	**−0.6**	**−0.9**
Beacon power sharing (db)	**−0.1**	**−0.1**
Power level at antenna input (dbm)	**31.0**	**28.8**
Antenna gain (db)	**18.5**	**15.0**
Single channel ERP (dbm)	**49.5**	**44.8**

a Based upon acceptance test and orbital measurements.

b Based upon worst case loss allowance, minimum acceptable TWTA power output and minimum allowable antenna gain at 9.5 of beam center for extreme test conditions.

despin motion of the radiating aperture. A rotary choke joint at the lower end of the MDA housing permits efficient transfer of energy between the spacecraft's fixed and despun waveguide sections. The radiating aperture is rugged and lightweight.

By providing high surface thermal conduction and by using radiation-resistant thermal paint, thermal distortion of the reflector, support, and horn is minimized. Heat transfer across the horn flange is held to a minimum to maintain near-constant temperatures in the MDA for all solar aspect angles.

The MDA employs a d.c. brushless torque motor with resultant high-torque efficiency and good starting characteristics (no special circuits or brushes are required). Vacuum-melt, angular-contact bearings, lubricated with Ball Brothers Research Corporation Vac-Kote lubricant impregnated in the bearing retainers and reservoirs, support either end of the drive shaft. A 2 to 3 lb preload is maintained through the use of a diaphragm spring. Careful attention to the mechanical/thermal design and an extensive bearing test program give high confidence of long orbital life.

A rotary transformer-type resolver, tachometer, and position feedback are used for despin control. Earth-horizon sensor data are compared in the AOCS circuits to maintain pointing accuracy to within $\pm 1.0^{\circ}$. If there is a failure of both redundant Earth-horizon sensors, a backup mode is provided in which simulated Earth pulse signals are generated and referenced to the ultra-stable oscillator of the communications-transponder timing generator. Drift is corrected by ground command; in the backup mode pointing error build-up is held to less than $\pm 2^{\circ}$ with a command update every 10 days.

Telemetry, Tracking and Command (TTC)

The TTC Subsystem performs the following functions:

1) Telemetry data are acquired and formatted for determining spacecraft attitude, equipment status, and command verification. Data are encoded in PCM format and transmitted in a phase-coherent carrier mode whenever the uhf command carrier signal is present. When the command carrier is absent, non-coherent transmission is maintained;

2) Two-way coherent doppler tracking is provided by coherently translating the command carrier for retransmission as

the telemetry carrier. Thirty-one FSK/PM commands are accepted and executed, controlling spacecraft functions and selecting equipment redundancy configuration.

The TTC subsystem block diagram (Fig. 9) depicts the following components: 1) two uhf transponders, each consisting of a telemetry and tracking transmitter, a command receiver, and a diplexer; 2) two uhf antennas, each associated with a specific transponder; 3) one telemetry unit consisting of a telemetry generator and a digital submultiplexer; 4) two command bit detectors and processors in switched redundant configuration, the outputs of each combine in a single command combiner from which control signals are sent to spacecraft equipment; and 5) one control and timing unit (CTU) consisting of a triply-redundant oscillator and timer chain with majority voting and parallel redundant power supplies.

The uhf transponder has two modes of operation: 1) in the absence of a command carrier, the transmitter operates independently of the receiver in a noncoherent mode, and its output carrier frequency is referenced to an on-board local oscillator, and 2) when the uplink carrier is present, the receiver acquires the signal with a phase-lock loop and provides a reference drive signal to the transmitter in the coherent mode. The transmitted carrier is then phase and frequency coherent with the received signal.

The command uplink signal is received by the selected antenna and coupled to the receiver via the diplexer. The two-tone FSK command message is demodulated in the command receiver.

The transmitter multiplies, in frequency, the coherent drive signal to produce an output at 400 MHz. The transmitted carrier is biphase modulated by the 1024-Hz telemetry subcarrier, which is, in turn, biphase modulated with telemetry data at 256 bps.

Isolation between transmitter output and receiver input is provided by the diplexer, permitting the use of a single uhf antenna for reception and transmission. Total spherical antenna coverage is achieved by employing two-monopole arrays at either end of the spacecraft, each providing more than hemispherical coverage. Each array consists of four $\lambda/4$ elements mounted on a 15-in. diam ground plane.

The telemetry unit receives analog and binary monitoring data from spacecraft subsystems. Analog signals are time-division multiplexed and converted into a seven-bit binary code. Encoded analog data are formatted with input digital data and frame synchronization signals to obtain a serial PCM bit stream. The multiplexer output frame is 64, eight-bit words with a 2 sec repetition rate. A digital submultiplexer increases the total data capacity by subcommutating 48 binary inputs on slowly changing spacecraft parameters.

The command bit detector demodulates the sequence of tone bursts (2000/2900 Hz) into a baseband PCM bit stream which is decoded in the command processor.

Discrete command functions are used for on-off switching operations. The proportional vector functions command the controls which require quantitative inputs (for example, SPS pulsing and MDA backup). Command verification is provided by using sequences of two or more command words to complete control operations. Typically, the execute vector in the first word directs the desired command vector into temporary storage for verification by telemetry. After verification, a second word is transmitted which contains the appropriate execute vector to enact the stored command.

The command function is of paramount importance in potential diagnostic and corrective orbital operations. Consequently, command equipment redundancy control is handled in a unique manner. Command receivers, bit detectors, processors, and associated d.c.-d.c. converters are automatically cycled by alternately switching power to redundant units. This power switching is controlled by the control and timing unit, which alternates the bus voltage to d.c.-d.c. converter No. 1 and d.c.-d.c. converter No. 2. A 30-min cycle is provided for launch and transfer orbit and a 32-hr period is allotted for on-station operations. When a satisfactory command link exists the command reset of the CTU timer inhibits the switchover.

Electrical Power

The 29.0-v bus supplies the subsystem d.c.-d.c. converters with the power which is needed for each subsystem. The converters, in addition to supplying positive and negative controlled voltages to associated equipment, accomplish two additional functions: 1) to limit the maximum current drawn from the bus, if an equipment malfunction raises current demands above set threshold, and 2) to highly attenuate noise and ripple accepted from or reflected onto the bus.

Selective load inhibition protects the essential functions from temporary undervoltages. If main-bus voltage falls below

25.5 v, telemetry and tracking are inhibited. If undervoltage persists for 0.5 sec, the communications transponder is replaced with a dummy load across the main bus. Automatic, sequential restoration of the operation occurs when the bus voltage returns to 27.5 v.

The solar array consists of eight curved panels forming a cylinder 54 in in diameter and 32 in high. Panel substrates are formed from aluminum honeycomb with fiberglass facing sheets. Cells are laid down in nominal 19- x 3-cell modules, but this arrangement is modified on panels having apertures for SPS nozzles, umbilicals, and sensors. Four series modules form strings 76 cells long, with four strings per panel. Cells are covered with a fused silica cover slide which has an antireflective coating and a blue filter. Bypass diodes are incorporated where cells are shadowed by SPS nozzles. Table 4 summarizes other solar array characteristics.

Table 4 Solar array characteristics

Solar cell	Cover slide		
Size: 2/2 cm x 0.014 in Type: silicon N/P, diffused junction Number: 7236 Efficiency: 11.2% (A.M.O.)	Thickness: 0.006 in Material: fused silica Coatings: antireflective, 410 mμ bluefilter		
Solar array			
Weight: 21.87 lb	Power (@30.35v)	Solstice	Equinox
Output voltage: 30.35 v	Start of life - 5 yr in orbit	97 w 86 w	105 w 92 w

The power control unit provides array output regulation, battery output regulation, battery charging control, automatic load control, and ABM ignition/heater control. A partial shunt regulator limits the array output to 29.4 v. This controls the current that is shunted from the lower half of the solar array. Extensive redundancy is employed in the regulator.

Battery voltage is always lower than the main bus voltage. When required, a pulsewidth-modulated boost converter is used to discharge the battery. There are two batteries on the satellite which are connected for independent operation. They are charged at a constant 600 ma until they reach a temperature-dependent voltage limit; each battery is charged alternately in 15-min intervals. Charging current is reduced if the battery temperature exceeds 105°F, or if solar array power output becomes marginal. If excessive current is drawn, battery charging is shut down. Undervoltage sensing prevents excessive discharge. A command override is provided for many functions.

Batteries deliver 77 w to the boost converter during eclipse. Each battery is constructed of 16 hermetically sealed cells, each with a minimum capability of 6 ah. Cells are assembled in two rows of eight each under a structural preload: 1) to provide structural rigidity and thermal conductivity, 2) to improve cell electrical characteristics by positive pressure, and 3) to prevent diaphragm cycling because of outgassing.

Attitude and Orbit Control Subsystem (AOCS)

The AOCS provides attitude sensing, orientation, and a stationkeeping. It incorporates the following components:

1) Sun-sensor assembly - two sun sensors and two sun-angle sensor heads and associated electronics;

2) Earth-sensor assembly - two IR Earth sensors and two sun-guard sensors;

3) AOC electronics unit (AOCE) - logic circuitry which processes sensor information; and a

4) Nutation damper.

A simplified block diagram is shown in Fig. 10.

Sensors

Sensors are used in the subsystem to determine attitude in the control circuits for spacecraft maneuvers.

The Earth-sensor assembly provides steering signals for the communications antenna and an indication of spin axis orientation relative to Earth. The assembly comprises: 1) two Earth sensors (sensitive in the 14-16 μ-band) each with a nominal 1 x 1° field of view, and 2) two 7.2 x 20° sun-guard sensors. The optical axes of the sensors lie in the plane of the spacecraft spin axis and are offset ±4° from the equatorial plane.

Two sun-pulse sensors provide a synchronization reference for the SPS thrusters, a measure of spin rate, and a backup reference for directing the communications antenna. Similar to the sun-guard sensors, except that their 10 x 120° fields of view overlap ±30° to provide 180° coverage, these sensors provide redundant coverage for normal on-station operation.

Sun-angle sensors provide a digital output accurate to an angular resolution of 0.75^{o} for total field of view equal to 180°. This attitude information is used primarily in the transfer orbit during the periods when the Earth sensors are not viewing Earth. In addition, the sun-angle sensors provide redundant sun-pulse information.

Nutation Damper

The damper removes nutational motion after booster separation and following ABM and SPS operation. The device is simple; it consists of a steel ball which is free to move in a curved nitrogen-filled aluminum tube. Because of small ball-tube clearances, energy dissipation primarily results from viscous drag. The time constant of the damper is approximately 6 minutes and the dead zone is less than 0.2^{o}.

Electronics Unit

The AOCE processes sensor information to supply a communications antenna steering signal, pulse synchronization for SPS thrusting, and outputs representing roll attitude and spin rate.

Earth-pulse switching circuits receive Earth-edge pulses from the Earth sensors and sun-guard sensor signals. The output of either Earth sensor may be used to control the communications antenna; however, when the sun appears in a sun-guard sensor's field of view, control is automatically switched to the alternate Earth sensor. Earth pulsewidth is measured to calculate spacecraft attitude. The time difference between positive and negative Earth-edge pulses is telemetered. From this data the roll attitude is calculated to an accuracy of ±0.3^{o}.

The sun-pulse threshold and combiner conditions and combines sun-sensor outputs. The sun pulse is forwarded to the command processor for synchronization of SPS thrusting. Sun-pulse logic provides a measure of spin rate which is telemetered with an accuracy of 0.1 r/min.

If the Earth sensor fails, the backup mode is initiated by command. The sun-pulse, with a variable delay time, is substituted for the Earth pulse. For communications antenna control, Earth-edge pulses are processed in the AOCE to form the error signal which controls the MDA.

Secondary Propulsion Subsystem (SPS)

The SPS (Fig. 11) consists of the following: 1) four thrust chamber assemblies (TCA) in redundant pairs aligned axially and radially with respect to the spin axis, 2) valve drivers, 3) manifolding, 4) fill and drain valves, 5) pressure transducer, and 6) tankage. The subsystem is nitrogen pressurized and fueled with 25 lb of hydrazine, used in a monopropellant reaction, and catalyzed by Shell 405 to provide nominal 5-lb thrust pulses for orbit velocity and orientation corrections.

A propellant control valve and a catalyst chamber/nozzle assembly, connected by a narrow gauge heat-isolating pipe, forms the TCA (Fig. 12). Earlier in the program a copolymer elastomer soft-seat valve was used, but leakage problems led to the adoption of hard-seat valves in a dual-seat configuration. The catalyst chamber is thermally isolated from both the spacecraft and the control valve. The spray head widely distributes propellant over the catalyst bed. Two grades of catalyst are used to obtain quick startup with a cold bed and with minimal pressure drop when operating in a steady-state. The length of the catalyst bed represents a compromise between incomplete decomposition and endothermic disassociation of the ammonia product, either of which result in lower specific impulse. The nozzle is sized to obtain the best tradeoff between insufficient expansion ratio and excessive boundary layer growth, again optimizing specific impulse. Individual valve drive amplifiers feed the torque-motor valves with pulses initiated in the TTC subsystem.

Tank pressure is telemetered from orbit, indicating, with appropriate temperature correction, the remaining fuel volume.

Titanium propellant tanks and tubing are welded to steel alloy TCA connections. Special transition joints are formed by friction welding the otherwise nonweldable materials. The use of transition joints creates a flexibility which permits arc welding for subsystem assembly.

Fig. 1 IDCSP/A satellite.

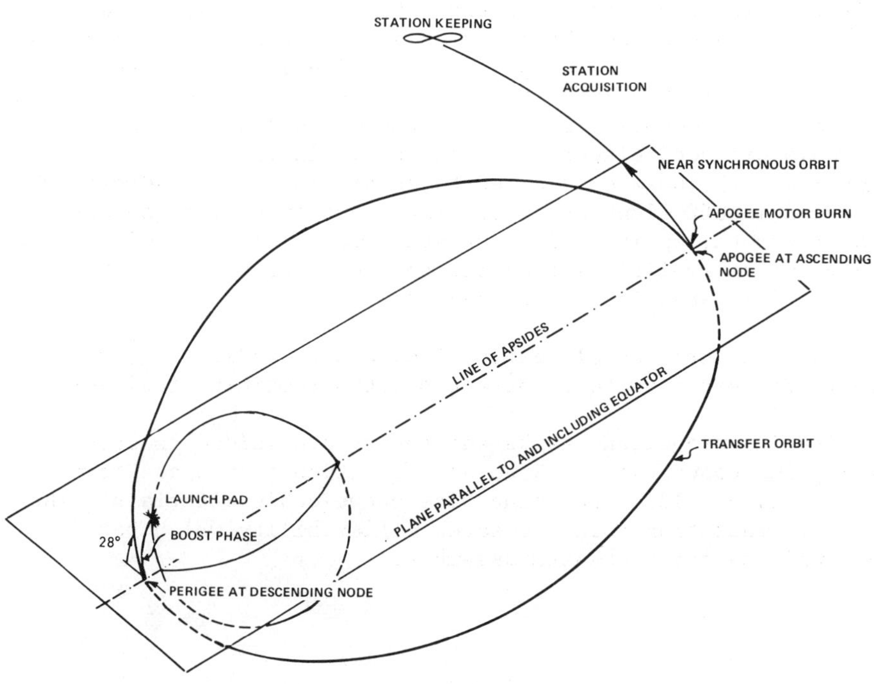

Fig. 2 Mission profile geometry.

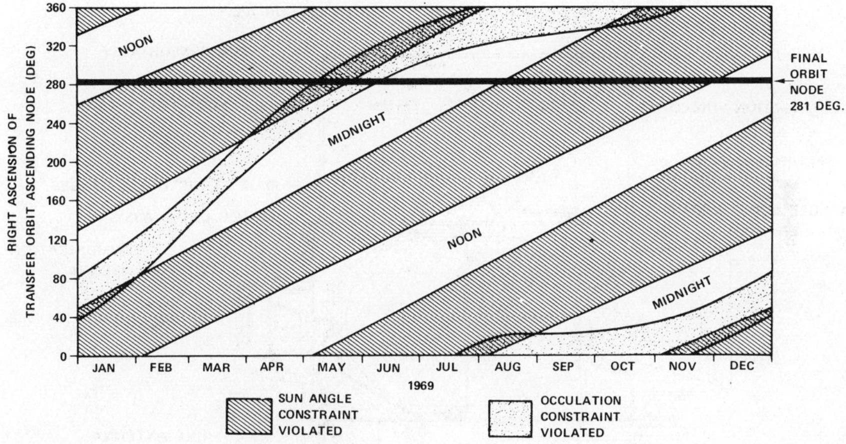

Fig. 3 Thermal/power constraints.

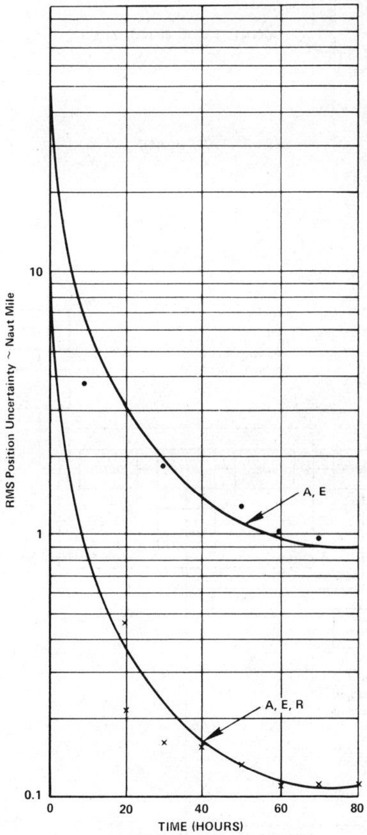

Fig. 4 Position uncertainty of transfer orbit for eight periods.

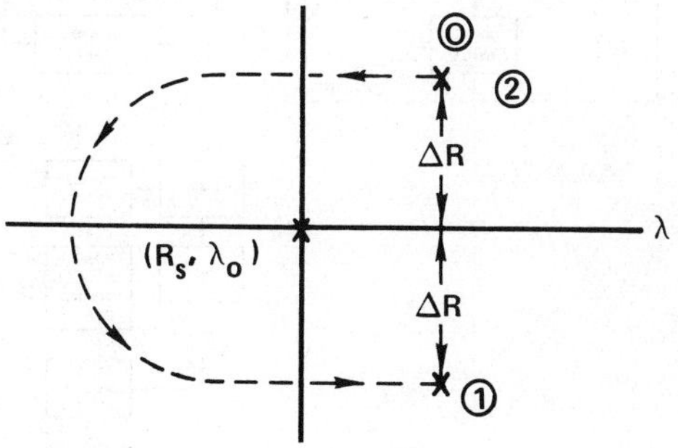

Fig. 5 Stationkeeping corrections.

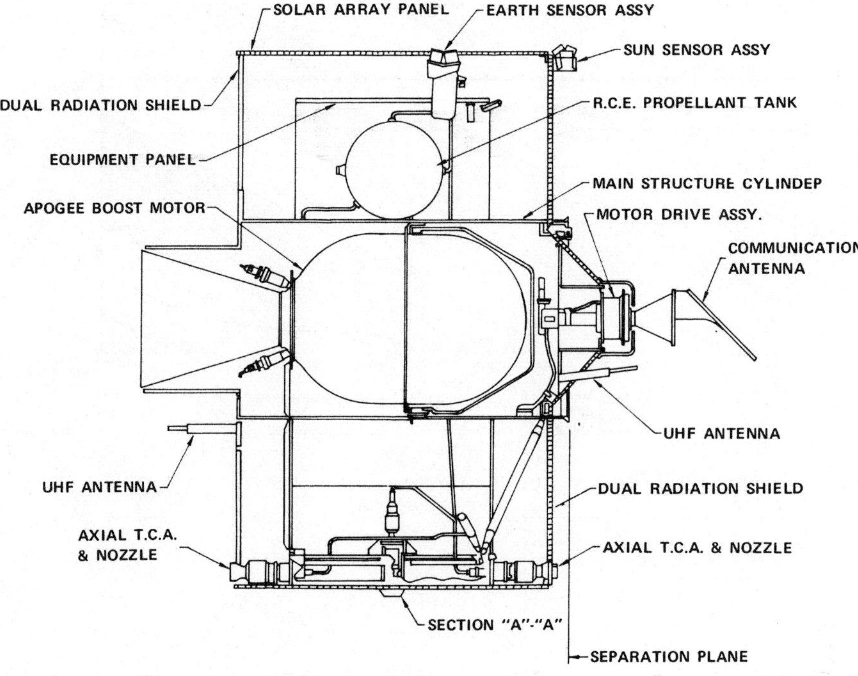

Fig. 6 Spacecraft layout.

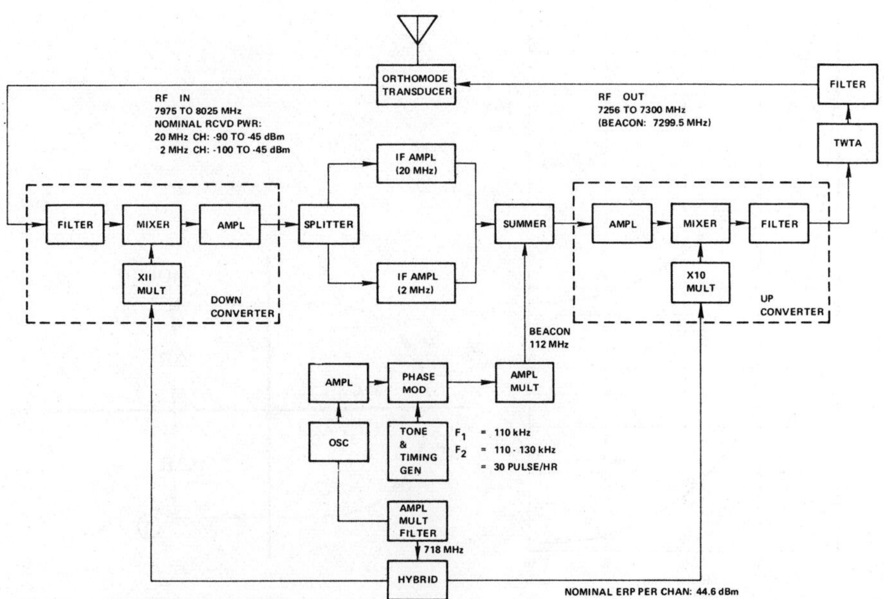

Fig. 7 Communications transponder.

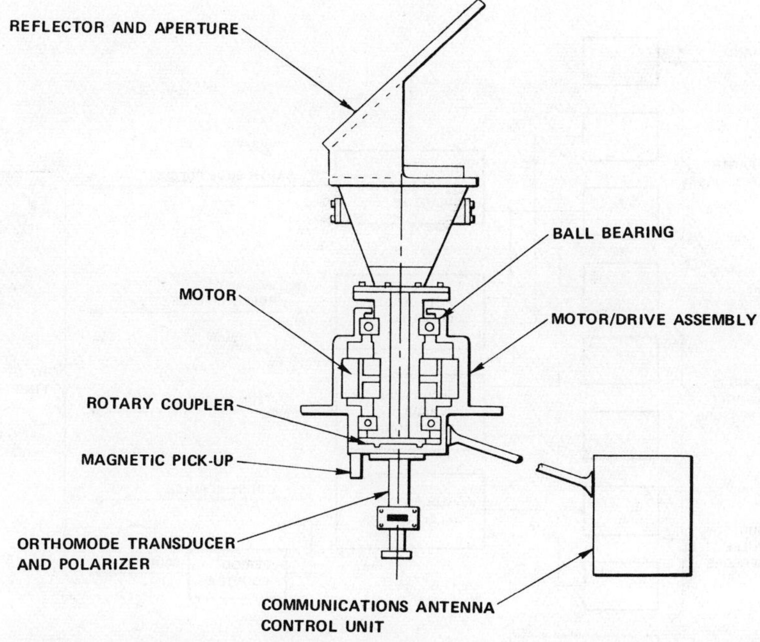

Fig. 8 Communications antenna.

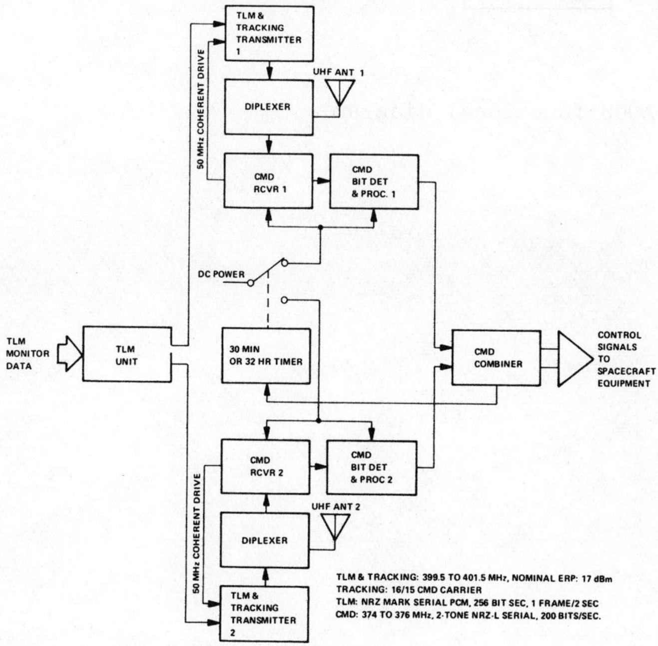

Fig. 9 TTC subsystem.

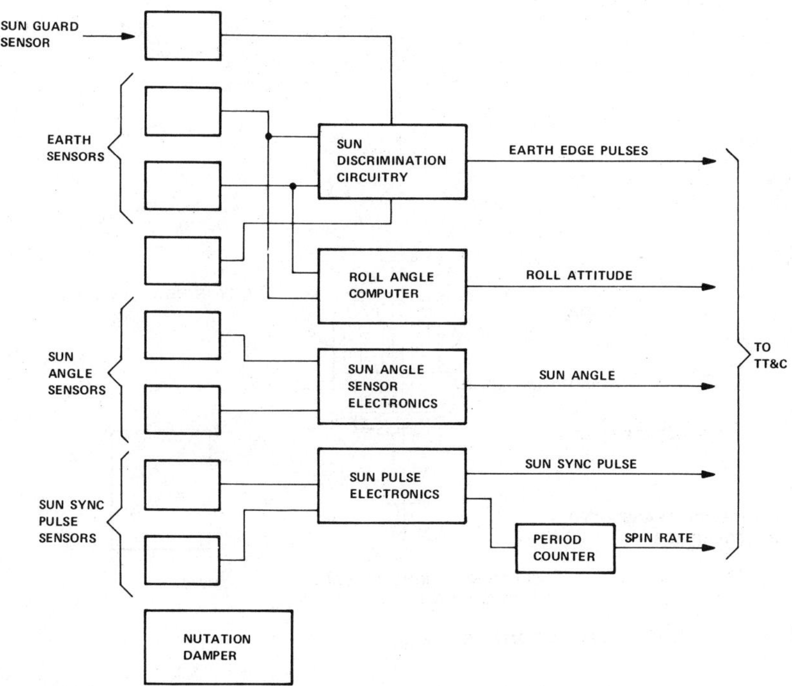

Fig. 10 AOCS functional diagram.

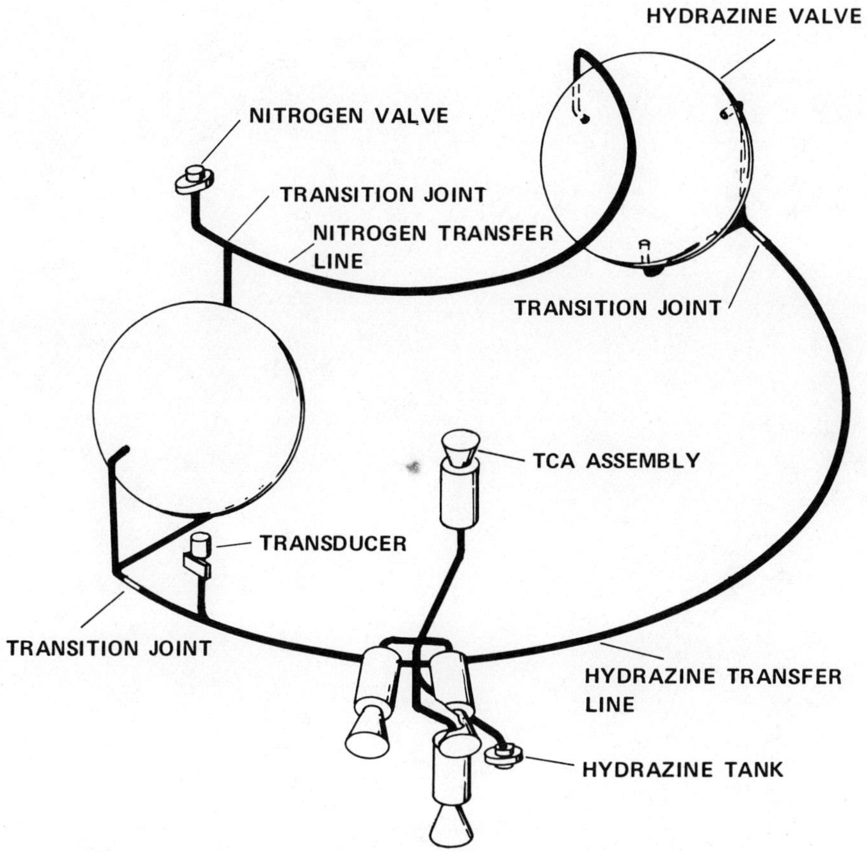

HYDRAZINE VALVE
NITROGEN VALVE
TRANSITION JOINT
NITROGEN TRANSFER LINE
TRANSITION JOINT
TCA ASSEMBLY
TRANSDUCER
TRANSITION JOINT
HYDRAZINE TRANSFER LINE
HYDRAZINE TANK

Fig. 11 Reaction control equipment schematic.

9.72
CATALYST BED
VALVE
THRUSTER
1.44 DIA. MAX
5.10

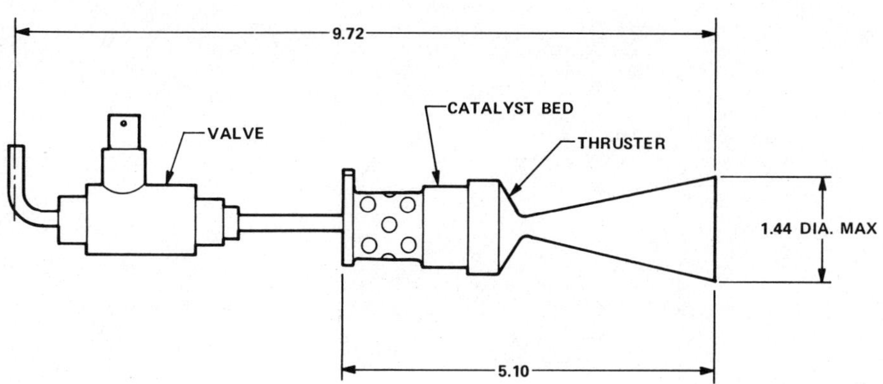

Fig. 12 Thrust chamber assembly envelope.

A PHASE II SATELLITE FOR THE DEFENSE SATELLITE COMMUNICATIONS SYSTEMS

A. T. Finney*

The Aerospace Corporation, El Segundo, Calif.

Abstract

This satellite is now being developed for the Defense Satellite Communication System. The satellite will provide communication capability via a directional earth coverage antenna and steerable narrow coverage antennas. The communication transponder design will provide separate earth coverage, narrow coverage communication channels, and a capability for crossover between channels. The satellite will operate at synchronous altitude and will have stationkeeping capability. It will be spin-stabilized and will utilize a despun platform on which will be mounted the earth coverage and narrow coverage antennas and the transponder. Earth orientation of the despun platform will be maintained by means of earth sensor IR data. Control of the narrow coverage antennas will be by ground command and by automatic on-board circuitry. Power will be supplied from a body-mounted solar array. Other design features include a hydrazine propulsion system and S-Band command telemetry equipment.

Introduction

Since its initial deployment in 1966, the Defense Satellite Communication System has been providing a vital military communication function. Originally intended as an interim system, the satellite segment already has exceeded life predictions by a factor of three.

Presented as Paper 70-493 at the AIAA 3rd Communications Satellite Systems Conference, Los Angeles, Calif., April 6-8, 1970

*Systems Engineering Director, Program 777.

Simplicity was a primary goal in the design of the Phase I system. Toward this end, the satellite contained a single channel X-band repeater together with a minimum state-of-health telemetry capability, but without a command capability. Power was provided by a solar array, but no batteries were utilized for eclipse operation. The Phase I satellites were deployed into drifting near-synchronous orbits in multiple payloads on the T-IIIC. Of the twenty-six satellites placed in orbit, twenty-three continue to operate satisfactorily and will continue to support the satellite communication needs of the DCS until the Phase II system is deployed in 1971.

Planning for the Phase II satellite system has been proceeding for the past several years and, during that period, the basic design concept has undergone several major iterations. After deployment of the Phase I system, it was planned that future military satellites would take advantage of the reliability improvements in command/control and stationkeeping capability which had ensued. Accordingly, both the TacSat and Skynet satellites were designed to incorporate provisions for geostationary operation. With the passage of time, additional state-of-the-art improvements were considered for the Phase II design. The most notable of these was the incorporation of narrow-beam antennas that would provide the potential for focusing rf energy over a small portion of the Earth's surface, providing such attendant advantages as increased channel capacity and privacy. In addition, it was planned to design a secure command system into the satellite up-link as well as down-link telemetry encryption. Other improvements included the provision of a capability for rapid repositioning of a satellite to a new operating point, as well as the incorporation of certain survivability features.

In mid-1968, a Phase II concept was finalized and approved for procurement. This concept embodies the satellite features previously discussed, as well as improved ground stations which would take advantage of increased satellite performance. Design, development, and test of the satellite are being performed by TRW Systems under a contract from the Space and Missile Systems Organization of the Air Force Systems Command. Effort on this contract has been under way since March 1969, with its first launch scheduled in 1971.

The following paragraphs describe some of the salient features of the satellite with particular emphasis on communications and attitude control, two areas which represent a particular amount of technical sophistication.

General

The basic satellite configuration is illustrated in Fig. 1. Structurally, the vehicle consists of two sections which have an angular velocity relative to each other. The outer cylindrical section is composed of a body-mounted solar array, an aluminum truss structure which bears the principal launch loads, and an equipment platform. The outer section spins at approximately 60 rpm, with the spin axis oriented normal to the orbital plane. The inner section of the structure consists of a despin platform which is isolated from the outer section by a despin bearing/motor assembly. By means of this assembly and its associated controls, the despin platform is driven so that a reference point on the platform is continuously pointed toward the center of the Earth. All of the communications equipment is mounted on the despin platform. A pair of earth coverage communication antennas, transmit and receive, are fixed rigidly to the platform and are aligned with the aforementioned reference point so that they are continuously pointed at Earth center. Also mounted on the despin platform are a pair of parabolic narrow-beam antennas, each of which can be independently driven in two axes.

Command and control of the satellite is achieved by means of an onboard tracking, telemetry, and command (TT&C) subsystem which is compatible with the Space-Ground Link System employed by the Air Force Satellite Control Facility located at Sunnyvale, Calif. Most of the components of this S-band equipment are located on the spinning platform; however, certain elements, chiefly those associated with the communications and attitude control functions, are contained on the despin platform. The TT&C antenna is deployed from the aft end of the satellite and is attached to the spinning section.

Attitude control and stationkeeping of the satellite are achieved by the operation of a hydrazine monopropellant system, the thrusters for which are mounted on the spinning section of the satellite in both axial and radial positions. In addition to its normal attitude control and stationkeeping functions, the onboard propulsion equipment will be capable of making at least one, and possibly more, major translations of a satellite to a new orbital position, depending upon the velocity at which these maneuvers are performed.

The primary source of power for the satellite will be a body-mounted solar array composed of N/P solar cells and attached to the spinning structure. Three nickel-cadmium batteries will provide a secondary source of power for use during eclipse.

Communications

The Phase II satellite will be tasked to accommodate a wide range of missions from long-haul trunking to intra-theater communications and will be required to be compatible with terminals varying in size between large ground stations and airborne command posts. To this end, the satellite communication system has been designed to provide maximum operational flexibility by utilizing four separate communication channels that can be interconnected to operate with earth coverage and narrow beam antennas in various modes.

Some of the outstanding features of the multichannel transponder are illustrated in Fig. 2. There are four possible paths or channels which an incoming signal can take, depending on its frequency. Channel 1 is a straight-through Earth-coverage channel which transmits and receives via the Earth-coverage antennas. Its nominal bandwidth is 125 mHz. Channel 3 is the straight-through narrow-beam channel which transmits and receives via the narrow-beam antennas and has a nominal bandwidth of 185 mHz. There are also two crossover channels: Channel 2, which receives on narrow-beam and transmits on Earth-coverage, and Channel 4, which receives on Earth-coverage and transmits on narrow-beam. Both of these crossover channels have a bandwidth of 50 mHz. Figure 3 is a graphic presentation of the frequency plan.

The effective radiated power through the Earth-coverage transmit beam is 28 dbw for any point on Earth from which the evaluation angle to the satellite is greater than 7.5°. The effective radiated power of a single narrow-beam antenna is 43 dbw for any point on Earth from which the line-of-sight to the satellite is within 1° of the desired position of the narrow-beam center. When utilizing the narrow-beam channels, it is possible to transmit in one of three modes. In the normal mode, signals can be received on one narrow-beam antenna and transmitted on the other; in another mode, signals can be received and transmitted on the same narrow-beam antenna; while the third mode of operation is one wherein a signal is received on one narrow-beam antenna and transmitted on both. Any one of these modes is selectable by commanding the position of a switch in the transponder. In the case where two narrow-beam antennas are being used for transmission, the effective radiated power from each one drops to 40 dbw.

Each of the four transponder channels is of the single conversion type, and each channel is redundant in its in-line functions. All of the channels are quite similar in design, the differences being in the receiver diplexer, the filters for receive and transmit, and also in over-all gain distribution. In view of this, the following discussion is addressed to some of the details of the Earth-coverage channel as being typical of the other three.

The Earth-coverage receiver section of the transponder contains a two-stage tunnel diode amplifier acting as a preamplifier. Following the preamplifier is a diplexer/equalizer assembly which routes incoming signals into the appropriate channel, depending upon their frequency. In the case of the Earth-coverage receiver section, there are two possible routes for a signal; the straight-through Earth-coverage channel and the Earth-coverage/narrow-beam crossover channel. Each of these two channels has an associated filter and an equalizer section. The output of the equalizer section (in this case, the straight-through Earth-coverage signal) is then transmitted to the tunnel diode amplifier limiter that has a gain setting, adjustable by ground command, which can vary over a 24-db range. Control of the gain of these devices is achieved by varying the bias settings or, for the lowest possible setting, by turning off a stage. The maximum gain setting can be approximately 47 db.

The next section of the transponder contains a mixer channel combiner which provides the necessary conversion from the receive to the transmit frequency. In the case of the Earth-coverage straight-through channel, this offset is 725 mHz. After frequency conversion, the signals for Channels 1 and 2 are summed in a hybrid and fed into the transmitter section.

At the input to the transmitter section is a level-adjusting pad which is used to produce a difference in signal of 6 db, favoring Channel 1 over Channel 2. This signal is then fed to a tunnel diode limiter summing amplifier that drives a low-level traveling-wave-tube (TWT) amplifier. This device has a gain of approximately 36 db and is used to drive a high-level TWT amplifier which has a nominal output of 20 w. Between the low-level and high-level TWTs, there is a variable attenuator which can be set by command to change the operating point of the high-level TWT amplifier.

All of the local oscillators in the transponder are driven from a common 5-mHz crystal oscillator that is contained in the frequency generator. The frequency generator also generates beacon signals for the Earth-coverage and narrow-beam channels. The beacons are bi-phase modulated by an appropriate code that emanates from a code generator within the frequency generator.

Earth-coverage communication will be achieved through two circularly polarized horns, one each for transmit and receive. The receive antenna is approximately 8% smaller than the transmit; however, in other respects, the horns are similar in construction. Both horns are rigidly mounted to the despun platform and aligned with the magnetic pipper.

The narrow-coverage antennas are parabolic reflectors having an aperture diameter of approximately 44 in. Each dish utilizes a feed which consists of a conical horn and also an ortho-mode transducer which facilitates the use of RHCP and LHCP senses.

Attitude Control

The attitude control system which has been designed for the Phase II satellite contains several concepts which merit further discussion since they have been evolved to meet an extremely tight pointing accuracy requirement of 0.3°. This accuracy, largely a result of the requirement for narrow-beam pointing, must be met with a minimum of ground command. In this regard, a program objective is the minimization of attitude control commands so that they need not be spaced closer than 3 weeks for normal operations.

As previously stated, the basic satellite configuration consists of a large spinning rotor and a despin platform, isolated from each other by a despin bearing assembly. When properly deployed and spun up, the satellite has a favorable moment of inertia ratio, at least 1.1/1 at end-of-life, regardless of the state of the despin platform. A simplified block diagram of the attitude control system is contained in Fig. 4.

The functions of the attitude control system can be summarized as follows: to maintain a reference point on the despin platform, and consequently, the Earth-coverage antennas, pointed at Earth center; to maintain the satellite spin axis within acceptable limits; and to point the narrow-beam antennas to selected areas on the Earth's surface and maintain their position within the required accuracy. In actuality, the last requirement has a considerable influence on the degree of accuracy to which the other two are maintained;

however, for clarity, these requirements are separately discussed in the following paragraphs.

The despin control system maintains the despin platform by operating upon an error signal based on Earth orientation. Earth orientation is derived from a pair of Earth sensors mounted on the spinning rotor. The sensors are mounted in a "V" configuration so that one sensor scans the northern hemisphere and the other scans the southern hemisphere. As each sensor rotates past the leading and trailing edges of the Earth, a pair of pulses is generated. These pulse pairs are used to form an Earth image signal. Orientation of the despin platform relative to the rotor is obtained by use of an electromagnetic pickup, one part of which, the magnetic pipper, is located on the platform in close alignment with the Earth-coverage antennas. This pickup produces a platform reference pulse which also is an input to the despin control system. The despin control system drives the despin platform so that the platform reference pulse is centered within the Earth image signal, thus ensuring that the Earth-coverage antennas are pointed at Earth center. Torque is provided by redundant two-phase synchronous motors. The despin controller is actuated by ground command after injection of the satellite into orbit. Before entering the normal mode, the controller will pass through several preliminary modes, each of which is ground-commandable. The despin controller also can be commanded into a standby mode in which all power is removed from the despin platform.

Data from the Earth sensor pairs also are transmitted to the ground via telemetry, where they are used to ascertain the position of the satellite spin axis. This can be accomplished by comparison of chord lengths if data from two sensors are available or by noting any changes in the chord length if only one sensor is utilized. Since the Earth sensor pair will only measure attitude deviation in one plane, it is necessary to make a pair of Earth sensor measurements 90° apart in orbit. This permits errors to commute to the proper axis, orbital roll, where they can be detected by the Earth sensors. When the satellite spin axis deviates from its desired position by an excessive amount, corrections can be applied by means of the onboard thrusters.

The third function of the attitude control system is pointing of the narrow-beam antennas. Each narrow-beam antenna can be driven independently in azimuth and elevation through gimbals mounted on the despin platform. The driving mechanism for the gimbals are stepper motors acting on harmonic drives which result in an incremental step of approximately 0.03°. Maximum angular rate at which an antenna can be driven is 0.12°/sec.

Commands for driving the narrow-beam antennas may emanate from three sources. These are a) slewing commands from the ground to change antenna position, b) nodding commands to compensate for the pointing errors introduced by orbit inclination, and c) bank angle compensation commands to correct for spin axis drift.

Commands for initial antenna position, as well as for any subsequent repositioning, are generated by the Satellite Control Facility using a computer program prepared for that purpose. These commands are transmitted to the vehicle via the command system and drive the antennas open loop through the gimbal/stepper motor assembly as previously described.

At present, it is planned that the operational orbit will be chosen so that the orbital inclination of the synchronous system will not exceed 3°. If uncorrected, the results of this inclination would be an apparent oscillation of the antenna line-of-sight having a maximum value of approximately 0.5° for a maximum inclination of 3°. To compensate for this effect, the onboard attitude control system will contain a separate programmer which generates a trapezoidal approximation to a sine curve having a period of one sidereal day. A separate program will be generated for each gimbal in the antenna drive assemblies. Since the actual inclination will be varying from 3° to 0 and then back to 3° over a period of 5 years, it is necessary to make periodic adjustments in the phase and amplitude of each command. These adjustments, as well as periodic updating of the programmer clock, are accomplished by ground command.

To minimize the ground commands required to maintain pointing accuracy, the attitude control system contains a bank angle compensator that automatically adjusts the narrow-beam antenna when the spin axis deviation exceeds a nominal value. The Earth image signal, described in a previous paragraph, is periodically examined by the appropriate logic circuitry, and any observed changes in bank angle greater than one gimbal step will occasion a command to the narrow-beam antennas.

An important feature of the attitude control system design are its damping characteristics. In this respect, the satellite will utilize a tuned pendulum damper mounted on the despin platform. The time constant of this damper will be less than 10 min for any operating extremes of the satellite design range.

System Deployment

The Phase II satellite is being designed for launch on a T-IIIC booster. The nominal launch configuration, as illustrated in Fig. 1, will consist of two satellites in a stacked configuration with the narrow-beam antennas of each satellite folded down in a stowed position. The T-IIIC will be launched from the Eastern Test Range and will inject the payload into a synchronous near-equatorial orbit. A typical launch and injection sequence is shown in Fig. 5. Each satellite is injected into orbit with the T-IIIC oriented normal to the orbital plane. After separation of a satellite from the T-IIIC, the despin platform is unlocked, the TT&C antenna is deployed, and solid propellant rockets spin the satellite up to the desired rotational rate. Despin of the platform is then accomplished by ground command, and the narrow-beam antennas are deployed, also by ground command.

After the proper amount of tracking data has been accumulated, each satellite will be placed in an orbital position where preliminary test and checkout can be performed. The satellites will then be deployed into their final operating position in the DCS network.

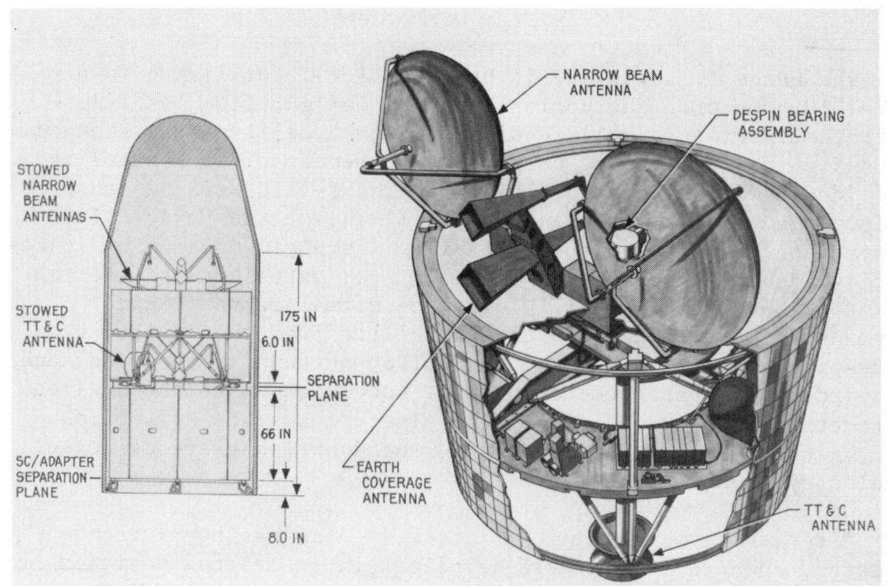

Fig. 1 Payload design.

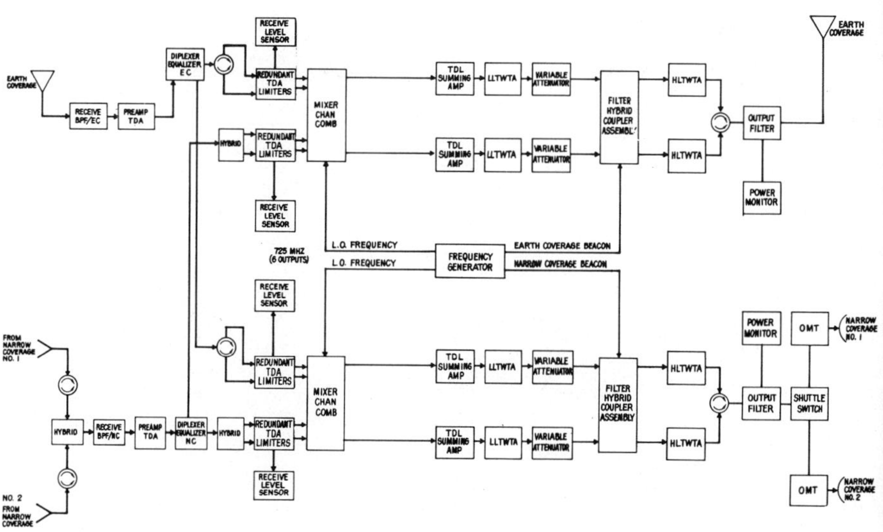

Fig. 2 Transponder block diagram.

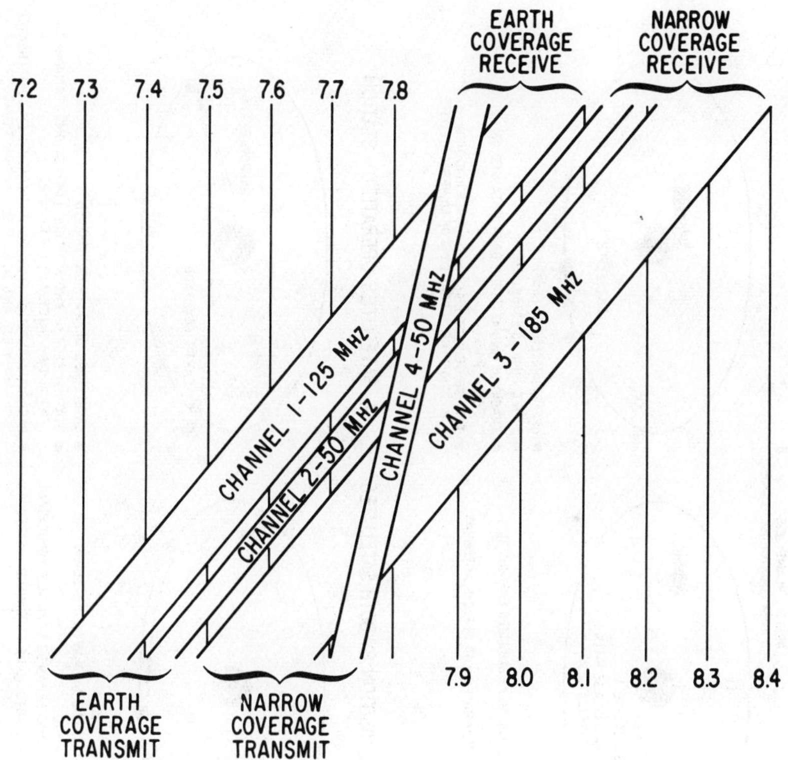

Fig. 3 Frequency plan.

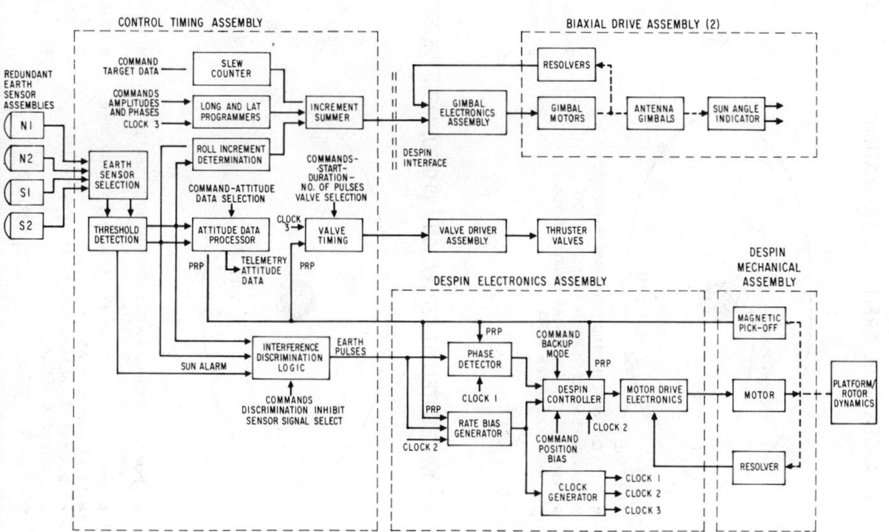

Fig. 4 Attitude control and station keeping subsystem.

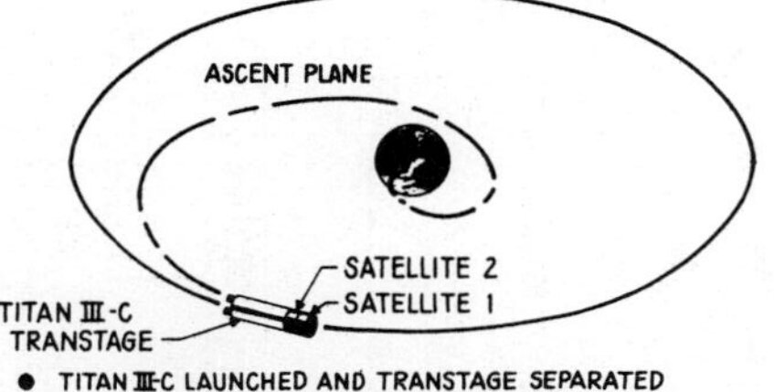

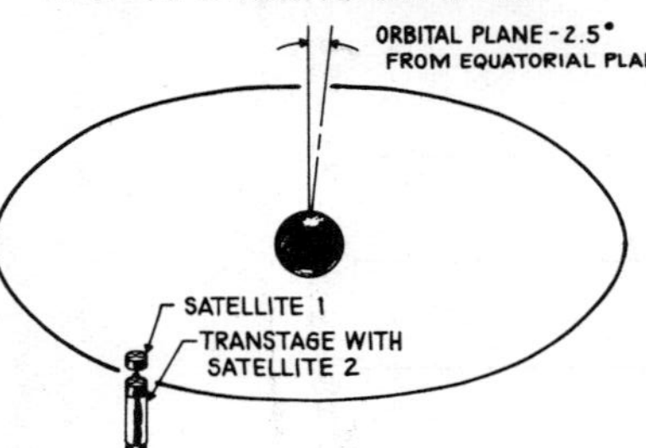

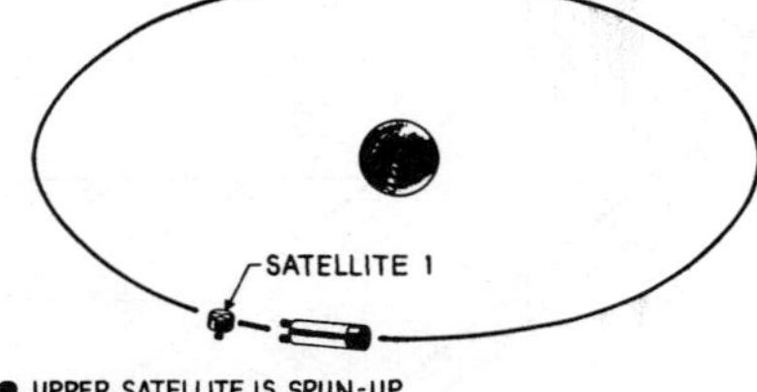

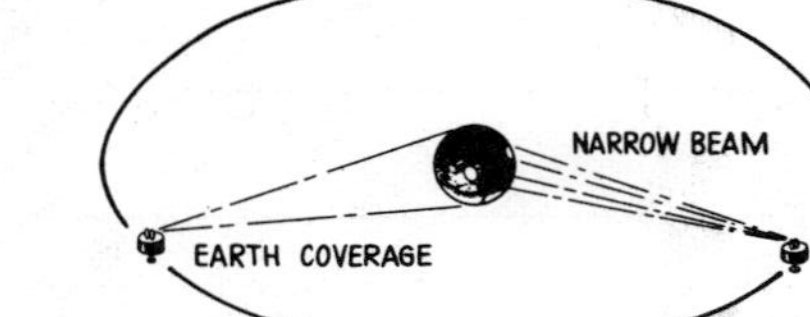

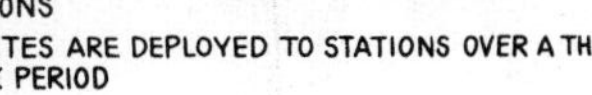

Fig. 5 Mission sequence.

LINCOLN EXPERIMENTAL SATELLITES 5 AND 6

D. C. MacLellan,* H. A. MacDonald,† P. Waldron,‡ and H. Sherman[δ]

Massachusetts Institute of Technology, Lexington, Mass.

Abstract

The paper describes the satellites and their use both as part of a demonstration of the feasibility of satellite communications at VHF/UHF to small mobile terminals and as test beds for experimental subsystems of interest to spacecraft designers. Both spacecraft employed highly efficient solid-state rf sources and circularly polarized antennas. The LES-6 antenna was electronically despun to realize about

---

Presented as Paper 70-494 at the AIAA 3rd Communications Satellite Systems Conference, Los Angeles, Calif., April 6-8, 1970. This work was sponsored by the Department of the Air Force. The success of LES-5 and -6 is due to the many people, inside and outside of Lincoln Laboratory, who worked the long hours, paid meticulous attention to details, and, motivated by professional pride, refused to settle for anything except a job done properly. Organizations outside the Laboratory deserving of particular thanks include: EOS, Walter Kidde Company, Aerospace Corporation, Western Microwave Laboratory, and Fairchild-Hiller, Republic Aviation Division. Inside Lincoln Laboratory, the authors must express their gratitude especially to the staff, technicians, designers, and draftsmen personnel of Groups 61, 62, 63, and 71 and to the machinists of Group 75 who constituted the indefatigable "satellite crew."

*Group Leader, Space Techniques and Equipment Group, Lincoln Laboratory.

†Associate Group Leader, Mechanical Engineering Group, Lincoln Laboratory.

‡Assistant Group Leader, Space Techniques and Equipment Group, Lincoln Laboratory.

δStaff Member, Communications Division, Lincoln Laboratory. Presently Technical Director of Ambulatory Care Program, Beth Israel Hospital, Boston, Mass.

8.5-db gain. Experiments emphasized autonomous satellite operations in stationkeeping and attitude control. The stationkeeping thrust was provided by a novel pulsed plasma thruster. Other experiments measured uplink rf interference in the band from 255-315 MHz, localized Earth albedo in the visible spectrum, and the effects of the synchronous orbit environment on solar cells. The on-orbit performance of the satellites and the experimental subsystems are included.

I. Introduction

LES-5 and -6 were built as part of a program to demonstrate the feasibility of using satellites that operate in the military UHF band (225-400 MHz)** for communications to and from small mobile terminals. Propagation measurements in this band had been made using a previous satellite in the program, LES-3,[1] which functioned as a simple beacon transmitter. This paper discusses LES-5 and -6, their principal subsystems and experiments, and reports their on-orbit performance. LES-5 and LES-6 are similar in many respects. Of the two satellites, LES-6 is the more complex and, from the standpoint of the experimental systems on board, the more interesting. Accordingly, the bulk of this report will be devoted to LES-6.

II. General Description, LES-5 and -6

Titan III-C Vehicle No. 14 placed LES-5 into a subsynchronous orbit on July 1, 1967. The orbit has a period of 1316.1 min and was inclined to the equatorial plane at 7°. Similarly, Titan III-C No. 5 placed LES-6 into a 2.9° inclined synchronous orbit on September 26, 1968. Final adjustment into stationary orbit was accomplished by use of an on-board cold ammonia thruster system.

LES-5 and -6 are shown in Figs. 1 and 2. Each is a right circular cylinder with a flat equipment platform dividing the cylinder near the midpoint. Above the periphery of this platform is a section called the view band through which sensors and thrusters view the outside world. Each satellite spins about its long axis which is held perpendicular to the orbit plane by an onboard attitude control system. Physical properties of both satellites are given in Table 1, and their subsystems, experiments, and rf characteristics are summarized in Tables 2 and 3. The physical similarity between LES-5 and LES-6 is shown in Fig. 3. The extensions beyond

**This band overlaps VHF and UHF but will be referred to as UHF.

Table 1 Physical properties

	LES-5	LES-6
Diameter, in.	48	48
Length, in.	64	66
Weight, lb		
Prime structure	22	29
Equipment	115	256.5
Solar array and antennas	63	82.5
Dispenser assembly	22.5	25.4
Trim weights	7	4.7
	229.5	398.1
I_{spin}, lb in.^{2}	72,032	113,766
$I_{transverse}$ (max) lb in.^{2}	58,050	96,359
I_s/I_t	1.24	1.18
Spin period, sec	5.784	7.353

the LES-5 solar panels were required to provide enough length for a set of cavity backed slot radiators which constitute half of the circularly polarized antenna system. On LES-6, this area was utilized by solar panels to meet the increased power requirements; the solar array was built in four discrete cylindrical assemblies. The sensor view band height was also increased from 4 in. to 6 in. to provide for the large complement of sensors and thrusters. The number of support struts below the deck was increased from 6 to 8 and they were extended to mate with the solar array support fittings to provide additional structural rigidity. Some growth capability had been built into the LES-5 structure and this fact, coupled with a nominal reduction of flight loads for LES-6, meant that only minor structural modifications were required to carry the increased payload weight.

III. LES-5

Because of the similarities between the two spacecraft, this section will be kept brief. Discussion will be limited to those areas which yielded unexpected results or results not found on LES-6.

A. Transponder

The LES-5 transponder[2] does a conventional frequency translation of the received signal to an intermediate frequency

Table 2 Subsystems and experiments LES-5 and -6

SUBSYSTEMS	LES-5	LES-6
Telemetry	100 bps, 8-bit accuracy; 75 channels main frame, 64 channels sub max	100 bps, 8-bit accuracy; 62 channels main frame, 168 channels sub max
Command	32 cmds, 42 bits each at 0.4 bps	60 cmds 47 bits each at 0.4 bps or 6.25 bps
Solar array	30-v series string, 10Ω-cm N on P cells, 6-mil cover slide, no cell contact protection. 136 w max start of life	26-v series string, same cells as LES-5, with cell contact protection 220 w max, start of life
Antenna despinning	Experimental logic flown, no actual switching	Operational electronic antenna despinning system
Stationkeeping	None	Autonomous synchronous orbit E-W stationkeeping $\pm 2^{\circ}$ accuracy in longitude
Attitude control	Autonomous magnetic stabilization system $\pm 2^{\circ}$ accuracy spin axis orientation	LES-5 type mag. stab. system. Autonomous gas thruster system, accuracy variable to $\pm 0.16^{\circ}$
Thrusters	None	Cold ammonia gas system--attitude control and stationkeeping. Experimental pulsed plasma thruster stationkeeping

Table 2 (continued)

SUBSYSTEMS	LES-5	LES-6
Temp. controlled oscillator	None	Dual primary oscillators; temp. control $\pm 0.004^{\circ}$C; frequency drift $<1\times10^{-9}$/day long term
EXPERIMENTS		
Solar cell degradation	5 cells, 2 different types, measure I_{sc} and V_{oc}	30 cells, 9 different cell types, 5 cover slip types; measure 20 point I-V characteristic each cell vs time, incidence angle; measure cover slip degradation vs time
Radiation measurement	None	Advanced LES-4 expt., measure trapped electron spectrum
Earth albedo	None	5 beams, $0.1^{\circ} \times 0.1^{\circ}$, scan Earth each measures Albedo in 6 spectral bands from 0.41μ to 1.00μ
Rf interference at UHF	Measure average rf energy and peak-to-average rf energy in 120 kHz steps from 225 to 280 MHz	Same as LES-5 but from 290 to 315 MHz
Precision spin period	None	Autonomous spin period measurement with 1 part in 10^{6} accuracy

Table 3 rf Characteristics – LES-5 and -6

	LES-5	LES-6
Frequency up	255.1 MHz	302.7 MHz
Frequency down	228.2 MHz	249.1 MHz
Frequency beacon	228.43 MHz	254.14 MHz
Transmitter power	30 w	120 w[a]
Antenna gain (net of filer and matching losses)	2.5 db circularly polarized toroidal pattern	9.5 db circularly polarized, electronically despun
E.I.R.P. (measured in in orbit)	45 w	890 w
Rec. noise figure	3.6 db	3.6 db
Rec. I.F. bandwidth (switchable)	100/300 kHz	100/500 kHz

[a] Varies with maximum available Solar Bus power. Power amplifiers operate in "optimized" mode, directly from Bus.

where it is amplified, limited, and translated again to the transmitting frequency. The signal is then amplified in a highly efficient solid state chain (η = 68%, dc to rf) and fed to the circularly polarized antenna through a triplexing filter. In the viewing period immediately after launch, the transponder was exercised extensively by Lincoln Laboratory and by agencies of all the military services. The ease and flexibility with which communications were established among aircraft, ships, and fixed and mobile ground terminals conclusively demonstrated the feasibility of this form of satellite communications.

Some eight months after launch, on March 18, 1968, the receiver sensitivity suddenly displayed a 17-db decrease as LES-5 emerged from Earth shadow. This sensitivity loss was attributed to an open circuit in the first rf preamplfier brought about as the average temperature of LES-5 decreased. (LES-5 average temperature is highest in December and lowest in June.) This hypothesis was borne out when, on November 14, 1968, the sensitivity returned to normal. Subsequently, the sensitivity has dropped and recovered in March

1969 and November 1969, respectively. In addition to the problem with sensitivity, one of the two transponder local oscillators (L. O.) exhibited a sudden change in frequency of 1700 Hz (about 8 parts in 10^6) in December of 1968. This oscillator had previously exhibited a long term stability of about 3 in 10^7. Capacitors similar to the type used in this L. O. have subsequently been observed to experience an abrupt change in value. Although this may be the cause, it is not certain. The L. O. offset shows up as a shift in translation frequency which causes no operational problem to ground users. With these two exceptions, the transponder has performed as designed. The sensitivity decrease demands more uplink power from a ground station but in practice has not hampered the usage of LES-5.

B. Command System and RFI Experiments[††]

The command system and Radio Frequency Interference (RFI) experiment receive timing information from the same source. Immediately after injection into orbit (19:23 GMT) timing signals were seen to be running at a rate almost twice normal. This precluded the satellite from accepting commands and the RFI experiment from stepping frequency at the required rate. Seven hours after ejection the problem spontaneously cleared up for some five hours and then reappeared. After several months, of mostly abnormal timing, it became clear that the period of normal timing occurred in the five hour interval around that point in the LES-5 orbit when it was opposite the sun. Although not proven, it seems highly likely that the source of the extra pulses is due to an unsuspected coupling from one of the logic systems which acts on Earth-sun inputs and changes its state at this point in orbit. This event reinforced the need for stringent system compatibility tests before launch.

Data have been collected from the RFI experiment as LES-5 moves around the Earth. Although these data are limited to the period around local satellite midnight, it has proven most interesting in that particular sources of interference can be detected and identified. Complete results will be presented in a forthcoming publication.[3]

††The RFI instruments flown on both LES-5 and LES-6 were built by the Laboratories Division of the Aerospace Corporation and were supplied to Lincoln Laboratory in a cooperative venture.

C. Solar Cell Array

The LES-5 solar cell array was designed to provide power for a five year lifetime on the assumption that degradation caused by trapped radiation would amount to no more than a 5%/yr decrease in maximum available power. In May 1968, however, the telemetered solar bus voltage was seen to have dropped to 22 v from its launch day value of 30 v. This was the first indication that the LES-5 array was experiencing the same accelerated degradation previously reported on other synchronous spacecraft in early 1968.[4,5] The best extimate is that the LES-5 array degraded 22% in the first year; 4% short term degradation caused by uv darkening of cover slides, 15% caused by proton edge effects, and 3% caused by other miscellaneous effects. Had this rate continued, the rf power amplifiers would shortly have shown a marked decrease in output power as their dc converters shut off as a result of low bus voltage. As of January 1970, this has not yet happened, leading the conclusion that after the first year or so the degradation rate decreased markedly.

In addition to difficulties caused by the high degradation rate, the LES-5 solar array experienced an open circuit in one series connected string on a panel. This had the effect of dropping the available power by an additional 13% once per satellite revolution when the affected panel is illuminated by the sun.

D. Magnetic Stabilization and Antenna Switch System

LES-5 carried an autonomous attitude control system to maintain its spin axis perpendicular to the orbit plane. Orientation is accomplished by a magnetic torquing system which has been described elsewhere.[1,6] The in-orbit operation of this system was degraded by a failure in one of four sun sensors which trigger the switching action necessary to maintain an inertially fixed magnetic moment. As a result, while the system is correcting attitude errors along one axis it produces small errors along a second, orthogonal axis. In effect this reduces the spin axis correction rate from the expected value of $3/4^{\circ}$/orbit to about $1/4^{\circ}$/orbit. The decrease in rate did not affect the ultimate accuracy, and the LES-5 spin axis has been maintained perpendicular to the orbit to within $2^{\circ} - 2.5^{\circ}$.

This sensor failure also affected the experimental antenna switching logic in that individual sections performed well but the system as a whole did not work. Sufficient information was obtained from the experiment to verify that the system could be used on LES-6 and, with changes to the sun sensor input logic, such a system was flown.

This section has emphasized the anomalies on LES-5. To put these in perspective it should be emphasized that the principal objective of LES-5 was met fully. It has served a very useful role as the pioneering satellite to operationally demonstrate tactical communications systems and continues to serve an operational need today.

IV. LES-6

The purposes of LES-6 were twofold: 1) to provide a greater capability than LES-5 to communicate with small mobile terminals, and 2) to provide a test-bed for a variety of new experiments and techniques for use in future programs in space communications. Particular emphasis was placed on the demonstration of autonomous attitude control and stationkeeping systems.

The LES-6 discussion begins with an event which occurred prior to or during satellite ejection from the launch vehicle. Immediately after ejection, telemetered information indicated that LES-6 was not spinning about the expected axis (the symmetry axis of the cylinder or Y axis in satellite coordinates). Analysis showed that the satellite Y axis was offset from the angular momentum vector by about 2.2° in the direction shown in Fig. 4. Another anomaly soon became apparent: one entire solar panel (of 32) was not delivering power. These events are related by the fact that the failed panel lay in the plane defined by the spin axis and the axis of symmetry.

Review of all factors led to the following conclusion: LES-6 either gained or lost weight sometime before it cleared the launch vehicle; to explain the in-orbit satellite unbalance this weight must have been at least 1.1 lb. It was further concluded that LES-6 had gained this weight since over-all satellite performance and telemetered information could reveal nothing missing and no structural parts of sufficient mass could have conceivably come free.‡‡ An hypothesis that the offset was there before satellite installation was ruled out. The dynamic balance procedures before launch would have detected < 3% of the in-orbit offset. (These were repeated on a LES-6 Structural Qualification Model and their accuracy borne out.)

‡‡ The failed power panel could not have come off without seriously degrading the antenna pattern. No pattern degradation is observed.

The resultant spin motion of LES-6 is such that the satellite equatorial plane appears to "wobble" $\pm 2.2^{\circ}$, as the nominal satellite spin axis nutates about the angular momentum vector. This has limited the operational use of the autonomous attitude control system.

A. RF Systems

Table 3 gives the rf characteristics of the LES-6 transponder and antenna. The LES-6 transponder is a double-conversion, uhf, bandpass-limiting repeater designed to operate with either a 100- or 500-kHz bandwidth. The receiver noise figure is $\sim$ 4 db. There are two transmitters in the transponder: one transmitter amplifies the communication signal to a level of 120 w, at a nominal start of life bus voltage; the second transmitter smplified an internally generated beacon signal to a level of 4 w . These two amplified signals, along with the received signal, are channeled via a triplexer to a common antenna system.

Since the launch of LES-6, the receiver noise figure and the beacon and communications signal power have been measured to be the same as before launch. The amplitude characteristic of the 500-kHz channel, which has been used exclusively since launch, is the same as the prelaunch characteristics. All telemetry data points in the transponder report readings in good agreement with ground measurements before launch.

Two unexpected changes in performance occurred after launch. The power of the transmitted communications signal varies with satellite rotation over a range of 0.9 db, as a result of a corresponding change in power generated by the solar cells because of the axis tilt and the lack of power from one panel (see Sec. IV H). The second change is a fortuitous decrease from the prelaunch measurements of the power generated in the receiver band by intermodulation between the beacon and communication signals in the common antenna system. This drop in intermodulation power was sufficient to permit simultaneious operation of beacon and communication transmitters with no loss in receiver sensitivity.

The LES-6 beacon and communications signals were chosen so that only a high-order intermodulation product (15th order) fell in the receiver pass band. Generation of bothersome intermodulation products at low order had been a problem on LES-5 and it was believed that the higher-order product would be acceptably low in power. This was not the case; sporadic noise bursts 10 — 15 db above the receiver noise level were commonly seen during ground tests. In both satellites, the intermodulation products were generated in spring finger contacts used at the edges of the slot antenna cavities behind the solar panels. Extensive efforts to minimize these products before launch were only partially successful. Cleaning, burnishing, and gold plating of the contact points did not eliminate the effect. The spontaneous disappearance of intermodulation products in orbit (in both satellites) led to speculation that some phenomenon such as vacuum welding had occurred but nothing is known for certain.

A noteworthy feature of the transponder is that the power amplifiers operated directly from the solar bus (limited to 26 v) in a "power optimized" mode,[7] i.e., the power amplifier load line is adjusted to be coincident with the locus of maximum power points as the output from the solar array varies with sun illumination and satellite life. This means that the power amplifiers are free to consume all available power beyond that demanded by the rest of the spacecraft. There is no dc-dc converter loss, or unutilized power for radiation degradation margins. In operation this amounts to almost 3 db more rf power at the start of satellite life than in a conventional design. At the end of satellite life there is still about 1/2-db advantage. The optimizer has performed well as attested by the above-mentioned variations in signal power with bus voltage as LES-6 rotates. These variations are the same in orbit as measured on the ground (Fig. 5) to within the 1/2-db measurement accuracy.

LES-6 carries the first electronically despun uhf antenna system.[8] Like LES-5, the LES-6 antenna system is circularly polarized to offset the effects of Faraday rotation and permit the use of linearly polarized Earth antennas. Dipole antennas radiate an axial polarization and slot antennas, excited in phase quadrature, radiate an orthogonal circumferential polarization to produce the circularly polarized wave. The radiating slots are located immediately adjacent to each of the solar panels (Fig. 6). Dipoles and slots are simultaneously excited two at a time to produce a beam 34° wide in the spin axis plane and 54° wide in the equatorial plane. As the satellite rotates, successive dipole/slot combinations are excited in either of two phases to produce sixteen overlapping beams.

Signals are fed to the antenna elements from a PIN diode switch matrix. The use of PIN diodes at the LES-6 power levels was made possible by paying particular attention to the diode thermal resistance and by proper heat sinking. In addition, it was necessary to increase the diode reverse bias to 100 v to avoid generation of intermodulation products. The switch matrix insertion loss is about 1/2-db.

Post launch measurement of satellite ERP indicated that antenna system gain is in the region of 8.5 – 9.0 db which agrees with prelaunch measurements.

B. Antenna Switch Logic

The antenna switching logic philosophy used on LES-6 is similar to that on LES-4 (Ref. 1). Each has two systems to despin the antennas; one is based on determining the Earth's center once per satellite revolution; the other is based on measuring the Earth-satellite-sun (ESS) angle at a specified point and, from sun sensor inputs plus a determination of the orbit period, updating the Earth center direction as the satellite moves in orbit. These are referred to as direct sensing mode (Mode I) and orbit interpolate mode (Mode II). Mode I accepts Earth center information from visible wavelength sensors and, as a consequence, is jammed for about 45° of orbit travel when the satellite is near opposite the sun. Mode II, because it needs visible light sensor inputs only at specific points of large ESS angle, does not have this limitation. LES-6 is also designed to operate in mixed mode, a combination of Modes I and II.

C. Radio Frequency Interference Experiment

The LES-6 RFI experiment[3] is similar to the LES-5 experiment in its operation. It scans from 290 to 315 MHz in 120 kHz steps, dwells at each step for 2.56 sec, and measures average rf power and the ratio of peak to average power at each step. (The LES-5 experiment operates from 255 to 280 MHz.) There are 256 steps and one scan is completed in 11 minutes. On every fourth scan the experiment is calibrated in frequency and amplitude by a set of markers generated from a crystal controlled oscillator. Because the LES-6 antenna system has higher gain and lower mismatch losses than the LES-5 system, the LES-6 experiment is sensitive to a ground station e.i.r.p. as low as 10 w, some 5 to 10 db lower than that required by the LES-5 experiment.

D. Attitude Control System

In addition to a magnetic stabilization system like that flown on LES-5, LES-6 also carries a cold ammonia gas attitude control system. Three thrusting nozzles are mounted at the view band: two tangential thrusters produce spin-up or spin-down torque, and one axial thruster precesses the spin axis for attitude control. The attitude control electronics is divided into three subsystems: Disaster System, Spin-Erection System, and Automatic Gas Attitude Stabilization System.

The Disaster System[δδ] automatically precesses the spin axis toward or away from the sun until the solar latitude is between ± 87° above the satellite equator. Satellite ejection from the launch vehicle was at nominal orientation and this subsystem has not been used.

The Spin-Erection System allows operation by ground command to orient the spin axis anywhere in inertial space or change the spin rate. Operating modes include spin-up, spin-down, and four precession directions of the spin axis — toward or away from the solar direction or either way perpendicular to the solar direction. The desired mode is selected first by gound command, then an initiate command is sent. The timing of thruster valve opening is controlled by the subsystem logic. For spin-up or spin-down the appropriate valve is opened for two satellite revolutions. For precession in a particular direction, the attitude thruster valve is opened for ~ 5° precession in 30 min time. A stop command may be sent to terminate operation at less than 256 revolutions. It has not yet been necessary to operate the Spin-Erection System.

The Automatic Gas Attitude Stabilization System was designed to hold the satellite spin axis to within ± 0.16°. This system uses the coincidence of narrow beam Earth and sun sensor inputs which occur at points in orbit defined by the relative placement of the sensors around the satellite view band. The system operates by precessing the spin axis to force the measured Earth latitude to fall within the adjustable deadband around the satellite equator (zero degrees). This occurs at four orbital positions spaced 90° apart, ± 45° and ± 135° away from the Earth-sun line. Thus, attitude errors along each of two axes are corrected once every twelve hours. The system design is based on the assumption that accumulated attitude errors in the twelve hour period are small compared with satellite requirements. This is true for spinning satellites of the LES-6 class.

Two Earth latitude sensors are used to allow the selection of dead-band widths by ground command; these are located 135° apart on the satellite view band. Because of the

δδ This system and its name were inspired by a previous vehicle difficulty on the LES-4 launch.

offset between the nominal and actual spin axes, these sensors report attitude errors in opposite directions, when the actual satellite spin axis is normal to the orbit plane. Full time automatic operation would thus result in wasteful gas expenditure four times per orbit. Since launch, the satellite attitude has been controlled by activating the automatic system for a short period when the satellite is near an appropriately selected fix point in orbit. The response during these operations (eight since launch) indicates that the control logic is performing as designed.

Between corrections, the satellite orientation has been allowed to drift under the influence of solar pressure. Figure 7 indicates how the actual spin axis orientation varies in time. The radial coordinate represents the magnitude of tilt from the orbit normal; the circumferential coordinate is the direction of the tilt. The long curving arcs correspond to spin axis precession due to solar pressure. The straight lines correspond to correction maneuvers. The rate of solar induced precession has varied with sun latitude during the year. At negative sun latitudes the center of pressure has been below the center of mass of the satellite and has produced drift rates of up to 0.05°/day. During the summer, the center of pressure was above the center of mass and drift rates up to 0.07°/day were produced. The sun latitude at which the center of pressure coincides with the center of mass is about -10° Now that typical solar torquing behavior has been determined, the spin axis will be positioned to allow for this and the humber of required corrections thereby reduced.

E. Thruster Systems

LES-6 carries a pressure regulated cold ammonia gas system and an experimental pulsed plasma thruster system.

The gas system is actually made up of two systems operating in parallel to provide stationkeeping and orbit adjustment capability. One of these systems also feeds the spin nozzles and the attitude control nozzle through a pyrotechnic shut down valve which can be fired on command to protect against a downstream failure. Each system has a 10 in. pitch diameter, 4 in. cross section diameter stainless steel toroidal tank (Fig. 8). An individual tank contains 5.7 lb of liquid ammonia, with 10% ullage at 70° F and a saturation pressure of 129 psia. Submerged in the liquid ammonia and thermally bonded to the tank wall are a 5/16-in.-diam cross-section toroidal preplenum chamber, and a 1 1/4-in. cross-section diameter toroidal plenum chamber. Each tank assembly has a pressure regulation system consisting of two redundant solenoid valves, each operated by a narrow differential pressure switch. The control points of the two switches are separated by approximately twice the operating differential of each switch. The tank mounted regulator assemblies draw fuel from the tank and introduce it to the submerged preplenum where it is flashed to 25 psia, converting its sensible heat to latent heat. The additional heat for vaporization and superheat is provided by heat transfer from the tank. The superheated ammonia gas then passes through a flow restriction to the plenum chamber where it becomes available on demand to the thruster nozzles. Flow to each nozzle is controlled by a close coupled solenoid valve. Heat given up by the fuel supply to the low-pressure gas is recovered from the satellite.

Thruster nozzle design parameters and duty cycles are contained in Table 4. Prelaunch measurements indicated that the cold ammonia gas was delivering a specific impulse in excess of 90 sec.

Table 4 Nozzle design parameters

	Nozzle		
	Station-keeping	Attitude control	Spin up/ Spin down
Throat diam, in.	.0675	.0135	.0265
Half angle, deg	30.	15°	30.
Area ratio	425.	235.	227.
Nominal thrust-lb	.106	.006	.017
Pulse length, sec [a]	.33	.75	5.78
Impulse/pulse, lb-sec	.035	.004	.098

[a] Based on 5.78 sec spin period.

The flight system was an upgraded version of a prototype system designed and built by the Space Systems Division of AVCO. The most significant modifications were the substitution of improved solenoid valves developed by Walter Kidde Corporation and the redesign of plumbing to facilitate cleaning and minimize the possibility of valve contamination.

The two systems have performed a combined total of 5000 stationkeeping firings and 1000 attitude control firings over a period of 16 months without incident, providing a high order of confidence in well executed ammonia cold gas systems.

The LES-6 pulsed plasma thruster[P] (PPT) contains, as basic elements, a 2 μf energy storage capacitor, two discharge nozzles, two blocks of solid teflon fuel and two discharge initiating spark plugs.[9,10] A negator spring for each teflon block keeps the fuel firmly pressed against a retaining shoulder in the discharge region. The ignition of a 500 v capacitor across a spark plug creates a micro-discharge which momentarily shorts the gap between the upper and lower electrodes of the nozzle. This triggers the 2 μf capacitor, which is terminated at these electrodes, and discharges two joules of energy in less than 3 μs across the face of the teflon are ablated. Gas dynamic and electromagnetic forces expel the plasma with a resultant impulse bit of 5.7 μlb-sec and a specific impulse of 300 sec. The energy storage and trigger capacitors are recharged and the process is repeated in the other nozzle.

A comparison of space and ground results have shown that the impulse bit (5.7 μlb-sec) as measured on the ground agrees closely with the value measured in space (5.35 μlb-sec). Life tests on the ground have shown that this thruster develops a firing intermittency after a perod of time. This is primarily caused by a fuel overhang which blocks the spark plug discharge from reaching the lower electrode. Intermittency has been noted in space although the time of onset was much longer than observed on the ground. As in the ground tests, thruster intermittency in space was interspersed with periods of self correction. Even with the intermittency the thrusters have maintained the satellite's position to the design goal of $\pm 2^{\circ}$ from station and as of January 19, 1970 each thruster has fired for more than 5000 hr.

F. Autonomous Stationkeeping System

LES-6 carried an autonomous control system to maintain the satellite position in orbit with no requirement for ground command inputs.

The principles and operation of the stationkeeping system are covered extensively elsewhere[11-13] so only a summary description will be given here. The principle is that the

P Built by Republic Division, Fairchild-Hiller Corporation.

time of an event (some given coincidence of sun and Earth sensor observations) is compared to the time at which that event should have happened as seen by an on-board clock. This time difference equates to a longitude position error which is corrected by firing thrusters as appropriate. The control strategy damps the satellite motion around its assigned station. The on-board clock time is corrected for the difference between sidereal time and solar time which is seen by the sensors.

A preliminary checkout of the LES-6 stationkeeping system was performed in early October 1968. In late December 1968 LES-6 was placed under the control of the automatic stationkeeping system which proceeded to bring the satellite towards the assigned station of 92°W longitude. This operation was interrupted in mid-July 1969 when LES-6 was moved on request to a position near 38°W longitude where it is now located. The stationkeeping system has once more resumed operation and will be observed through the next year.

During the first operating period, thrusting was done exclusively with the pulsed plasma thrusters. Because of their low average thrust level, LES-6 moved to its assigned station very slowly and the in-orbit damping behavior of the control system was never tested. The ammonia gas thrusters will be used at the new station location to demonstrate the damping operation. Proper operation of all other principal parts of the system has been verified.

G. Sensors

LES-6 carries a large complement of visible wavelength sensors[14] which provide inputs to the antenna switching system, stationkeeping system, and attitude control system. Sensors of this type have been used extensively on previous LES spacecraft. Easy to build and check out, such sensors have proven to be simple, reliable, and inexpensive. The detectors provide large output signals and, operating on reflected sunlight from the Earth, are capable of providing attitude measurements to accuracies of better than $\pm 0.25^{\circ}$.

The chief drawback of such sensors is their inability to operate well when the sun is near their field of view. With moderate baffling, a narrow beam visible wavelength Earth sensor is capable of working to within $10^{\circ} - 15^{\circ}$ of the sun. At smaller angle than this, scattered sunlight overrides the reflected light from the Earth and jams the sensor.

In spite of this, these sensors can be readily employed on those circular orbit spinning spacecraft which, because of stored angular momentum and/or appropriate system design, can maintain attitude and perform their function with only periodic inputs from Earth sensors.

H. Solar Power Degradation

The problem of rapid solar plant degradation near synchronous altitude was mentioned in discussing LES-5. In an attempt to minimize this degradation, the LES-6 solar panels were treated by coating the exposed solar cell contact areas with RTV 3140 to protect against proton bombardment. The solar cell edges were not coated because of physical limitations

The on-orbit performance of the LES-6 array is somewhat uncertain, but the present best estimate of degradation in maximum available power is 12% ± 4% . These results will be refined by further analysis but they do cast doubt on the efficacy of the coating "fix" that was limited to the exposed contacts only.

These doubts are reinforced by the results of the LES-6 Solar Cell Experiment.[15] This experiment measures solar cell current at twenty different voltage points for each of thirty test cells. Four of these cells are the same as those on the LES-6 main array; two of them have edge and contact protection; two have none. At the maximum power point, the protected cells show markedly less degradation than the unprotected cells (10% vs 18% after one year). Detailed results of this experiment are being reported elsewhere at this conference.[16]

I. Spin Rate Measurements

As a by-product of a special clock rate generator on LES-6, a high accuracy measurement of the spin period was available. This system has a capability of measuring spin periods up to 40 sec; resolution of the measurement is 1.6 μ sec .

Operation is quite simple. An accurate sun crossing is generated by a relatively simple sun sensor. By counting 10 MHz clock pulses between sun crossings, the measurement is actually made to a resolution of 0.1 μ sec, but the four least significant bits in the counter are not telemetered.

In orbit, the system has noise amounting to a spread of about 10 μ sec; by smoothing, the data is readable to about 2 μ sec . The noise may be inherent in the sensor, but the problem has not yet been analyzed. With this accuracy, however, effects on the spin rate from temperature variations, misalignment of thrusters, and possibly spindown from magnetic drag are easily observable. For example, the firing of the pulsed plasma thrusters causes a spin period decrease of about 70 μ sec/hr and is quite apparent. Firing a gas jet yields a period decrease rate of about one millisecond per hour.

V. Summary

LES-5 and -6 have helped demonstrate the feasibility of uhf satellite communications between small, mobile terminals. The principles of autonomous satellite attitude control and stationkeeping have been realized in hardware and demonstrated in orbit. The program has further demonstrated the utility of visible wavelength sensors and the operation of pulsed plasma thrusters for extended periods in orbit. Data from the RFI experiments will be of continuing value in measuring uhf background noise levels; data from the solar cell experiment has shed valuable light on important parameters for the design of satellite solar arrays.

VI. References

[1]Sherman, H. et al., "Lincoln Experimental Satellite Program (LES-1, -2, -3, -4)," Journal of Spacecraft and Rockets, Vol. 4, No. 11, pp. 1448-1552.

[2]Ward, W. W., Nochols, B. E., "Lincoln Experimental Satellite-5 (LES-5) Transponder Performance in Orbit," TN-1968-18, Nov. 1, 1968, Lincoln Lab., Lexington, Mass.

[3]Ward, W. W. et al., "The Results of the LES-5 and LES-6 Experiments," TN-1970-3, July 2, 1970, Lincoln Lab., Lexington, Mass.

[4]ATS Power Subsystem Radiation Effects Study, Phase I/ Final Report, Hughes Aircraft Co., Space Systems Division, SSD 80099R, NASA Contract 5-3823, Feb. 1968.

[5]Curtin, D. J., "Solar Cells at Synchronous Altitude," Summary Rept. of Symposium held at Comsat Corp. March 15, 1968, Washington, D.C.

[6]Black, W. L., Howland, B., Vrablik, E. A., "An Electromagnetic Attitude Control System for a Synchronous Satellite," Journal of Spacecraft and Rockets, Vol. 6, No. 7, pp. 795-798.

[7]Braga-Illa, A. and Snider, D. "Transmitted Power Maximization in Communication Satellites," Journal of Spacecraft and Rockets, Vol. 6, No. 2, Feb. 1969, pp. 173-177.

[8]Assaly, R. N. et al., "LES-6 Antenna System," TR-465 March 10, 1969, Lincoln Lab., Lexington, Mass.

[9]Guman, W. J. and Nathanson, D. M., "Pulsed Plasma Microthruster Propulsion System for Synchronous Orbit Satellite," Journal of Spacecraft and Rockets, Vol. 7, No. 4, April 1970, pp. 409-415.

[10]Vondra, R. J., Thomassen, K., and Solbes, A., "Analysis of Solid Teflon Pulsed Plasma Thruster," AIAA Paper 70-179, New York, N. Y., 1970; Journal of Spacecraft and Rockets, Vol. 7, No. 12, Dec. 1970, pp. 1402-1406.

[11]Braga-Illa, A., "Automatic Satellite Stationkeeping," Journal of Spacecraft and Rockets, Vol. 6, No. 4, April 1969.

[12]Braga-Illa, A., "Preliminary Report on the Orbital Operation of the Automatic Stationkeeping System of LES-6," AIAA Paper 69-934, Princeton, N. J., 1969.

[13]Braga-Illa, A., "The Future of Self-Contained Control of Synchronous Orbits," AIAA Paper 70-479, Los Angeles, Calif., 1970, to be published in AIAA Progress in Astronautics and Aeronautics: Communications Satellite Technology for the '70's, edited by N. E. Feldman and C. M. Kelly, Vol. 26.

[14]Trudeau, N. R., Sarles, F. W., Jr., and Howland, B., "The Design of Visible Light Sensors for Use in Circular Near-Equatorial Orbits," AIAA Paper 70-477, Los Angeles, Calif., 1970, to be published in AIAA Progress in Astronautics and Aeronautics: Communications Satellite Technology for the '70's, edited by N. E. Feldman and C. M. Kelly, Vol. 26.

[15]Sarles, F. W., Jr., Stanley, A. G., and Burrowes, C., "The Solar Cell Calibration Experiment on LES-6," Proceedings of the 7th Annual Photovoltaic Specialists Conference, Nov. 1968, pp. 262-268.

[16]Sarles, F. W., Jr., "Preliminary Results from the LES-6 Solar Cell Experiment," AIAA Paper 70-600, Los Angeles, Calif., 1970, to be published in AIAA Progress in Astronautics and Aeronautics: Communications Satellite Technology for the '70's, edited by N. E. Feldman and C. M. Kelly, Vol. 26.

Fig. 1 LES-5.

Fig. 2 LES-6.

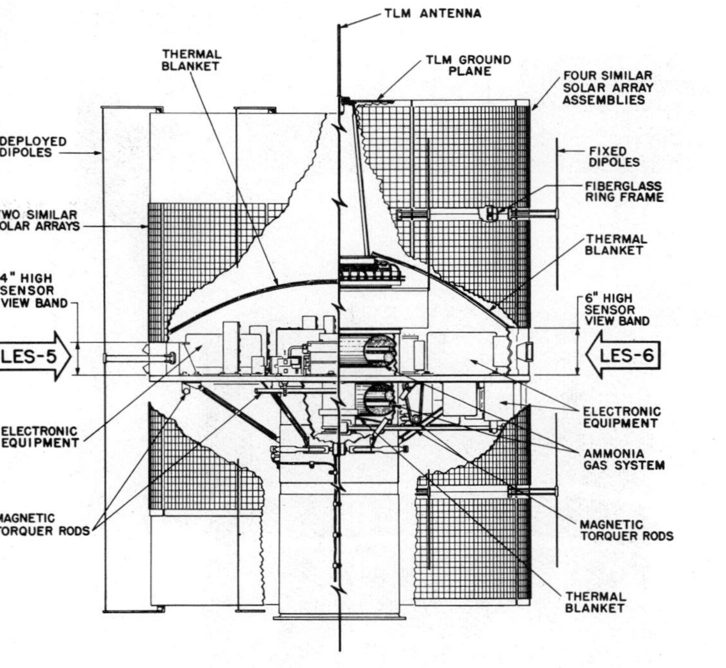

Fig. 3 LES-5/LES-6 comparison.

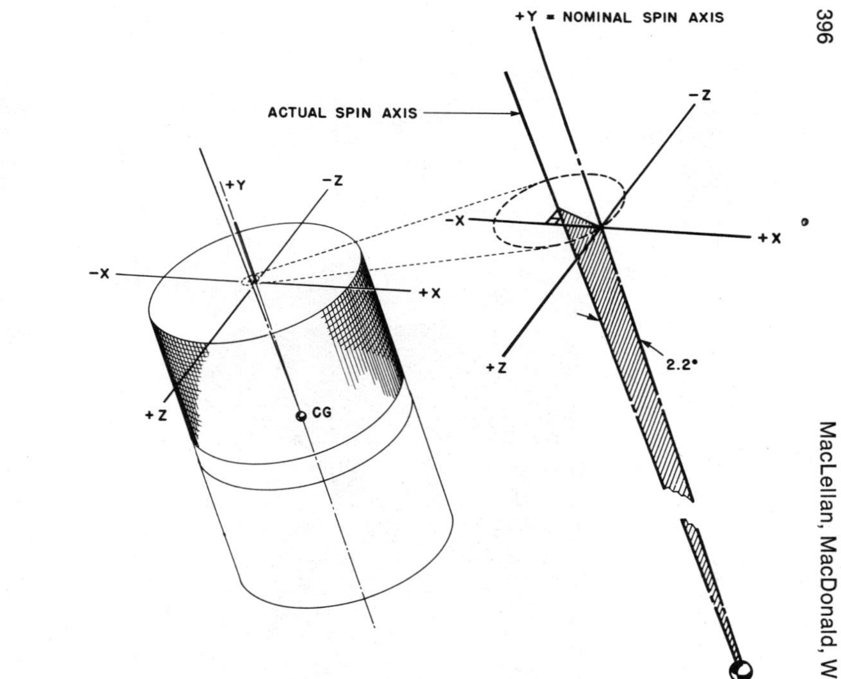

Fig. 4 LES-6 spin axis misalignment.

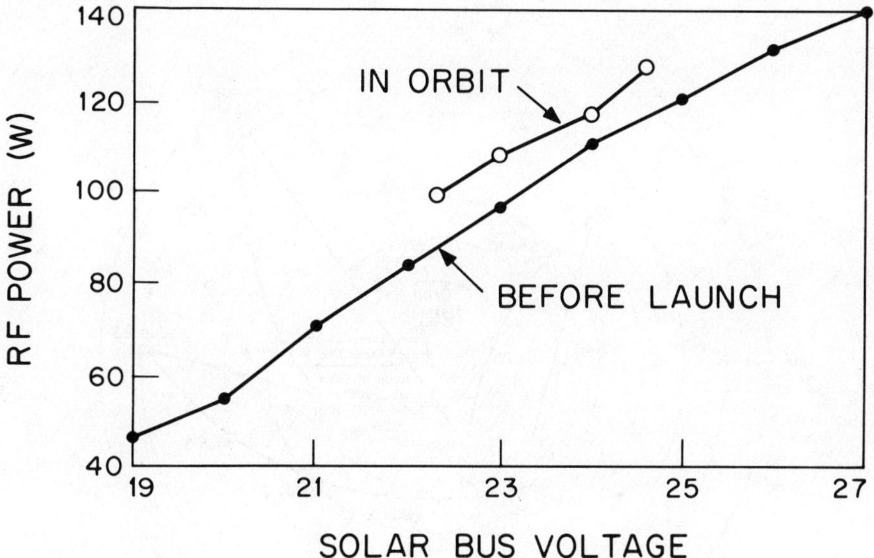

Fig. 5 LES-6 PA power out vs bus voltage.

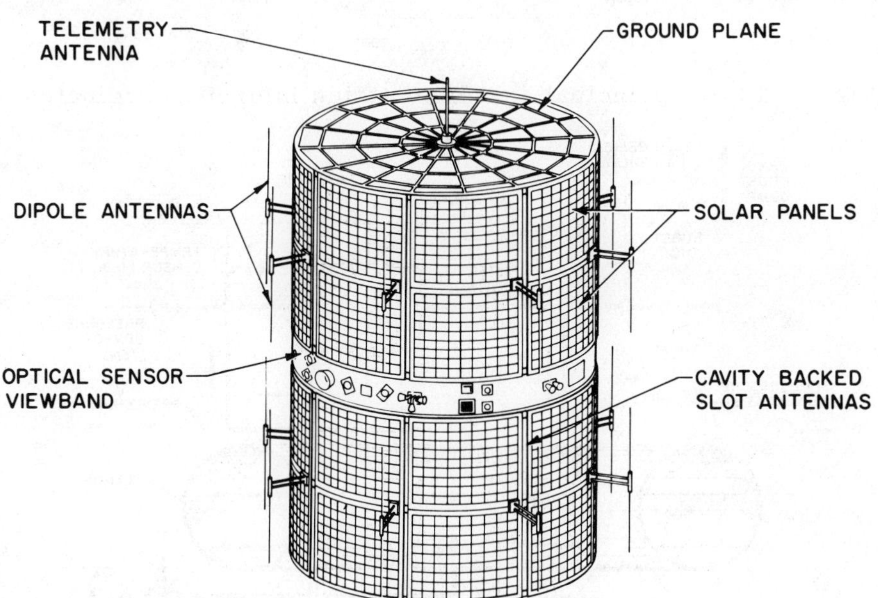

Fig. 6 LES-6 antenna configuration.

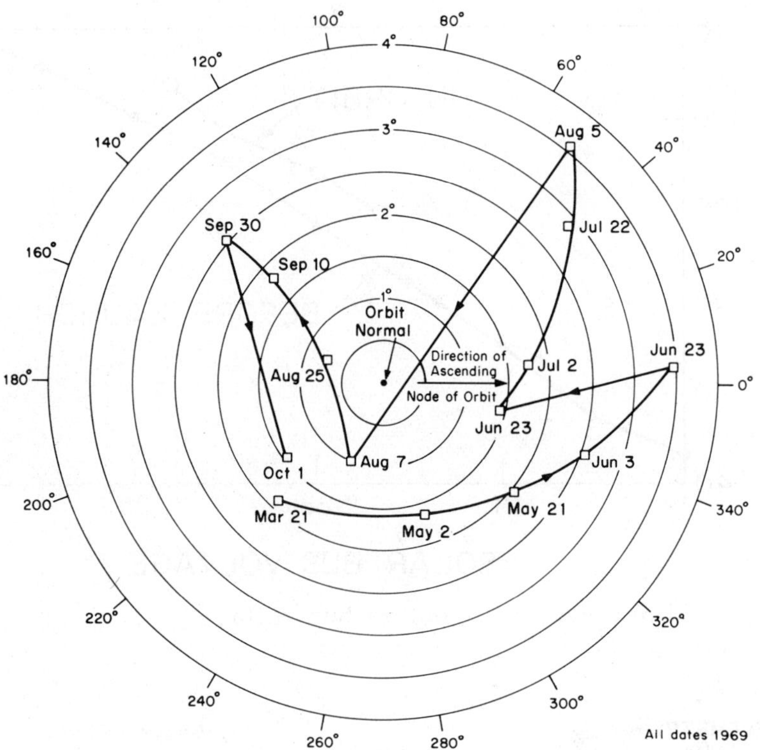

Fig. 7 LES-6 principal axis orientation in orbit coordinates.

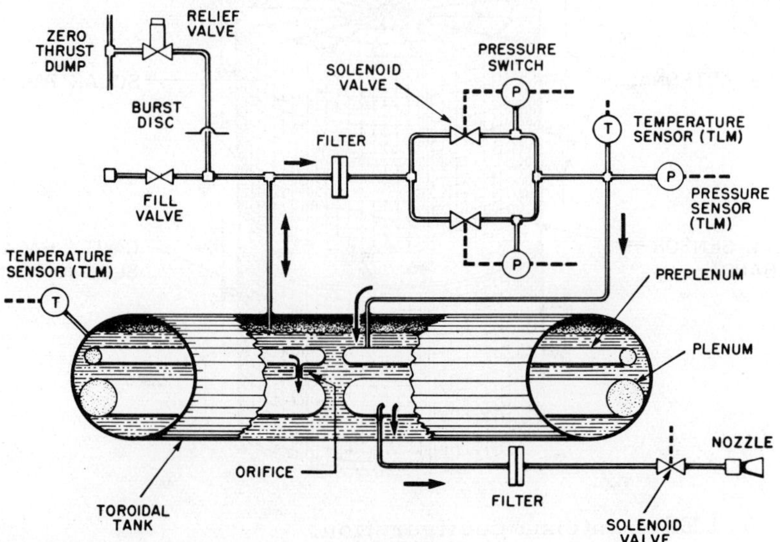

Fig. 8 Gas system - simplified schematic.

DIGITAL TECHNIQUES IN MILITARY COMMUNICATION SATELLITE SYSTEMS

Joseph A. Buegler*
U. S. Army Satellite Communications Agency, Ft. Monmouth, N. J.

Richard S. Davies+ and D. T. Magill#
Philco-Ford Corporation, Palo Alto, Calif.

Abstract

This paper reviews the development of several digital techniques having potential for use in military communication satellite systems. A high-speed quadriphase modem has been developed which can transmit up to 40 Mbps with economical use of rf bandwidth. Up to 10 Mbps have been transmitted through the Initial Defense Satellite Communication System (IDCSP) satellite. A convolutional encoder operating with a threshold decoder has demonstrated performance gains of almost 2 db. Another convolutional encoder operating with a sequential decoder has demonstrated performance gains of over 5 db. A time-division multiple-access (TDMA) experiment has been initiated which will permit determination of those technical and operational parameters that are essential in allowing many different Earth terminals to transmit digital data simultaneously through the same satellite transponder with high efficiency and without mutual interference. These developments will demonstrate the feasibility of high-speed digital transmission through communication satellites.

Presented as Paper 70-495 at the AIAA 3rd Communications Satellite Systems Conference, Los Angeles, Calif., April 6-8, 1970.

*Systems Engineer.
+Manager, Communication Systems.
#Manager, Communication Sciences.

Introduction

The objective of this paper is to discuss some of the digital techniques, now in the experimental stage, which are expected to permit efficient use of the high-power, wideband, multi-purpose second-generation satellites, now in the developmental stage, of the Defense Satellite Communication System (DSCS). These satellites will usher in Phase II of the DSCS.

Phase I of the DSCS, known as the Initial Defense Communication Satellite Program (IDCSP), provided the capability of transmitting high-quality photographs from Vietnam to Washington[1] using digital techniques. The picture transmission system known as Compass Link gave the Defense Communication System (DCS) its first sampling of the "digital world"[2] on a wideband global basis. Phase II of the DSCS eventually will allow the complete introduction of this digital world to the DCS on a system basis.[3]

Pulse code modulation (PCM) will be added to the DCS standard arsenal of hardware to buffer the voice analog interface to the new digital DCS interface. Time-Division Multiplexing (TDM) will combine those digital channels which can time-share a common transmission circuit. Both PCM and TDM will incorporate conventional techniques similar to those used in existing hardware for both commercial and military systems.

Similar conventional techniques will also be used in portions of the ground terminal communication subsystem which will match the DCS digital interface to the Phase II satellite links. However, before these communication subsystems can be fabricated to become a part of the DCS standard arsenal of hardware, critical design parameters must be determined and appropriate techniques and procedures must be devised to allow optimum interface of a multitude of ground terminals with a single satellite repeater.

The time-division multiple-access (TDMA) technique[4] is well suited for efficient handling of many digital signals through a single satellite repeater. TDMA allows greater flexibility in mixing various size terminals and in changing network configurations. Areas not yet fully resolved for TDMA include determination of network timing and user slot rates, synchronization of all terminals into a master frame rate and buffer storage required at each ground terminal operating in the network.

Each terminal in the TDMA network is allotted one or a number of time slots during which it transmits its data at a high burst rate to other terminals. The modem used must allow for efficient utilization of available satellite repeater bandwidth. The demodulator must be capable of responding rapidly to the transmission directed to its terminal. The modem must work coherently with each transmitted carrier to operate with maximum efficiency.

Experimental tests[5] of the digital military satellite channel determine that it closely approximates a binary symmetric channel having binomial distribution of errors making it a practical candidate for error correcting coding to improve the transmission efficiency. Convolutional encoding with threshold decoding[6] can provide a low-cost capability of correcting for burst errors to provide the bit error rates of 10^{-10} or less required in the transfer of computer-to-computer data.

Considerable improvement in error performance can be obtained using sequential decoding[7-9] techniques at the expense of increased complexity and possible added memory. Sequential decoding has been used successfully in deep-space telemetry systems and by Lincoln Lab with its 16-ary frequency-hopped modulation scheme in the Lincoln experimental (satellite) terminal. Both of these applications involved operation at low data rates. Considerable effort has been expended[8] to apply sequential decoding techniques to the high-speed data rates envisioned in the Phase II DSCS.

This paper will describe some of the experiments and tests that will be performed by, or for, the U. S. Army SATCOM Agency to provide the techniques required in the areas of TDMA, high-speed digital modems, and error coding, and will prepare for the full-fledged "digital world" in military communication satellite systems.

Quadriphase Signaling

High-speed digital signaling can be accomplished by many techniques. The desirability of quadriphase signaling is established by the following arguments: first, the satellite channel characteristics, e.g., peak power limitation, favor angle modulation because of its low ratio of peak-to-average powers; second, digital phase modulation offers the advantages of greater bandwidth efficiency and simple, flexible implementation when compared to alternate modulation systems such as FSK.

Quadriphase signaling represents a desirable compromise between bandwidth efficiency and energy efficiency. Quadriphase and biphase signaling have the same error rate performance energy-per-bit-to-noise-power-spectral-density ratio (E_b/N_o); however, quadriphase signals require only one-half the bandwidth of biphase signals. Extension to higher-order phase modulation further reduces the bandwidth requirements, but the E_b/N_o requirement for a given error rate is increased as well as the complexity of the modulator and demodulator. Thus, quadriphase signaling is used as a high-speed digital modulation technique since it possesses both good energy and bandwidth efficiencies. These performance criteria are particularly important for a satellite link characterized by limited power and bandwidth.*

Differential encoding with the phase ordering shown in Fig. 1 was used to resolve phase ambiguity introduced by the carrier phase-lock loop. The characteristics of this modem are as follows:

5 data rates - 2.5, 5, 10, 20 and 40 Mbps;
70 MHz carrier;
Carrier reconstruction by X4 multiplication;
Coherent matched filter detection; and
Differential encoding and decoding to resolve phase ambiguity.

Figure 2 shows the basic elements of the modem. The modem can communicate at rates of 2.5, 5, 10, 20, and 40 Mbps. Figure 3 illustrates the measured error rate performance at a data rate of 40 Mbps when differential coding is employed. Note that the measured performance lies very close to theory. Test results over an IDCSP satellite link with a similar modem are shown in Fig. 4. The downlink capacity quotient of 69 db (Hz) available with this IDCSP satellite limited testing to data rates of 10 Mbps and below and to quite high error rates. These results and other experiments have confirmed that quadriphase signaling is very effective over satellite links.

*A variable data rate quadriphase modem was developed by Philco-Ford under contract DAAB07-68-C-0351 with the U. S. Army Satellite Communications Agency.

Sequential Decoding

A convolutional encoder hard-decision sequential decoder has been built by Codex Corporation for the USASATCOM Agency. It operates at input data rates of up to 5 Mbps. Rate-1/2 coding is used. A parity check is generated for each information bit transmitted. The check bit is computed on the basis of information bits that enter a multi-stage shift register with many taps. These taps feed a parity check network which forms the sum modulo-2 of the inputs. Each parity check is interleaved between each information bit.

The sequential decoder (Fig. 5), after deinterleaving, directs the information bits to the same type of multi-stage shift register as is used in the encoder. Check bits from the decoder are modulo-2 added to the deinterleaved parity bits to produce syndrome bits. Information bits, delayed by the shift register and syndrome bits, are entered in cadence to the core memory. Once the core memory is full, the central decoder logic examines the syndrome bits and systematically searches and tests for compliance with a hypothetical sequence. If a sequence within preset limits of acceptable distance is found, a decision is made as to the correctness of the data bit under consideration. If at first no acceptable sequence is found, alternative sequences are generated, altered, and examined until one alternative becomes more probable than the other. Finally then, a decision as to the correctness of the bit under consideration is made. A correction bit (0 or 1) is generated for each delayed information bit. Correction is made by modulo-2 addition immediately prior to presentation as a data output bit to the user.

The convolutional encoder-sequential decoder has been tested back-to-back at the U. S. Army SATCOM Agency[11]. The performance with added noise provided the equivalent gain in error performance shown in Fig. 6. It provided gains in error performance of from 4.3 to 5.2 db over straight PSK at a P_e of 10^{-5}. Maximum gain was achieved at 9600 bps, minimum at 5 Mbps. Coding gain is primarily a function of decoder memory length and search depth.

This same error encoder/decoder was tested with a differentially encoded PSK modem at rates up to 900 Kbps and again showed coding gain of 5 db. This gain is realized only if the differential encoding is done before the convolutional encoder and differential decoding after the sequential decoder. Operation with the above described QPSK modem showed coding gains of only 3.5 db. At the time of test, however, the differential decoding was performed before the sequential decoding. Equipment design did not readily allow tests otherwise. Based on experience with the PSK modem, it can be said that the full 5 db gain can be realized with the QPSK modem providing the error coding is done inside the differential encoding/decoding function.

Threshold Decoding

A threshold decoder (Fig. 7) has been built which can correct up to two random errors or an error burst of up to 8 bits received in a block of 11 bits. Rate-1/2 convolutional encoding is employed; that is, a parity bit is generated by the convolutional encoder and interleaved between each data bit.

The received data bits are used to generate parity in the same identical manner used in the encoder. These parity bits are modulo-2 added to the received parity bits to produce syndrome bits S(i), which are read into an 18-stage shift register. For the decoder illustrated, four logic functions are generated from the stored syndrome bits from which a majority vote threshold decision is taken, T(i), as to whether the present output bit is in error. If the decision is that an error has been detected, T(i) is used both to invert the received data bit and to invert certain of the syndrome bits stored in the syndrome shift register, thus eliminating the effects of the error from those bits affected.

The results of measurements made in the laboratory are shown in Fig. 8. These results lead to the following relationship between the input bit error probability P_i and the output bit error probability P_o (for random bit errors):

$$P_o = 153\ P_i^3$$

The threshold decoder provides a 1.3 db gain over an uncoded data link for an output probability of 10^{-5}. A 2 db gain is obtained when the output probability is 10^{-10}. While these gains are smaller than those obtained for the sequential decoding, the reduced complexity and cost of the threshold decoder makes it an attractive technique for application where the output error probability must always be less than 10^{-5}. Furthermore, the threshold decoder is capable of operating at data rates above 50 Mbps.

Time-Division Multiple Access

Time-division multiple access (TDMA) is a multiple-access technique that offers both excellent energy and bandwith efficiencies[10] and is ideally suited for handling digital traffic.

A TDMA modem consists of the following major subsystems (see Fig. 9): 1) data buffers, 2) data modem, 3) timing subsystem, and 4) interface and control subsystems. These units are required at each terminal to permit TDMA communication.

The data buffers perform the functions of converting a relatively low-speed continuous-data stream to a very high-speed burst-data stream (at the transmitter) and the inverse operation (at the receiver). Output and input burst rates on the order of 100 Mbps are anticipated. Furthermore, the buffer must be flexible and provide a variety of input and output rates. As a result, buffer costs may be quite high unless the buffer size is kept small (less than 1000 bits) by employing short frame times.

The data modems perform the function of transmitting and receiving the data transmissions to and from other terminals sharing the same satellite transponder. Due to the burst character of TDMA transmissions, the burst rate of each terminal is much higher than the input data rate. As a result, a bandwidth efficient modulation technique, such as quadriphase signaling, is recommended. The short-frame time requirement implies that, if the multiple-access efficiency is to be kept high, no time can be allotted at the beginning of each time slot for carrier reconstruction or bit sync loop acquisition. As a result, an inertial-type operation is recommended where carrier and symbol clock phase coherence is maintained from one frame to the next.

Figure 10 shows how the terminal costs vary with time efficiency. The inertial synchronization technique employs multiple carrier/bit synchronizers at each terminal. One synchronizer is required for each terminal in the network being received. For example, in a 10-terminal network, 9 signals could be received during a single frame period, thus requiring 9 separate synchronizers.

The reacquisition synchronizing technique employs only one carrier/bit synchronizer at each terminal. This synchronizer must lock onto the carrier and bit frequencies at the beginning of each time slot prior to data burst transmission. The time required to perform this reacquisition function accounts for the loss of time efficiency compared to the inertial synchronizer.

To obtain a high time efficiency when using the reacquisition synchronization technique, the data burst duration must be large compared to the signal reacquisition time. The longer burst duration causes the buffer size to be large, whereas an inertial synchronization technique with the same time efficiency requires a shorter data burst duration and, hence, a smaller data buffer. This difference in buffer size accounts for the inertial synchronization technique being less expensive.

Preliminary experiments performed both at 70 MHz and at 7 GHz have demonstrated the feasibility of this approach. Figure 11 illustrates the analog waveform prior to matched-filter detection with a frame time of 1 msec and a duty factor of 1%. Note that no significant phase acquisition or distortion occurs at the leading edge of the time slot.

The timing subsystem performs two major functions. First, it establishes all the local time bases (both transmit and receive) at each terminal. Second, it performs the network timing function of coordinating all terminals so that their transmitting time bases are properly aligned at the satellite. This is accomplished by a long loop control system utilizing a central timing transmission and local timing transmission from each terminal. For preliminary demonstration purposes, both central and local transmissions are allotted separate time slots.

The interface and control subsystem permits: 1) coordination of data, timing, and buffer subsystems; and 2) operator selection of data rates, burst rates, frame rates, and time slot locations. Frame rates of 600, 1200, 2400, 4800, and 9600 frames per second may be selected. Burst rates can be set at 2.5, 5, 10, 20, and 40 Mbps. Data rates have been selected to be of the form

$$75 \times 2^N$$

All the subsystems use digital techniques extensively, which gives them the ability to accommodate rapidly changing network operation. Variable data rates are achieved simply by changing digital divider ratios. Stable time bases and time slot locations are obtained by dividing the frame into 4096 discrete quanta which act as fundamental building blocks.

Two TDMA terminals have been constructed and tests are now under way to evaluate the performance of this equipment while operating with the IDCSP satellite and AN/MSC-46 Earth terminal.

Conclusion

The feasibility of high-speed digital transmission through communication satellites has been established. The development and demonstration of the high-speed quadriphase modem, the sequential and threshold decoders and the time-division multiple-access modem is a step in providing the equipments necessary to implement digital transmission systems using communication satellites.

References

[1]Foster, J. S., Jr., address to the Dallas chapter of the Association of the U. S. Army and Dallas Council on World Affairs, Dallas, Texas, Dec. 13, 1967.

[2]Backes, R. G., "Potential Future Roles of Satellite Communication in Support of National Security," AIAA Paper 68-409, San Francisco, 1968.

[3]"Introduction of TDM into the DCS," Rep. 405600-1-3, Jan. 1969, Defense Communications Agency by Communications and Systems, Inc.

[4]Davies, R. S. and Stephenson, J. M., "A Communication Satellite System for Many Users," AIAA Progress in Astronautics and Aeronautics: Communication Satellite Systems Technology, Vol.19, Edited by R. B. Marsten, Academic Press, New York, 1966, pp. 681-694.

[5]Langelier, R. M., Quagliato, D., and Quigley, J., "Characterization and Coding of DCSP Satellite," presented at the Electronics and Aerospace Systems Convention, Washington, D. C., Sept.1968.

[6]Massey, J. L., Threshold Decoding, MIT Press, Cambridge, Mass., 1963.

[7]Wozencraft, J. M., "Sequential Decoding for Reliable Communication," 1957 National IRE Convention Record, Vol.5, Part 2, pp. 11-25; also TR 383, MIT Research Lab. of Electronics, Cambridge, Mass.

[8]Forney, G. D., Jr. and Langelier, R. M., "A High-Speed Sequential Decoder for Satellite Communications," IEEE International Conference on Communications, 1969, pp. 39-9 to 39-17.

[9]Forney, G. D., Jr., "High-Speed Sequential Decoder," Final Report on Contract DAAB07-69-C-0051, June 6, 1969, submitted by Codex Corporation to U. S. Army Satellite Communications Agency, Fort Monmouth, N. J.

[10]Magill, D. T., "Multiple-Access Modulation Techniques," AIAA Progress in Astronautics and Aeronautics: Communication Satellite Systems Technology, Vol.19, Edited by R. B. Marsten, Academic Press, New York, 1966, pp. 667-680.

[11]Stark, R. J., "Coding for Satellite Communications Channels," presented at the Fifteenth Meeting of the Working Panel, Military Space Communications of Sub-Group, Technical Cooperation Program (TCP); Feb. 16-18, 1970, Nel Center, San Diego, Calif.

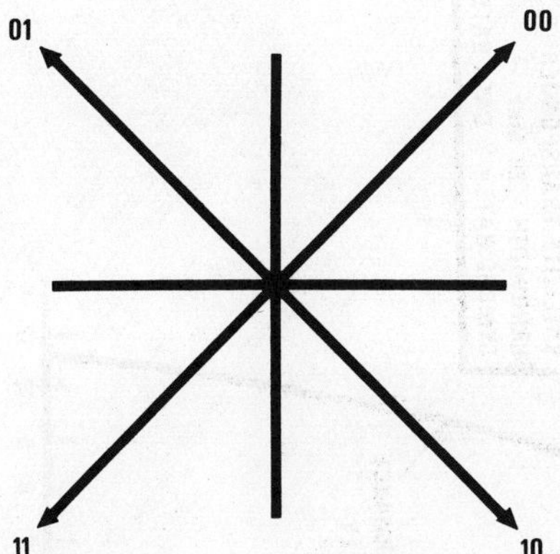

Fig. 1 Quadriphase signaling. Advantages: 1) reduced bandwidth (1/2 biphase PSK BW); 2) good energy efficiency (same as biphase PSK).

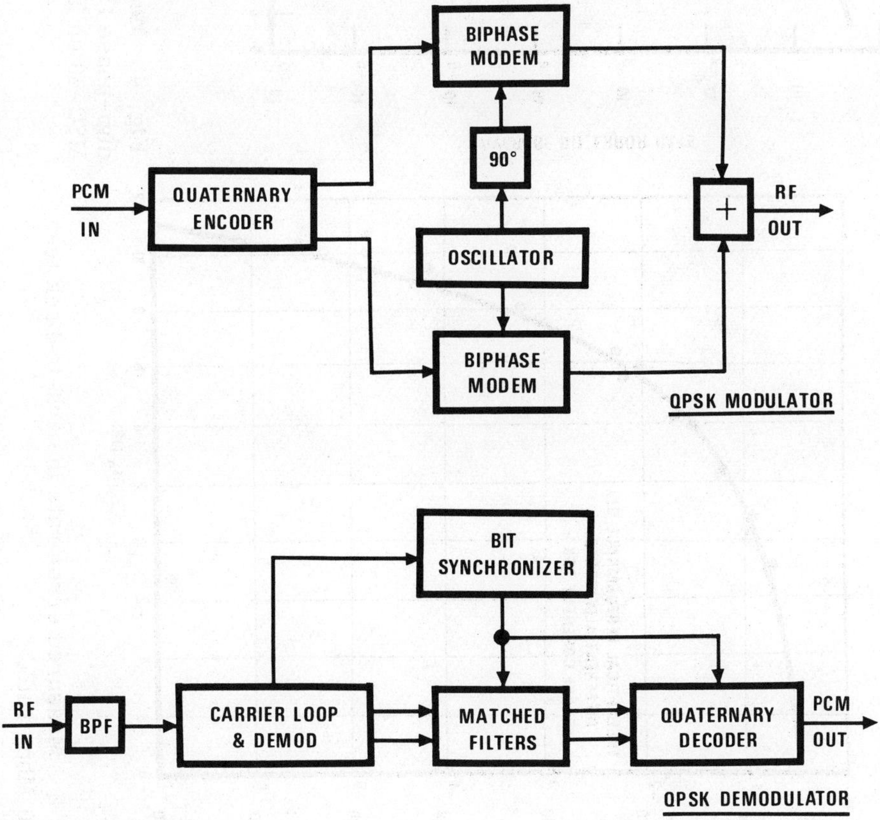

PCM IN
QUATERNARY ENCODER
BIPHASE MODEM
90°
OSCILLATOR
BIPHASE MODEM
+
RF OUT
QPSK MODULATOR

RF IN
BPF
CARRIER LOOP & DEMOD
BIT SYNCHRONIZER
MATCHED FILTERS
QUATERNARY DECODER
PCM OUT
QPSK DEMODULATOR

Fig. 2 Simplified block diagram of quadriphase modem.

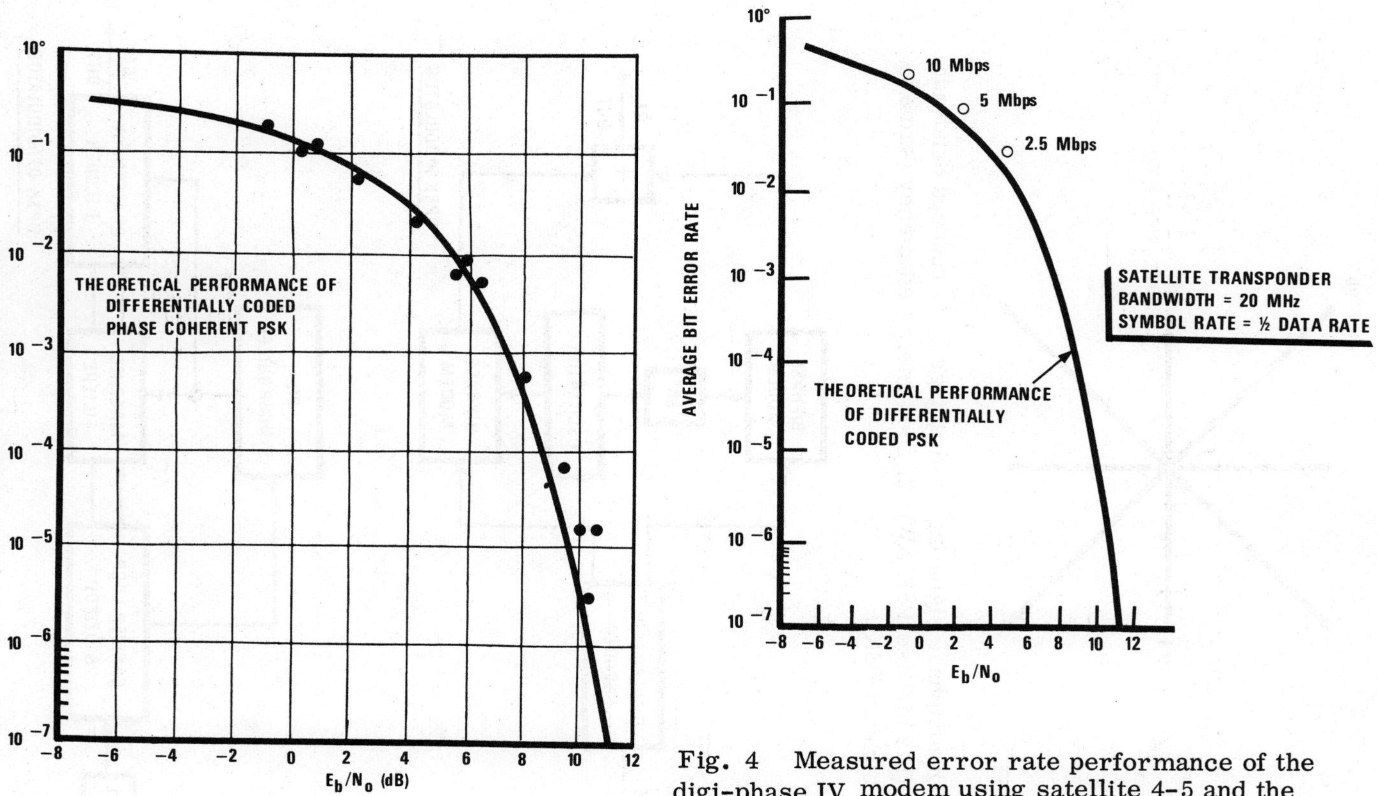

Fig. 3 Measured error rate in back-to-back test at 40 Mbps rate.

Fig. 4 Measured error rate performance of the digi-phase IV modem using satellite 4-5 and the MSC-46 on Thursday, August 22.

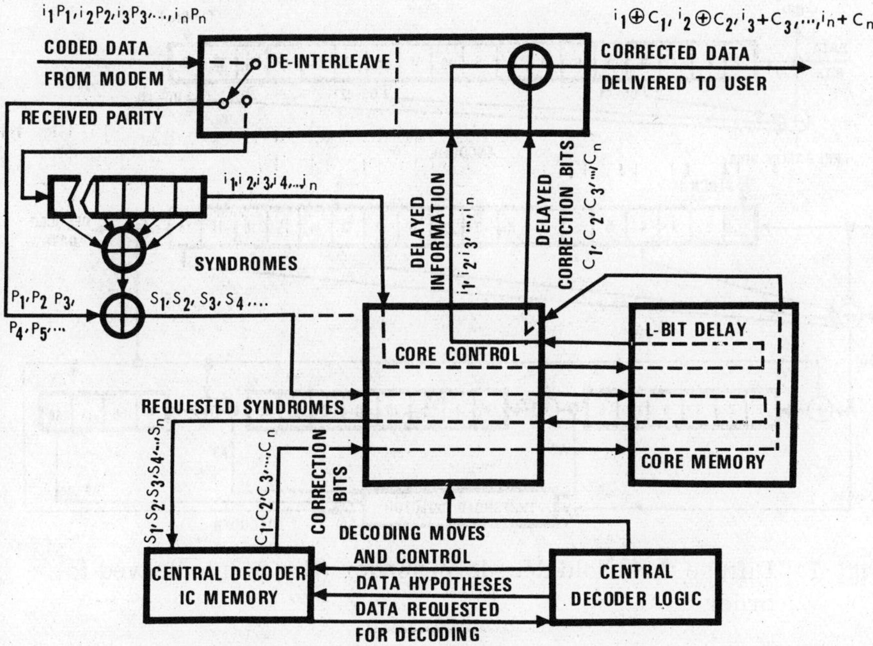

Fig. 5 Simplified flow diagram of high-speed sequential decoder.

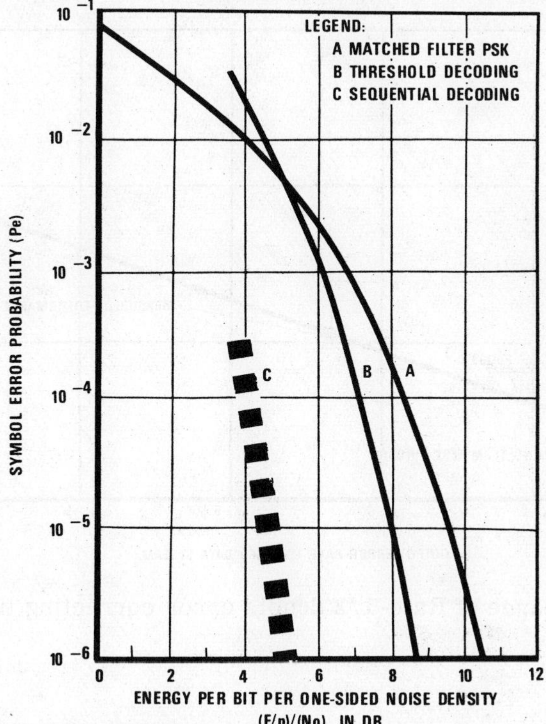

Fig. 6 Measured performance of sequential decoder.

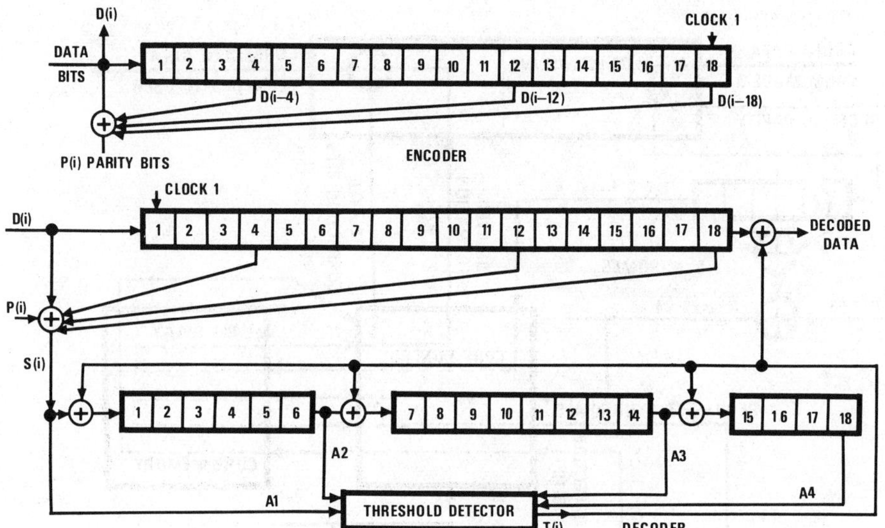

Fig. 7 Diffuse threshold decoder/encoder (N = 2) Interleaved to order 2.

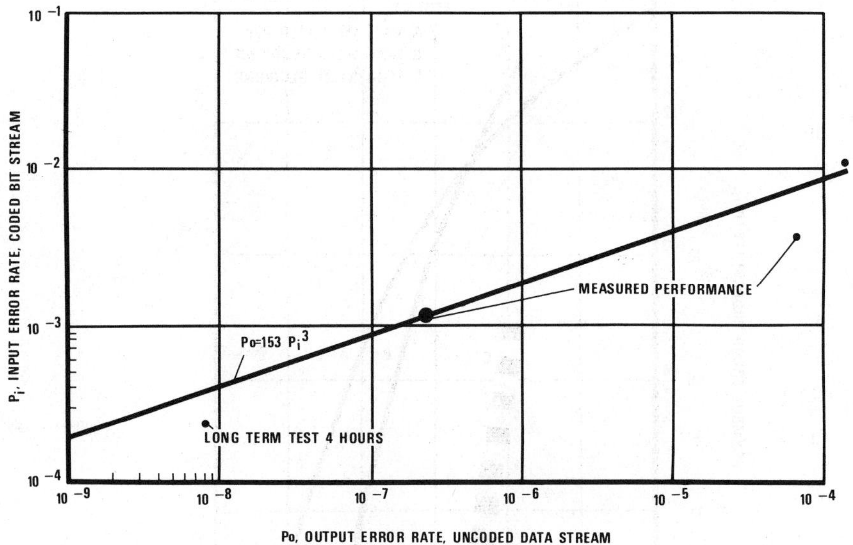

Fig. 8 Performance of Rate-1/2 double error correcting threshold decoder.

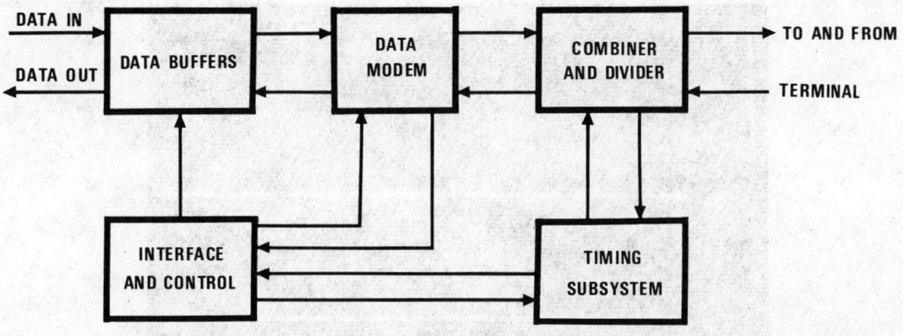

Fig. 9 Block diagram showing the four major subsystems (plus the combiner and divider) of a TDMA Modem.

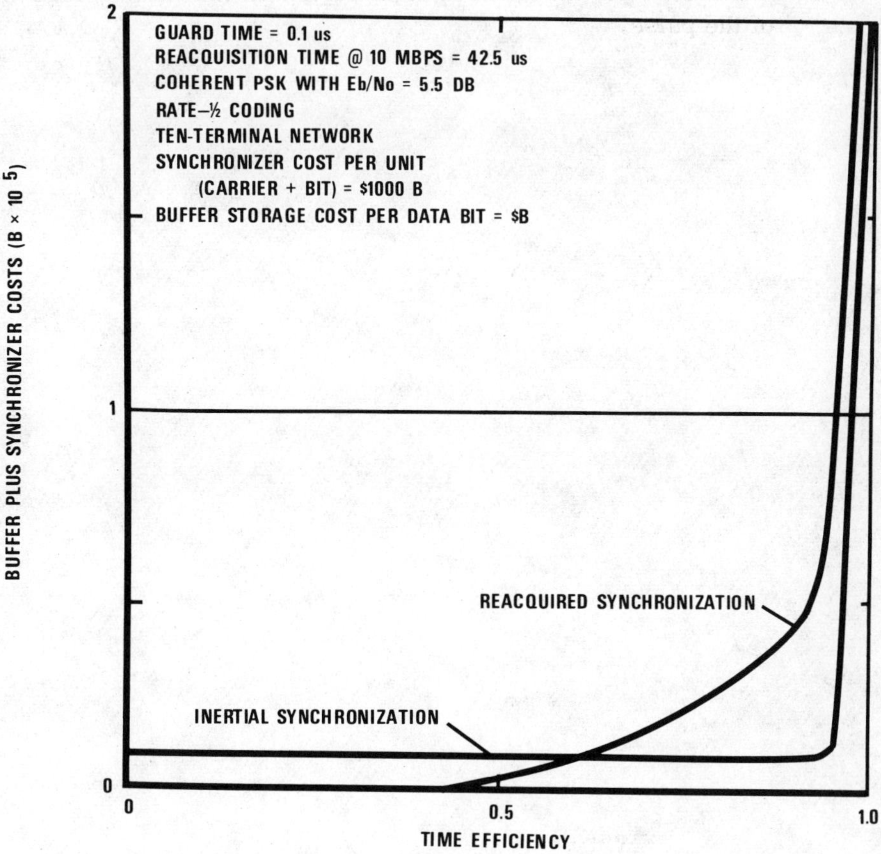

Fig. 10 Comparison between TDMA synchronization techniques.

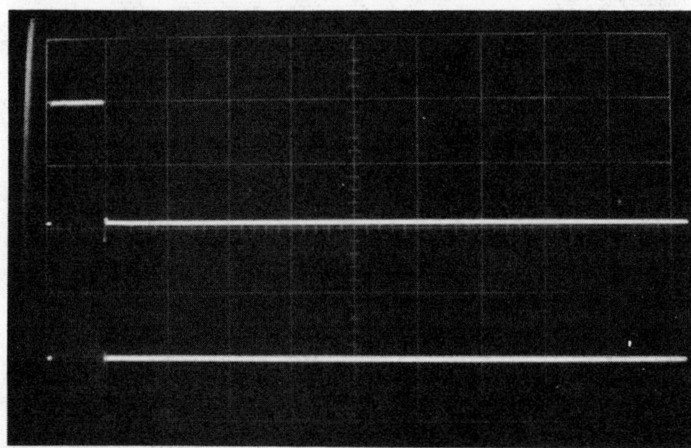

Fig. 11 Matched filter input illustrating frame-to-frame coherence with a time slot of duration 10 μ sec and a frame time of 1 M sec. Note the absence of distortion at the leading edge of the pulse.

PERFORMANCE OF SPREAD-SPECTRUM SYSTEMS IN A MULTIPATH ENVIRONMENT

Ivan Kadar* and Heinz H. Schreiber*

Grumman Aerospace Corporation, Bethpage, N.Y.

Abstract

The performance of pseudo-noise phase shift keying (PN/PSK) and frequency hopping (PN/FH) modulation techniques is analyzed for a low-altitude satellite receiving (VHF) transmissions from a network of Earth coverage synchronous relay satellites. The interference environment assumed may be caused by simultaneous illumination of the user by two relay satellites or by a specular ground reflection. A time-varying channel response is derived, and estimates of group delay and fade frequency are given. The PN/PSK system utilizes nonoptimum coherent demodulation and matched filter chip detection; the PN/FH system utilizes matched filter envelope chip detection. The signal-to-interference (SIR) ratio performance of the PN/PSK is found to be a monotonically increasing function of the chip time-relative doppler shift product, while the SIR for the PN/FH system is shown to be directly proportional to the number of channels available to the hopping code.

Introduction

A system has been suggested to provide continuous communication, dual tracking, and ranging between low-altitude Earth-orbiting spacecraft (user) and Earth coverage synchronous satellites operating in the VHF band. The dual tracking function requires hemispherical coverage (23° beamwidth), antenna patterns on the relay satellites, and omnidirectional antennas on the users. A severe interference problem exists in the user-relay satellite link as a result of the simultaneous illumination of a user by two relay satellites and by multipath propagation (off the Earth's surface).

This paper models and analyzes the simultaneous illumination and the (specular) multipath propagation channels. The models

---

Presented as Paper 70-496 at the AIAA 3rd Communications Satellite Systems Conference, Los Angeles, Calif., April 6-8, 1970.

* Electrical Engineer.

are characterized by fade frequency (differential doppler) and differential time delay. In addition, the time varying channel response is derived for the simultaneous line-of-sight (LOS) illumination. The modulation techniques selected (influenced by the channel characteristics) must have the capability of separating the simultaneous LOS illumination from each other and from the multipath signal. If the system is to be operated at a common carrier frequency, spread-spectrum code division multiplexing (CDM) schemes will provide a basis for separating the signals. Code division multiplexing may be achieved by proper phasing of a common code among the relay satellites and users or by separate codes. The former approach is selected for the system.

The performance of nonoptimum (post-detection correlation) pseudo noise phase shift keying (PN/PSK) and frequency hopping (PN/FH) modulation techniques are analyzed for noiseless receivers. The performance of each technique is evaluated in terms of the signal-to-interference (SIR) ratio at the matched filter detector output at the chip sampling time. The multiple access problem on the user to relay satellite link is not considered in this study, but has been treated extensively elsewhere.[1] A nonoptimum demodulation technique was selected for the PN/PSK system since it lent itself to convenient mathematical analysis and implementation, and provided a basis for comparison to the frequency hopping model on a chip error basis. The optimum demodulation technique (predetection) is discussed extensively in extant literature and is not considered here.

System Geometry

The relay satellite-user deployment geometry showing all propagation paths is given in Fig. 1. The relay satellites are assumed to be in equatorial synchronous orbits with users at 100, 200, and 300 naut. mile altitudes and angular separation between any two relays of 40°, 80°, and 120°. The common carrier frequency was assumed to be 150 MHz.

LOS Channel

Refer to Fig. 1. The relay satellites are in circular equatorial orbits, user altitude is h, the distances to the satellites from the user are R_1 and R_2, respectively, user relay satellites separation angle is x, and relay satellite separation angle is Q. It is assumed for simplicity that the user is at the same longitude as a satellite. Let a be the Earth's radius.

From the law of cosines,

$$R_1^2 = d_1^2 + d_2^2 - 2d_1d_2\cos(\chi) \qquad (1)$$

$$R_2^2 = d_1^2 + d_2^2 - 2d_1d_2\cos(Q-\chi) \qquad (2)$$

In order to compute the relative doppler shifts we require the rate of change variation in differential path lengths

$$\dot{R}_1 = (\dot{\chi}d_1d_2/R_1)\sin(x) \qquad (3)$$

$$\dot{R}_2 = -(\dot{x}d_1d_2/R_2)\sin(Q-x) \qquad (4)$$

$\dot{x}$ is the angular velocity of the user.

Equating the centripetal and gravitational forces, the angular velocity of the user is given by

$$\dot{\chi} = 2\times 10^7(d_1)^{-3/2} \text{ radians/sec} \qquad (5)$$

d_1 is measured in meters. We define the differential time delay as

$$\tau = (R_1 - R_2)/c \text{ sec} \qquad (6)$$

where c is the speed of light, in m/sec.

Similarly, the differential doppler is defined as

$$F = ((\dot{R}_1 - \dot{R}_2)/c)f_0 \qquad (7)$$

where f_o is the carrier frequency, assumed to be 150 MHz.

Differential doppler, defined above, is also known as the fade frequency.[2] Because the path length difference changes with time as a result of user motion, the relative phase of the two LOS waves incident at the antenna constantly changes. This causes the net received signal strength to pass through a series of maxima and minima, and the rate at which the two signals "beat" against one another in this manner is called the fade frequency. Therefore, the terms fade frequency and differential doppler will be used interchangeably throughout the paper.

The individual and differential doppler shifts and differential time delay as a function of angular separation x with user altitude as a parameter have been computed. The doppler shift for the relay to user path as a function of x is given in Fig. 2. It should be noted that the relay satellite orbital velocity has been neglected in the model since its contribution is not significant. The differential doppler shifts and differential time delays as a function of normalized longitudes are given in Fig. 3 for the two LOS paths (R_1 and R_2) to the two relay satellites for a user altitude of 100 naut. miles. As expected, the differential doppler shift and differential path time delay reach a maximum of 7.125 KHz and 34 msec, respectively. The rate of change of relative doppler (as determined from the slope of the relative doppler curve) is 20 Hz/sec. The variations in doppler and delay for other user altitudes (e.g. 200 and 300 naut miles) are negligible.

Multipath Channel

Differential doppler shift and time delays as a function of the angular separation (x) for one multipath channel have been evaluated exactly using the equations given in Ref. 3. Referring to Fig. 1, the law of cosines yields

$$R_1^2 = d_1^2 + d_2^2 - 2d_1 d_2 \cos(\chi) \quad (1)$$

$$d_2^2 = L_2^2 + 1 + 2L_2 \cos(\theta) \quad (8)$$

$$d_1^2 = L_1^2 + 1 + 2L_1 \cos(\theta) \quad (9)$$

$$R_1^2 = (L_1 + L_2)^2 - 4L_1 L_2 \cos^2(\theta) \quad (10)$$

Note that all dimensions have been referenced to Earth's radius a. If θ is chosen as the independent variable, then L_1, L_2, R_1, and x may readily be solved for from Eqs. (9, 8, 10, and 1), respectively; i.e.,

$$L_1 = -\cos(\theta) + \sqrt{d_1^2 - \sin^2(\theta)} \quad (11)$$

$$L_2 = -\cos(\theta) + \sqrt{d_2^2 - \sin^2(\theta)} \quad (12)$$

$$R_1 = \left[(L_1 + L_2)^2 - 4L_1 L_2 \cos^2(\theta)\right]^{1/2} \quad (13)$$

$$\chi = \cos^{-1}(d_1/2d_2 + d_2/2d_1 - R_1^2/2d_1 d_2) \quad (14)$$

The differential time delay is given by

$$\tau = a(L_1 + L_2 - R_1)/c \quad (15)$$

Differentiating Eqs. (1, 11, 12, and 10) yields

$$\dot{R}_1 = [(d_1 d_2/R_1)\sin(\chi)]\dot{\chi} \quad (3)$$

$$\dot{L}_1 = \{\sin(\theta) - [\sin(\theta)\cos(\theta)]/\sqrt{d_1^2 - \sin^2(\theta)}\}\dot{\theta},$$

$$\dot{L}_2 = \{\sin(\theta) - [\sin(\theta)\cos(\theta)]/\sqrt{d_2^2 - \sin^2(\theta)}\}\dot{\theta} \quad (16)$$

$$\dot{R}_1 = (L_1 + L_2)(\dot{L}_1 + \dot{L}_2)/R_1 - (2/R_1)(\dot{L}_1 L_2 + L_1 \dot{L}_2)\cos^2(\theta) + {} + (4L_1 L_2/R_1)[\cos(\theta)\sin(\theta)]\dot{\theta} \quad (17)$$

$\dot{R}_1$ is solved for from previously calculated values of x and R_1. Substituting Eq. (16) into Eq. (17) and equating to this $\dot{R}_1$ allows one to solve for $\dot{\theta}$ and hence also for $\dot{L}_1$ and $\dot{L}_2$.

Differential doppler is calculated from

$$F = (\dot{L}_1 + \dot{L}_2 - \dot{R}_1)\ a f_o / c \tag{18}$$

A computer program was written to solve for x, and F as a function of Θ. The results are plotted in Figs. 4 and 5. For large x they are appreciably different from those given by flat earth models.[4] In addition, variations in fade frequency and differential delay are very dependent upon altitude.

Comparison with the LOS characteristics shows that much smaller differential doppler and time delays exist on the multipath channel, indicating that separating the desired and interference signals on the multipath channel may be more difficult to solve than the simultaneous illumination problem.

It should be noted that in the multipath model the scattered power levels from the Earth's surface have not been computed. Reference 3 gives estimates on the upper bound of the probability density of the scattered power density for various user altitudes at VHF and S-band.

A specular multipath model has been used throughout the analysis over smooth Earth that is believed to be valid over low grazing angles. If rough reflecting surfaces are encountered, especially at high grazing angles the reflected ray will have a highly diffuse character. However, the analysis in this paper is restructed to bounding the problem to the specular case.

LOS Channel Frequency Response

With reference to Fig. 1, the unmodulated received waveform s(t) at the user from the two relay satellites may be written as

$$s(t) = \sin \omega_o t + \alpha \sin [\omega_o (t-\tau) + \theta + \Delta\omega t] \tag{19}$$

where ω_o = carrier frequency, rad/sec; α = relative attenuation of paths; τ = differential time delay, sec; θ = random variable, phase; and $\Delta\omega$ = relative doppler, rad/sec. The envelope squared of the received signal is given by

$$\{E_{nv}[s(t)]\}^2 = 1 + \alpha^2 + 2\alpha \cos[\omega_o\tau - \theta - \Delta\omega t] \tag{20}$$

If we let $\alpha = 1 - \epsilon$, where $\epsilon \to 0$, then Eq. (20) becomes

$$\{E_{nv}[s(t)]\}^2 = 4\cos^2[(\omega_o\tau - \theta - \Delta\omega t)/2] \tag{21}$$

Let Eq. (21) define $|H(w,t)|^2$, the time-varying channel frequency response. The argument of H(w,t) may be obtained from Eq. (19) as

$$\varphi(\omega,t) = -\tan^{-1}(\alpha \sin(\omega_o\tau - \theta) - \Delta\omega t)/(1 + \alpha\cos(\omega_o\tau - \theta - \Delta\omega t)) \tag{22}$$

The channel group delay is defined as

$$\tau(\omega,t) = -(\partial/\partial\omega)\,\varphi(\omega,t) \tag{23}$$

The instantaneous frequency $w_i(t)$ of the received wave is related to the group delay as

$$\omega_i(t) = -(\partial/\partial\omega)\varphi(\omega,t) = -[\Delta\omega\,\tau(\omega,t)]/\tau \tag{24}$$

Therefore, in calculating the group delay we readily obtain

$$[\tau(\omega,t)]/\tau = (\underset{\sim}{\alpha}\cos[\cdot] + \underset{\sim}{\alpha}^2)/(1 + \underset{\sim}{\alpha}^2 + 2\underset{\sim}{\alpha}\cos[\cdot]) \tag{25}$$

For $\underset{\sim}{\alpha} = 1 - \underset{\sim}{E}$ as before, Eq. (25) becomes

$$[\tau(\omega,t)]/\tau = (\cos^2(1/2)[\cdot] - E)/(2\cos(1/2)[\cdot] + E^2/2(1-E)) \tag{26}$$

where $[\cdot] = \omega_0\tau - \underset{\sim}{\theta} - \Delta\omega t$ and $E \sim 0$

Plots of $[\tau(\omega,t)]/\tau$ and $|H(w,t)|^2$ are given in Fig. 6. The graphs clearly depict the time varying nature of the channel indicating severe nulls in instantaneous frequency and time varying amplitude frequency response. The null spacing, as shown in Fig. 6 is a function of the differential time delay. The rate at which nulls occur at a given frequency is a function of differential doppler.

Suppressed Carrier System

The nonoptimum PN/PSK system considered is based upon the premise that an uncorrupted carrier can be reconstructed from the incoming waveform consisting of desired and interference terms; it utilizes post-detection correlation.

The transmitted waveform is made up of data, synchronization and PN code signals, combined in a balanced modulator with a coherent reference and upconverted to the desired carrier frequency. The received signal y(t) is applied to the loop shown in Fig. 7, where the squaring operation in reconstructing the reference carrier separates the desired and unwanted terms in the presence of interfering modulation spectra. At the input to the squarer for analytical simplicity either one of the simultaneous LOS illumination signal or the specular multipath reflected wave is assumed to be present, not both. Following the coherent detection process, the output signal is detected in an integrate and dump matched filter, where the decisions are based on the individual chips. The synchronization, data, and PN codes are recovered in subsequent operations as shown in Fig. 7, based upon the work of Springett.[5]

Waveforms

The user received waveform y(t) is of the form

$$y(t) = \cos[\omega_o t + br(t)] + \alpha(t)\cos[(\omega_o + \Delta\omega)t + br(t-\tau) + \theta] \quad (27)$$

where b $= \pi/2$ - modulation index, rad; r (t) = combined ranging code, sync, and data assumed to take on values of ± 1; α (t) = relative attenuation of paths; θ = random variable, phase U (o, 2π); $\Delta\omega$ = relative doppler shift, rad/sec; ω_o = carrier frequency, rad/sec; and τ = differential time delay, sec.

The received signal y (t) is applied to the loop shown in Fig. 7 where the carrier reference is extracted and r (t) is recovered. Referring to Fig. 7, the output of the squarer neglecting d.c. terms and expanding the trigonometric functions can be written as

$$\begin{aligned} y^2(t) = &-(1/2)\cos 2\omega_o t - (\alpha^2(t)/2)\cos[2(\omega_o + \Delta\omega)t + 2\theta] + \\ &+ \alpha(t)\cos[(2\omega_o + \Delta\omega)t + \theta + r(t) \pm r(t-\tau)] + \\ &+ \alpha(t)\cos[-\Delta\omega t - \theta \pm r(t) \mp r(t-\tau)] \end{aligned} \quad (28)$$

The approximate spectrum of y^2 (t) is shown in Fig. 8. We will assume that the loop filter, centered at $2\omega_o$ and of rectangular bandwidth $2B_L$ is narrow enough such that $2[\omega_o + \Delta\omega]$ is outside $2\omega_o \pm B_L$; in other words $2[\omega_o + \Delta\omega] > |2\omega_o + B_L|$ Since the filter was assumed to be rectangular, the maximum possible interference power in $2B_L$ bandwidth occurs when the interfering term is centered in the filter bandwidth, and is given by

$$\begin{aligned} P_{int} &= (\alpha^2/R)\int_0^{B_L} [(\sin \pi f/R)/(\pi f/R)]^2 df = \\ &= (\alpha^2/\pi)\int_0^{\pi B_L/R} [(\sin^2 x)/x^2]\, dx \end{aligned} \quad (29)$$

where R = bit rate of code. However, the interference spectrum is essentially flat in the band $2\omega_o \pm B_L$. Therefore, P_{INT} is bounded by

$$P_{int} < (\alpha^2/\pi)\int_0^{\pi B_L/R} (1)\, dx = \alpha^2 B_L/R \quad (30)$$

The reference carrier power P_c is $P_c = 1/8$. Therefore

$$P_c/P_{int} \geq R/8\,\alpha^2 B_L \quad (31)$$

Phase Lock-loop Interference

The effect of interference on a narrow-band phase lock-loop (PLL) has been considered in Ref. 6. It has been shown there that for CW interference sweeping across the input signal, loss of lock (cycle skipping) will occur at

P_c/P_{int} = -3 db, with the loop signal-to-noise ratio with no interference present being in the order of 6 db. In addition, it was found that if the interfering signal is separated from that of the desired signal by several multiples of the loop noise bandwidth of the PLL or if the interference sweeps through this bandwidth at a sweep rate of the order of 10 times the square of the loop bandwidth the interference effects are negligible. Clearly, P_c/P_{int} in this model will always be greater than unity, and by the squaring operation the interfering carriers fall outside the loop noise bandwidths, provided that the doppler shifts are greater than the one-sided loop noise bandwidth. Therefore, the probability of losing lock under the above assumptions with the system constraints considered is small.

Matched Filter Output

The output of the integrate and dump filter is

$$y_o(t) = \int_0^T \{\cos[\omega_o t + b r(t)] + \\ + \underset{\sim}{\alpha} \cos[(\omega_o + \Delta\omega)t + br(t-\tau) + \underset{\sim}{\theta}]\} \sin \omega_o t \, dt$$

Neglecting $2\omega_o$ terms,

$$y_o(t) = \pm T/2 + (\underset{\sim}{\alpha}/2)\int_0^T r(t-\tau)\cos(\Delta\omega t + \underset{\sim}{\theta})\,dt \qquad (32)$$

Define signal-to-interference ratio as

$$SIR = E^2[y_o(T)]/\{E[y_o^2(T)] - E^2[y_o(T)]\} \qquad (33)$$

The SIR ratio represents the signal-to-interference output of the matched filter without considering thermal noise. Note that $y_o(t)$ is not gaussian as in the case of thermal noise. To evaluate the interference term,

$$I = (\underset{\sim}{\alpha}/2)\int_0^T r(t-\tau)\cos(\Delta\omega t + \underset{\sim}{\theta})\,dt$$

we will make the following definitions and assumptions: Let $\tau = kT + \underset{\sim}{\gamma}$ where $\underset{\sim}{\gamma}$ is U(o,T) and $\underset{\sim}{\theta}$ is U(o, 2π) as defined before. During the interval τ, $r(t-\tau)$ becomes

$$r = (t - kT - \underset{\sim}{\gamma}) = \begin{cases} \underset{\sim}{a}_1 & 0 \le t < \underset{\sim}{\gamma} \\ \underset{\sim}{a}_2 & \underset{\sim}{\gamma} \le t < T \end{cases} \qquad (34)$$

where $a_{1,2} = \pm 1$ with probability $\frac{1}{2}$. Therefore

$$I = (\underset{\sim}{\alpha}/2)\int_0^{\gamma} \underset{\sim}{a}_1 \cos[\Delta\omega t + \underset{\sim}{\theta}]\,dt + (\underset{\sim}{\alpha}/2)\int_{\gamma}^{T} \underset{\sim}{a}_2 \cos[\Delta\omega t + \underset{\sim}{\theta}]\,dt \qquad (35)$$

$$y_o(t) = (\underset{\sim}{d}/2\Delta\omega)\{\underset{\sim}{a}_1[\sin(\Delta\omega\underset{\sim}{\gamma}+\underset{\sim}{\theta}) - \sin\underset{\sim}{\theta}] + \underset{\sim}{a}_2[\sin(\Delta\omega T+\underset{\sim}{\theta}) - \sin(\Delta\omega\underset{\sim}{\gamma}+\underset{\sim}{\theta})]\} \pm T/2 \quad (36)$$

The expectation of $y_o(t)$ is

$$E\{y_o(t)\} = E\{\cdot\} \pm T/2 \quad (37)$$

But since $E\{a_{1,2}\} = 0$, and the terms in the braces are products of the independent random variables, the expectation of the product equals the product of expectations; hence $E\{\cdot\} = 0$.[7] Therefore, $E\{y_o(t)\} = \pm T/2$. The variance of I is equal to

$$\mathrm{var}\,[I] = (\underset{\sim}{d}^2/4(\Delta\omega)^2)\,E\{\underset{\sim}{a}_1^2[\sin^2(\Delta\omega\underset{\sim}{\gamma}+\theta) + \sin^2\underset{\sim}{\theta} - 2\sin(\Delta\omega\underset{\sim}{\gamma}+\underset{\sim}{\theta})\sin\underset{\sim}{\theta}] + \underset{\sim}{a}_2^2[\sin^2(\Delta\omega T+\underset{\sim}{\theta}) + \sin^2(\Delta\omega\underset{\sim}{\gamma}+\underset{\sim}{\theta}) - 2\sin(\Delta\omega T+\underset{\sim}{\theta})\sin(\Delta\omega\underset{\sim}{\gamma}+\underset{\sim}{\theta})]\} \quad (38)$$

Evaluating the expectations in the variance expression term-by-term, the variance of I becomes

$$\mathrm{var}\,[I] = (\underset{\sim}{d}^2/2(\Delta\omega)^2)(1 - \sin(\Delta\omega T)/\Delta\omega T) \quad (39)$$

Substitution of Eq. (39) in Eq. (33) yields

$$\mathrm{SIR} = (T\Delta\omega)^2/2\underset{\sim}{\alpha}^2[1 - \sin(T\Delta\omega)/(T\Delta\omega)] \quad (40)$$

The expression for SIR is only valid when $\tau \neq 0$ (excludes the case when the codes overlap), $\Delta\omega \neq 0$ and $T \neq 0$. This means that the model breaks down for the overlapping code case, i.e., when the differential delay vanishes. The SIR, Eq. (40), was evaluated on the computer and the result is plotted as a function of chip time - doppler shift product in Fig. 9. Equation (40) highlights some of the existing incompatibilities in system design caused by the interference. The chip time - doppler shift product should be large to minimize the effect of the interference (see Fig. 9). However, the chip time T is determined by the minimum differential time delay between the user and the relay satellites and by the coarse range accuracy requirement of the system. Therefore, for a given relative doppler (determined by carrier frequency and orbital parameters) the chip time-doppler shift product cannot be made arbitrarily large. This implies that improving range accuracy by selecting shorter chip times gives poorer SIR per chip. However, the analysis assumed detection of the combined code bit-by-bit (integrating over the chip time). In correlating the ranging code, the integration time is over the entire code or component codes (code length is determined from the path lengths and unambiguous ranging considerations), which is much longer than the chip time; and in data recovery the integration is over each data bit, which is also longer than T. Therefore, an enhancement of SNR will be realized over the chip detection when the component codes (ranging, synch and data) are recovered.

Some Results

The analysis indicates that the nonoptimum PN/PSK will work in the simultaneous LOS illumination environment as well as for the specular multipath channel except for the case when the codes overlap.

If we assume a coarse range accuracy of 200 ft, the corresponding chip time is approximately T = 1 sec, and the rf bandwidth is approximately 2 MHz. Even with the maximum doppler shift in the LOS model, the chip time-doppler shift product is much less than one. However, with the PN code of 0.25 sec (as required for unambiguous two-way ranging) the processing gain of 54 db offsets the SIR loss.

Noncoherent Frequency Hopped Carrier

Frequency-hopping (FH) may be used as a modulation technique to provide multiple-access capability, combat multipath, and to alleviate the multiple illumination problem in the system. Synchronization of a locally generated system code-word may be used to provide ranging, probably with an accuracy comparable to that of the coherent binary PSK system discussed here and in extant literature. Lack of coherence implies that range-rate is not readily obtained.

Figure 10 is a block diagram of the FH-binary FSK model used in the analysis. The received signal $S^{(r)}(t)$ and the interference $S^{(i)}(t)$ are down converted with a frequency-hopped local oscillator, assumed to be in synchronism with $S^{(r)}(t)$. An IF filter passes the dehopped signal and interference to a set of matched-filters, the outputs of which are envelope detected, subtracted, and sampled at signal chip-times, mT. A decision may be made for each chip, using a majority vote to decide on the message bit, or a decision may be made after the chip samples are linearly combined. Results are obtained in terms of chip signal-to-interference ratio and may be used in either case.

Waveforms

Mark and space waveforms in the IF, after mixing with the synchronized and hopped local oscillator, define the matched filters

$$S_1(t) = \begin{cases} \cos \omega_1 t & 0 \le t \le T \\ 0 & T < t < 0 \end{cases}$$

$$S_0(t) = \begin{cases} \cos \omega_0 t & 0 \le t \le T \\ 0 & T < t < 0 \end{cases}$$

The matched filter inpulse responses are $h_1(t) = S_1(T-t)$ and $h_0(t) = S_0(T-t)$. The received signal waveform in the IF is $S^{(r)}(t) = \cos [(w_k + \delta)\, t + \underset{\sim}{\theta}]$ $0 \le t \le T$ where $k = 0,1$; δ = doppler shift; $\underset{\sim}{\theta}$ = random variable, phase; and T = FH code chip time. The interfering wave is a delayed replica of the signal

$$S^{(i)}(t) = \begin{cases} \underset{\sim}{\alpha}_k \cos [(\underset{\sim}{\omega}_k + \Delta)t + \underset{\sim}{\varphi}_k] & 0 \le t \le \underset{\sim}{\gamma} \\ \underset{\sim}{\alpha}_\ell \cos [(\underset{\sim}{\omega}_\ell + \Delta)t + \underset{\sim}{\varphi}_\ell] & \underset{\sim}{\gamma} \le t \le T \end{cases}$$

where $\underset{\sim}{w}_k$ ($\underset{\sim}{w}_1$) is the frequency of the interfering wave in the IF after mixing with PN-FH local oscillator; $\underset{\sim}{\varphi}_k$ ($\underset{\sim}{\varphi}_1$) is the random phase of the interference in the IF; $\underset{\sim}{\Delta}$ is the doppler shift of the interfering wave; $\underset{\sim}{\alpha}_k$ ($\underset{\sim}{\alpha}_1$) is the relative amplitude of the interfering wave in the IF; $\underset{\sim}{\gamma}$ is a random variable the delay between the two codes.

To perform the analysis of system performance we have assumed that 1) chip sync is established with $S^{(r)}(t)$; 2) $\underset{\sim}{\theta}$ and $\underset{\sim}{\varphi}$ are independent and uniformly distributed over $(0, 2\pi)$; and 3) $\underset{\sim}{\gamma}$ is an independent random variable uniformly distributed over $(0, T)$.

The output of each matched filter (prior to envelope detection) is given by

$$e_0(t) = \begin{cases} \int_0^t [S^{(r)}(\tau) + S^{(i)}(\tau)]\, S_j(T+\tau-t)\, d\tau & 0 \le t \le T \\ \int_{t-T}^T [S^{(r)}(\tau) + S^{(i)}(\tau)]\, S_j(T+\tau-t)\, d\tau & T \le t \le 2T \\ 0 & 2T < t < 0 \end{cases} \quad (j = 0,1)$$

For the special case of a mark transmission, the outputs are readily found:

Mark transmitted/Mark filter output

$$e_1(t) = \begin{cases} (t/2)\left|\dfrac{\sin(\delta t/2)}{(\delta t/2)}\right| \cos\left[\omega_1(t-T) + \underset{\sim}{\theta} + \delta t/2\right] & (41) \\ \qquad 0 \le t \le T & \\ \dfrac{(2T-t)}{2}\left|\dfrac{\sin\frac{\delta}{2}(2T-t)}{\frac{\delta}{2}(2T-t)}\right| \cos\left[\omega_1(t-T) + \underset{\sim}{\theta} + \frac{\delta}{2}t\right] & (42) \\ \qquad T \le t \le 2T & \end{cases}$$

We see that the relative doppler shift causes a loss in the sampled envelope output of

$$\rho = \left|(\sin \delta T/2)/(\delta T/2)\right| \qquad (43)$$

As a result of doppler, there is also a cross-talk term, i.e., the response of the space filter to a mark signal. This is readily shown to be

Mark transmitted/space filter output

$$e_0(t) = \begin{cases} \dfrac{t}{2}\left|\dfrac{\sin\frac{\omega_1-\omega_0+\delta}{2}t}{\frac{\omega_1-\omega_0+\delta}{2}t}\right| \cos\left[\dfrac{\omega_1+\omega_0+\delta}{2}t - \omega_0 T + \underset{\sim}{\theta}\right] & (44) \\ \qquad 0 \le t \le T & \\ \dfrac{2T-t}{2}\left|\dfrac{\sin\frac{\omega_1-\omega_2+\delta}{2}t}{\frac{\omega_1-\omega_2+\delta}{2}t}\right| \cos\left[\dfrac{\omega_1+\omega_0+\delta}{2}t - \omega_0 T + \underset{\sim}{\theta}\right] & (45) \\ \qquad T \le t \le 2T & \end{cases}$$

The signal to cross-talk ratio, which will determine the lower bound on signal-to-noise ratio, is proportional to

$$\left[(\sin(\omega_1-\omega_0+\delta)T/2)/(\omega_1-\omega_0+\delta)T/2\right]^{-1}$$

We will assume that the mark-space frequency spacing $(\omega_1 - \omega_0)$ is large enough so that under all doppler conditions the cross-talk ratio does not exceed 1/10. In addition, we also will assume that the signal loss resulting from doppler is less than 1 db. The requirements may be written as

$$20 \text{ LOG}_{10}\left|(\sin \delta T/2)/(\delta T/2)\right| > -1 \qquad (46)$$

from which we determine the maximum chip time

$$T < 1.7/\delta \qquad (47)$$

From the cross-talk limitation we find that

$$|\omega_1 - \omega_0| > 20/T + \delta \qquad (48)$$

Substitution of Eq. (48) into Eq. (49) yields (approximately),

$$|\omega_1 - \omega_0| > 13\,\delta \qquad (49)$$

A penalty in bandwidth is exacted because of the doppler. Without it, a frequency spacing of $2\pi/T$ would (theoretically) suffice; however, we see from Eq. (48) that doppler uncertainty requires almost 3 times as much bandwidth.

The interference term is a delayed version of the received signal of almost the same rf amplitude. However, since it is not in synchronism with the frequency-hopped local oscillator, its occurrence in the IF will be described statistically over the received signal chip-time. For each interference term, the matched filter outputs may be written as

Mark-filter output.

$$e_1(t) = \begin{cases} (\alpha_1/2)(t-\gamma)\cos[\omega_1(t-T)+\theta+\varphi_1-\theta-\gamma\omega_1] & (50) \\ (\alpha_2/2)(T-t+\gamma)\cos[\omega_1(t-T)+\theta+\varphi_2-\theta-\omega_1\gamma+\omega_1 T] & (51) \end{cases}$$

Equation (50) applies if the interfering chip begins in (0,T); Eq. (51) applies if the interfering chip begins in (-T,0).

Envelope detector output. Let ψ denote the uniformly distributed variable

$$\psi = \begin{cases} \varphi_j - \theta - \omega_k \gamma \\ \varphi_j - \theta - \omega_k \gamma + \omega_k T \end{cases}$$

where j=1, 2, 3, 4; k=0, 1; all additions (subtractions) are taken modulo 2π. On the assumption that a mark is sent, the envelope of mark filter output resulting from both the signal and the interference at time t = T is

$$e_1(T) = (T/2)\sqrt{1+2\alpha_1(1-\mu)\cos\psi+\alpha_1^2(1-\mu)^2} \qquad (52)$$

$$e_1(T) = (T/2)\sqrt{1+2\alpha_2\mu\cos\psi+\alpha_2^2\mu^2} \qquad (53)$$

We have defined $\mu = \gamma/T$, where μ is a uniformly distributed random variable over the interval (0,1). Clearly Eqs. (52) and (53) are symmetrical in μ since μ and $(1-\mu)$ are both uniformly distributed random variables over (0,1).

The space filter output is solely due to interference; at time t = T it is given by

$$e_0(T) = \begin{cases} (T/2)\alpha_3(1-\mu) & (54) \\ (T/2)\alpha_4 \mu & (55) \end{cases}$$

These equations are also symmetrical in μ. The difference output may be written as

$$(2/T)\,e(T) = \sqrt{1 + 2\underset{\sim}{\alpha}_1 \underset{\sim}{\mu} \cos \underset{\sim}{\psi} + (\underset{\sim}{\alpha}_1 \underset{\sim}{\mu})^2} - \underset{\sim}{\alpha}_3 \underset{\sim}{\mu} \quad (56)$$

Output signal-to-interference ratio. To evaluate the expected values we recognize that e(T) is a function of the random variables $\underset{\sim}{\alpha}$, $\underset{\sim}{\mu}$, and $\underset{\sim}{\psi}$, defined earlier, and proceed to apply the conditional expected value theorem,[7]

$$E[g(\underset{\sim}{\alpha}, \underset{\sim}{\mu}, \underset{\sim}{\psi})] = E\{E[g(\underset{\sim}{\alpha}, \underset{\sim}{\mu}, \underset{\sim}{\psi})/\underset{\sim}{\alpha}]\}$$

To simplify the analysis, we assume that $\underset{\sim}{\alpha}_{1,3}$ can take on the values either 0 or 1. We further assume that mark and space interference (each occurring with probability of ½) are mutually exclusive. For the case of M possible frequencies available to the FH code word, the probabilities that the delayed and received signals coincide in frequency are

$$P\{\underset{\sim}{\alpha}_k = 1\} = 1/M, \; k = 1, 3$$

$$P\{\alpha_1 = 1, \alpha_3 = 1\} = 0$$

$$\left.\begin{array}{l} P\{\alpha_1 = 1, \alpha_3 = 0\} \\ P\{\alpha_1 = 0, \alpha_3 = 1\} \end{array}\right\} = 1/M$$

The probability of no coincidence is

$$P\{\alpha_1 = 0, \alpha_3 = 0\} = (M-1)^2/M^2$$

Then we can write for expected value of the normalized receiver output at t=T

$$E\{(2/T)\,e(T)\} = (M-1)^2/M^2 + (1/M)\int_0^{2\pi} (1/2\pi)\,d\psi \int_0^1 d\mu \sqrt{1+\mu^2+(2\cos\psi)\mu} + (1/M)\int_0^1 (1-\mu)\,d\mu \quad (57)$$

which reduces to

$$E\{(2T)\,e(T)\} = (M-1)^2/M^2 + 1/2M + (2/M)\int_0^{\pi} (1/2\pi)\,d\psi \int_0^1 d\mu \sqrt{1+\mu^2+(2\cos\psi)\mu} \quad (58)$$

The integral on μ may be obtained from Petit-Bois tables; yielding

$$F(\psi) = \int_0^1 d\mu \sqrt{1+\mu^2+(2\cos\psi)\mu} = \cos^3(\psi/2) - (1/2)\cos\psi - (1/2)\sin^2(\psi)\ln\left[\frac{\sin(\psi/4)}{\sin(\psi/2)}\;\frac{\cos(\psi/2)}{\cos(\psi/4)}\right] \quad (59)$$

The last term of Eq. (59) has been integrated over $(0, \pi)$ using numerical techniques. The endpoint indeterminacy of this term, i.e., $\sin^2 \pi \ln [(\cos \pi/2)]$, was readily solved for and found to be zero. The numerical integration was performed by breaking the integral into two parts, $(0, \pi/2)$ and $(\pi/2, \pi)$; a spacing of $h=\pi/10$ was used and calculations were made by slide rule. The result is .21/M and Eq. (58) reduced to

$$E\{(2/T)\,e(T)\} = (M-1)^2/M^2 + 0.71/M \tag{60}$$

The mean-square output is given by

$$E\{[(2/T)\,e(T)]^2\} = E\{1 + 2\alpha_1\mu\cos\psi + (\alpha_1\mu)^2 + (\alpha_3\mu)^2 - 2\,\alpha_3\mu\sqrt{1 + 2\alpha_1\mu\cos\psi + (\alpha_1\mu)^2}\}$$

Introducing probabilities and their densities yields

$$E\{[(2/T)\,e(T)]^2\} = (M-1)^2/M^2 + (1/M)\int_0^{2\pi}(1/2\pi)\,d\psi\int_0^1 d\mu\,[1 + 2\mu\cos\psi + \mu^2] + (1/M)\int_0^1 d\mu\,[1+\mu^2-2\mu]$$

which reduced to

$$E\{[(2/T)\,e(T)]^2\} = (M-1)^2/M^2 + 5/3M$$

The variance is given by

$$\sigma^2 = (1/M)\,[(1-1/M)^2\,(0.58-1/M) + (5/3)\,(1-3/10M)].$$

Then the signal-to-interference ratio as a function of M is given by

$$SIR = \frac{1/2 + M(1-1/M)^4 + 1.42(1-1/M)^2}{(1-1/M)^2\,(0.58-1/M) + (5/3)(1-3/10M)} \tag{61}$$

A plot of this function is given in Fig. 11. To a good approximation for all M,

$$SIR \sim M/2.24$$

Some Numerical Results

The analysis indicates that the noncoherent binary FSK-FH modulation technique can work well in the simultaneous illumination environment (as long as the relative delay exceeds the chip time) if sufficient channel bandwidth is available. For the VHF case the maximum doppler shift in each leg of transmission is the order of 3.85 kHz (Fig. 2). From this and Eq. (49) we determine that the matched filter spacing must be at least 13 x 3.85 = 50 kHz and that the chip-time must be less than 70 usec; this corresponds to a chip bandwidth of about 28.5 kHz. As a result, the nominal dehopped IF bandwidth requirement is 50 + 28.5 = 78.5 kHz; but, in the 136- to 138-MHz channel, e.g., there are only 25 such channels available (i.e. M=25). Furthermore, if a data rate of 1000 bits/sec is desired, then there will be about 1000/70 = 14 chips per bit (i.e., integrate 14 chips to decide on each data bit); thus, we can expect a processing gain of 10 log (14) = 11.5 db. For an extension of this work, and for comparison with extant analyses, we consider the results given by Wittman[8] for FH-multiple access. We apply his results for average chip error probability to the case of 2 accesses:

$$\bar{P}_e = \sum_{m_M=1}^{2} \sum_{m_S=0}^{1} \frac{m_S}{m_M + m_S} P\{m_M, m_S\}$$

where

$$P\{m_M, m_S\} = \binom{1}{m_{M-1}} \left(\frac{1}{M}\right)^{m_M - 1} (1 - 1/M)^{2 - m_M} \cdot \binom{1}{m_S} \left(\frac{1}{M}\right)^{m_S} (1 - 1/m)^{1 - m_S}$$

Solving for average chip error rate yields

$$P_e = (3M - 1)/6M^2$$

A plot of this function is included in Fig. 11.

If simultaneous illumination as well as ground reflections must be considered (i.e., N=4 accesses) this work may be applied directly to estimate performance; however, slightly pessimistic results will be obtained because Wittman's work assumes all accesses have the same power level, whereas the reflected waves will usually be less than the direct waves. For large M, the average bit error probability reduces to $(2M)^{-1} = (4.48 SIR)^{-1}$. We note that this is the same type of variation that is observed for slow, nonselective Rayleigh fading, viz;

$$\bar{P}_e = 1/(2 + \gamma_o)$$

where γ_o is the mean SNR.

Conclusions

The performance of PN/PSK and PN/FH modulation techniques has been analyzed for a low altitude satellite receiving transmissions from a network of Earth coverage synchronous relay satellites. Considering a specular multipath model and/or LOS illumination the noiseless receivers used, namely postdetection correlation for the PN/PSK and noncoherent envelope chip detection for the PN/FH indicates that the modulation techniques are suitable to combat specular multipath type interference. The SIR has been found to be a valuable tool in bounding the system performance.

This analysis represents a beginning to a larger effort. Further work is needed to evaluate chip error probability and receiver noise. The model should be extended to include specular and diffuse reflections as well as simultaneous illumination.

References

[1]Wittman, J. H., "A comparison of Satellite Multiple Access Techniques," IEEE EASTCOM Conference, Oct. 1967.

[2]NAS9-8166, "An Investigation of the Lunar Multipath Problems Associated with the Apollo Spacecraft S-band Tracking Antennas," May 1969.

[3]Ott, R. H., and Herman, J. E., "Preliminary Study of Satellite Multipath Interference Characteristics," ERLTM-ITS 176, ESSA Research Labs, June 1969.

[4]Bond, F. E., and Meyer, H. F., "Fading and Multipath Considerations in Aircraft/Satellite Communication Systems," AIAA Progress in Astronautics and Aeronautics: Communication Satellite Systems Technology, edited by R. B. Marsten, Vol. 19, Academic Press, New York, 1966, pp. 193-213.

[5]Springett, J. C., "Telemetry and Command Techniques for Planetary Spacecraft," 32-495, Jet Propulsion Lab., Pasadena, Calif., Jan. 1965.

[6]"Effects of Interference on Narrow Band Phase-Lock-Loops," CCIR Rep. 219-1, Annex I, 1966.

[7]Papoulis, A., Probability, Random Variables, and Stochastic Processes, McGraw-Hill, New York, 1965.

[8]Wittman, J. H., "Analysis of a Hybrid Frequency - Time Hopping Random Access Satellite Communication System", IEEE Convention Record, 1967.

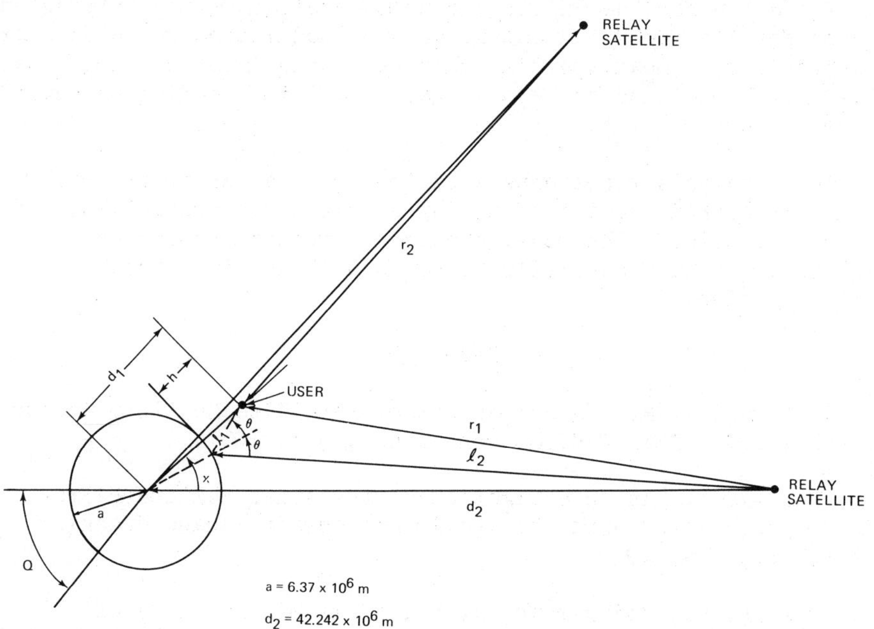

Fig. 1 System geometry.

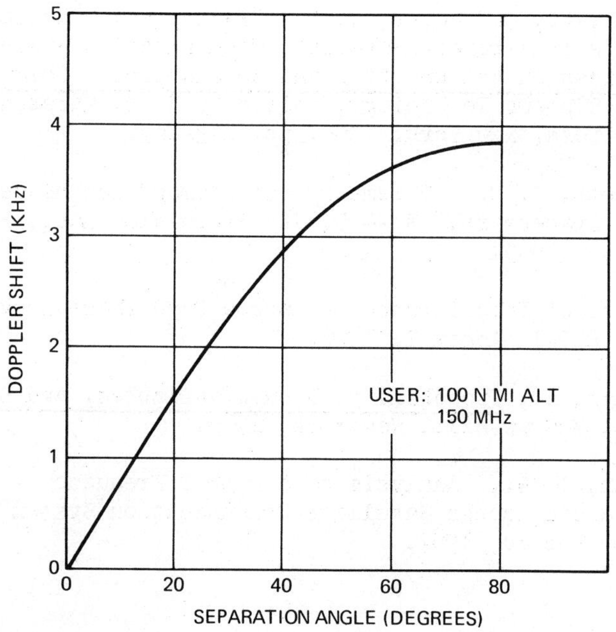

Fig. 2 Doppler shift for LOS between relay and user satellites (VHF).

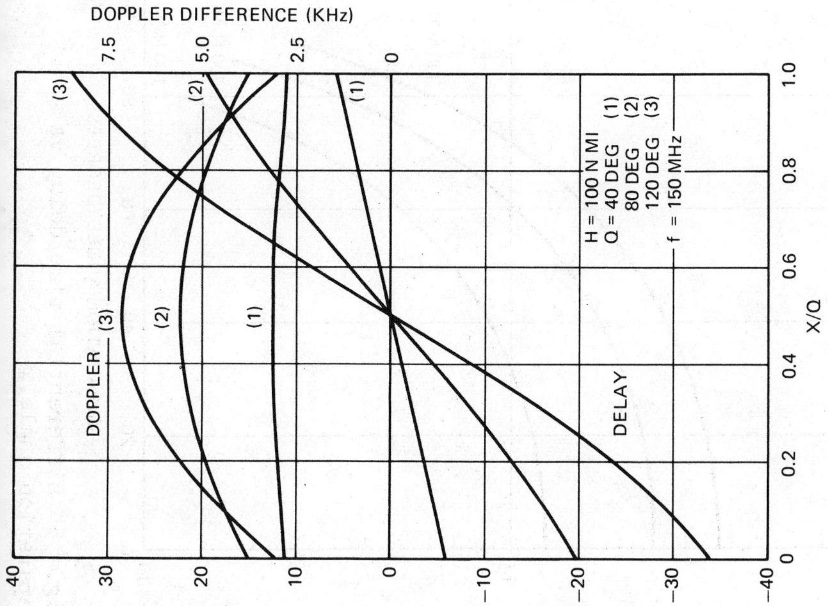

Fig. 3 Difference in doppler and delay for relay and user satellites.

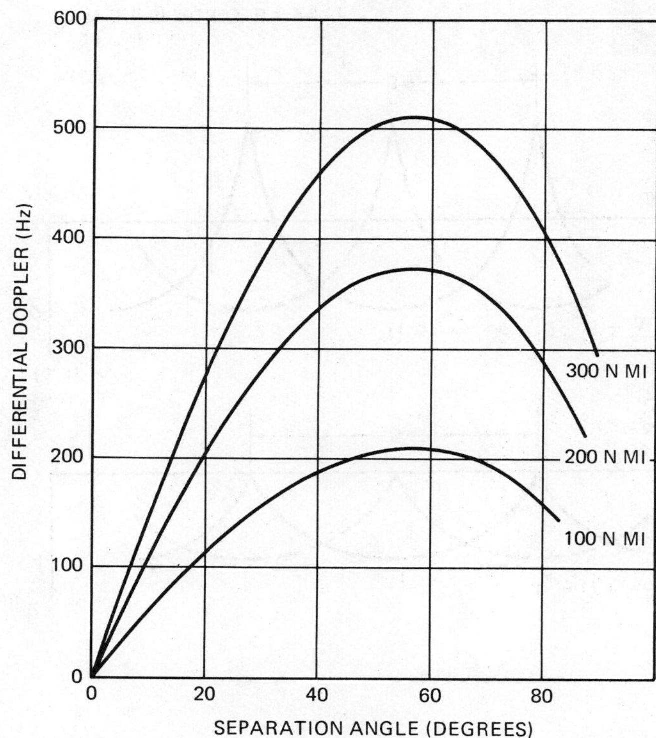

Fig. 4 Differential doppler vs separation angle.

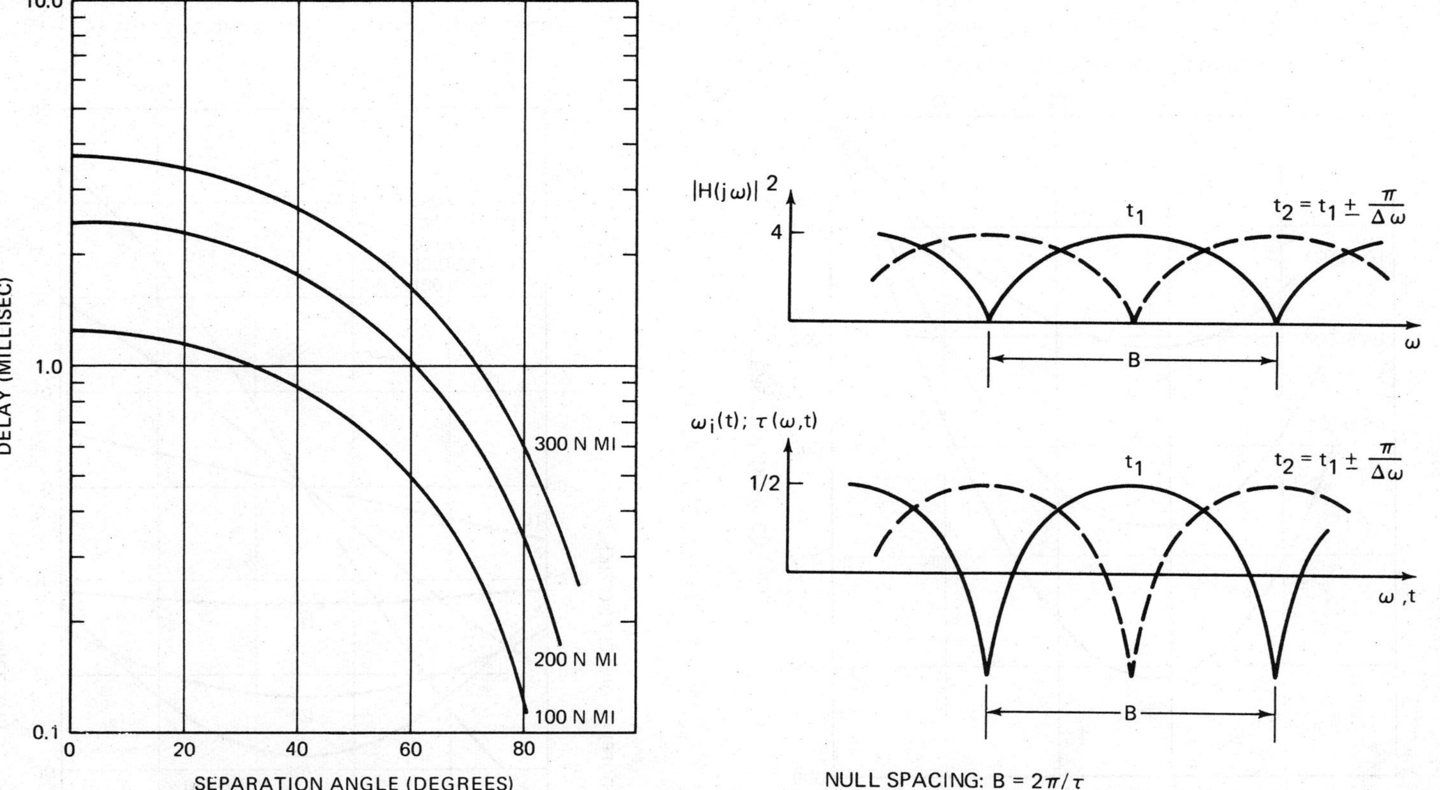

Fig. 5 Differential time delay vs separation angle.

Fig. 6 Time varying channel amplitude and instantaneous frequency response.

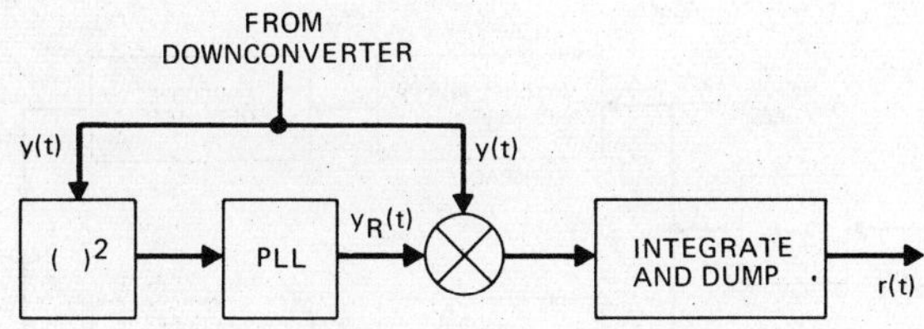

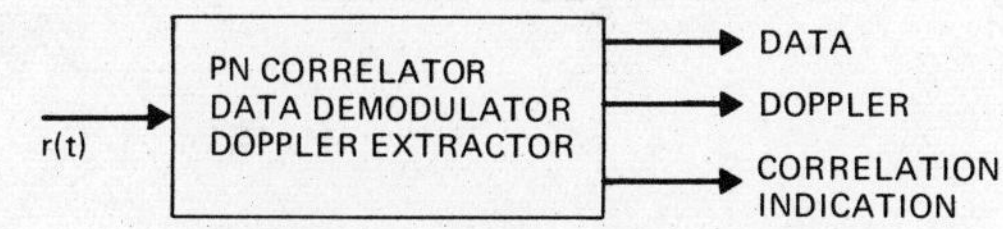

Fig. 7 Receiver implementation.

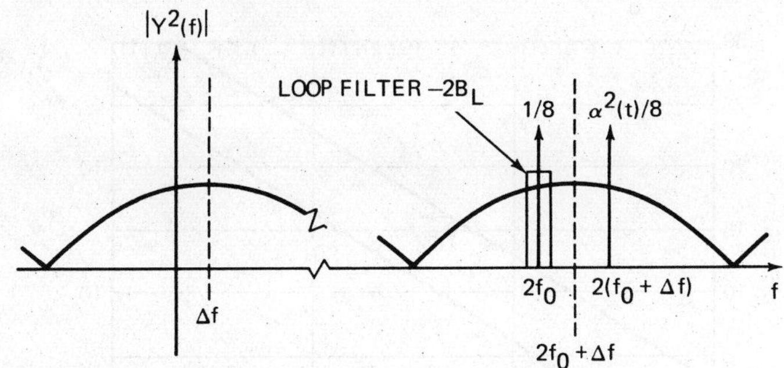

Fig. 8 Spectrum of squarer circuit output.

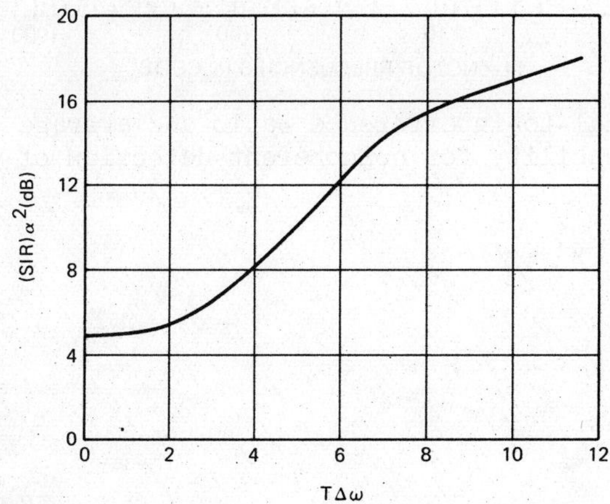

Fig. 9 Signal-to-interference ratio vs chip time-differential doppler product.

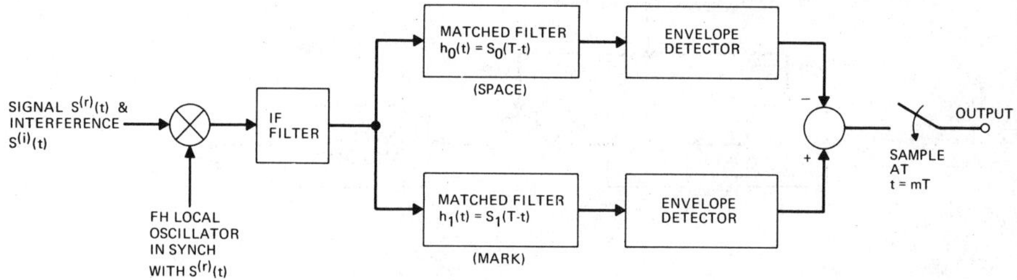

Fig. 10 Model for FH-Binary FSK system.

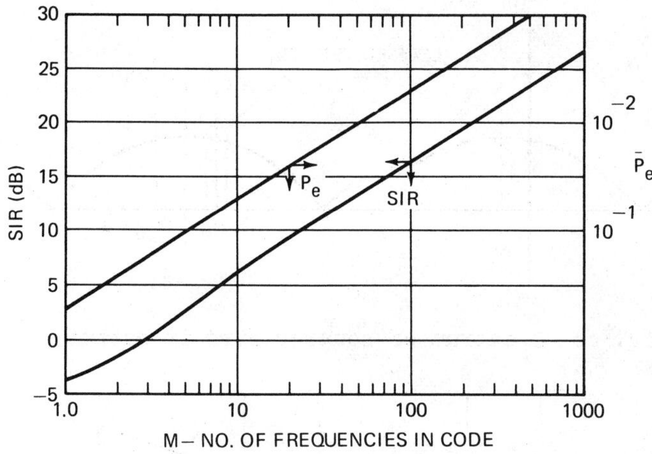

Fig. 11 Signal-to-interference ratio and average chip error probability for noncoherent detection of FH signal.

SATELLITE SYSTEMS FOR INTEGRATED COMMUNICATIONS, NAVIGATION, AND IDENTIFICATION

P. M. Diamond*

The Aerospace Corporation, El Segundo, Calif.

Abstract

Integrated communications, navigation, and identification (ICNI) implies two objectives: 1) implementation of a worldwide available command/control system and 2) equipment and signal integration. Satellite systems can provide navigation capability and communication capability. Tactical satellite communications experiments have demonstrated feasibility; however, ICNI requirements dictate a wideband, high-power, and large, expensive satellite. This satellite can be developed during the latter 1970s. The navigation capability requires development and demonstration in the near future. A suitable system can be developed by the 1974-1975 period. The principal problems of equipment integration relate to differences among requirements for "navigation-only" users and ICNI users, and ICNI equipment differences between air-air/ground-air functions vs the satellite mode. It appears feasible to resolve these differences via a multifunction satellite system combining, for example, an shf wideband communications and navigation capability for ICNI subscribers and a uhf narrower band navigation capability for all military operators.

I. Introduction

Integrated communications, navigation, and identification activities[1] are focused on the achievement of a means of providing an inclusive system to support those functions which require a radio-media interface between tactical autonomous elements, especially aircraft, and external information sources and sinks. The mission purpose of those functions is to enable effective command and control of those otherwise

---

Presented as Paper 70-503 at the AIAA 3rd Communications Satellite Systems Conference, Los Angeles, Calif., April 6-8, 1970.

*Associate General Manager, Office for Development Planning.

autonomous elements in carrying out their assigned missions. Since those functions have proliferated to excess in the last two decades, another purpose of the ICNI activities is to enable a systematic reconfiguration of the equipments in question to achieve a reduction in the device plethora, and to thereby realize savings in equipment, maintenance, and logistics. The approach utilized with respect to the latter purpose is directed to the ultimate reduction of these devices to a suitably configured common transmitter-receiver-processor at all system terminals in association with an appropriate common modulation of a signal at a single center frequency. Thus, to implement ICNI requires addressing two basic objectives: 1) implementation of a worldwide available command/control system which nets together a central facility and a complex of autonomous elements, as well as netting the autonomous elements among themselves, and 2) implementation of an equipment and signal structure which is common to all users and which enables a minimal number of devices, ultimately a single, integrated device.

The functions which correspond to the objectives can be characterized simply as communications and navigation, with each user having the capability to actively participate in the network. Communications, in this instance, is defined in the most general fashion to include all data: status, position, velocity, mission reports, and mission redirection data, etc. The major emphasis excludes the specific control of the terminal phase of mission execution because of the short-time constant involved and the uniqueness of this phase for each user; and the consequent technical inability to mechanize systems to accomplish the objectives (or conversely, the desirability of intelligent autonomous control by the element commander in the field). It focuses on the dispatch, enroute, rendezvous, and force recovery phases. That is, the focus is on basic force allocation control. For these purposes, it appears feasible and desirable to utilize digital techniques, capable of interference resistance, and data rates as low as possible, commensurate with the functions. A representative situation might be summarized as follows: 1000 users simultaneously participating in an objective area of 1000-naut-mile diam requiring a communications capability of 1000 bits each every 10 sec for a total data rate on the order of 100 kbps. A band spreading to, e.g., 100 MHz will provide 30-db interference protection. Pseudorandom modulation with a spread-spectrum and a time slot user message assignment structure is a representative technique. The navigation requirement is somewhat different in that a relatively coarse capability may be adequate for much of the enroute phases while an extremely accurate capability in three dimensions of position and velocity is required (and is feasible to deliver) for such activities as weapon delivery and blind landing. Furthermore, the navigation capability, in all its aspects, is required by many users, not all of whom will be participants in the ICNI network and who will not require the complex and expensive devices associated with ICNI. Additionally, the optimum frequency for a navigation-only user will not, in all probability, be the same as the optimum frequency for ICNI. Thus, it is useful to require that the radio navigation data be arranged so that it is conveniently available to all users while still integrated for the ICNI application. Fortunately, the navigation requirement can also be most effectively satisfied with a digital, pseudorandom, spread-spectrum system although the information data rate is considerably smaller, say 20 - 100 bps, and the spectral occupancy required is smaller, say, 20 MHz. These data can be utilized in a ranging scheme to satisfy the navigation requirements.

It is obvious that a highly ordered communications signal as has been described contains within it the properties required of a navigation signal, providing the number of ranging signals and the geometry of their points of origin can be arranged to satisfy the constraints. Explicitly considering the role of satellite systems in ICNI, the above discussion implies a worldwide satellite network providing dedicated communications and navigation to an ICNI network, as well as general DOD navigation, operated in conjunction with objective area users and objective area ground facilities netted with the users through ground-air and air-to-air relay. This general concept is shown in Fig. 1. Although the ICNI communications satellite capability is formidable and requires a large and powerful satellite, the technology of communications satellite can support the requirements. Considering all the problems, however, that capability appears to be a late 1970s implementation. The crucial, nearer term, satellite system issue appears to be that of developing and demonstrating a navigation satellite system capable of providing the accurate data to all DOD users and, especially, high-speed maneuvering aircraft. Extensive studies[2-5] have led to an understanding of the question sufficient to conclude positive feasibility with high confidence.

II. Navigation Satellite

A set of system properties which will satisfy the requirements of a broad spectrum of users of navigation systems, including ICNI, is 1) wide area or global coverage, 2) continuous availability, 3) three-dimensional position and velocity fixes, 4) real-time measurements, 5) accuracy to a few tens of feet in position, 6) accuracy to a fraction of a foot/second in velocity, 7) usable on rapidly maneuvering vehicles, 8) no radiation from user—no system saturation (excluding ICNI communications), and 9) equipment adaptable to many classes of users.

A variety of navigation systems using satellites can be configured. Analysis indicates that only systems based on measuring the ranges between satellites and the user, or the range difference between links to pairs of satellites, can meet the accuracy goals while being available to users in high-speed aircraft. Furthermore, the desire to accommodate high-speed users requires all of the measurements needed to establish a navigation fix to be made simultaneously.

Measurement of range can be accomplished by comparing the time of arrival of a signal with a clock at the user synchronized to a clock at the satellite. To maintain clock synchronization with any clock presently feasible for inclusion in the user's equipment, one additional measurement must be made. Since a three-dimensional fix is required, signals from four satellites need to be received. By measuring the time of arrival of the four signals relative to the user's clock, three-position coordinates and a clock correction thus can be determined. If, in addition, the frequency shift is measured, the three-compoment velocity vector of the user can be found. Since a clock correction is calculated, the user effectively has an extremely accurate time standard, and frequency standard, available to him.

Geosynchronous orbits (generally not circular or equatorial) are preferred from the standpoint of providing good coverage and the ability to deploy less than a full global system. A particularly attractive configuration employs one satellite in a synchronous, near-circular, equatorial orbit in conjunction with three or four satellites in inclined, elliptical carefully phased orbits, having the property that their ground traces follow a common near-circular path around the trace of the first (center) satellite. Such a constellation provides continuous regional geographic coverage with a nearly ideal geometry for providing accurate navigation. Four such constellations can provide global coverage with high redundancy. The satellite configuration which has just been described is shown from two points of view in Fig. 2. On the left, the satellites are shown in orbit as viewed by an observer in space who rotates with the earth. In this set of coordinates, the satellites do not appear to be in their orbital planes.

In the right-hand view, the same series of satellites for an inertially fixed observer are seen to be inclined elliptical orbits. This constellation array is also highly compatible with the ICNI communications requirements in that the central satellite of each constellation can be designed to provide the dedicated ICNI support. Four such satellites appear as a rotating Y in the sky, and five as an X. These satellite constellations provide a near-optimum geometry for navigation.

A system using the measurement technique just described will be termed a one-way pseudorange and pseudorange rate system. Such a system makes measurements equivalent to one-way range and range rate measurements made by the use of a synchronized user clock. It is also equivalent to (except for the computation algorithm) a hyperbolic system in which three range differences are measured to four satellites.

A conceptual diagram of a system based on the foregoing considerations as deployed for regional coverage is shown in Fig. 3.

Since the satellite positions form the reference from which navigation is performed, it is essential that their positions be accurately determined in geodetic coordinates. A master tracking station containing a tracking antenna sequentially acquires the four satellites and, for a period of several minutes, measures range and range rate using the two-way path from the master station to the satellite. From these data and a model of the Earth's gravitational potential, the satellite ephemerides are determined.

Figure 3 illustrates the process by which the user obtains a position and velocity fix. An arbitrary set of coordinates X, Y, and Z are introduced in which the satellite positions are assumed to be known and in which the user wishes to obtain his position and velocity. Each satellite transmits a signal to the user. These signals contain identifiable range codes modulated upon the carrier, typically by biphase modulation. The signals also contain the equivalent of satellite ephemerides modulated at a low data rate. The signals are either at different carrier frequencies or are modulated by orthogonal codes in order that the signals may be distinguished by the user. In addition to the primary signal, which is the wideband interference protected signal and which provides the most accurate and precise capability, another signal may be projected which offers somewhat poorer navigation performance to a user equipped with a minimum-cost set of equipment operated under a less demanding environment. This user could be a civil operator. The auxiliary signal would have a smaller bandwidth and would not be as protected from interference as the primary signals.

Figure 4 illustrates the method by which a position fix is determined from these signals. A typical signal from one satellite is shown together with the same signal as generated by the user's unsynchronized clock. This clock typically would be a high-quality quartz crystal oscillator of the type commonly incorporated in high-grade field equipment. By means of a correlation detector, the time shift between a satellite signal and the user clock is determined. This time shift (T in Fig. 4) consists of the speed of light transit delay from the satellite to the user and the lag of the user's time reference relative to system time. Four T_j's as measured from the four satellites together with the positions of the four satellites can be expressed as a set of four nonlinear algebraic equations in four unknowns (three components of user position and clock bias). Since the signals from the four satellites are synchronized as one of the ground station functions, the user clock bias is the same in each of these equations. The equations can then be solved for the user position and the clock bias.

If, at the same time as the pseudoranges are measured, the rate of change is measured by normal Doppler extraction techniques, four equations with user velocity as unknown may be written and solved for user velocity in three dimensions. The computation of user position and velocity is normally made in a digital computer of the sort used in an integrated avionics system. If less than full accuracy is required, many simplifications are possible which result in more modest user equipment.

The desired accuracy of a few tens of feet in position and a fraction of a foot/second in velocity can be conservatively guaranteed by a technique requiring a more careful system calibration than has so far been implied. The same navigation signal which is received by the user is received at the master station and at two or more calibration stations. Pseudorange measurements at each station are made and the results of these measurements are returned using a data link through the center satellite to the master station. At the master station, the implied position of each station is computed in the same manner as a user would compute his position. These are compared with the surveyed position of each station and corrections and clock biases in order to eliminate the difference between apparent and surveyed station locations.

The error performance of the systems already discussed has been analyzed, considering the error sources listed in Table 1, and the results of an analysis based on a representative five-satellite constellation are summarized. The results are based on the use of the four satellites with the highest elevation angles; better accuracy may be achieved by using all satellites of a four-satellite constellation (at the expense of some geographic coverage) or by using signals from all five satellites of a five-satellite constellation when they are visible.

Table 1 Major contributions to system error

Satellite tracking, errors, and geometric effects
Tracking station location
Geopotential model
Satellite-user/satellite-tracking station geometry
Signal propagation
Ionosphere uncertainties
Troposphere uncertainties
Multipath delayed signals
Instrumentation errors
Receiver/transmitter delay
Receiver resolution
Receiver noise
Tracking station clock drift
Satellite clock drift

Satellite tracking accuracy is determined primarily by the number and location of ground stations taking tracking measurements, the degree of complexity and sophistication of the algorithm used to transform tracking measurements into satellite ephemeris estimates, and the interval of time over which the satellite ephemerides are predicted. In general, satellite position errors are smallest in the radial direction and largest in the in-track direction; errors in the cross-track direction typically lie between these two extremes. To facilitate a more general description of the effect of satellite tracking accuracy on navigation accuracy, a typical satellite error ellipsoid shape is assumed: the principal axes of the ellipsoid are in the radial, in-tracking, and cross-track directions and have relative magnitudes of 1, 8, and 4, respectively; an intersatellite radial correlation of 0.6 also is assumed. A representative set of values are (1σ), 50 ft, 400 ft, and 200 ft, respectively.

Of particular concern is the area of error sources due to propagation delays because of ionospheric and tropospheric uncertainties. This follows because of the unique association these factors have with frequency choice. There is clearly another major factor involved, that of satellite rf power, which, for natural noise limited design, is obviously a strong function of frequency ($f2.5$ considering Doppler tracking requirements as a function of frequency). However, for designs in which artificial radio interference is the dominant constraint, the rf power level required is very nearly the same over a wide frequency range from a few hundred MHz to a few GHz. This range includes the most desirable frequencies for navigation.

Propagation delay errors are those associated with prediction or measurement of the signal delay during its passage through the ionosphere and the troposphere. The range increment because of the ionospheric delay, for frequencies above approximately 200 MHz, is inversely proportional to the square of the frequency and a direct function of the propagation path total electron content, as shown in Fig. 5. For frequencies below C band, it is necessary to make a correction in order to establish user position to the required accuracy. This correction may be made by utilizing a dual frequency measurement scheme which enables a real-time computation, by the user, of the proportionality factors which are variable in space and time, and exhibit marked variation at the dawn and dusk periods. The disadvantages of this approach are that users require a dual set of microwave receiver equipment, and at least five and possibly as many as eight radio channels as compared to four for a single-frequency technique. Furthermore, the satellite is penalized since both channels require equivalent electrical power during their transmission period. Another technique with a single frequency is to utilize field calibration stations which, since their positions and velocities are known, enables complete nulling of all errors for a local region. In the L band, it appears that statistical modeling of the ionosphere, without explicit local user measurement, may suffice. Figure 6 gives the calculated ionosphere range uncertainty utilizing statistical modeling of available data. In addition, a modification of the dual-frequency technique at one frequency is possible if the signal bandwidth is sufficiently large. A correction might be made by the user which would bound the error, even for such extreme solar/ionospheric events as the May 1969 event, to the order of 60-70 ft. Finally, at frequencies in the range above approximately 3 GHz, no correction whatsoever is necessary for a single-frequency system. The troposphere correction at L band is indicated in Fig. 6 and can be seen to be relatively small and easily made.

Figure 7 summarizes the geographic variation of accuracy for a particular configuration of constellation and ground stations and also shows the coverage limits for a single constellation. These results are presented for the worst case over a 24-hr span resulting from slight variations in the system properties which occur. The nominal (mean) performance can be characterized, as before, by the statement "a few tens of feet." Similar performance characteristics are exhibited for vertical position. The geographical variation of velocity is also similar in form with smaller variation and can be characterized by 1/4 - 1/2 fps. Time synchronization variation is similar and can be characterized by several tens of nsec uncertainty.

III. Communications and an ICNI Satellite

Achieving the requisite communications performance to the ground from a satellite is not simple. Indeed, as in the navigation case, the aircraft user tends to dictate the system design constraints. In the navigation case, consideration of such factors as tracking loop bandwidth, integration time, signal/noise ratio, and aircraft antenna gain (which for maneuvering aircraft means essentially an omnidirectional antenna or a multiple beam automatic search and pointing antenna of higher gain) is necessary. The complex, high-performance antenna, of course, very seriously presses technological limits as well as weight, volume, and installation characteristics. Similarly, communications are constrained by physical limitations and functional requirements. These are summarized in Table 2. Each of these factors impacts the nature of the satellite system. In general, for example, it seems desirable to utilize the uhf region, but the wideband requirements force consideration of higher frequencies such as shf. This, in turn, forces consideration of narrower beamwidths than Earth-coverage, perhaps 1 - 3°, which enables coverage of a "nominal" objective area, but utilization of shf may require significant development in the area of aircraft antennas. Thus, the classical tradeoff analyses among bandwidth, antenna factors, interference resistance, receiver noise, modulation factors, as well as new factors such as navigational accuracy and propagation questions, must be accomplished.

Present capabilities, and also the demonstrated technological basis for future development, are represented by the Tactical Communications Satellite, TACSAT I.[6-8] TACSAT I, the largest satellite ever placed into geosynchronous orbit by the United States (Fig. 8) was launched on February 9, 1969.

Table 2 Tactical satellite - aircraft communications

Constraints/problems/objectives

- Transmitter power
- Receive antenna gain
- Interference resistance/security
- Spectrum conservation/frequency allocation
- Increased capacity

Development directions

- Higher frequencies
- Digital processing, encryption
- Spread-spectrum
- Narrow transmit/receive antenna beams
- Multiple beams
- System discipline/ordering

Its performance was verified as nominal and operational concept testing by the military services is proceeding. Based upon the data gathered, acquisition of an operational system may be initiated. Following the current testing program, TACSAT I will be made available for limited operational support. The TACSAT I operates to uhf and shf terminals and has a very limited capacity. This capacity is illustrated in Table 3.[9] While there is a moderate capability at shf, only very limited service at uhf is provided. This is an illustration of the complexities of the problem serving to underline the assertion that thousands of channels for small users are not easy to come by. The capacity represented by the 132 vocoded voice channels is generally in accord with the ICNI need. However, this is achieved with a satellite of nearly one kilowatt of raw power and with a stationary ground terminal utilizing a 4-ft-diam paraboloid antenna, and with over-all bandwidth limited by comparison with ICNI requirements. A significantly higher-capacity system cannot be expected in the near future because of the high cost and necessity for implementing improved subsystem techniques.

Such techniques are motivated by consideration, among others, of interference. Interference resistance, mitigation, and avoidance are principally a problem in the user receiving system caused by ground (or air) based sources. As mentioned previously, signal processing is an approach leading to partial solution of the problem. In addition, highly directional, high-gain, satellite antennas will assist, leading towards the consideration of narrow-beam systems and multiple-beam systems.

Table 3 TACSAT I capacity

Repeater	Receive Terminal	Modulation	Accesses
uhf	1-1/4 ton truck (4-ft-diam antenna)	FM voice	8
		Encrypted voice	8
	Aircraft (blade antenna)	FM voice	5
		vocoded voice	20
		teletype (100 wpm)	630
shf	1-1/4 ton truck (4-ft-diam antenna)	FM voice	52
		Encrypted voice	23
		Vocoded voice	132

One may speculate on the nature of the satellite which satisfies the diverse requirements posed by the navigation and communications problems discussed previously. In particular, suppose a satellite system designed as implied earlier, i.e., a central satellite of the stationary constellation carrying the ICNI communications function, in concert with an outer ring of satellites performing the ICNI navigation function as well as the general military navigation function. This combines the general support, with Earth coverage, with the ICNI system to service one objective area at a time. This concept is illustrated in Fig. 9. The assumed total bandwidth includes provision for a dedicated and integrated ICNI capability. Since other DOD users are presumed not to be interested in participating in the wideband ICNI network, a separate, but adjacent, band for providing general precision navigation service, is provided within the total bandspread. The frequency chosen for this communications/navigation signal is L band. No attempt here is made to argue the degree to which the design is optimum. Indeed, it is presented only in an exemplary manner.

The design of the satellite (Fig. 10) is dominated by the required power-generating capability dictated primarily by the communications function. Transmitted radiated power on the order of 400 w for the objective area communications function and 100 w for the navigation service is required. A total raw power input to the transmitters of 1.4 kw is required. Taking into account the power requirements of the rest of the satellite and in-orbit degradation over the satellite lifetime, it is estimated that a raw generated power capability of 2.5 kw is necessary to perform the mission. A 20-ft dish antenna is required to provide 36-db gain with a 3° beamwidth for the satellite to aircraft objective area link. A 3-ft-diam parabolic antenna, providing 19-db gain, is utilized for transmitting the navigation signal with Earth coverage. The onboard electronic equipment consists of the L-band transceivers, S-band and C-band equipment for satellite ephemeris determination and ground support, power management and control equipment, and data demodulators and modulators. Since the antenna beamwidth is relatively narrow, a method of pointing the beam with high accuracy (on the order of a degree or less) is required. This requirement implies a spacecraft stabilized about an Earth-centered orbital planar reference system.

Recent developments in deployable solar arrays have included roll-up panels which have a figure of merit of 30 w/lb compared to 9 w/lb for conventional foldout arrays. Since the advantages to be gained over more conventional arrays were obvious from a weight and space standpoint for the powers involved, roll-up arrays were chosen for this design. These arrays consist of rolled up plastic Kapton sheets upon which the solar energy converter materials are deposited. Deployment is by extensible booms to a rectangular configuration 10 ft wide × 17 ft long. The configurations discussed are being built for test in the early 1970s and should be available for systems implementation by 1975.

The parabolic antenna design is based upon a design utilizing a flexible wound rib concept. In this scheme, the rotation of an electric motor deploys flex ribs upon which an antenna reflective fabric is stretched. The antenna is fixed, when deployed, with small beam direction changes effected by moving the antenna feed. Attitude control is accomplished in the suggested design by cold gas reaction jets with the system reference slaved to the earth horizon and orbit plane. This concept was chosen because it is a well-known technique and provides performance and reliability equaling or exceeding any other method for the weights and inertial distribution under consideration and provides no development risk. Other concepts such as the dual spin concept are worthy of consideration. The satellite would weigh on the order of 2700 lb and require an improved T-IIIC for launch or could be launched by the projected Space Transportation System.

The foregoing satellite is the center satellite for the constellation. The outer satellites must provide the navigation support in conjunction with the center satellite. These satellites would not require the large antenna, and corresponding power system and communications equipment, and would be more conventional in appearance. They would require a solar array power supply of approximately 650 w and would weigh on the order of 800 lb.

Further growth potential is offered through the application of a multibeam antenna system, shown conceptually in Fig. 11. Space-fed, passive lens techniques appear to offer great advantages for the future pending further development. Frequency allocation problems are eased, coverage and capacity increased, resistance to interference improved, and flexibility of operations is provided.

IV. Conclusion

Some notions of satellite systems for an ICNI system have been described and it has been shown that there is a demonstrated basis for providing the communications capability required. This capability, dominating the satellite design for a dedicated ICNI satellite, is sufficiently complex that it should not be expected to be developed until the late 1970s. The navigation capability for ICNI as well as for general military application is crucial in that 1) it is a necessary capability for an ICNI system, 2) it is necessary for improvement of overall operational effectivness, and 3) it is appropriate and feasible to demonstrate the technology in the immediate future. Finally, an integrated satellite system complex can be conceived which would support the desired objectives.

References

[1]Gobelman, I. J., "Improvement in CNI - Will They Help Create a New Air System?," Astronautics & Aeronautics, Vol. 7, Feb. 1969, pp. 62-69.

[2]Woodford, J. B. and Soule, P. W., "Precise Navigation over Wide Areas Using Satellites," TR-0066(5521-01)-1, Aug. 1969, The Aerospace Corporation, El Segundo, Calif.

[3]Woodford, J. B., "A Satellite System to Support an Advanced Air Traffic Control Concept," paper presented at IEEE Conference, New York, Oct. 1969.

[4]Leonard, B. P. and Woodford, J. B., "Characteristics of a Satellite Navigation System Operated in Conjunction with a User Inertial System," paper presented at Guidance and Control 9th Meeting, Advisory Group on Aeronautics Research and Development, Department of Defense, NATO, Sept. 23, 1969.

[5]Woodford, J. B., Melton, W. C. and Dutcher, R. L., "Satellite Systems for Navigation Using 24-Hour Orbits," paper presented at IEEE Conference, New York, Oct. 27-29, 1969.

[6]"United States Aeronautics and Space Activities," rept. to the Congress from the President of the U.S., National Aeronautics and Space Council, 1968.

[7]Miles, M., "Big Satellite to U.S. World Defenses," Los Angeles Times, May 14, 1968.

[8]Wilford, J. N., "U.S. Launches a Test Satellite to Aid Military Communication," New York Times, Feb. 9, 1969.

[9]Mattics, T. J., "Tactical Satellite Communications Requirements," Signal, Vol. 24, No. 1, Sept. 1969, pp. 65-69.

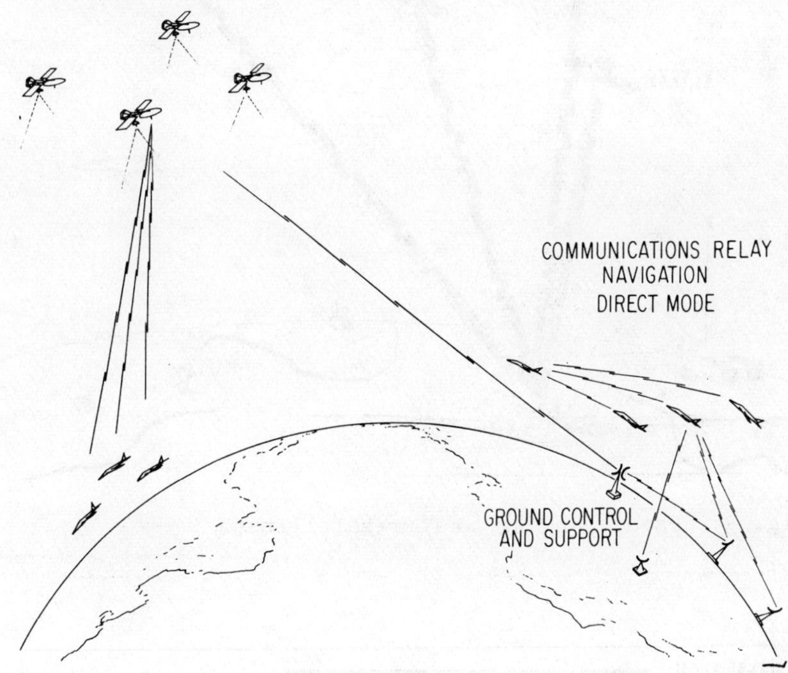

Fig. 1 ICNI concept.

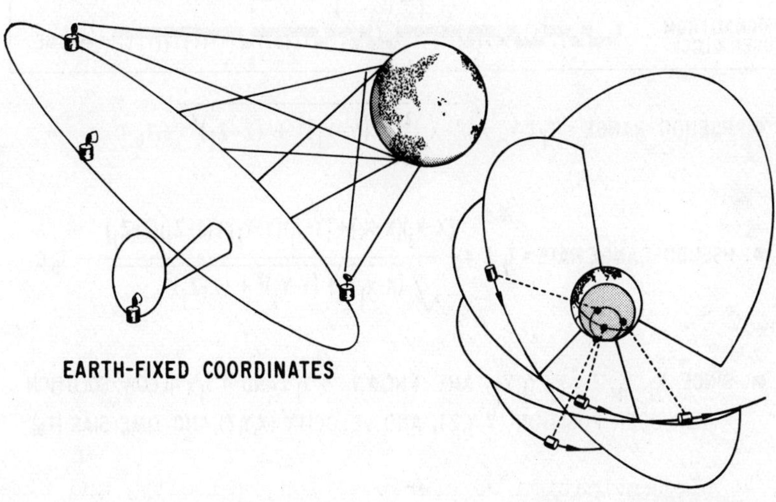

Fig. 2 Orbital deployment-two views.

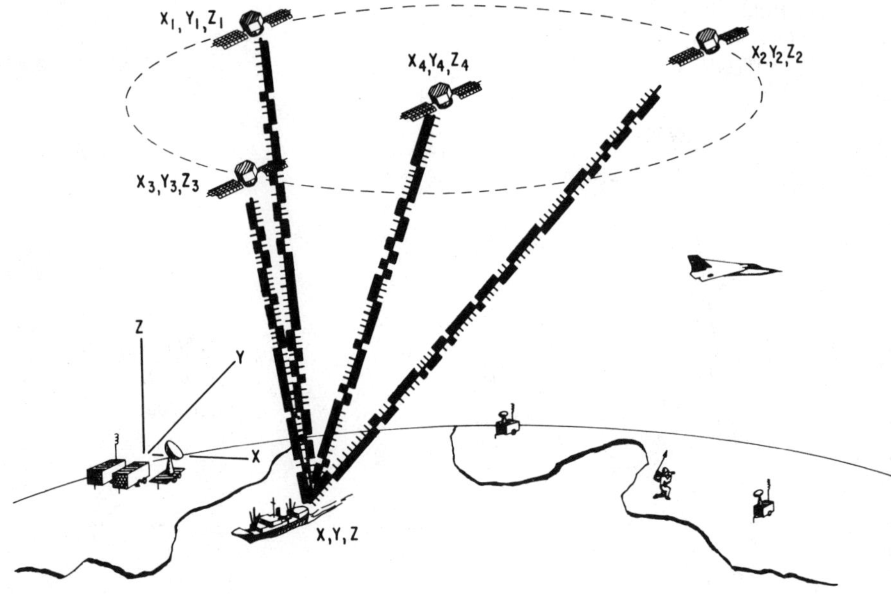

Fig. 3 Pseudoranging to four satellites.

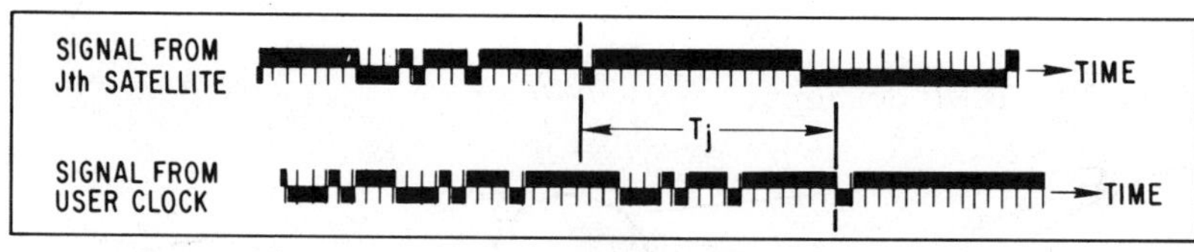

- PSEUDO-RANGE = $T_j \cdot C = \sqrt{(X-X_j)^2 + (Y-Y_j)^2 + (Z-Z_j)^2} - T_B \cdot C$

- PSEUDO-RANGE RATE = $\dot{T}_j \cdot C = \dfrac{(X-X_j)(\dot{X}-\dot{X}_j) + (Y-Y_j)(\dot{Y}-\dot{Y}_j) + (Z-Z_j)(\dot{Z}-\dot{Z}_j)}{\sqrt{(X-X_j)^2 + (Y-Y_j)^2 + (Z-Z_j)^2}} - \dot{T}_B \cdot C$

- SINCE $X_j, Y_j, Z_j, \dot{X}_j, \dot{Y}_j, \dot{Z}_j$, ARE KNOWN, 4 T_j's AND 4 $\dot{T}_j$'s ALLOW SOLUTION FOR USER POSITION (X,Y,Z), AND VELOCITY ($\dot{X},\dot{Y},\dot{Z}$), AND TIME BIAS (T_B)

Fig. 4 Determination of user position, velocity, and time.

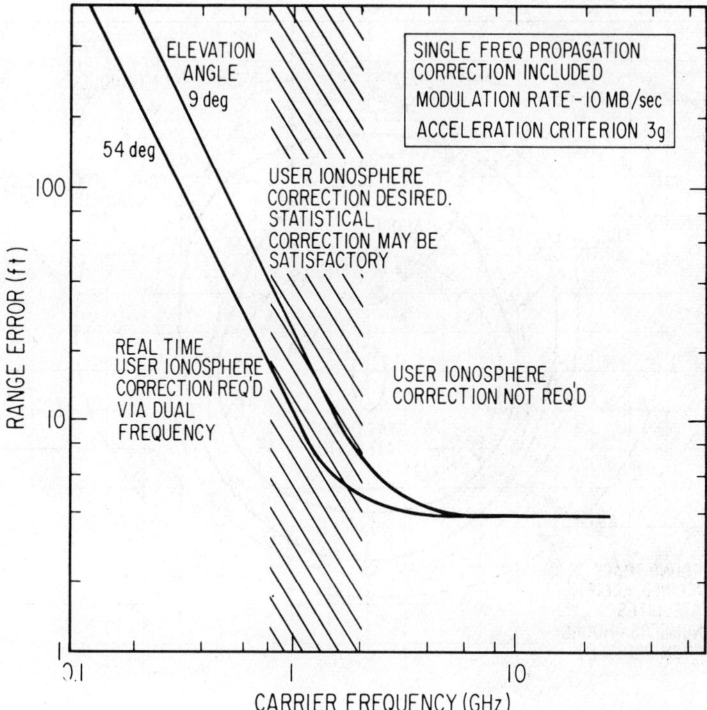

Fig. 5 Frequency dependency of range error.

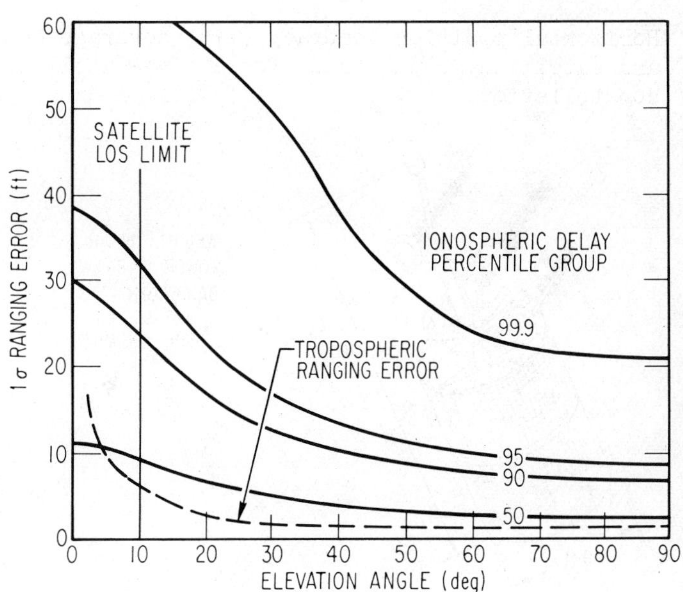

Fig. 6 Propagation ranging error.

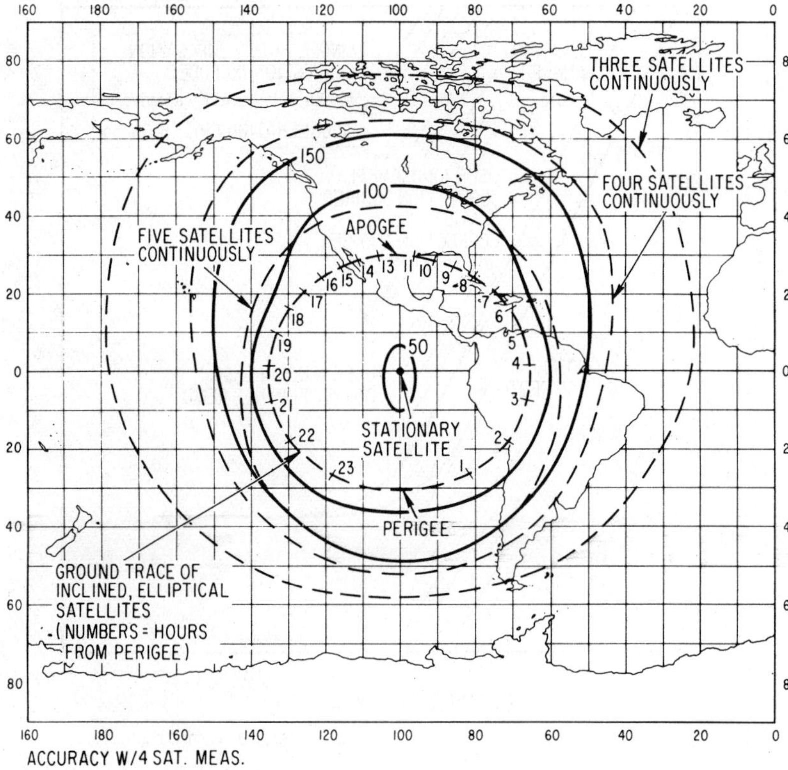

Fig. 7 Horizontal position accuracy (1σ), coverage contours, and satellite ground trace for a five-satellite constellation.

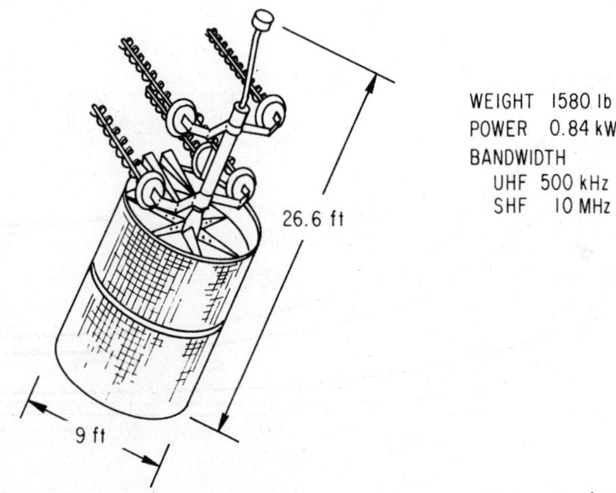

Fig. 8 Tactical communications satellite 1 (TACSAT 1).

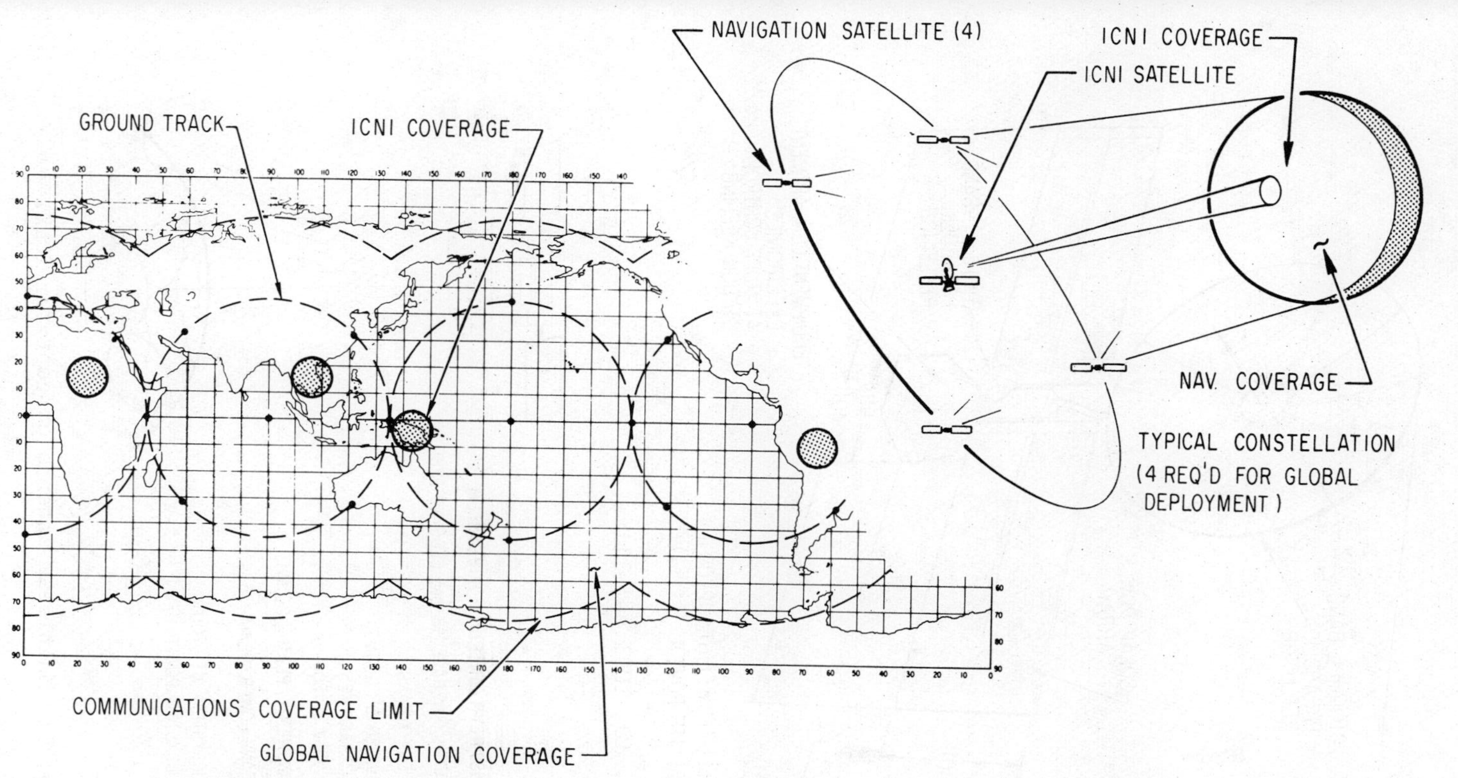

Fig. 9 ICNI satellite system concept.

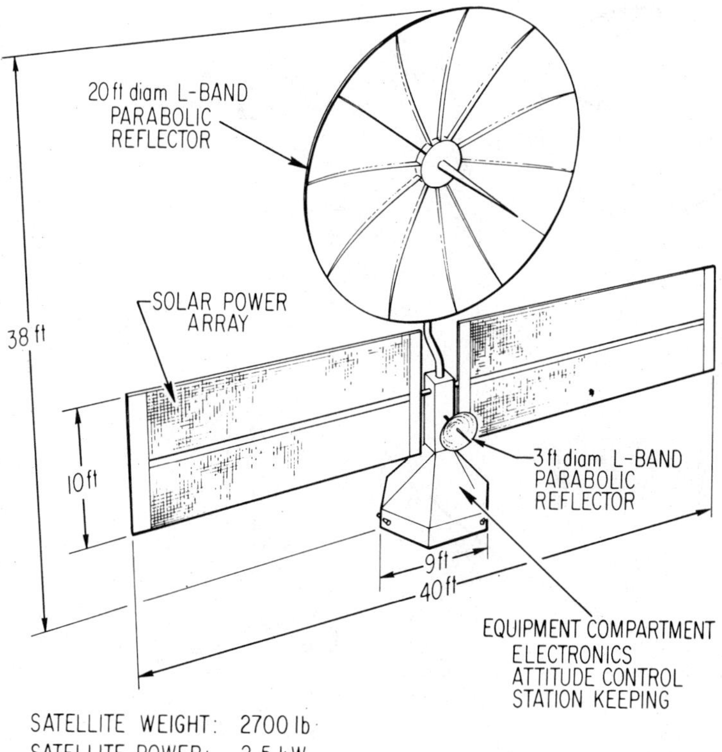

Fig. 10 ICNI satellite.

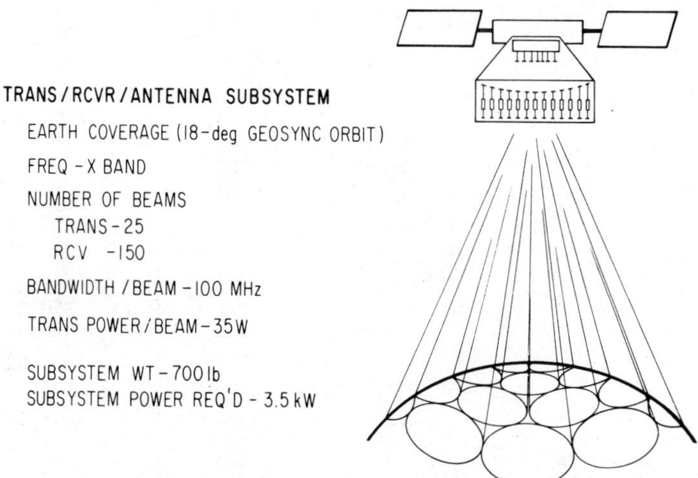

Fig. 11 Multibeam antenna concept for ICNI.

VIII. SYSTEM ENGINEERING

INFORMATION TRANSFER IN 1990

Roger W. Hough*

Stanford Research Institute, Menlo Park, Calif.

Abstract

Results of a recent study of potential demand for electronic information transfer services in 1970-1990 are presented. Projections are made for new services such as electronic mail, remote library browsing, checkless society transactions, video telephone, and others, as well as conventional services such as telephone, telegraph, and network program transmission. The principle method used was one of subjectively selecting a reasonable estimating parameter specific to each service; using past trends in these parameters—including analogous logistic growth curves in certain cases—to project potential demand for the services; then combining the requirements into common categories such as voice, video, record and data, and written material. Major conclusions indicate that, other estimates to the contrary, voice traffic will continue to dominate in terms of input to a total (both local and long distance) "U.S. information system," and that data will constitute only a small fraction of that input.

Introduction

This paper examines one aspect of the "communications revolution" in the United States—demand for new services. The study was commissioned by NASA as part of a series of investigations into information transfer via satellite. The research cited[1] did not concentrate on satellites, since it was an

---

Presented as Paper 70-443 at the AIAA 3rd Communications Satellite Systems Conference, Los Angeles, Calif., April 6-8, 1970. This research was supported by NASA Contract NAS2-5369. The author wishes to thank the many individuals who contributed to this study. Among them are Carolyn Fratessa, Virginia Holley, A. H. Samuel, and L. J. Wells, who are also co-authors of the final report. Other individuals provided much helpful advice and assistance during the research, particularly L. R. Parkinson, E. B. Shapiro, and L. Feldner, Consultant.

*Senior Industrial Economist, Electronic Industries Research.

investigation of telecommunications needs and services without regard to the means of transmission. Other studies in the series[2,3] have addressed the question of possible satellite missions.

The method used in the study was to identify potential as well as present-day services, classify them into major categories, and establish quantitative estimates of message traffic occurring within the United States during the period 1970-1990. In order to aggregate diverse services such as voice, video, and record and data transmission, traffic estimates were converted to the common denominator of bits/yr. (It is important to note, that the study did not consider, nor do the conclusions imply, estimates of circuit requirements.)

Scope

The term "information transfer," as used in this paper, is intended to convey something broader than present-day electrical communication, that is, radio, television, telephone, and telegraph. For example, the study considers present-day written communications, such as postal services, newspapers, books, and magazines, as well as the rapidly growing area of man-computer and computer-computer communications.

An enormous number of new services using digital data communications are technologically possible in the future. During the research approximately 400 such services were identified and cataloged, including remote library browsing, cashless society transactions, law enforcement systems, patent searches, airline reservations, cultural and sports reservations, and so on. It is impractical to present all of these in this paper. Mention is made of them, however, to indicate the breadth of the study.

Economic Framework

One of the most common techniques used in long range forecasting is the very simple method of extrapolating past trends, using time-series data plotted on semilog paper. Many indicators of economic and technological growth increase at a relatively constant rate each year, leading to straight lines on such log-linear scales. Thus, when used expeditiously, projection of constant compound annual growth rates into the future is an entirely acceptable method of forecasting.

There are, however, many important limitations in projecting constant rates of growth. One of these results from the fact that many products and industries go through a life cycle that is characterized by increasing rates of growth in early history and by decreasing rates of growth as maturity is approached.[4] In these cases, a plot of the growth variables resembles an S-shaped curve, which is characteristic also of biological growth. Such curves are commonly called logistic or Gompertz curves.

The fact that products and groups of products exhibit such life cycles is very important. For example, one of the required tasks of the Ref. 1 study was to estimate demand for information transfer services that have not yet come into being. Traditional trend projection methods are of little value in this case, because historical data are completely lacking.

To provide a solution to this problem, a technique called the method of historical analogy was borrowed from the field of technological forecasting.[5,6] This method encourages one to use information from past historical developments and apply it to future events of a similar nature.

The Telephone

The history of telephone service in the United States is shown graphically in Fig. 1. Both number of telephones and number of conversations are plotted in Fig. 1, to illustrate the close correlation between the two series.

To quantify the changing rates of growth, we calculated the average compound annual growth rate for the first year, then for the first two years, the first five years, and so on, up to the average annual rate for the first twenty years. These calculations are shown on the figure. Between the first and second years (shown in the chart as "Year 1"), the increase in number of telephones was 300%.[†] For the first two years of service the average increase was 200%/yr. For the first five years of service the growth was about 80%/yr; for 10 years, it was 50%/yr; and for 20 years, 28%/yr.

[†]Clearly, this calculation and those that follow are very sensitive to the values given for "Year 1". The data used in the study were the "best available," that is, Government statistics where possible and acceptable industry information elsewhere. In general, subscribers or users for Year 1 represent those recorded for the first (recorded) year of operation. Thus, the curves necessarily start at a point other than zero.

Telegraph Service

There is quite a difference between the history of telephone service and the history of telegraph service, which is illustrated in Fig. 2. The most obvious characteristic of this series is, of course, its absolute decline in recent years. Since 1945, when 236 million telegrams were sent, the service has declined approximately 4% each year. In 1966 only 93 million messages were sent; if the trend continues 80 million will be sent in 1970, and about 53 million will be sent in 1980.

Television

Of the five technological developments illustrated, the history of television, shown in Fig. 3, exhibits the most regular as well as one of the highest growth rate patterns. Receivers installed grew by more than one thousand percent in the five year period 1946-1951.

Automobiles and Computers

The growth histories of automobiles and computer installations in the United States are shown in Figs. 4 and 5. In each of these cases growth proceeded rapidly from the beginning, and at almost constant rates throughout the period illustrated.

Summary

The growth trends presented in this section are summarized in Table 1. These trends give a picture of the manner in which the growth of some developments has actually taken place in the past. In the next section this evidence is used to provide a basis for projecting new services that have no historical base at the present time.

Table 1 Average annual rates of early growth

		Growth Rate (%/yr)				
		First year	First 2 yrs	First 5 yrs	First 10 yrs	First 20 yrs
Telephone	1876-1896	300%	200%	80%	50%	28%
Telegraph	1867-1887	10	17	12	13	11
Television	1946-1966	75	370	320	190	58
Automobile	1900-1920	85	70	60	50	41
Computers	1951-1971	700	400	300	210	84 (est.)

Projections of Individual Services

The major quantitative results of the study were derived by considering approximately thirty existing and potential information transfer services in such a way as to estimate demand volume, local, nonlocal, and total, in terms of bits/yr. This was accomplished by selecting a representative sample from the list of services developed early in the study; estimating a projection parameter for each service based on number of calls, messages, transactions, hours of operation, and so on; projecting this variable over the time period of interest; then converting all demand to the common denominator of bits per year.

In this paper we illustrate five of the services so analyzed, providing a brief account of the method used to project the transaction variables of interest. A table is then given that summarizes the information developed during the study for all of the services. It is important to note that the projections given do not in all cases imply a real demand for information transfer, even after the conversion to bits/yr. What they imply is simply potential demand, or, in other words, an upper bound on demand. Thus, no attempt has been made to predict time scales for development of new electronic services to replace present-day nonelectronic means of information transfer, and the actual development of such services is not implied.

Telephone

For the purpose of extrapolating into the future, more data are available for telephone service than for any other. Statistics have been collected since the beginning of the service in 1876; and at an early date separate records were started for long distance, business and residential use, and for revenues by type of service. These data show that both the number of telephones and total telephone conversations have been increasing at a constant rate (about 5%/yr) for many years.

For a number of reasons, some forecasters in the past have taken the position that this rate of increase must decline before too many more years pass. They say, for example, that it is impossible to talk on the telephone more than X number of minutes per day; and since conversations are increasing faster than people, saturation must begin to take place soon. As a matter of fact, projections have been made in the past that assumed saturation, only to find that it did not occur.

For this reason, telephone conversations are projected at present rates of increase, as shown in Fig. 6. Over a span of five years or more, both local and total calls exhibit average annual growth rates of about 5%/yr. Long distance calling is growing faster, however, especially in recent years. Taking account of this, we have projected an increase in long distance conversations of 8.5%/yr from 1970-1990.

Videotelephone

Videotelephone or Picturephone[≠] has not yet begun as an offered public communications service, except on an experimental scale. Consequently, there are no historical data upon which to base a projection or forecast. On the other hand, various pieces of data may be fitted together to provide a composite picture of the potential growth of this service.

According to recently published information,[7] Picturephone service is soon to be introduced on a regular basis in several cities. During experimental tests in 1968 and 1969, 40 sets were used at Westinghouse Electric Corporation, communicating regularly between New York and Pittsburgh. In addition, some 60 sets have been in use within the Bell System for several years, making a total of about 100 sets in operation now, on an experimental basis.

In 1876, the first year of operation of the telephone system, there were 3000 telephones in service. Also, since 1880, calls per day per telephone have consistently averaged between four and eight, and the greatest number of calls/day occurred during the earliest years of telephone service.

Using these guidelines, we established the following assumptions:

1) During the first year of operation (nominally 1970), approximately 1000 videotelephones will be installed.

2) Each subscriber will place on an average four calls/day, making a total of about 1.2 million calls during the first year.

3) Calls will increase at an average annual growth rate of about 40% between 1970 and 1990. This rate falls approximately half-way between that for telephone and television in the first twenty years of their existence (28% and 58%, respectively) and is slightly less than that for the first twenty years of the automobile, at 41% growth per year. (See Table 1.)

[≠] Registered trademark of the Bell System.

Using these assumptions, projected videotelephone conversations are plotted as shown in Fig. 7.

Network Television Transmission

The question of future television activity is of major interest to the country. As a recent report points out,[8] greatly expanded cable facilities (CATV) should be encouraged; viewers need a wider range of programming, including public, community, and educational programs; and so on. Other qualitative indicators of growth and change are available also. However, for quantitative purposes, even the method of historical analogy and logistic growth curves is of little use in this situation. Therefore, we project several alternative future possibilities regarding the number of national networks in operation during the time period of interest, as shown in Fig. 8.

The alternative possibilities depicted for 1990 are five, ten, and twenty networks. We assume that each of these networks will broadcast approximately the same number of hours per year as each of the present three (3600 hrs),[8] resulting in the three forecasts of 18,000 hrs, 36,000 hrs, and 72,000 hrs of broadcasting in 1990.

Value Transfer—Checks and Credit Transactions

The checkless society has been much discussed in the literature, as well as being the subject of several conferences and a number of studies. To establish potential demand for such services, we turn to data on the number of transactions involving checks and credit purchases over the last several years, shown in Fig. 9.

These data indicate that the number of checks written in the United States is growing at a rate of approximately 6%/yr. Credit transactions are growing considerably faster, at about 12%/yr. All of these transactions are directly replaceable with some form of electronic funds transfer system.[9]

First Class and Airmail

The last projection to be illustrated concerns what we have referred to as "written material," in this case first class mail and airmail. In this application we assume that an image of the letter, or the letter's contents alphanumerically coded, will be transmitted electronically to a distant point where the letter will be reconstituted.

As shown in Fig. 10, mail service will increase in the years to come at a projected rate of 3.5%/yr. This results in a first class mail and airmail volume of 50 billion, 70 billion, and 100 billion pieces over the twenty-year period of interest.

Summary of the Projections

Table 2 summarizes the projections shown on the previous figures, and includes estimates for the remaining services that were also investigated. In total, 28 services were analyzed, including voice service (telephone), 2 video services (videotelephone and network broadcast transmission), 22 record and data services (telegraph plus 21 data transmission applications), and 3 written services (mail, books, and newspapers). For each of these a projection parameter unique to that service was subjectively established. This is given in the column "Units" in Table 2. The values of the projection variable are then given for the years 1950 and 1960 where historical data are available and for 1970 to 1990 in all cases.

Conversion to Bits/Year

The rationale and conversion factors used to derive information transfer volume in bits/yr, resulting from the projected transactions given in Table 2, are illustrated in the examples below. Bits/yr are calculated for the five services described previously, using data for 1990.

Telephone: $(482 \times 10^9$ calls/yr$) \times (6$ min/call$) \times$
$(60$ sec/min$) \times (64{,}000$ bits/sec$) = 1.1 \times 10^{19}$ bits/yr

Videotelephone: $(1.0 \times 10^9$ calls/yr$) \times (6$ min/call$) \times$
$(60$ sec/min$) \times (6.3 \times 10^6$ bits/sec$) = 2.3 \times 10^{18}$ bits/yr

Television Transmission: $(72{,}000$ hrs/yr$) \times (3600$ sec/hr$) \times$
$(64 \times 10^6$ bits/sec$) = 1.7 \times 10^{16}$ bits/yr

Checks and Credit Transactions: $(340 \times 10^9$ transactions/yr$)$
$\times (50$ char/trans$) \times (8$ bits/char$) = 1.4 \times 10^{14}$ bits/yr

First Class and Airmail: $(100 \times 10^9$ letters/yr$) \times$
$(3 \times 10^5$ bits/letter$) = 3.0 \times 10^{16}$ bits/yr

Similar calculations were made for all other services, using the following basic set of assumptions:

Table 2 Summary of demand trend projections

Service	Units	1950	1960	1970	1980	1990
Voice						
Telephone	calls/yr x 10^9	62	105	174	286	482
Video						
Videotelephone	calls/yr x 10^6	0	0	1	36	1000
Television transmission	hrs/yr x 10^3	11	11	14	32[a]	72[a]
Record and Data						
Telegraph	messages/yr x 10^6	180	125	80	53	35
Stolen vehicle information transfer	cases/yr x 10^3	160	320	820	1950	4600
Facsimile transmission of "mug shots," fingerprints, and court records	cases/yr x 10^6	2	4	7	13	25
Stolen property information transfer	cases/yr x 10^3	430	880	1700	3500	7000
Motor vehicle registration	items/yr x 10^6	49	74	110	164	245
Driver's license renewal	items/yr x 10^6	38	48	60	75	90
Remote library browsing	accesses/yr x 10^6	0	0	neg.	5	20
Remote title and abstract searches	searches/yr x 10^6	0	0	neg.	8	20
Interlibrary loans	books/yr x 10^6	...	...	neg.	40	100
Remote medical diagnosis	cases/yr x 10^6	0	0	20	60	200
Remote medical browsing	accesses/yr x 10^6	0	0	20	60	200
Electrocardiogram analysis	cases/yr x 10^6	0	neg.	20	60	200
Patent searches	searches/yr x 10^6	6	6	6.5	7	7
Checks and credit transactions	trans/yr x 10^9	11	25	56	135	340
Stock exchange quotations	trans/yr x 10^9	0	0	1	2	4
Stock transfers	trans/yr x 10^6	290	580	1200	2500	4900
Airline reservations	pass/yr x 10^6	19	62	193	500	1400
Auto rental reservations	reserv/yr x 10^6	0	neg.	10	20	40
Hotel/motel reservations	reserv/yr x 10^6	...	...	25	50	100
Entertainment reservations	reserv/yr x 10^6	...	...	100	140	200
National Crime Information Center	trans/yr x 10^6	0	0	6	20	70
National legal information center	trans/yr x 10^6	0	0	neg.	5	30
Written						
Mail (first class and airmail)	letters/yr x 10^9	25	35	50	70	100
Books (remote printing)	new books/yr x 10^3	11	15	34	60	105
Newspapers (facsimile transmission	no. of newspapers using service	0	1	2	10	20

[a]Assumes the projection to 20 networks in 1990.

1) Conversion of analog voice signals to digital form requires 64,000 bits/sec, assuming current methods of Pulse Code Modulation (PCM).[10] Similarly, videotelephone requires 6.3 megabits/sec, and television requires 64 megabits/sec.

2) Alpha coded text, that is, coding of each character into bits, requires about 30,000 bits/page. This convention assumes approximately 70 characters/line, 50 lines/page, and 8 bits/character.

3) No-gray-scale facsimile, similar to the quality of conventional copying machines, requires about 300,000 bits/page, based on the assumption of a 700 x 500 dot matrix.

4) High quality facsimile having sufficient resolution for halftone photographs, requires about 10 times the number of bits for no-gray-scale facsimile, or approximately 3 million bits/page.

Summary

Combining the results of the projections into a composite picture of demand is illustrated in Fig. 11. Note that these findings represent amounts of information that might be input to a conceptual "black box" information system, specifically, a connected U.S. information network. Thus, retransmission of the information by broadcasting it to many viewers or subscribers is not included.

In Fig. 11, video communications are shown at a projected compound annual growth rate of 35%, while data communications are projected at approximately 25% growth/yr. By contrast, voice communications are projected at only 8 %/yr.

On the other hand, the amount of information transfer represented by voice communications may be expected to exceed that in all other categories by a considerable margin. This conclusion contrasts markedly with the popular and widely held view that volume of data will equal that of voice sometime in the 70's.[11]

The reason for this apparent contradiction lies in the fact that different assumptions have been made in the past regarding the number of bits represented by voice conversations. One of these assumptions is based on the speed of a "human channel," which is about 180 to 300 words/min. Removing redundancies in the voice reduces this to about 40 bits/sec. Thus, in the past some statements about voice versus data, including Ref. 11 have assumed this conversion factor for voice. Alternatively, a conversion factor of 2000 bits/sec has been used. This rate represents the approximate digital transfer rate that can be achieved on a typical voice circuit without special conditioning of the line.

The assumption made in this study is that the most appropriate conversion factor to assume for an overall comparison is 64,000 bits/sec. However, this is not meant to imply that all voice calls might be digital in the near future, for the most likely transition to digital transmission in the next twenty years will be in the carrier plant, that is, between toll, tandem, and central offices, but not in the "local loop." Thus, at least those calls confined to a single central office are likely to remain in analog form, as well, probably, as some proportion of those between central offices.

References

[1]Hough, R. W. et al., "A Study of Trends in the Demand for Information Transfer," Stanford Research Institute, Menlo Park, Calif., Feb. 1970.

[2]Sedlacek, W. C. et al., "Information Transfer Systems Requirement Study," Lockheed Missiles and Space Company, Sunnyvale, Calif., March 1970.

[3]Bergin, P. A. et al., "Information Transfer Satellite Concept Study," General Dynamics/Convair, San Diego, Calif. (to be published).

[4]<u>Growth Factors: Compound Interest Tables</u>, Chemical Information Services, Stanford Research Institute, Menlo Park, Calif., 1967, p. 1.

[5]Bright, J. R., ed., <u>Technological Forecasting for Industry and Government, Methods and Applications</u>, Prentice-Hall, Englewood Cliffs, N. J., 1968.

[6]Jantsch, E., <u>Technological Forecasting in Perspective</u>, Organization for Economic Cooperation and Development, Paris, 1967.

[7]"Videotelephone Service—One Step Closer," <u>Bell Laboratories Record</u>, Vol. 47, No. 3, March 1969, p. 69.

[8]Rostow, E. V., "President's Task Force on Communications Policy: Future Opportunities for Television," Staff Paper 6, Clearing House for Scientific and Technical Information, U.S. Dept. of Commerce, Washington, D. C., June 1969.

[9] Cox, B. et al., "A Technoeconomic Study of Methods of Improving the Payments Mechanism," Stanford Research Institute, Menlo Park, Calif., Dec. 1966.

[10] Pierce, J. R., "The Transmission of Computer Data," Information, W. H. Freeman, San Francisco, 1966, pp. 96-112.

[11] Kappel, F. R., "New Developments in Communications," unpublished manuscript, Aug. 1964, American Telephone and Telegraph Company, New York.

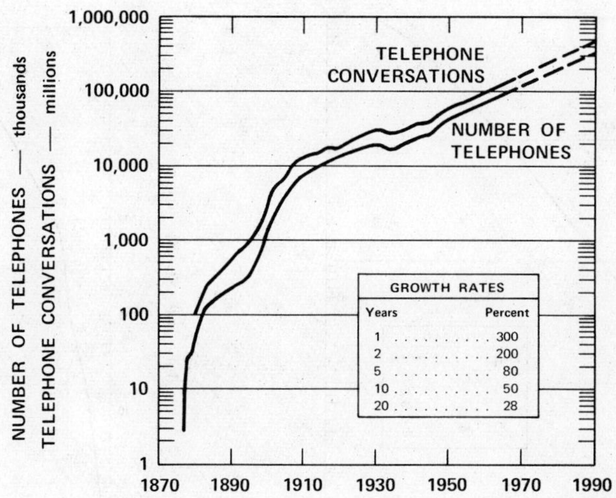

Fig. 1 U.S. telephone growth 1870-1970.

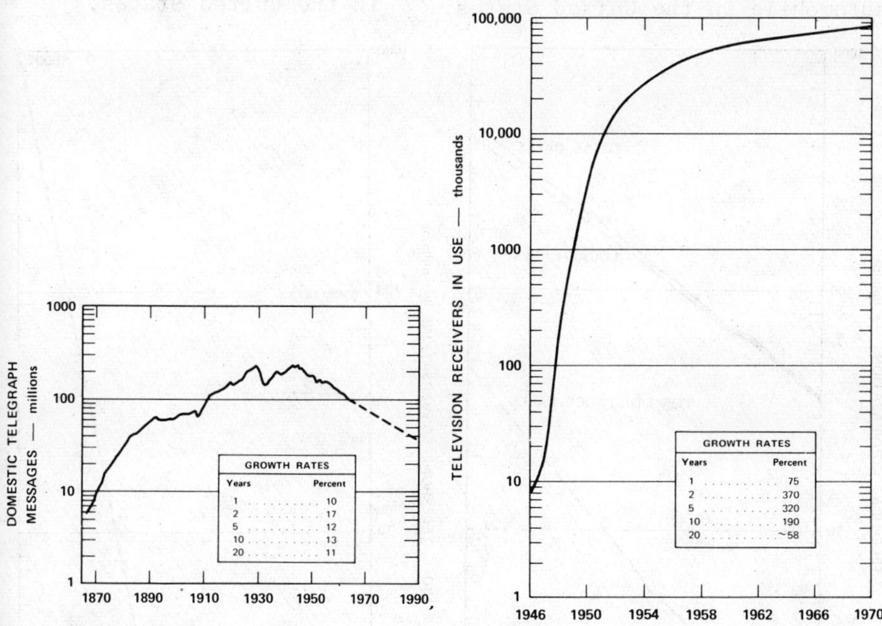

Fig. 2 U.S. telegraph service 1867-1970.

Fig. 3 Early growth of television in the United States.

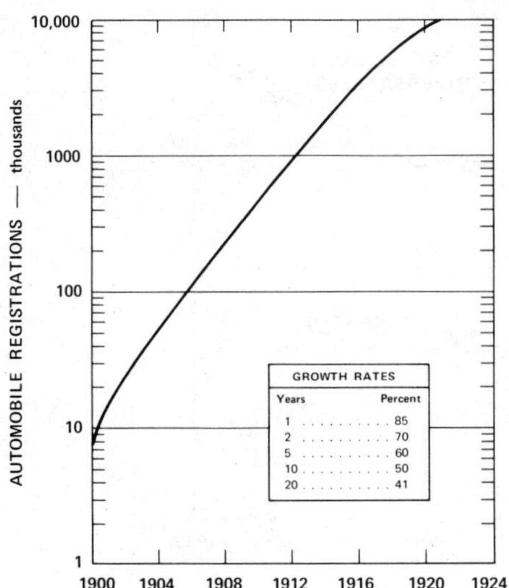

Fig. 4 Early growth of the automobile in the United States.

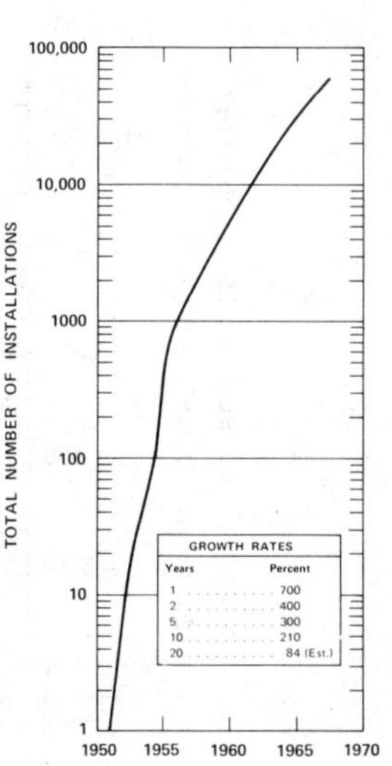

Fig. 5 Growth of computers in the United States.

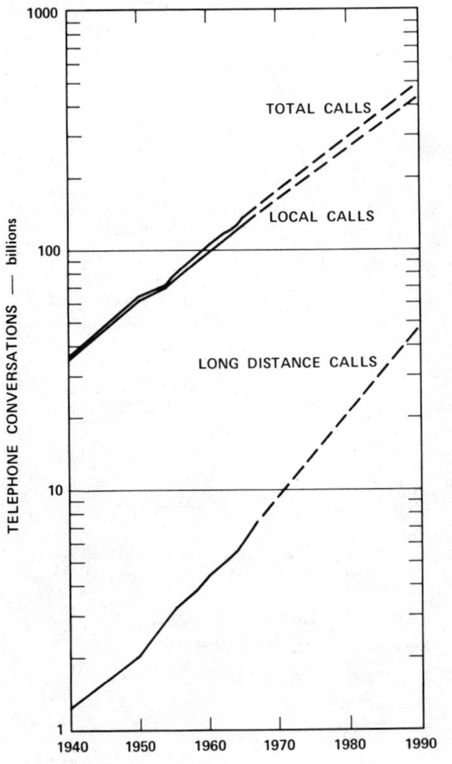

Fig. 6 Telephone service.

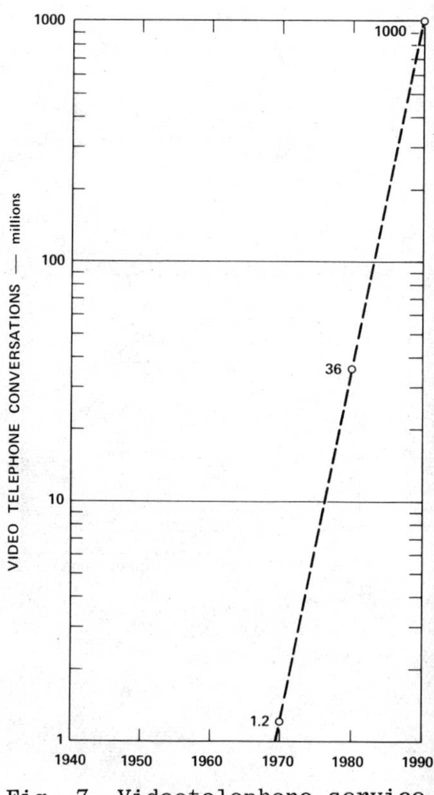

Fig. 7 Videotelephone service.

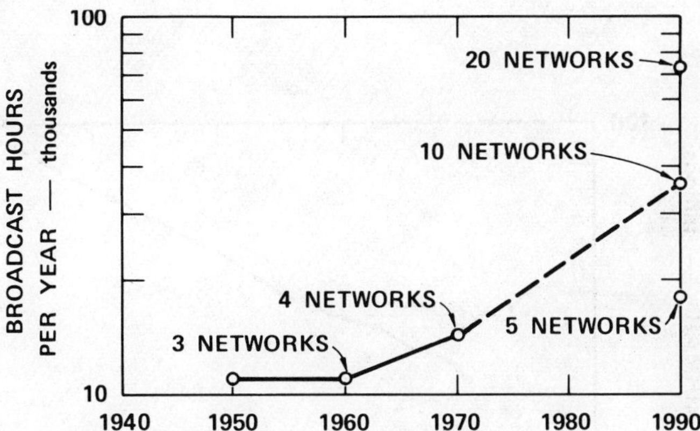

Fig. 8 Network television transmission, based on number of networks, 1960-1990.

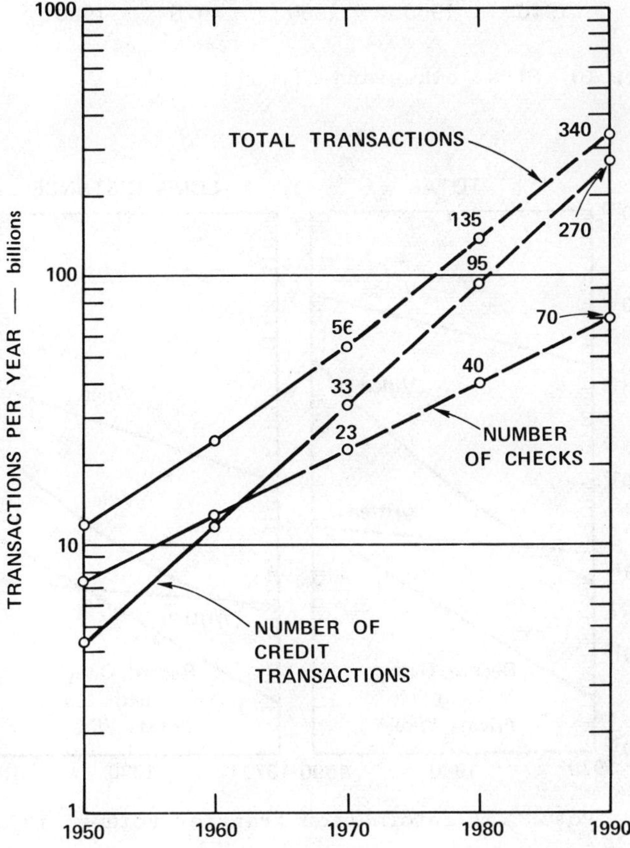

Fig. 9 Checks and credit transactions.

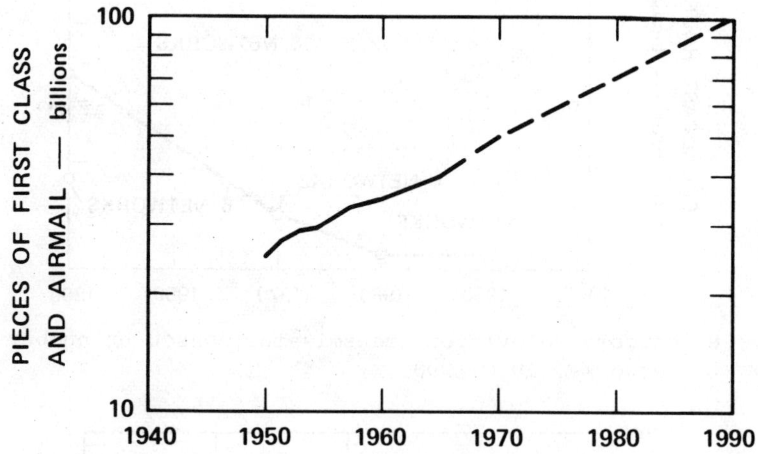

Fig. 10 First class and airmail.

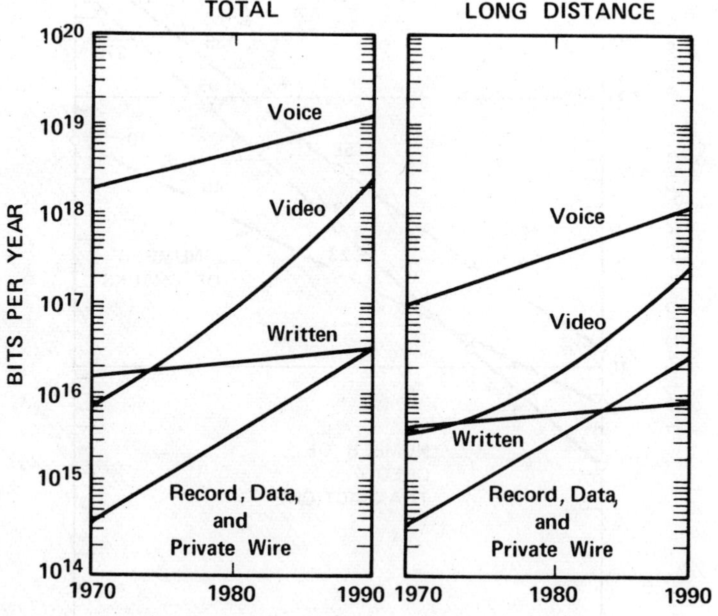

Fig. 11 Potential information transfer volume, 1970-1990.

COMPARATIVE EVALUATION OF MODERN TRANSMISSION MEDIA FOR GLOBAL COMMUNICATIONS

B.M. Dawidziuk * and H.F. Preston ≠

Standard Telephones and Cables Limited, London, England

Abstract

Submarine cables and satellites provide the wide-band communications media to meet today's (1970) traffic demands and are creating new dimensions for the future. The principal parameters influencing decision making by operating administrations when planning investment in international telecommunications are discussed. A formula is developed for evaluation of economic trade-offs in terms of circuit density and distance for both submarine cables and satellites. Analysis of traffic by continental area indicates a well-defined, two-level pattern of heavy interzone trunks and light, widely dispersed streams. The relationship between these patterns and the application of cables and satellites to each is considered. A review of the ownership and technical and operational aspects of a global communications network tends to the conclusion that to achieve the extensive diversity necessary to maximize security of communications, worldwide, both cables and satellites are essential elements of global network development. In this environment both technologies will enjoy unrestricted development to assure economic and reliable facilities in the future.

Presented as Paper 70-444 at the AIAA 3rd Communications Satellite Systems Conference, Los Angeles, Calif., April 6-8, 1970. The authors wish to express their gratitude to the management of Standard Telephones and Cables Limited for their encouragment and for permission to publish this paper.

* Market Development & Planning Manager.

≠ Market Planning Manager.

Submarine Systems Marketing Division, London E.16, England.

1. Introduction

Since the first application of submarine cables (1956) and communication satellites (1965) to transoceanic communications, the development of the global network has rapidly progressed. Both cable and satellite links now span the globe and provide operating administrations with a choice of wide-band facilities for their international communications. Each is capable of satisfying all the demands for communications traffic.

Before making their final decision on investment in either medium of communications, the operating administrations need to consider a wide variety of aspects concerned with traffic, economics, and system operation. This paper sets out to evaluate the principal aspects influencing decision making by the administrations.

2. Global Communications Network Environment

The environment of international communications at any point in time is governed by 1) technological progress; 2) the traffic mix; 3) Traffic growth, foreseeable and future; and 4) politico - economic influences.

In order to appreciate more fully the implications behind a given administration's decision to invest in and develop specific media for the distribution of their international traffic, it is worthwhile to consider briefly the current global network with its associated traffic patterns.

2.1 Development of Global Communications Network

2.1.1 Submarine cables. Modern, high-quality, and reliable transoceanic communications date back to 1956 when TAT-1 cable was opened. This 48-channel system, small by present (1970) standards, represented a major technological achievement at that time, laying the cornerstone for the expansion of world communications up to their present-day status.

The immediate success of TAT-1 resulted in a world-wide interest in submarine cables so that by 1964 both the North Atlantic and Pacific Oceans had been bridged by networks of cables, making available high-quality international outlets to more than 90% of the world's telephone users.

Figure 1 portrays milestones in the growth of oceanic cable system capacity, showing that systems have rapidly grown from the early 48 circuit to the 1840 circuits scheduled for 1971.

The significance afforded to submarine cable systems by administrations world wide is evidenced by the variety of systems constructed and laid from 1968 - 1970.

Table 1 Submarine cables laid or under construction in the three-year period, 1968-1970

Year	No. of cable systems laid	Total route mileage	Added capacity (channels)	Added circuit miles[a] (000)
1968	8	10308	2880	3610
1969	7	2892	2776	1199
1970	6	7876	2896	3971

[a] Note: total circuit miles to end of 1967 = 6.4 million; total circuit miles to end of 1970 = 14.2 million; increase, 1967-1970 = 122%

Additions to the network, in 1968-69, together with projects scheduled for completion in 1970, are shown in Table 1 and represent an increase of 122% in the available circuit mileage in just three years.

This is all the more significant because the period coincides with the peak of activity in establishing the global satellite communication system. In fact, in the five years since commercial satellite communications commenced, the cable network has almost trebled in circuit-mile capacity. Figure 2 demonstrates this and traces the growth of the cable network since the first repeatered systems were laid.

The principal cable routes in the Atlantic, Pacific and Indian Ocean areas are shown on Figs. 3-5 respectively. Although each of these cables is owned by at least two administrations, many have in fact come to fruition through multinational partnerships under Indefeasible Right of User (IRU) agreements. Overall, the submarine cable contribution to the global network will, at the end of 1970, amount to 94 cable systems, 85,000 naut miles of route, and 14.2 millions of circuit miles.

2.1.2 INTELSAT system. Figure 1 also shows that since the launch of INTELSAT-1 (Early Bird) in 1965, rapid developments in satellite technology have taken place. Today the commercial network consists of a series of INTELSAT III satellites, each with an approximate capacity of 1200 voice grade circuits and multiple access capability.

This series has established the global satellite system extending over the three major oceanic areas. It brought the subcontinent of India into a modern communication network enjoyed by all other principal world trading areas.

Development is proceeding on INTELSAT-IV satellites capable of some 6000 circuits capacity, through their 12 separate transponders, each capable of TV transmission.[1]

Great progress has also been made in the design of ground terminals, and 32 stations were in operation at the beginning of 1970 in North and South America, Europe, Australia, and Asia. Many more are planned, and it is anticipated that by 1972 some 62 sations will be available for satellite communications. The location of these stations and their relation to the cable network for each of three oceanic areas are shown in Figs.3-5. These figures highlight a very significant point. The nations which generate the largest volumes of international traffic are seen to have invested already in <u>both</u> cables <u>and</u> satellites. Of equal significance is the fact that the Indian Ocean area is so far not served by submarine cables, because the trunk "highway" concept that submarine cables provide does not lend itself readily to the low-density traffic routes and wide geographical distribution encountered in this region.

A close examination of Figs.3-5 reveals a definite pattern of global communications development and network topography based upon traffic patterns, and the state of local network development in given areas.

2.2 Traffic Patterns and Growth

The principal forces shaping the dimensions of the global communications network are those of traffic pattern and traffic growth. Both tend to be ill-defined, though their motivating parameters are well established and common to any telecommunications network.

<u>2.2.1 Traffic patterns.</u> Telecommunication traffic is generated by a complex interdependence of political, economic, trade, social, and cultural relations between any two communities. In the international sphere, it is these relationships which define the principal world traffic streams. Adequate facilities, both within national networks and between the countries concerned, are a prerequisite for traffic growth within a specified stream.

If we look specifically at the U.S.A., recent statistical evidence[2] reveals that 80% of her total international telephone traffic can be assigned to three specific streams (Europe, West Indies, and Hawaii), with the remaining 20% distributed in very small measures over the rest of the Earth's surface (Table 2 and Fig.6). Further analysis of the U.S. traffic streams to Europe, Asia, and Oceania indicates that this traffic is concentrated in highly industrialized countries and consequently requires heavy trunk route facilities. Corresponding traffic density patterns are evident for most of the developed countries, and Table 3 shows an additional example of traffic distribution to and from Australia.[3]

Table 2 Distribution of U.S.A. telephone and telex traffic 1967

Traffic stream	% of Total traffic	
	Telephone	Telex
U.S.A. - Europe	33.2	34.6
- West Indies	25.1	27.5
- Hawaii	18.6	17.1
- Asia and Oceania	11.8	9.5
- North, Central and S.America	10.6	10.5
- Africa and Near East	0.7	0.8

Unless dramatic changes occur in the world economic environment, communication traffic is likely to maintain its current established pattern. The industrialized nations of the world will continue to require, and expand, a basic network of high-density trunk routes as the communications mainstay, with an increasing number of light streams as an interstitial network.

Table 3 Distribution of telephone and telex traffic - Australia (1968)

Traffic stream	% of Total traffic	
	Telephone	Telex
Australia - Europe	32.2	39.4
- New Zealand and Oceania	27.8	11.9
- North America	25.2	24.5
- S.E. Asia	5.7	3.2
- Japan	3.9	14.3
- Rest of world	5.2	6.7

2.2.2 Traffic mix and growth. World-wide traffic statistics indicated that the traffic mix is critically dependent upon the stage of national network development and the quality of international connections.

Telephone traffic supported by telex are the main types for highly developed countries. Telegraph traffic is predominant in the case of developing countries. Whenever good facilities are provided, telephone traffic takes over the lead at the expense of telegraph traffic.

As the global network becomes more sophisticated, new facilities will be introduced, such as International Subscriber Dialing (ISD), wide-band data, picture-phone, video conference facilities, etc. When introduced, each of these new facilities will generate its own rapid rate of growth.

Any attempt to develop traffic forecasts is a subject in itself and outside the scope of this paper. However, it is worthy of comment that, in the authors' judgement, based on their independent research, a total of 10,000-20,000 equivalent voice circuits might be required in 1980 for Transatlantic traffic alone.

2.2.3 Potential revenues. The revenue potential of a modern, high-quality international communications network is very considerable. A clear example is given in figures published by the American Telephone and Telegraph Company (AT&T) and record carriers [2] for an 11-yr period 1957-1967. Figure 7 shows that the total revenue derived by the U.S. Common Carriers from overseas services has more than trebled in this period, from $82 million in 1957 to $276 million in 1967. Revenues derived from international services in some other countries in 1967 are shown in Table 4.

The distribution of the revenue derived from originating telephone and record traffic in the U.S.A. during 1967 is portrayed in Fig.8. The sharp differential in traffic pattern, referred to in a previous paragraph, between those regions with well-developed communications and the developing areas is clearly apparent.

With $276 million collected by the U.S. carriers in one year alone, and a further $303 million by the six sample operators listed in Table 4, it is quite evident that international communications is an industry of considerable size. Any investment in expansion of facilities (for example a trans-oceanic cable costing $70-80 million) should be viewed in this perspective.

Table 4 Operating revenues from international services in 1967 (from related annual reports)

International carrier	Total revenue $ million
U.S. Common Carriers	276
British Post Office	162
Kokusai Denshin Denwa (Japan)	57
Cable and Wireless Group (British Commonwealth and Middle East)	46
Overseas Telecommunications Commission of Australia	26
Compania Telefonica Nacional de Espana (Spain)	8
New Zealand Post Office	4
TOTAL - FOR 7 OPERATORS	579

2.3 Other significant factors

Although the traffic pattern growth and mix are the principal forces that shape the development of the global network, there are many other significant factors that influence the decision of given administrations on the choice of media to satisfy the traffic demand. These factors can be listed under the following broad headings: 1) economic issues; 2) political aspects; 3) technical factors; and 4) ownership and legal issues.

It is within this environmental framework that communication agencies undertake their own evaluations leading to a choice of cable or satellite systems for expansion of their international/intercontinental communications.

3. Comparative economics

After five years of commercial use of satellites and considerably longer operating experience of submarine cables, it is now possible to formulate a meaningful structure of systems costs for comparative purposes.

A greatly simplified analysis is possible, when all the costs are reduced to the common denominator of annual costs per circuit. This approach was suggested by the authors in earlier papers [4,5] and has been used since by others.[6] It is the most straightforward approach for the case of satellite systems for it by-passes the complex issue of investment in the space segment by INTELSAT, merely taking into account its end product, the charge for the use of the space segment. It is this parameter, after all, which concerns administrations in computing operating costs.

Further simplification is effected by considering only "standard" Earth stations (G/T of 40.7 db), since these stations will form the mainstay of the system until the tradeoffs of increased satellite power vs lower G/T ratio are proven more fully.

3.1 Capital investments

3.1.1 Submarine systems. The capital costs vary directly with system capacity and route length. For transocenic routes in excess of about 1500 n.m. however, per-mile capital costs tend to become substantially constant as a function of the cost of the undersea equipment. Table 5 shows some typical costs for modern British long-haul systems taken between Supergroup distribution frame at the terminal stations, while Fig.9 presents the same data graphically.

Table 5 Typical installed costs for oceanic submarine cable systems (over 1500 naut miles)

Approximate system bandwidth in MHz	1	5	14
System capacity (3-kHz circuits)	160	640	1,840
Cost per system naut miles in dollars	9,000	13,000	20,000
Cost per 3-kHz circuit naut mile in dollars	56	20	11

Once a cable system has been installed and commissioned, no further capital outlay for the system is involved. It is clear that after opening a cable to traffic, additional circuits could be routed through this facility until saturation at no incremental cost.

Table 5 indicates that modern wide-band cable systems offer economy of scale. A typical transatlantic circuit 3500 naut miles in length in 14-MHz cable would cost only $38,000. With an expected lifetime of 20-25 years, this investment level represents a very attractive proposition to the Administrations, since recovery of the circuit capital cost can be easily effected in less than one year of its exploitation, even at the reduced tariff scales in force and envisaged for the future.

3.1.2 Satellite system. Apart from a relatively small capital investment in INTELSAT, any given administration is required to provide its own Earth station. Current capital cost for a standard station is of the order of $3.5 million. This represents some $1.5-2.5 million reduction from the levels prevailing some three years ago. The $3.5 million does not include the terrestrial link from the station to the international gateway exchange.

3.2 Annual costs

The annual costs for transoceanic links may be readily assessed in terms of the capital investment.

3.2.1 Cable systems. Typical values for cable systems are shown in Table 6. They cover all the costs.

Table 6 Annual costs for transoceanic cable systems

Item	Annual cost as % of system capital investment
System operation and maintenance	2
depreciation. (over 20 year life)	5
Interest	5
Total	12

An allowance of 2% for operation and maintenance is rather generous for modern, wide-band cable systems. The estimated maintenance costs of the TAT-5 system for example, are only 1.2% of capital cost,[7] or of the order of $1000 per circuit per annum.

Cable systems are designed for a minimum lifetime of 20 years. Accordingly, depreciation of their capital value may be extended over this period, effecting a low contribution to the total annual costs. Using the parameters shown in Tables 5 and 6, it is possible to derive a complete economic profile for submarine cable systems in terms of associated annual cost per circuit, Ac, by use of the expression

$$A_c = \frac{0.12L.C.}{N}$$

where L is the route length, C is the cost per naut mile, N is the number of working circuits.

A range of Ac parameters for different rate of fill of the currently available British cable designs is shown in Fig.10 and summarized in Table 7.

Table 7 Annual costs per cable circuit, Ac (for fully loaded system)

System capacity, 3-kHz channels	Ac for cable length as stated in dollars 1000 naut miles	4000 naut miles	8000 naut miles
160	6,700	27,000	54,000
640	2,400	9,750	19,500
1,840	1,300	5,200	10,400

As previously indicated, annual costs are very low for modern wide-band cable systems. A typical value of annual costs for an estimated IRU circuit in a 14-MHz transatlantic cable 3500 naut miles long would be 12% x $38,000, or only $4600 per annum.

Like satellite communications, cable technology is still rapidly developing. Indications are that further economies of scale are achievable in the future when systems of 3000 circuits or more become available.

3.2.2 Satellite systems. Several authors recently have considered economics of satellite communications.[5,6] The paper by Lutz [6] introduced subdivision of annual costs for the ground segment, which is accepted for the present analysis, in which the following notation is used:

C_s = capital cost of Earth station

T_o = noncircuit annual station cost defined as 25% of C_s made up by

- operation and maintenance, 10% of C_s
- depreciation (over 10 years), 10% of C_s
- Interest, 5% of C_s

T_1 = per circuit annual cost [6];

T_r = per route annual cost [6]; and

S = space segment rate (annual).[8]

The authors consider that the annual costs of an Earth station must have a direct relation to the associated capital investment. For this reason, T_o is related to station capital costs, C_s. Also, in the development of satellite systems there is a different time frame of investment, since some of the early stations constructed prior to say 1967 had much higher capital cost than stations added more recently to the global network. Therefore, it is meaningless to consider only low-cost stations of the future since by now some stations are in operation, and their depreciation and other costs are already fixed at a much higher level. Moreover, earth station lifetimes are reckoned as 10 years' average, resulting in higher rates of depreciation (10%) than encountered in cable systems.

To reflect this time dependence of annual costs, three cases are considered in Table 8. These cover stations constructed in 1967, stations constructed in 1970, and new additions to the system at a future date, arbitrarily selected as 1975.

Table 8 Annual cost parameters for satellite circuits, thousands of $

Item	Case 1 station commissioned in 1967	Case 2 station commissioned in 1970	Case 3 station to be built soon (1975?)
C_s	6000[a]	3500	2500
T_o at 25% of C_s	1500	850	625
T_1	3	3	3
T_r	10	10	10
S	20	16	5

[a] This $6 million is an average value for a number of stations,[4,6] in the U.S.A. specifically, C_s in 1967 was nearer $9 million.

To simplify analysis, assume that additional costs per circuit resulting from T_r are small. Also assume that the number of channels at station A and station B, N_A and N_B, are approximately equal, i.e., $N_A = N_B = N$. Total annual cost per satellite circuit A_s between stations A and B is

$$A_S = 2x\left(\frac{T}{N} + T_1 + S\right)$$

The annual costs for satellite circuits (A_S) as based upon parameters stated in Table 8 are shown in Fig. 11, and are summarized for a selected number of channels in Table 9.

Table 9 Annual costs for satellite circuit A_S, using standard INTELSAT Earth stations and N circuits, thousands of $

No. of circuits N	Station commissioned in 1967	Station commissioned in 1970	Station to be built in 1975
10	346	208	141
20	196	123	77
50	106	72	41
100	76	55	28
500	52	41	18
1000	49	38	16

The absolute timing (Table 9) is of little importance since the analysis is of a general nature and indicative of trends and approach to economic comparison. The reader can readily use his own system cost parameters to establish his own specific economic profile. Similarly, the effects of any future adjustments in space segment charges S, can be readily estimated using the approach outlined in Table 8.

3.3 Comparative Economics

Using annual costs per circuit as the best available parameter for economic comparison of submarine cable and satellite alternatives, it is possible to establish relative costs for each alternative medium as a function of system loading (number of circuits, N), and break-even distance.

The resulting economic profiles, showing competitive standing of 160, 640, and 1840 circuit cable, respectively, for different cable circuit loading conditions, are indicated in Figs. 12-14.

A break-even distance, defined as a point of indifference between cable and satellite annual costs, (Ac = As) is presented in Figs. 15-17 for the three British cable systems under discussion. Typical break-even distances as shown by these curves are summarized in Table 10.

Table 10 Break-even distances resulting in equal cable/ satellite annual circuit costs (cable systems operating at 50% fill)

Cable system capacity (circuits)	Terrestrial distance between terminal stations (naut miles) In 1970	In 1975
160	4,800	2600
640	8,900	4100
1840	15,000	6700

A definite common trend is evident from curves shown on Figs. 12-17. Satellite systems offer cost advantages for relatively light traffic routes. They find particular application to traffic patterns requiring diverse, low-density long-haul circuits.

Cable systems on the other hand, are more attractive economically for transoceanic routes with heavy trunk traffic requirements, as typified by the transatlantic and transpacific routes (Fig.6). Because of this fact, further transatlantic and transpacific cables are expected to mature. Advance plans for additions in both areas have already been announced by the U.S. Common Carrier.[9] There is a considerable "grey" area in the curves where both systems offer similar cost profiles. In such cases the final choice of the medium is influenced by issues other than economics.

4. Political, Legal and Ownership Issues

4.1 Political Aspects

The majority of the large administrations throughout the world, who contribute most of the volume of international traffic, have applied balanced judgment to the relative roles of submarine cables and satellites in satisfying their specific requirements for traffic mix and distribution.

In the United States, the existence of a Carrier's Carrier with a sole interest in satellite communication provides a situation not found elsewhere.

For many developing nations, it is politically expedient to cultivate the satellite technology because it is something new, and therefore has prestige appeal. Additionally, it provides facilities for distribution of their relatively small volume of international traffic at a reasonably economical cost, but already some are planning to consolidate their satellite facilities through provision of cable outlets.

Any country with a reasonable traffic volume can no longer afford not to have both satellite and cable circuits to economically maximize network reliability and traffic distribution facilities.

4.2 System Ownership

As shown in Sec. 3, the economics of submarine cable operation favors their choice for high-density point to point routes. Submarine cables can demonstrate an added advantage over their satellite counterparts in the area of system ownership and control.

The most commonly employed method to afford ownership in major international systems is that of Indefeasible Right of User (IRU).

The IRU principle of system ownership offers possiblity of cooperative participation of many administrations in a specific transoceanic cable achieving three very important end results: a) underwriting of capital investment by many partners; b) immediate use of IRU circuits, thus ensuring high initial cable fill and economic viability of cable project (Figs. 12-17); and c) extensive diversity of routing, since traffic streams of N circuits are now routed in several separate cables.

By contrast, participation in the global satellite network does not, under the present infrastructure, allow for ownership in its fullest sense. Unlike cables the establishment of a point-to-point circuit in a satellite system inevitably involves a third party (INTELSAT).

The majority of the large administrations who form the membership of INTELSAT also have extensive interests in worldwide cable circuits. They well understand the advantages offered by the total legal right of ownership in cable systems, and the necessity for balanced network development.

5. Technical Factors

Recent progress in space communications technology is well documented.[10] Similarly, developments in submarine cable systems have been widely described. Detailed technical comparisons are outside the scope of this paper. However, the following are some broad observations in the technical area.

Although earlier satellites were power limited, new generations will be bandwidth limited. To overcome this limitation, the use of frequencies higher up the microwave spectrum (18 or 30 GHz) are likely to be introduced for space communication in future years. Their use will introduce new problems in securing stability of the transmission path during heavy rain conditions, the solutions to which will be found, but inevitably at a cost.

Following the removal of constraints upon satellite power, there seems a strong case, advanced in several quarters, for smaller and cheaper Earth stations.[6, 11] It remains to be seen whether this concept will be accepted by INTELSAT who currently forsee only standard (26 meter) Earth stations for some years to come.

Considerable interest has recently been shown in development of multiple access demand assigned systems. Realization of this concept should result in added economics for relatively light traffic Earth stations with increased operational flexibility. No major economic advantages are foreseen for heavy traffic stations, since in such cases most of the circuits are in preassigned groups.

Availability of highly reliable semiconductor devices has opened up new bandwidth dimensions for submarine cables and development work is proceeding on cable systems with at least 3500 circuit capacity.

Theoretical studies indicate that this is still far from the limit of cable technology. Progress and timing will be dependent upon the market demand for wide band, point-to-point facilities requiring 10,000 circuits or more. The economics of such wide band cables will be very attractive, since the current trend line is likely to be maintained (Fig.9).

Development of wide band cables has rendered the reasons for using the original TASI-A (Time Assignment Speech Interpolation) obsolescent, but the policy of using efficient 3-kHz voice circuits is likely to be retained for long-haul ocean cables.

Although cable systems provide convenient interface with conventional terrestrial systems, a number of phenomena inherent in satellite communications have necessitated the development of peripheral techniques, all of which, in the end, add to the cost.

Complex tracking, telemetry, and control systems are required in special control stations to maintain satellite position keeping, and to insure correct and constant transponder loading. A deviation in either could result in serious loss of service.

Satellite systems of the geostationary type must inevitably be subject to transmission delay of the order of 250 msec. By contrast, the longest terrestrial circuit likely to be encountered (Europe-Australia, 16,000 miles) has a corresponding delay of some 130 msec.[12]

Although the transmission delay on a single-hop satellite link is considered by most to be tolerable for telephone traffic, the general concensus of opinion among many administrations is that a two-hop link would not be satisfactory.

Where domestic, or regional satellite communications systems mature, it will be necessary to provide sufficient terrestrial circuit capacity to insure that under no combination of domestic plus international connections will a tandem satellite link be established.

In a country such as the U.S.A., this criterion dictates the assignment of cable circuits in the major traffic streams (Fig. 6) of at least 50% of the total requirements for the route. Moreover, special switching arrangements at the International Exchanges will be required to safeguard against tandem satellite connection, either on a "barred access", or split GO (cable)/RETURN (satellite) basis, as embodied by AT & T in the design of their TASI-B equipment.[9]

6. Operational Issues

The advance towards more machine-oriented communications (inter-register signalling, real-time data, etc.) opens up a new dimension in system capacity requirements. In this developing environment, the accent is upon providing increasingly greater circuit capacity to absorb traffic peaks whilst maintaining reasonable grades of service without causing congestion in common equipment at terminal switching centers.

Not only are large circuit groups implied for routes of major traffic density, but also implied is the maximum diversity of routing to cater for overflow or path failure conditions. With the future prospect of fully programed traffic assignment by computer, the need for comprehensive triangulated facilities by both terrestrial and spatial means becomes essential.

Controversial issues have from time to time been raised regarding the relative operational reliability of cable and satellite systems. Advances in system design and laying techniques have effectively eliminated the risk of damage to submarine cables.[13] Similarly with the experience gained from Intelsat-3 satellites, reliability precautions are being built into future satellite systems in the same way as with cables. With the expansion of an integrated cable/satellite network, the reliability of any individual link may not be so paramount in the future as hitherto.

For the reasons given above network reliability will be secured without excessive deployment of funds to insure absolute reliability of one specific medium.

Long term traffic growth has been shown [14] to follow a well defined exponential pattern with no tendency yet to saturation.

New classes of service such as videophone, extra wideband data etc., are foreseen to maintain the pattern so ensuring an increasing demand for bandwidth.

A broad estimate of the 1980 requirements for a transatlantic route offer a potential equivalent to some 10,000-20,000 voice circuits. To meet and satisfy this demand, and to meet the increasingly stringent requirements of quality, diversity and security, it is considered that a fully integrated global network, in which both satellites and cables form a balanced part, is the necessary and only prerequisite.

7. Conclusion

Satellite systems and submarine cables are the two modern media capable of providing wide-band communications facilities to meet expanding traffic demands. Each of these two technologies offers specific technical, operational, and economic advantages to the system user/owner.

Cable systems continue to maintain their competitive standing for heavy traffic density routes. Traffic analysis indicates that a large part of global communications requirements consist of such heavy trunks.

Satellite systems provide economical solutions for multiple paths of relatively small circuit capacity. They also offer alternative routing facilities for high-density traffic streams.

Both techniques are complementary, and their unrestricted growth is essential to meet the ever increasing demand for economical and reliable telecommunication facilities.

References

[1]"Report to the President and the Congress, for the Calendar Year 1968," COMSAT, Washington, D.C., March 1969.

[2]"Statistics of Communications Common Carriers" Years ended Dec. 31, 1964-1967. Federal Communications Commission, Washington, D.C.

[3]"Annual Report and Balance Sheet," Overseas Telecommunication Commission, Sydney, Australia, March 31, 1969.

[4]Mackay, A.D., Dawidziuk, B.M., and Preston, H.F., "Economic, Operational and Technical Aspects of Modern Global Communication Systems," Electrical Communication, Vol.43, No.1, April, 1968, pp. 70-82.

[5]Dawidziuk, B.M. and Preston, H.F., "Current Trends in Global Communications," Paper presented at the XVI International Congress on Communications, Genoa, Oct. 1968.

[6]Lutz, S.G., "Economic Factors Influencing the Break-Even Relations between Satellite and Terrestrial Point to Point Communication," Telecommunication Journal, Vol.36, No.7 July 1969, pp. 317-328.

[7]"Response to FCC Enquiry regarding future Atlantic Area Communication Facilities," American Telephone and Telegraph Co, New York, N.Y., Nov. 1967.

[8]"Sixteenth Report to the ICSC, Advisory Subcommittee on Finance", ICSC, Washington, D.C., Jan. 1969.

[9]"Submission to FCC on International Facilities Requirements", American Telephone and Telegraph Co. New York, N.Y., Aug.1969.

[10]Metzger, S. "The Commercial Communications Satellite System 1963-78", Astronautics and Aeronautics, Vol.6, No.4, April 1968, pp. 42-51.

[11]O'Nians, F.A. and Blonstein, J.L., "Some aspects of the economics of satellite communications", Telecommunication Journal, Vol.35, No.12, Dec. 1968, pp. 662-669.

[12]Halsey, R.J. "British Commonwealth Ocean Cables," Transactions of IEEE Communication Technology, Vol. COM.12 No.3, Sept.1964, pp. 6-16.

[13]Duncan, C.C., "Plowing Cables under the Sea", IEEE International Conference on Communications, Philadelphia, Pa., June 1968.

[14]Deloraine, E.M., "Telecommunication Traffic", Electrical Communication, Vol.44, No.1, April 1969, pp. 4-13.

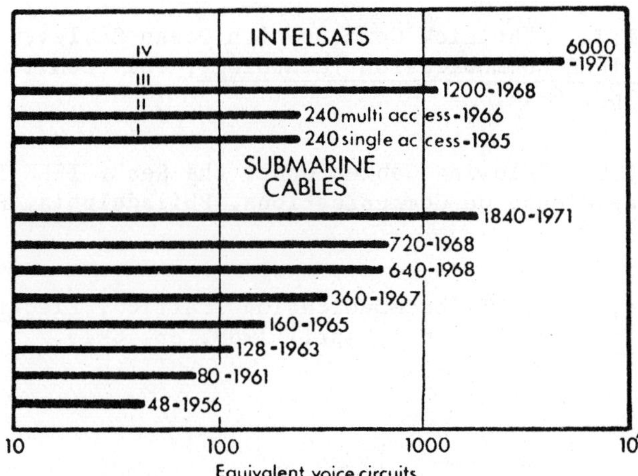

Fig. 1 Development milestones in intercontinental communications media.

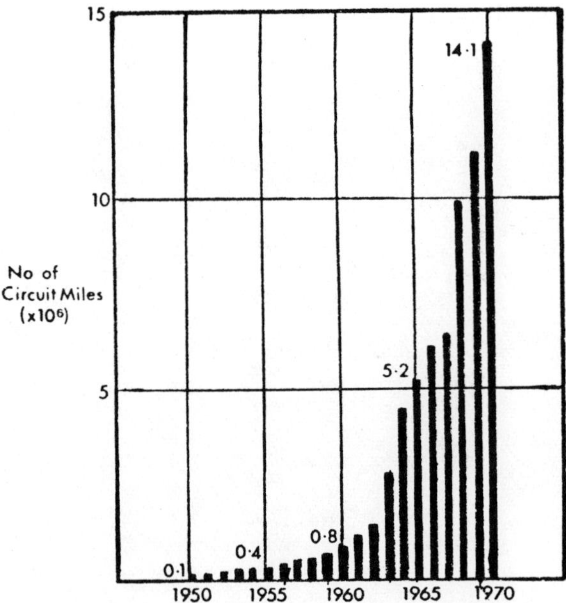

Fig. 2 Growth of worldwide submarine cable network.

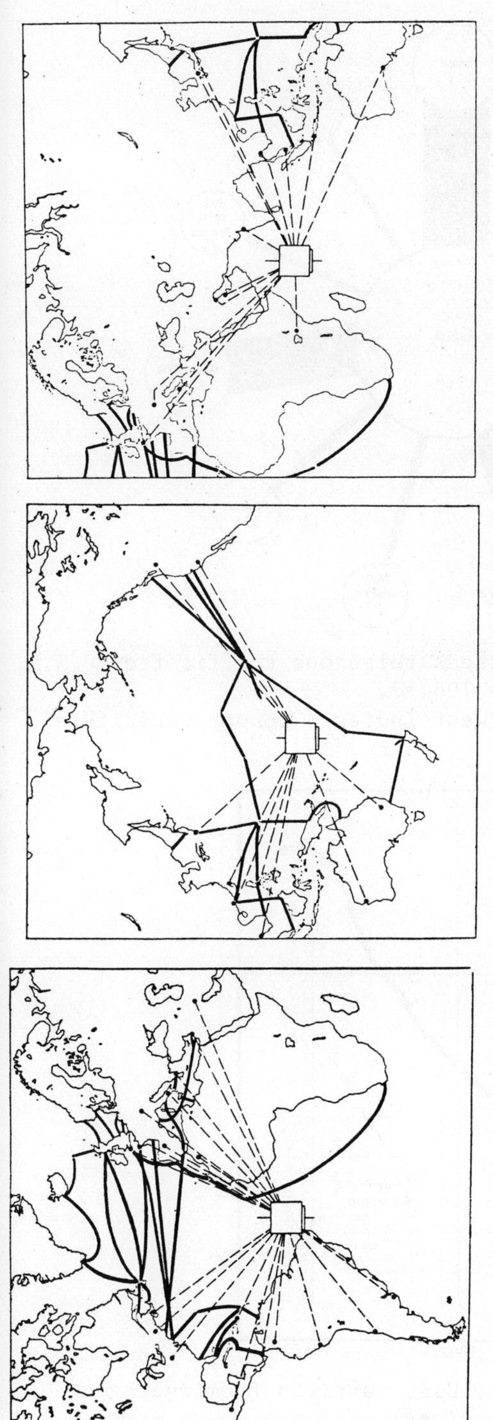

Fig. 3 Principal communication arteries - Atlantic Ocean.

Fig. 4 Principal communication arteries - Pacific Ocean.

Fig. 5 Principal communication arteries - Indian Ocean.

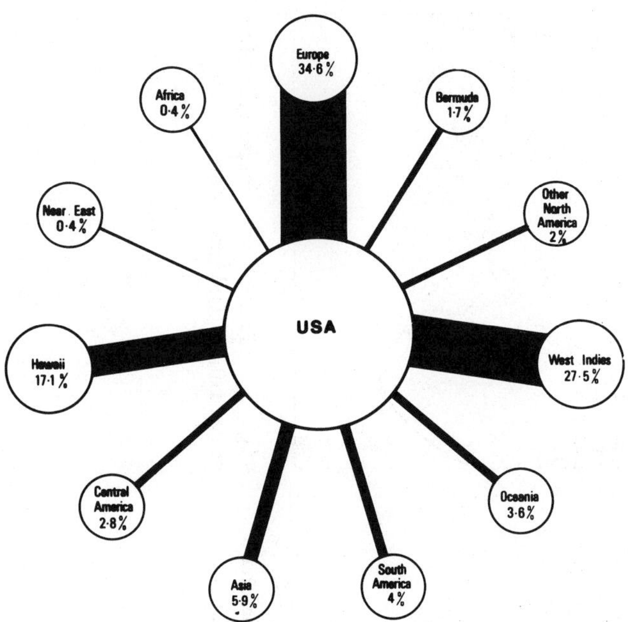

Fig. 6 Distribution of overseas telephone traffic from U.S.A. average over 4 years 1964-7.
Note: U.S.-Hawaii + West Indies + Europe constitute 80% of total.

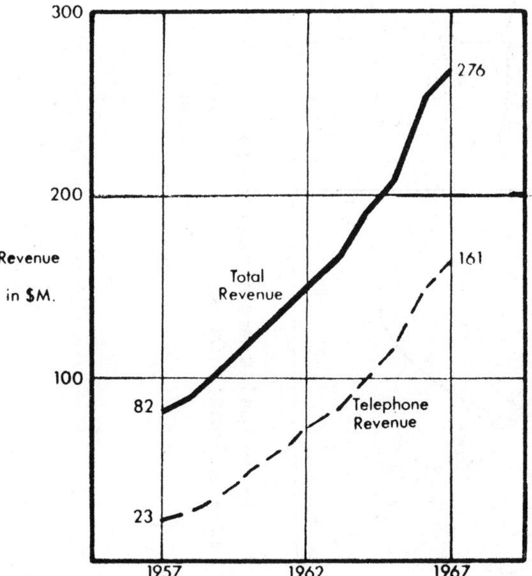

Fig. 7 Revenue derived by U.S. carriers from overseas communications services.

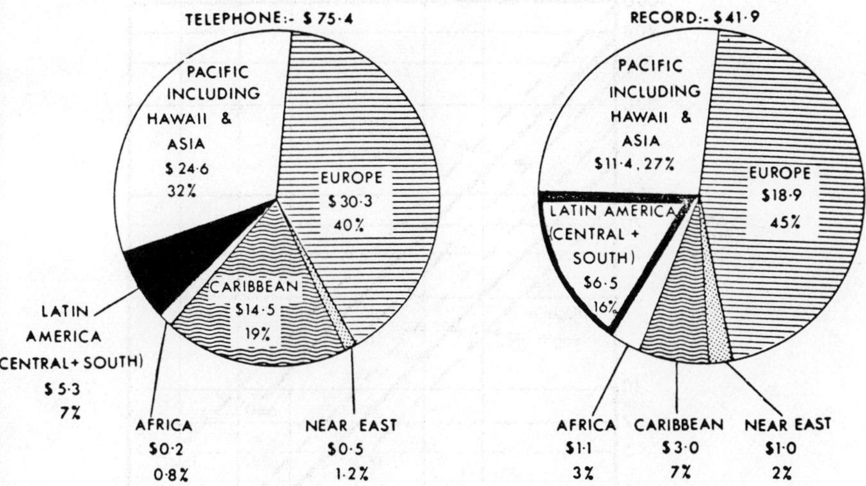

Fig. 8 Total revenue and distribution by region U.S. originating traffic - 1967. (millions of $).

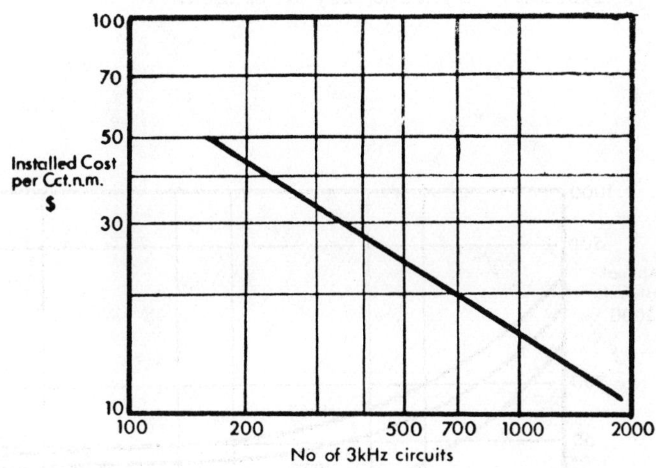

Fig. 9 Typical installed costs for submarine cable systems.

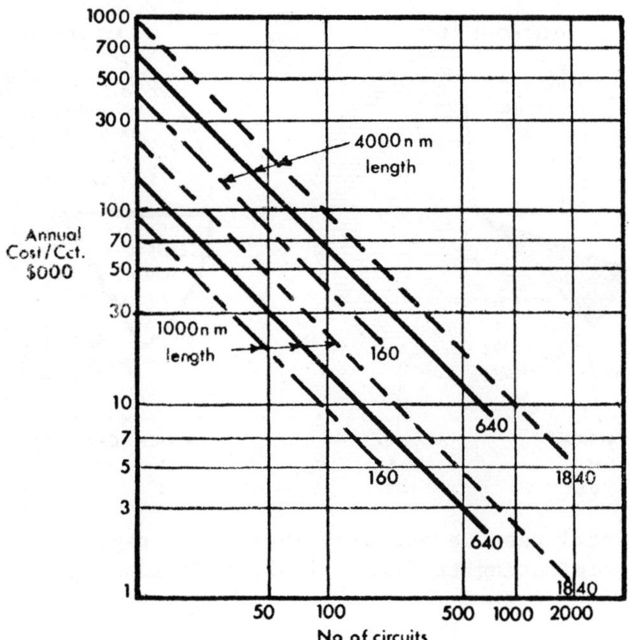

Fig.10 Annual costs (A_c) for British transoceanic cable systems - by capacity & length.

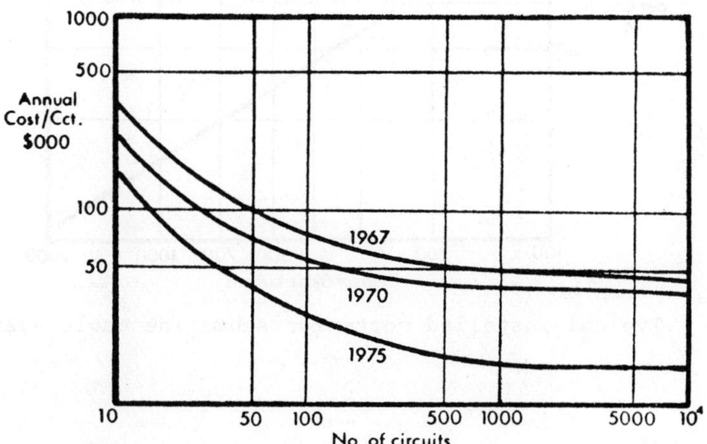

Fig.11 Annual costs (A_s) for satellite circuits - using standard 40.7 db earth stations.

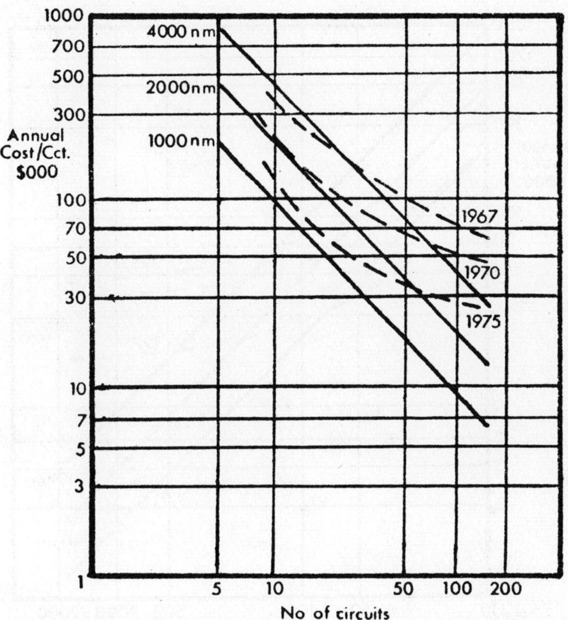

Fig.12 Comparison of annual costs for 160 circuit submarine cable (—) and satellite systems (---).

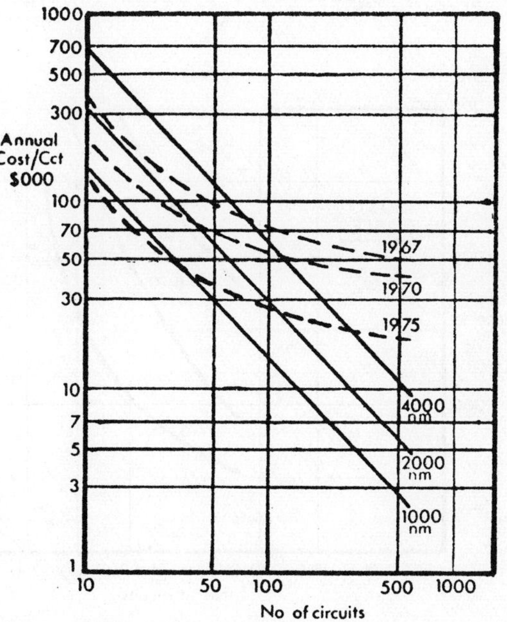

Fig.13 Comparison of annual costs for 640 circuit submarine cable (—) and satellite systems (---).

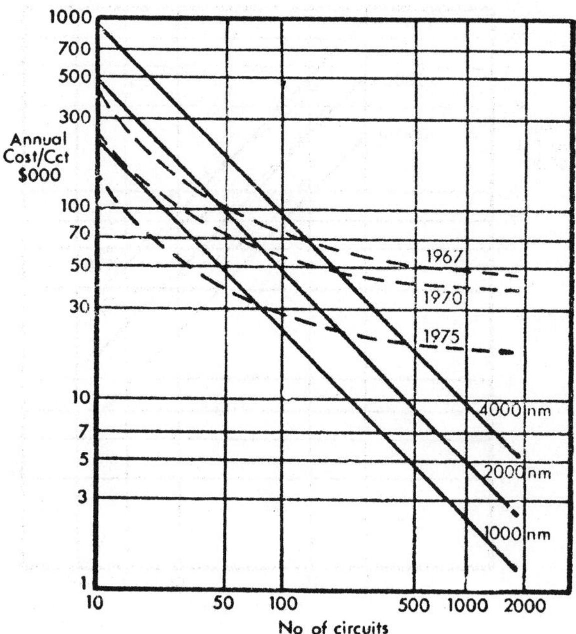

Fig.14 Comparison of annual costs for 1840 circuit submarine cable (—) and satellite systems (---).

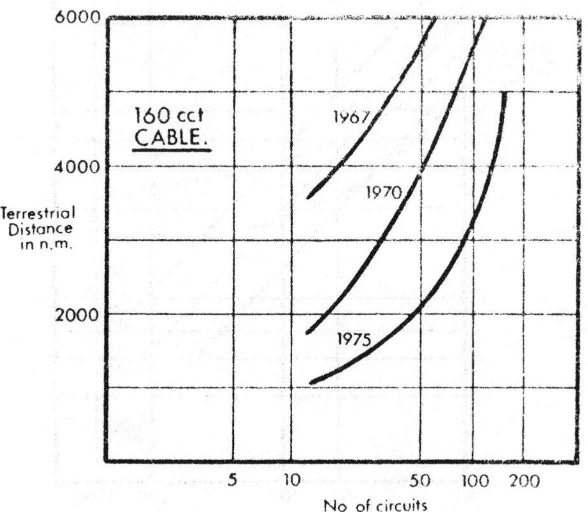

Fig.15 Break-even conditions to achieve same annual costs for submarine cable and satellite circuits.

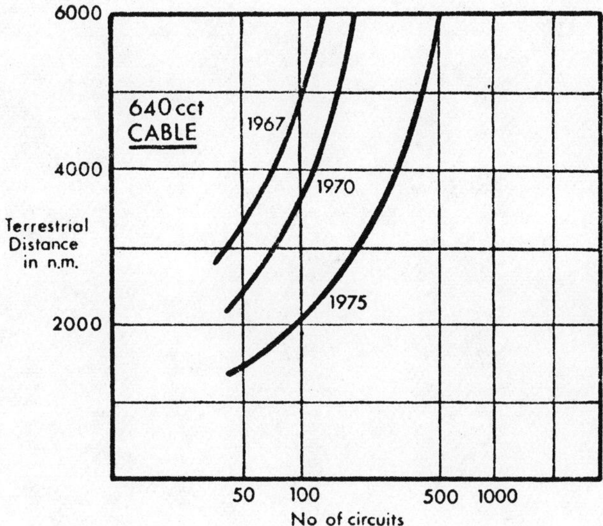

Fig.16 Break-even conditions to achieve same annual costs for submarine cable and satellite circuits.

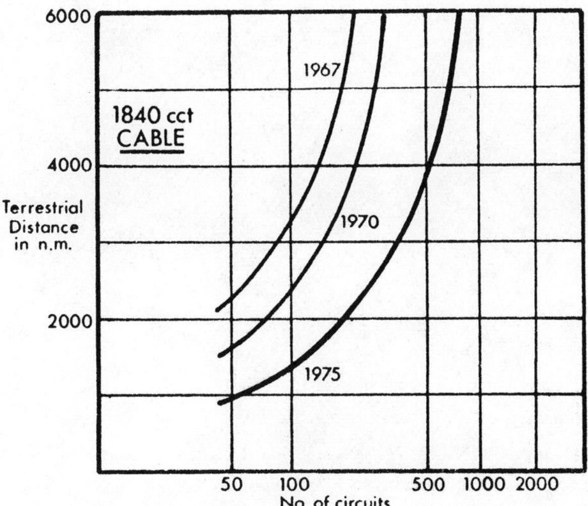

Fig.17 Break-even conditions to achieve same annual costs for submarine cable and satellite circuits.

USE OF AN EMPIRICAL PRODUCT-ORIENTED COST DATA BANK FOR PREDICTING AND NEGOTIATING FUTURE SATELLITE SYSTEMS COSTS

Hyman Silver*

North American Rockwell Corporation, Downey, Calif.

Abstract

This paper outlines a method for estimating, negotiating, and reporting costs within the aerospace industry. In the approach presented, the traditional system of identifying costs by function is replaced by one that also identifies costs with the product or end item. Specifically, the end-item cost system used on an Air Force communications satellite program is described. The paper discusses the manner in which this system was used to form a comprehensive cost data bank and the subsequent use of this data bank for estimating new programs. Also described are the benefits derived by both the procuring agency and the contractor when contracts are negotiated on the basis of product-oriented cost estimates.

Introduction

Since the early 1960's, a great deal of attention has been given to methods of estimating, collecting, controlling, reporting, and negotiating costs within the aerospace industry. Because the various phases of the procurement cycle involve many levels, interest in the topic is far reaching, ranging from contractor financial and program personnel to the senior planning and budgeting levels of DOD. At issue are the large sums of money that change hands during

---

Presented as Paper 70-445 at the AIAA 3rd Communications Satellite Systems Conference, Los Angeles, Calif., April 6-8, 1970.

*Director, Business Operations, Advanced Programs, Space Division.

the development, acquisition, production, and deployment of military systems. If these transactions are to be accurately and efficiently monitored, costs must be identified in a manner that will optimize the functions of cost estimation, collection, etc.

Traditionally, the primary cost-identification method used by the industry is the functional system. Aerospace firms are normally organized along the lines of major departmental functions, e.g., engineering, manufacturing, testing, tooling, etc. A functional cost-identification system is one that identifies costs in these terms. Such a system has certain advantages. For one, costs can be organized with relative ease for estimating purposes. For another, control is facilitated because the estimated expenditures of a given function can be readily compared with the actual expenditures. In addition, it is acceptable to report expenditures to the Government on the same functional lines. Finally, most aerospace firms believe that organizing along functional lines is a virtual necessity for maximum personnel control.

There are, however, certain disadvantages to viewing costs along functional lines only. The most serious of these is the loss of end-item visibility during estimating, pricing, negotiating, and billing. This could result in a complete or partial invisibility of end-item costs (or end-item subdivisions) below the gross levels of program product breakdown.

This paper describes a cost system in which the traditional method of identifying costs by function is replaced by one that also identifies costs by product or end item. In addition to describing an actual end-item cost system, the paper discusses the benefits derived from such a system, including the formation of a comprehensive data bank and its use in estimating new programs.

IDCSP End-Item Cost System

In 1963, the Space Systems Division (SSD) of the Air Force Systems Command (AFSC) began a project that evolved into a series of military communications satellite (ComSat) programs. The first ComSat program was designated the Initial Defense

Communications Satellite Program (IDCSP). At the time of its establishment, a cost-effectiveness analysis was performed of DOD's acquisition of self-owned communications satellites. This analysis indicated the favorable cost-effectiveness aspects of the basic IDCSP. It also appeared that DOD's acquisition of communications-type satellites would continue in a follow-on or modified-version sense. Later, some of the potential situations matured, resulting in several additional ComSat programs that ran concurrent with and subsequent to the original program.

With these possibilities in mind, the Air Force decided to build a reliable, detailed, and comprehensive data bank based upon actual expenditures of the IDCSP. The responsibility for this effort was assigned to the author of this paper. In addition to its desire to develop a sophisticated data bank for estimating future ComSat programs, the program office had a strong interest in counteracting some problems inherent in identifying costs solely on the basis of function. Coincident with this attempt to upgrade the cost-identification posture of the program, the AFSC, among others, initiated serious work on the development and use of the DOD/NASA PERT/Cost system.

Work Breakdown Structure

The PERT/Cost concept had a very attractive feature: it introduced, and placed general emphasis on, end-item cost identification. This concept did not call for any less activity in estimating and collecting costs along conventional lines. Neither did it change any time-honored organizational or functional cost-charging systems. However, it did require that information collected functionally be recapitulated into discrete end-item, or product, and product-subdivision categories. The purpose of this recapitulation was to permit the generation of a complete end-item-oriented data base for all contractual purposes (Fig. 1). To effect such a system, the contractor would develop an end-item-oriented matrix of cost-estimating, collection, and reporting centers. Under the matrix, all program costs would be successively divided into lower end-item subdivision levels. The process would start at the entire program level and develop lower and lower levels of detail along strict product/end-item lines. Eventually, the entire end item would be subdivided into units having a nominal value ($200,000). At this point, the next level of indenture would return to functional orientation. This final level, called the work-package level, normally had a value of no more than $50,000-$100,000/work package.

This entire matrix would often encompass up to eight levels of indenture (seven end-item-oriented levels and one work-package level, functionally oriented). It was termed then, as it is today, the Work Breakdown Structure (WBS) (Fig. 2). The WBS is composed of work packages (tasks) that identify all costs for the program. It is end-item-oriented; i.e., it is directed to collecting the total costs (design, manufacturing, test, support, etc.) of each major end item and its identifiable subassemblies. It is matrixed to the normal functional or organizational cost-collection system.

Under the PERT/Cost concept, the contractor was asked to provide, for every entry on the WBS, an individual resource estimate, including all labor, material, other direct costs (ODC), burden, G&A expenses, etc. Thus, each work package on the lowest level and each hardware subdivision (summary item) had a complete unit cost estimate and represented the selling price for the unit minus profit. Later, this WBS, with completed estimates, should form the basis for negotiation, budgeting, cost collecting, and reporting. As mentioned earlier, the newness of the PERT/Cost technology prompted a careful investigation of the feasibility of using the PERT/Cost system relative to its cost and technological risks. In the final analysis, a key factor in the decision to contractually employ PERT/Cost on the IDCSP was the system's ability to provide detailed end-item-oriented, cost-identification information. The system appeared to be quite useful to SSD for both program control and data bank development.

SSD decided to develop the first WBS (through cost level 4) in house and to present it to the contractor as a relatively firm guideline for the type of end-item breakdown desired. This was done, and the WBS was forwarded to the contractors who had been selected to compete in the definition stage of the program (Phase IB).

A very significant difference existed between this WBS and some earlier versions sponsored by various agencies. This was the presence of a balanced mixture between hardware and so-called level-of-effort items. In this regard, the WBS departed from some of the interpretations of the DOD/NASA PERT/Cost manual then in effect, because no attempt was made to segregate all program costs into the various hardware end-item categories. Rather, the WBS was designed to accommodate conventional accounting procedures which do not attempt to identify level-of-effort items with particular hardware end-item subdivisions.

This feature resulted in level-of-effort end items being placed on the WBS in direct relationship to their dollar ratios on the program. Consequently, some 25% of the summary items on the WBS were level-of-effort end items. If it were impossible or impractical to obtain realistic hardware end-item association on a detailed end-item subdivision basis, the cost was assigned to a level-of-effort end item. This resulted in "clean" hardware end-item categories limited to identifiable hardware end-item tasks and realistically segregated level-of-effort costs, identified and collected in the most practical manner.

The two Phase IB competing contractors then expanded the four-level WBS provided by the program office by adding two or three more levels. About 100 summary items were given to each contractor, and the fully developed contractor WBS ran from 400 to 500 summary items and work packages. Both contractors used the four levels provided in the RFP with minor deviations. Program office approval was required for the fully developed WBS, and the existing WBS technology was fragmentary. Consequently, a number of changes in this WBS were negotiated with the contractor during Phase IC of the program definition. The final WBS, therefore, represented a fully coordinated program office/contractor effort (Fig. 3). Ultimately, a single contractor was selected for the acquisition phase of the program.

Advantages of an End-Item Cost System

In general, both Phase IB contractors responded well to the PERT/Cost concept, and both found the level-four WBS proposed by SSD acceptable with only minor modification. Neither contractor had attempted to use a PERT/Cost system before the IDCSP proposal. Of course, as the program matured toward the actual contract start date, many problems arose in the contractor's development and implementation of the PERT/Cost system. A coincidental factor that ultimately helped the PERT/Cost effort was an unusually long Government-review period (Phase IC). During a series of program start delays, both the contractor and SSD had ample opportunity to exercise the system. Thus, when the program started (Oct. 1964), the PERT/Cost system was running smoothly.

Toward the end of the Phase IC period, our collective confidence in the PERT/Cost system mounted, and other potential advantages of an end-item, cost-identification system began to manifest themselves. Since the end-item structure formed an excellent lowest-common-denominator framework for the entire program, a decision was made to introduce a unique type of contract work statement—one aligned with the WBS on a one-to-one basis. That is, for every summary item and summary-item subdivision through cost level five, a separate and autonomous line-item paragraph would appear in the work statement.

This scheme had several advantages (Fig. 4). First, a contract would ensue that was specifically correlated to the manner of collecting contract costs on a detailed basis. The subsequent effect of contract change on contract price, and vice versa, would become readily apparent. It would be far easier to make tradeoff decisions affecting costs, make-or-buy decisions, etc. Disputes regarding contract coverage are minimized when coverage is provided on a specific dollars-per-piece/part basis. In general, the increased correlation between costs and contract terms provides visibility and a legal foundation not often found in an aerospace contract. For these and other reasons, therefore, the contract work statement was closely aligned with the WBS.

Another advantage of detailed end-item cost identification is improvement in the manner and substance of contract negotiations. Employing the detailed estimates of the more than 400 summary items and work packages, both sides had a common meeting ground for discussing, challenging, justifying, and ultimately negotiating the costs of the program. Although an end-item data bank for previous similar work was not available for this particular negotiation, it was a great advantage to deal with costs in this manner. An important contributor to this advantage is the improvement in understandability. When dealing with specific detailed parts of the program, individually priced, the negotiators could quickly grasp the characteristics of the item discussed, review the physical makeup of the item, and relate these details to individual costs. The phenomenon of understanding through simplification is well understood and well documented.

Another advantage is that cost details can be instantly assessed for a given situation. Because the PERT/Cost system provides for lowest-common-denominator information (including labor category, material, ODC, travel, computer expenses, per hours per work package, etc.), the negotiator instantly had available a detailed pool of information for every item discussed, however small (Fig. 5).

During the life of the program, few difficulties were encountered in running the PERT/Cost system. As situations evolved, the WBS underwent minor changes as any cost structure should. Only one significant problem arose; this was in the area of estimating and collecting material costs for discrete work packages. The contractor's normal material accounting system was not oriented to a detailed, end-use measurement. Normal buying practices (e.g., buying material in gross amounts for "economical lot buy" considerations, buying in advance, common material purchase orders, etc.) conflicted with the work package detailed estimating requirements. Further, the practice of charging common parts and common raw materials to large across-the-board purchase orders conflicted with the cost-collection requirements of the PERT/Cost system. However, it was considered essential to retain the PERT/Cost characteristics of total costs per end item for purposes of developing a sound basis for re-estimating. Therefore, the contractor was asked to make the necessary adjustments to his internal system to accommodate subject requirements. The contractor complied, and the subsequent adjustments to his systems proved to be minor.

Use of the IDCSP Data Bank

Toward the latter part of the IDCSP program, when the contractor's costs were well understood, the assembled end-item information was first used as a data bank. Under the system that was set up, actual costs were recorded each month against every work package. In addition, a forecasted completion was made each month for every work package. As a result, a total price per work package was available each month, with the degree of confidence increasing monthly as a greater and greater share of the total (forecast) price became the actual expenditure. Thus, use of the data for estimating follow-on or similar programs became increasingly feasible as the program progressed.

Follow-On Programs

As had been anticipated at the outset of IDCSP, a number of follow-on or related programs became a reality. First, at about mid-point in the IDCSP, the ComSat multiple-payload program was undertaken. This involved the procurement of four follow-on IDCSP satellites from the same contractor for a subsequent combined launch with other spacecraft. As the IDCSP neared completion (after the first two launches) and with costs of the program 99% consumed, a new and significantly different program was initiated. This was for the development of a much larger communications satellite—a tactical communications satellite (TCS). This contract was awarded in Jan. 1967. Finally, still another program was initiated to procure two modified IDCSP satellites. Contracts for these latter programs were awarded during the first quarter of 1967. At this time, the actual expenditures were virtually complete on the original IDCSP. The military ComSat family eventually numbered five distinct and separate programs. The configuration of any one program was sufficiently related to the others so that a complete cost data bank for one program could be usefully employed in estimating the costs of the others. As events unfolded chronologically, the basic IDCSP was (and still is) used to estimate the costs of the others. This was possible for the following reasons: 1) all program costs were known; 2) the program was uncomplicated by a high incidence of changes or significant, nonrecurring overrun; 3) the PERT/Cost system was successfully employed throughout the life of the program; 4) the WBS was employed in an end-item mode; and 5) the level of cost data received by SSD was sufficiently low (level 7) to permit full visibility of detailed particles by the SSD cost analysts.

The basic IDCSP cost data bank was used extensively in preparing the Government in-house estimate for all of the programs listed. Although tailored to meet individual program characteristics, the techniques used for each program were quite similar from a methodology standpoint.

The estimation of the two identical follow-on programs was performed with the greatest ease and accuracy. Interestingly, however, estimation of the TCSP, although more difficult and less accurate, was more useful to the Government, because a different-configuration situation contains a larger range of uncertainty than does a like-configuration situation. With or without an effective data bank, an identical satellite poses no great problem to an estimator; his main concern is with learning curves, start-up costs, labor and material escalation, etc. In contrast, an unfamiliar configuration poses a substantial challenge to the estimator—a challenge significantly less formidable when a detailed end-item-oriented cost data bank is used.

Estimating the TCSP

For the purposes of this paper, it is useful to describe briefly the methodology employed by the Government in estimating the TCSP. The task, in brief, was this: to estimate the total costs of a ComSat program involving one large space vehicle that was to be placed in synchronous orbit and positioned on station by a ground-controlled command station. The satellite, which would be boosted aloft by a Titan IIIC launch vehicle, was to be extremely heavy and, dimensionally, one of the largest ever conceived by DOD. It was to have high power output, to be solar powered with batteries, and to contain a solid-propellant engine for maneuvering capability.

The cost data-bank program (IDCSP) involved 24 satellites; these were to be placed eight at a time in nearly synchronous orbits by three Titan IIIC launch vehicles. The satellites were quite small and lightweight. They were solar powered with no batteries, had relatively low power output, and contained no command or maneuvering system.

Obviously, these two programs were very different. Yet both were military communications satellites procured by the same Government system project office, with similar funding and similar schedule situations.

The first step in the estimating procedure was to divide the new program into end-item elements down to the work-package level of the WBS (levels 5 and 6). This procedure was accomplished by both the program control and technical personnel involved in the preparation of the program RFP. As far as possible, the specific logic involved in the selection of summary items, work packages, etc., was patterned after the existing cost data bank. As events turned out, the data-bank WBS and the one created for the new program had a high degree of commonality with respect to descriptive titles of hardware and level-of-effort end items. In fact, the two level-of-effort end-item WBS's were almost identical.

After the new WBS was formed, it was compared in detail with the data bank WBS for the purpose of defining the exact differences in the content of the two programs. An engineer was assigned to document the differences of each hardware and systems end-item breakdown at the component level. Subsequently, a management specialist documented the probable change in scope of each level-of-effort task. This was done after the hardware had been defined because of the obvious direct correlation between the specific hardware effort and the necessary level-of-effort support.

The estimating task, therefore, involved the following: 1) comparing the technical and production tasks of the deliverable end product with the actual costs known to apply to the base or data-bank program and 2) normalizing the data-bank costs for a single production unit.

The first step was to perform a detailed, cost-correlated "indicator" study for each component, subsystem, and system task or integration task. Following are the more important parameters selected for analysis. Although not complete, the list serves to indicate the tenor of the analysis: 1) component or task complexity; 2) component weight potential; 3) component dimensional-size/miniaturization requirements; 4) reliability and mean time to failure (MTTF); 5) material considerations; 6) integration characteristics (thermal, structural, etc.); 7) launch environment; 8) specific requirements (power output, etc.); 9) function or mission; and 10) state-of-the-art—design and production.

After the parameters had been established, the new items were rated against the data-bank items. The two WBS's were highly suitable for this comparative study because, as mentioned, they were quite similar. This similarity permitted the use of identical parameters in rating the existing and proposed WBS tasks. For both old and new tasks, the same parameter grading system was used. With this system, each parameter was assigned a scaled numerical rating. The rating was obtained by creating factors for each aspect of the parameter, scoring the factor, and then multiplying the score by an assigned weight per factor. An over-all scale of 100 points/parameter was established for the data-bank baseline. The newly evaluated Program A component could have the same rating/parameter or could fall above or below the 100 mark. When combined, the 100 points/parameter represented the actual total costs incurred for a component or task on the base ComSat program.

A bridge between the normalized actual data-bank costs by work package and the parameters was established by performing a detailed parameter analysis for each component or technical task of the program. This was done at the resource level of direct labor, direct material, and other direct costs collected on the base program, per work package.

As stated earlier, information per resource, per work package, per component had been collected throughout the program. Since this information was an integral part of the cost data bank, it was possible to employ any level of detailed cost data necessary to carry out the evaluation. For example, the data bank included labor skill classification/month by work package expressed in labor hours, direct dollars, or total dollars. However, as stated, it was decided to perform the estimate/work package on the basis of total direct labor, total direct material, and total other direct costs. This level of investigation was considered to be cost effective and to contain the accuracy of detail consistent with the purpose of the estimate. Although lower-level information was employed in specific situations, it was not generally used across the board. During subsequent fact finding and negotiations, however, every level of detail was used.

The problem, then, became one of relating the new program parameter scores to the known costs—in other words, to assess the impact of the varied parameter on a known cost base. This was not unduly difficult because of the preciseness of the study. Using the above detailed comparative technical and cost information, technical and production experts in the Government program office made individual judgments in the labor, material, and ODC areas for every work package/component on a single-product-unit basis. Justification and explanation of each expert's estimate had to be broken down to at least the total labor, material, and ODC level for every component planned for the program. In every case, there was sufficient data-bank information to generate good estimates. A possible exception was the case of totally new components or tasks (10% of the total work load). In the case of new items or tasks, however, estimators found it beneficial to study the data bank of related components and, at times, the bank of dissimilar components. Once a trained estimator established a sound, detailed understanding of the electronic end items in question, he could usually develop rather sophisticated estimates for new components. Thus, the entire hardware and system estimates were derived on generalized levels 5 and 6 of the WBS.

The next problem was the software end-item or level-of-effort areas under "Administration" on the WBS. In this area, there were about 70 software end items, most of which were level-of-effort entries. In a level-of-effort situation, this type of cost detail was as unique to administrative costs collection as the hardware end-item collection system previously discussed. The job of preparing the estimate for the new program was an order of magnitude easier than that for the hardware effort. This was due to the more generalized or less precise relationship of administrative tasks to the program end product. The over-all effect or cost "swing" of level-of-effort tasks is not as great as that of hardware; therefore, it is easier to re-estimate.

The Government management experts established criteria for evaluating the tasks. For the most part, these criteria were related to the results and criteria of the technical analysis. A listing of all the criteria used in the level-of-effort areas is not appropriate here due to the myriad detail involved. Because of the diverse types of level-of-effort items, it was necessary to create area-peculiar criteria and, in some cases, single-item-peculiar criteria to deal with the varying parameter. As indicated, the level-of-effort estimates, although less uniform, were usually simpler to make because of the greater degree of subjectivity necessary for characteristically nonprecise administrative and system-oriented work.

Using the normalized existing cost data bank in the same manner as did the technical estimators, the management estimators derived an estimate for the new program for each work package by direct labor, direct material, and other direct costs. None was estimated more grossly than by the labor, material, and ODC/work package previously established as the program-estimate justification level. After the raw estimates were completed, they were subjected to a tactical updating by universal parameters. This was necessary because general situations had changed during the time period in question (1963-67), and certain basic assumptions regarding the program's cost makeup had also changed. For example, differences in quantity were considered as a specific parameter in the basic re-estimate, but other universal parameters were not. Some of the more important universal parameters were as follows: 1) material cost increases; 2) labor cost increases; 3) burden and G&A trends, industry-wide; 4) start-up costs, as applicable; 5) quantity-difference impact, as applicable; 6) general technical state-of-the-art increases; and 7) availability of high reliability and exotic parts.

As accurately as possible, a quantitative judgment was made of the effect of each of the above parameters (plus miscellaneous items), and the basic estimates were appropriately adjusted. For example, information on material, labor, and burden increase trends was derived directly out of measured trends from established cost analysis data and subsequently used as an additive item in the basic estimates.

The remaining items on the foregoing list, although more subjective or judgment oriented, were also assessed in quantitative terms and applied to the basic estimates as appropriate (both increases and decreases as necessary). In some cases, individual universal adjustments were applied to individual components by cost element; in other cases, they were applied across the board. Every effort was made to apply these parameters to the lowest level of detail consistent with best use of the data, accuracy, and appropriateness of application. For example, the potential increase in burden rates was applied across all the engineering labor categories. In contrast, the adjustment for the current availability of quality solar cells was applied only to the power subsystem.

Finally, after all universal parameter adjustments had been made to the basic estimates, the system program office officially released the Government estimate to the necessary offices. Completed during preparation of the RFP, this estimate superseded a planning-purposes estimate generated in a much simpler manner some months earlier (Fig. 6).

When the contractors' bids were received, the system program office noted with great interest that the Government-generated estimates were well within 10% of the arithmetic mean of the six bidders' estimates. This knowledge permitted the Government to maintain a sophisticated negotiating position through the lowest levels of the contractors' cost proposals. Armed with this knowledge, the negotiators were able to handle the fact-finding sessions in a direct and enlightened manner. The subsequent selection of the winning contractor was also simplified and expedited. Moreover, from the cost standpoint, the detailed estimate made possible a markedly improved source-selection justification. Both sides found the final cost settlement equitable, free of loose ends, and streamlined. Both positions were soundly established and comprehensible to each other on both a gross and detailed level.

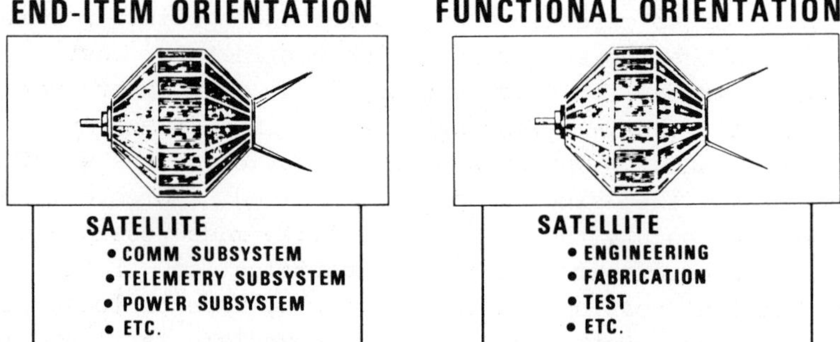

Fig. 1 End-item vs functional data base.

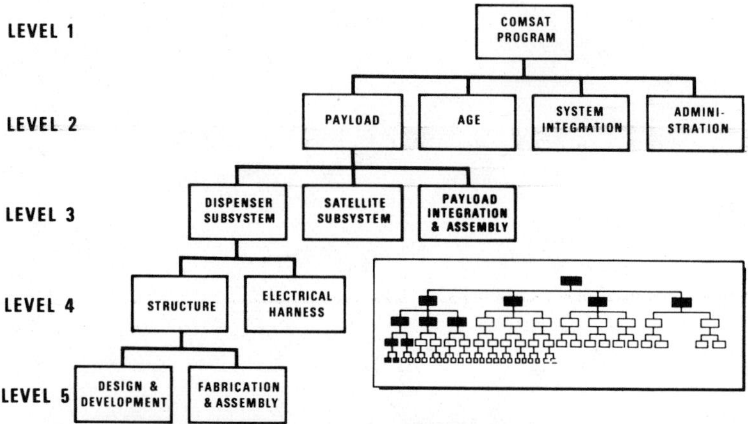

THE PROGRAM BREAKDOWN STRUCTURE IS:

1. **COMPOSED OF WORK PACKAGES (TASKS) IDENTIFYING ALL DIRECT COSTS ON PROGRAM**
2. **END-ITEM ORIENTED... DIRECTED TOWARD COLLECTING THE TOTAL COSTS (I.E., DESIGN, MFG, TESTING, SUPPORT COSTS, ETC.) OF EACH MAJOR END ITEM & ITS IDENTIFIABLE SUBASSEMBLIES**

Fig. 2 Program breakdown structure.

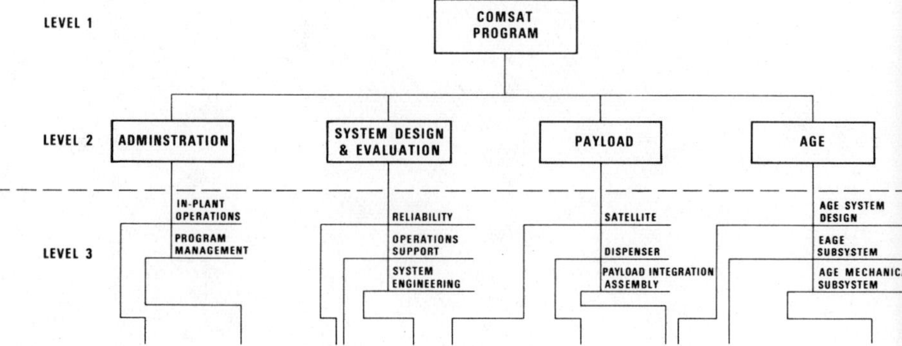

Fig. 3 Operational program breakdown structure.

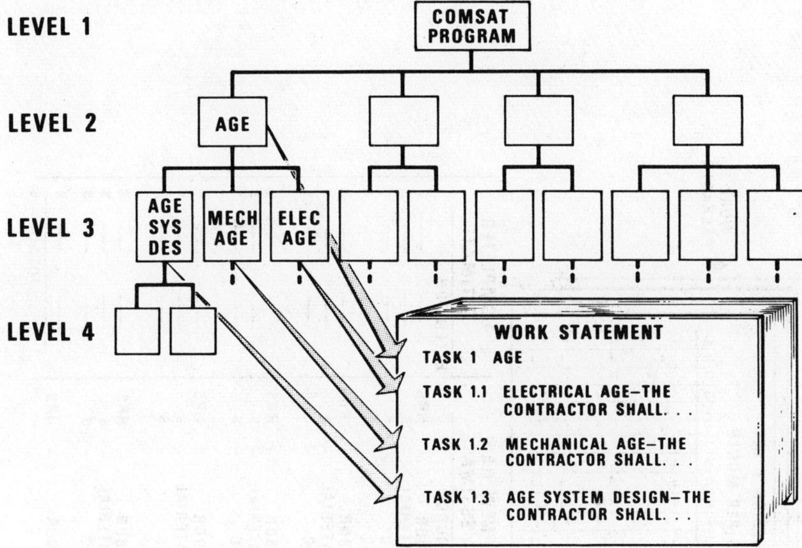

Fig. 4 Program breakdown structure/work statement relationship.

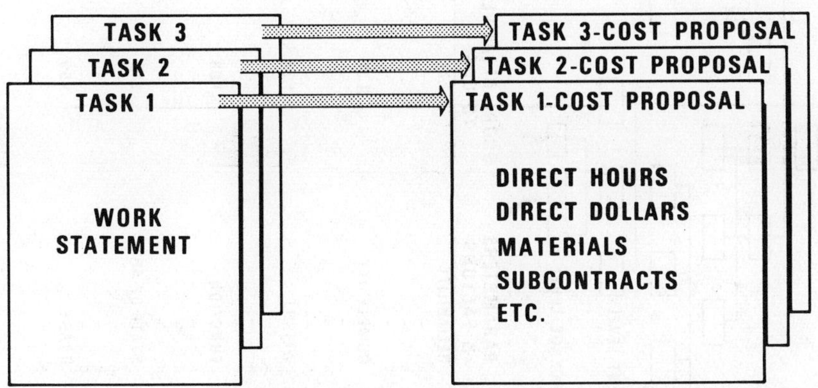

Fig. 5 Work statement/cost proposal relationship.

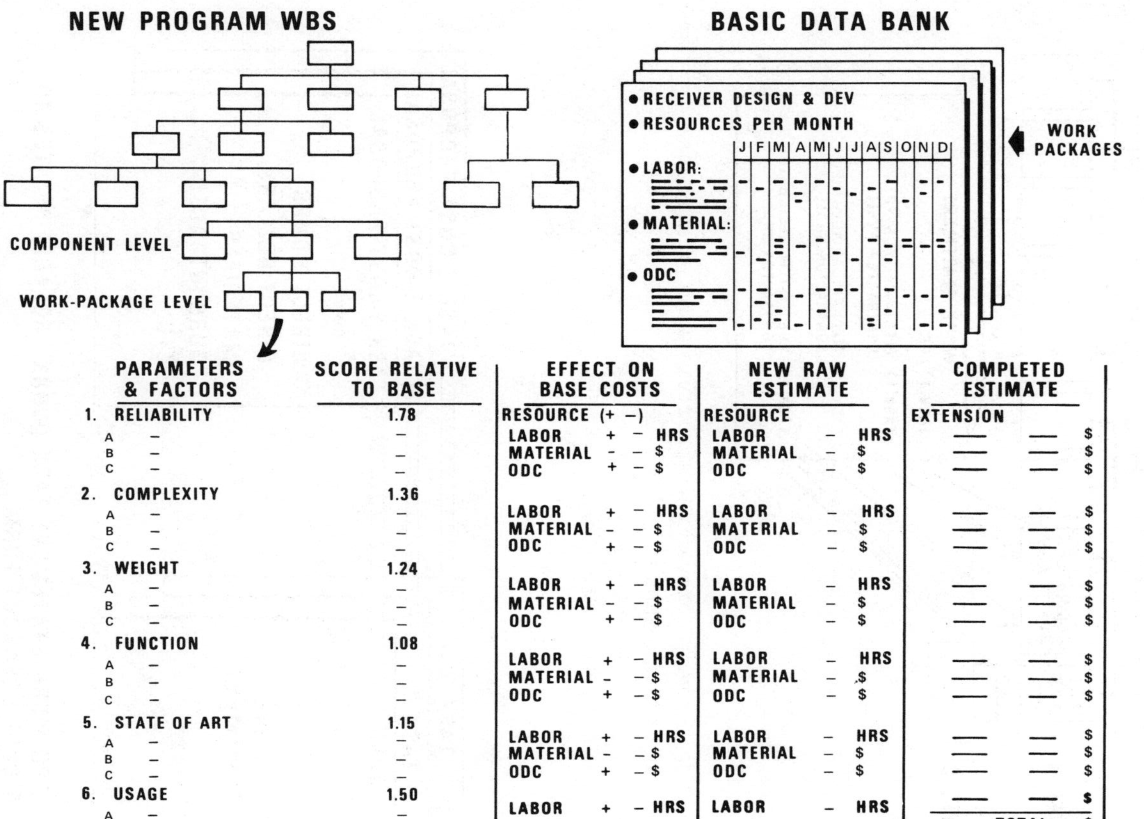

Fig. 6 New program estimate model.

INFLUENCE OF RAIN AND CLOUD ATTENUATION ON THE DESIGN OF A 20 TO 30 GHz SPACECRAFT COMMUNICATIONS REPEATER

Walter A. Koenig* and Clor W. Merle+

Lockheed Missiles & Space Company, Sunnyvale, Calif.

Abstract

Rain and cloud attenuation impose a formidable problem for communication satellite systems operating at 20-30 GHz. To study this problem a model is postulated for FM transmission of 525 line color television via a satellite repeater in a synchronous orbit. A 30-GHz up-link ground transmitter with a 104-dbw ERP is assumed. A 20-GHz down-link ground receiver with a G/T of 43.2 db/°K also is assumed. An analysis is made to determine design requirements for repeater ERP, dynamic range, and antenna gains for a 45 MHz system bandwidth to achieve adequate signal-to-noise ratios above threshold at the Earth receiver with varying rain and cloud attenuation on the up- and down-links. The analysis shows that the up-link attenuation affects the repeater design differently than the down-link attenuation. Calculated data supporting this conclusion are given in the figures.

I. Introduction

The decade beginning in 1970 will bring a greatly increased demand on a world-wide basis for commercial communication satellite services. This increased demand will require considerably greater channel capacity than presently exists in satellite communications, particularly to accommodate the high traffic volume which will develop between certain ground terminals such as New York and London. Therefore, it appears that the presently allocated bands below 11 GHz for satellite communications will be inadequate to furnish the required channel

Presented as Paper 70-498 at the AIAA 3rd Communications Satellite Systems Conference, Los Angeles, Calif., April 6-18, 1970.

*Chief Development Engineer, Communications Satellite Programs, Space Systems Diyision.

+Senior Staff, Engineer, Communications Satellite Programs, Space Systems Division.

capacity, especially if they are permitted for domestic use with hundreds of associated Earth stations. In addition the number of terrestrial microwave stations operating below 11 GHz will continue to increase, thus competing with satellite systems for increased spectrum assignments to meet their increasing needs.

It may be assumed that the present worldwide commercial satellite system operating in the 4 and 6 GHz bands will continue to do so in the 1970's; however, to meet projected needs for increased channel capacity indications point towards extensive use of the 20-GHz and 30-GHz frequency bands for future generations of communication satellites. The present posture of the FCC in developing the U.S.A. position for the next World Administrative Radio Conference is a major factor supporting this view.[1]

If we assume that the 20- and 30-GHz bands will be dedicated to satellite communications exclusively, then no statutory limitation on e.i.r.p. will exist for satellite communications in these bands.≠ Satellite e.i.r.p. may then be maximized to the extent possible within on-board power availability limits and by narrow antenna beams limited by satellite attitude control and beam pointing errors.

A formidable problem exists when one considers satellite communications systems at 20 and 30 GHz. This problem arises because of the severe attenuation by liquid water of electromagnetic waves with wavelengths shorter than 3 cm. In a satellite communication system, this attenuation results from the presence of cloud formations and rainfall in the signal paths between ground stations and the satellite repeater. Therefore it is logical to study this problem to determine in what manner, the attenuation by rain and clouds will affect the design of a 20- to 30-GHz spacecraft communications repeater.

In the following sections, rain and cloud models for temperate and tropical climates are described, followed by calculated values for up- and down-link attenuations at projected worldwide ground station locations. The use of space diversity at Earth terminals is not considered in this paper. A model of

≠The present CCIR limitation of flux density on the Earth's surface is necessary to control interference with terrestrial communications systems. This limitation is stated as $(-152 + \theta/15)$ dbw/4 kHz/square meter, where θ is the satellite elevation angle. Thus, the allowable satellite e.i.r.p. varies with ground station location and must account for relationships between peak flux density in a 4 kHz band and the total r-f power in a transmission channel.

a 20- to 30-GHz satellite transmission link is then analyzed using the attenuation factors predicted by the rain and cloud models. The satellite repeater design characteristics which are primarily affected by the link conditions are the receiving antenna gain and receiver noise figure (or G/T), repeater dynamic range, and transmit e.i.r.p. Ranges of values for these parameters are derived for the predicted ranges of up- and down-link attenuations at various ground terminal locations in the international communications satellite network. The results show that the link is feasible and indicate the effects of up-link and down-link rain and cloud attenuation on the design of the spacecraft repeater system.

II. Atmospheric Attenuation Models

In this section the cloud and rainfall attenuation models and the data required for calculating the values of attenuation of 20 and 30 GHz are taken from a paper by W. Holzer.[2]

Temperature Climate Atmospheric Model

A two-dimensional representation of a temperature climate atmosphere during a rainstorm is shown in Fig. 1. This model applies to rainstorms typical of the winter months, but not to a summer thunderstorm. The thunderstorm is similar in characteristics to the tropical model described later.

Cloud attenuation. The attenuation in clouds is given by

$$A_c = k\rho r \quad \text{db} \tag{1}$$

where k = coefficient of attenuation in db/km/g/m^3, ρ = liquid water content of cloud in g/m^3 and r = path length through the cloud in km. From Fig. 1 of Holzer's paper,

$$k = 0.37 \text{ db/km/g/m}^3 \text{ at 20 GHz}$$
$$k = 0.82 \text{ db/km/g/m}^3 \text{ at 30 GHz}$$

Rain attenuation. From Fig. 1 the horizontal extent E of the rainfall area is given by the empirical relationship[3]

$$E = 41.4 - 23.5 \log_{10} p \text{ km} \tag{2}$$

where p = the rainfall rate in mm/hr.

The Washington, D. C. rainfall rate that is not exceeded more than 0.1% of the time is 12 mm/hr. This rate is typical of many temperature climate areas. Using this value for p in Eq. (2) we find $E = 41.4 - 23.5 \log_{10} 12 = 16$ km.

The signal attenuation in rain is given by

$$A_p = qpr \quad \text{db} \tag{3}$$

where q = coefficient of attenuation in db/km/mm/hr, p = rainfall rate in mm/hr, and r = path length through the rain in km. From Fig. 2 of Holzer's paper,

q = 0.09 db/km/mm/hr at 20 GHz
q = 0.14 db/km/mm/hr at 30 GHz

Gaseous absorption. The remaining component of attenuation is due to gaseous absorption. At centimeter wavelengths the diatomic oxygen and water molecules in the atmosphere cause molecular absorption of microwave energy. The amount of absorption is a function of temperature, pressure, and humidity. Based on measurements made in radio astronomy[4] an estimate of 2-db loss at 20 and 30 GHz for a 10° elevation angle appears reasonable for this quantity.

Total attenuation. The total attenuation is the sum of the attenuation due to clouds, rain and atmospheric gases, i.e.,

$$A_T = A_c + A_P + A_G \tag{4}$$

where A_G = attenuation due to gaseous absorption, estimated to be 2 db. As an indication of the magnitude of attenuations to be expected, total attenuation for a temperate location with a 10° elevation angle to the satellite, which is a reasonable limit for these frequencies, is calculated as follows:

20 GHz: $A_T = 3.83 + 17.55 + 2 = 23.38$ db

30 GHz: $A_T = 8.49 + 27.3 + 2 = 37.79$ db

Tropical Climate Atmospheric Model

A two-dimensional representation of a tropical climate atmosphere is shown in Fig. 2.

Cloud attenuation. For the tropical climate atmosphere shown in Fig. 2, the cloud densities used are $\rho = 1.0$ g/m^3 in cloud columns and $\rho = 5.0$ g/m^3 in lower level clouds. These are cloud densities at Manila, Phillipines.

Rain attenuation. We assume that the rain rate p is constant over a given area. If we use p = 22 mm/hr and the empirical relationship given in Eq. (2) for horizontal extent E of the rainfall, $E = 4.14 - 23.5 \log_{10} 22 = 9.8$ km.

This extent agrees quite well with observations on thunderstorms in the United States[5], and is probably typical of a convective cell system for this magnitude of rainfall rate. It should be noted that the foregoing formula is not valid for very high rainfall rates such as are observed at the center of a thunderstorm.

Total attenuation. As in the case of temperature climate, the total attenuation is the sum of the attenuation resulting from clouds, rain, and atmospheric gases, i.e., $A_T = A_c + A_p + A_g$. The magnitude of attenuations which might occur for elevation angles of 25° are

$$A_t = 17.1 + 4.7 + 2.0 = 23.8 \text{ db at } 20 \text{ GHz}$$

$$A_t = 35.7 + 7.3 + 2.0 = 45 \text{ db at } 30 \text{ GHz}$$

It is obvious that tropical locations present a very difficult situation.

Comparison of Temperature and Tropical Climates

It is apparent from examination of the calculated values of total attenuation for the temperate and the tropical climates that for the former the rainfall attenuation predominates, whereas for the latter the cloud attenuation predominates. This is, of course, the result of selecting a 10° elevation angle to the satellite in the temperate climate and a 25° elevation angle in the tropical climate. Calculations made for a 10° elevation angle using the tropical climate model show the predominance of rainfall attenuation as in the case of the temperate climate model. However, the resultant total losses for a 10° elevation angle in a tropical climate are so large as to preclude use of so small an elevation angle. It appears reasonable to limit operational modes for a tropical location to elevation angles to the satellite which are compatible with feasible systems.

III. Attenuation Design Ranges

The attenuation factors derived earlier were based on specific values of rainfall rate. It is of interest to derive the attenuation for each model as a function of the probability that the attenuation exceeds a given level. Figure 3 shows this function for Washington, D. C. Similar curves can be derived for other locations where adequate statistical data on rainfall rates exist.

Because the structure of the atmosphere during storms is extremely variable, and little actual data exist on typical structures, it is of interest to compare the results obtained with the models to actual experimental data. Figure 4 is a plot of data taken at Crawford Hill, N.J. by the Bell Telephone Research Laboratories using the sun as a microwave source.[6] Data were taken at 16 and 30 GHz; the curve shown for 20 GHz is an estimate based on the other two curves. The 20 GHz curve indicates that the attenuation will exceed 16.5 db 0.1% of the time. This value is in fairly good agreement with the 23 db obtained with the temperate climate model in Sec. II, considering that the Crawford Hill measurements are of necessity taken over a range of elevation angles. However, several characteristics of the calculated curves in Fig. 3 suggest further examination. The calculated curves of attenuation decrease at a much slower rate than the Crawford Hill data indicates. This may be due in part to the assumption of a constant rainfall thickness in the temperate climate model of Fig. 1. The atmospheric structure is undoubtedly much more complex, and the foregoing factors are probably related to the rainfall rate. Hence the slope of the curves should be steeper, and more like the Crawford Hill data. It is concluded that the models represent a worst case condition, and hence a satellite design based on these models will be conservative.

The elevation angle to a satellite will vary from location to location, and hence the rainfall and cloud attenuation values also will vary. Some 70 satellite communications ground terminals around the world are anticipated to be in operation by 1972 using the 4 and 6 GHz bands.[7] The implementation of links at 20 and 30 GHz will undoubtedly use these same terminal locations to a great degree, hence it is of interest to examine the attenuation anticipated for these locations. Values of attenuation were calculated for each location using the models and equations of Section II, assuming a three satellite system, and using rainfall rates obtained by correlating total annual rainfall and rainfall rates. Figure 5 is a scatter plot of attenuation factor at 20 GHz vs. elevation angle to the satellites within view of each ground station. The attenuations represent values which would be expected to be exceeded for 0.1% of the time. A similar plot has been made for a frequency of 30 GHz. These plots indicate that the limiting values of attenuation calculated in Section II will be valid for approximately 90% of the projected Earth station locations. It is also apparent that a few tropical locations would have extreme difficulties at 20 and 30 GHz, and hence would be best served by appropriate expansion in the 4- and 6-GHz bands. Large numbers of temperate locations will have significantly lower attenuation values, so it is of interest to examine the complete-ground-to-satellite-to-ground link as it is affected by up-link and down-link attenuations.

IV. Link Analysis of a 20- to 30-GHz Communication Satellite System

Communication System Model

The satellite repeater model used in the analysis consists of a hard-limiting repeater receiving at 30 GHz and transmitting at 20 GHz. A 45-MHz repeater bandwidth is assumed for FM transmission of 525 line color television signals having a 4.2-MHz baseband bandwidth. The ground receiver threshold in this case is approximately 10 db. It should be noted that the following analysis is general for a 45-MHz repeater bandwidth and hence applies also to other forms of modulation. The following system parameters are also assumed: 1) ground transmitter power = 5 kw; 2) 40' ground transmit antenna, 67-db gain at 30 GHz; 3) satellite receiver noise figure = 8 db (overall); 4) ground receiver antenna gain = 65 db at 20 GHz; 5) ground receiver noise temperature of 400°K (under conditions of rain) at 20 GHz; 6) satellite antenna pointing errors may be neglected.

The remainder of this paper is devoted to a study of the effects of rainfall and cloud attenuation on the system described by the parameters stated above. The results of the analyses are presented in Figs. 6-12 which follow. We note, however, that the system behavior shown in these figures is not unique for the system parameters selected. That is, with appropriate changes of scale, these figures are applicable to other systems of the same type but with different parameter values such as ground transmitter power, satellite transmitter e.i.r.p., etc.

The basic equation which applies to either the up-link or the down-link is $W + G/T - L = C/T$ where W is the transmitter e.i.r.p., G/T is the ratio of receiver antenna gain to receiver noise temperature, L is the free-space loss between receiver and transmitter and C/T is the ratio of received carrier power to receiver noise temperature, all in db. Therefore on the up-link, values of noise figure (effective noise temperature) other than the value assumed may be accommodated simply by adjusting the values of receiving antenna gain to maintain G/T constant. On the down-link, values of satellite e.i.r.p. other than the value assumed may be accommodated by appropriate adjustment of the down-link rain loss values shown in the curves.

Hard Limiting Operation of the Repeater

Figures 6 and 7 show the variation of ground station carrier-to-noise ratio (C/N) as a function of satellite receive antenna gain for two values of up-link rain loss and with down-link rain loss as a variable parameter.# The plots shown assume that the repeater always operates in a hard-limiting mode, regardless of the rainfall and cloud attenuation of the received signal on the up-link of the repeater. Hard limiting is necessary because of the very wide variations in up-link losses which will occur. Inspection of these plots shows that if the down-link rain and cloud losses become large enough, the carrier-to-noise ratio at the ground receiver IF output does not vary appreciably over a wide range of satellite receiving antenna gains, nearly independent of the up-link rain and cloud signal attenuation. Thus, the choice of satellite receiving antenna gains is not extremely critical in the operation of the link and is hardly influenced by the up-link rain and cloud losses, once the down-link losses due to rain and clouds become sufficiently large. This result is somewhat surprising, but the reason for it is as follows: as the down-link rain and cloud losses increase, the noise on the signal received from the satellite contributes less and less to the total noise in the ground receiver IF output. A point is reached where the output SNR of the ground receiver is determined almost entirely by its own internal noise temperature. At that point the gain of the satellite receiving antenna (and hence the transmitted SNR of the satellite repeater) has almost no effect on the SNR at the ground receiver IF output provided the up-link C/N ratio exceeds 13 db. It should be kept in mind, however, that this is true only if the repeater satellite output power (e.i.r.p.) is a constant, regardless of the signal power received by the satellite from the ground transmitter. This in turn implies that the repeater has sufficient dynamic range to drive the output TWT to saturation for all levels of signals received from the ground.

#The derivation of the equations for the curves of Figs. 5, 6, 7 & 8 and the derivation of Eqs. (5), (6) & (7) which follow in the remaining Sections of this report are shown in the Appendix to AIAA Paper 70-498.

Repeater Operation with Limited Dynamic Range

Possibly a more practical case for study is that where the repeater has a limited dynamic range. In this case as the up-link rain and cloud attenuation increase, a point is reached where the signal received by the satellite is below the dynamic range of the repeater and limiting no longer occurs. At that point the TWT in the repeater is no longer driven to provide maximum saturated power output, hence the satellite e.i.r.p. is no longer constant, but begins to fall off, decreasing as the level of received signal from the ground decreases. If we assume that at that point the repeater output power becomes a linear function of the input received signal, then the repeater operates as a constant power gain device. This is in comparison with operation in the saturated power output mode where the repeater operates as a variable power gain device.

Figures 8 and 9 show the variation of ground station C/N as a function of up-link rain loss for two widely separated values of down-link rain loss. A satellite receiving antenna gain of 40 db was selected for purposes of analysis. On each figure there are two curves, one for a 150°K ground receiver and one for a (more realistic) 400°K ground receiver. The solid curves indicate link operation for the repeater operating under hard limiting conditions over a 37.5 db range of up-link rain and cloud loss. The broken lines on each figure show link performance with the repeater dynamic range limited to 20 db as an example. For instance, on each of these Figures, the repeater output power is constant over the range of up-link rain and cloud loss from 0 to 20 db. Beyond this range of loss, the satellite repeater power output decreases and the broken lines, rather than the solid curves, depict link performance.

The effect of limited dynamic range in the repeater is to cause the received C/N ratio to drop to the threshold value at a lower value of up-link loss, for a given value of down-link loss. Values of repeater dynamic range other than 20 db are, of course, possible. The effects of different values may be studied by drawing lines on Figs. 8 & 9 parallel to the broken lines depicting the 20 db value, intersecting the solid curves at the value on the abscissa (i.e., the up-link rain and cloud loss) corresponding to the assumed value (in db) of repeater dynamic range.

Repeater Dynamic Range Requirements

To obtain a quantitative measure of the requirements for repeater dynamic range, the curves of Fig. 10 were drawn from data obtained by graphical construction from a set of curves similar to Figs. 8 and 9. In deriving Fig. 10 it was arbitrarily assumed that a 3-db degradation in received C/N caused by limited repeater dynamic range was acceptable at an up-link rain loss of 36 db.

To illustrate the method used, consider the solid curve on Fig. 8 for a 400°K ground receiver. At a 36 db up-link loss, the ground station C/N ≅ 13 db. We move down vertically from this point on the curve to a new point 3 db lower on the vertical scale. From this new point we draw a line parallel to the broken lines on the figure until it intersects the solid curve for the 400°K receiver. This intersection occurs at an up-link rain and cloud loss of 25.5 db, hence this is the value of required dynamic range, for an expected down-link loss of 10 db caused by rain and clouds. Other points to obtain the curves of Fig. 10 were derived in a similar manner.

We note that both curves of Fig. 10 reach a value of 33 db required dynamic range for 36-db expected down-link rain loss. A set of curves similar to those in Fig. 10 can be derived for expected values of up-link rain loss less than 36 db. In general the latter will decrease in slope and range of ordinate values as the values of up-link rain loss are decreased. This is illustrated for a 22-db up-link rain loss condition, shown on the lowest curve of Fig. 10. However, it would seem that the expected 36 db up-link rain and cloud loss would be the more realistic value to use in determining dynamic range requirements for the repeater, at least when considering the total losses at 30 GHz which were calculated earlier in Section II for the extremes of climate models considered there.

Repeater ERP Requirements

In the preceding parts of this study, we have assumed that the effective radiated power of the satellite (ERP) was fixed at 47 dbw. We now look at the requirements that must be placed on $(ERP)_s$ in order to achieve the minimum C/N = 10 db threshold requirement at the ground receiver. Figure 11 depicts the repeater ERP requirements, assuming the following parameter values for the system:

ground receiver noise temperature = 400°K (total)
ground receiving antenna gain = 65 db

ground transmitter ERP	= 104 dbw
ground receiver C/N	= 10 db
satellite repeater noise figure	= 8 db
satellite repeater receive antenna gain	= 40 db

Figure 11 is a plot of the equation

$$(ERP)_s = 10 \log_{10} \left[10N_T(R+1)/(R-10) \right] + 145 - L_{RD}, \text{ dbw} \qquad (5)$$

where N_T = thermal noise power due to ground receiver noise temperature (400°K), R = carrier-to-noise ratio at the satellite repeater output, and L_{RD} = down-link rain loss in db.

$(ERP)_s$ as given by Eq. (5) is that power required to give a C/N = 10 db at the ground receiver. The value for R in Eq. (5) is calculated from the relationship

$$R = (S/N)_s = \text{antilog}_{10}(5.12 + 0.1\ L_{RU}) \qquad (6)$$

When the up-link rain loss L_{RU} is 28 db or less, the quantity (R+1)/(R-10) in Eq. (5) is close to unity and the lowest curve of Fig. 11 applies for all values of up-link rain loss up to 28 db. When the up-link rain loss reaches 38 db, then

$$R = \text{antilog}_{10} \left[5.12 + 0.1 \times (-38) \right] = \text{antilog}_{10} 1.32 = 20.8$$

i.e., the carrier-to-noise ratio at the satellite is 13.2 db and the satellite transmitted noise power is beginning to contribute significantly to the total noise power at the ground receiver. This is the middle curve of Fig. 11. The upper bound is given by the top curve of Fig. 11. Here the up-link rain loss is 41 db and the signal-to-noise ratio at the satellite is

$$R = \text{antilog}_{10} \left[5.12 + 0.1\ (-41) \right] = \text{antilog}_{10}\ 1.02 = 10.47$$

i.e., it is only slightly higher than the 10 db carrier-to-noise ratio at the ground station. Hence if the up-link rain loss increases beyond 41 db, the satellite ERP requirements increase without bound. This is seen by inspection of the term (R-10) in Eq. (5). R must be larger than 10, otherwise (R+1)/(R-10) becomes indeterminate.

From the point of view of practical system design, it is reasonable to select a minimum carrier-to-noise ratio at the satellite of 13 db as a design criterion. Then a 10 watt TWT power output tube in the satellite repeater, together with

3 db line losses and a 1° beamwidth transmitting antenna (44 db gain at 20 MHz) provide an ERP = 7 + 44 = 51 dbw. Using the middle curve of Fig. 11, this ERP will allow a maximum 38-db up-link rain loss and 20-db down-link rain loss to occur simultaneously while still achieving the threshold C/N = 10 db at the ground receiver. If we design the link for operation over 99.9% of the time, then the probability of link outage (due to excessive up-link or down-link rain loss) is 10^{-3}. Then if we assume that excessive rain on the up-link and the down-link are independent events, the probability they will occur simultaneously is $10^{-3} \times 10^{-3} = 10^{-6}$, i.e., the up-link rain loss will be greater than 38 db, and simultaneously the down-link rain loss will be greater than 21 db on the average of approximately 32 sec./yr. The lowest curve of Fig. 11 shows that with $(ERP)_S$ = 51 dbw the maximum down-link rain loss that can be tolerated is 23 db, even with no up-link rain loss.

Finally, it is worthwhile to note that the two lower curves of Fig. 11 are separated (vertically) by 2.5 db in $(ERP)_S$. Thus at a cost of increasing the repeater ERP by 2.5 db, one moves from the lower curve to the middle curve, gaining an increase of 10 db in the maximum allowable up-link rain loss. This appears to be a fairly profitable tradeoff to make. However, comparison of the middle and the upper curves of Fig. 11 shows that to obtain an additional 3 db in maximum allowable up-link rain loss, an increase of approximately 11.5 db in repeater ERP is necessary. This is obviously a very poor tradeoff to make in system design. One may therefore conclude that anything above the middle line of Fig. 11 represents the point of rapidly diminishing returns, as far as gaining increased tolerance to up-link rain loss in exchange for increased repeater ERP is concerned.

Repeater G/T Requirements

The preceding section illustrated the influence of up-link rain loss in determining design requirements for repeater ERP. To complete this study of the effects of rain and cloud attenuation on repeater design, we now study the requirements for repeater G/T as this quantity is influenced by up-link and down-link rain and cloud attenuation.

Figure 12 depicts the G/T requirements for the satellite repeater, assuming the following parameter values for the system:

ground receiver noise temperature	= 400°K (total)
ground receiving antenna gain	= 65 db

ground transmitter ERP = 104 dbw
ground receiver C/N = 10 db
satellite repeater ERP = 47 dbw

Figure 16 is a plot of the following equation:

$$G_{RS}/T_e = \frac{10kB}{P'_{RS}} \cdot \frac{N_T - \text{antilog}_{10}(0.1\ L_{RD} - 9.6)}{\text{antilog}_{10}(0.1\ L_{RD} - 9.6) - 10N_T} \qquad (7)$$

where G_{RS}/T_e = ratio of repeater receiving antenna power gain to receiver effective noise temperature P'_{RS} = antilog$_{10}$(-10.9 +0.1 L_{RU}), k = Boltzman's constant, B = repeater receiver bandwidth (45 MHz), L_{RU}, L_{RD} = up-link and down-link rain loss, respectively in dB (negative values), and N_T = as defined for Eq. (5).

A study of Fig. 12 reveals immediately the repeater design tradeoffs possible. For example, the 0 db and 18 db down-link rain loss curves are separated by only a 3 db difference in G/T, whereas the 18 db and 20 db curves are separated by a 19 db difference in G/T. Thus one might logically select a repeater G/T based on a maximum allowable down-link rain loss of, say, 18 db since Fig. 12 shows that to increase this to 19.5 db would necessitate an increase of 6 db in G/T, or to increase it to 20 db would necessitate an increase of 19 db in G/T over the value required for 18 db down-link rain loss. It is clear that in this case the point of diminishing returns in trading off G/T for increased allowable down-link rain loss is reached at approximately 18-db down-link rain loss. If we utilize the previous result of 38-db maximum allowable up-link rain loss, we find from Fig. 12 a very nominal G/T requirement of approximately -1.5 db/°K for a maximum allowable down-link rain loss of 18 db. This is, of course, based on an assumed 47 dbw for the satellite repeater ERP. It appears that this latter assumption is quite reasonable since, as indicated earlier, an ERP of 51 dbw is achieved with a 10-w amplifier, 3-db line losses, and a 1° antenna beam.

The comments in Section IV concerning the small probability that the maximum rain losses would occur simultaneously on the widely separated up-link and down-link apply equally well here.

V. Conclusions

The purpose of this paper was to consider the magnitude of signal losses which may be encountered in various parts of the world as a result of rainfall and cloud effects at 20 and

30 GHz, and then to examine the input and output requirements which must be placed on satellite repeaters operating at these frequencies.

A comparison of rain and cloud losses calculated from atmospheric models and data available from suntracker measurements at Crawford Hill, N.J., correlate quite well. The calculations indicate that rainfall and cloud margins of the order of 20 db at 20 GHz and 38 db at 30 GHz will be required. Examination of the effects on satellite G/T, e.i.r.p. and dynamic range to accommodate these values of losses indicate that the design of a satellite repeater is quite feasible even though the losses are much larger than normally encountered in the present 4- to 6-GHz bands. A typical range of design values for critical satellite parameters might be as follows:

Receive G/T	-1.5 to + 2.5 db/°K
Antenna gain	40 db
Noise figure	13 to 16 db
Transmit e.i.r.p.	+47 to + 51 dbw
Antenna gain	40 to 44 db
Power output	+ 7 dbw (5w)
Dynamic range	33 db

None of these parametric conditions exceed the present state of the art, and in fact, it is believed that they can be exceeded by comfortable margins with a reasonable expenditure of effort.

The question of space diversity on the ground-to-satellite links is not considered herein due to the lack of adequate experimental data. For many stations in temperate climates, diversity will not be required provided the satellite design results in adequate link margins. Diversity may be used to alleviate the difficult problems in tropical regions, or alternatively to improve the link reliability beyond the 99.9% figure used in this paper.

References

[1]FCC Docket 18294, Fifth Notice of Inquiry, Aug. 27, 1969, Washington, D. C.

[2]Holzer, W., "Atmospheric Attenuation in Satellite Communications," Microwave Journal, Vol. 8, Mar. 1965, pp. 119-125.

[3]James, W. J., "The Effect of the Weather in Eastern England on the Performance of X-Band Ground Radars," TN 655, Royal Radar Establishment.

[4]"Atmospheric Attenuation of Space Communications," Special Rept: Space/Aero Electronics, Space/Aeronautics Magazine, Vol. 31, Jan., 1959, pp. 74-77.

[5]Byers, H. R., and Braham, R. R., Jr., The Thunderstorm, Washington, D. C., U. S. Department of Commerce, Government Printing Office, 1949.

[6]Wilson, R. W., "Sun Tracker Measurements of Attenuation by Rain at 16 and 30 GHz." Bell System Technical Journal, Vol. 48, May-June, 1969, pp. 1383-1404.

[7]"Intelsat Satellite Communications System," COMSAT Corp. International Development Division, Mar. 15, 1969.

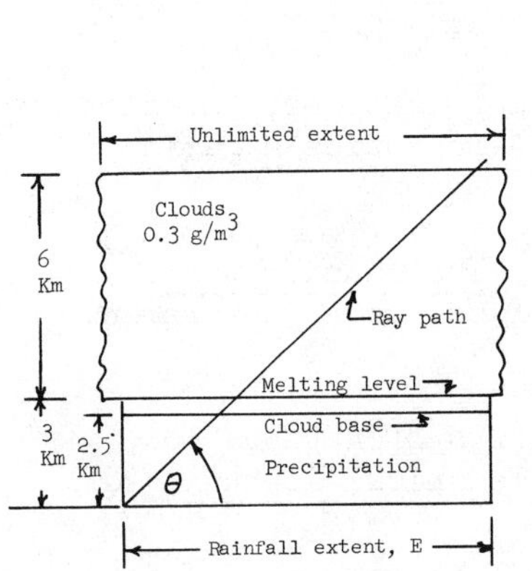

Fig. 1 Model atmosphere for temperate regions.

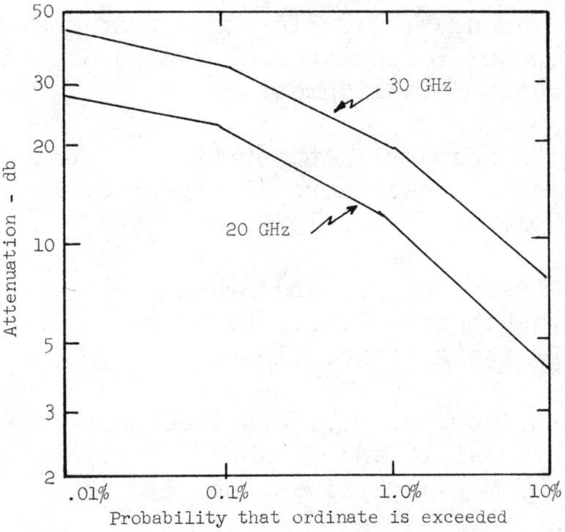

Fig. 3 Predicted total attenuation for 10° elevation angle at Washington, D. C., based on measured rainfall rates.

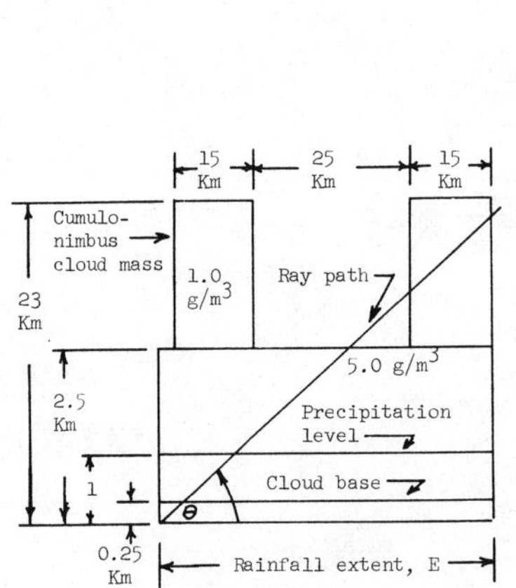

Fig. 2 Model atmosphere for tropical regions.

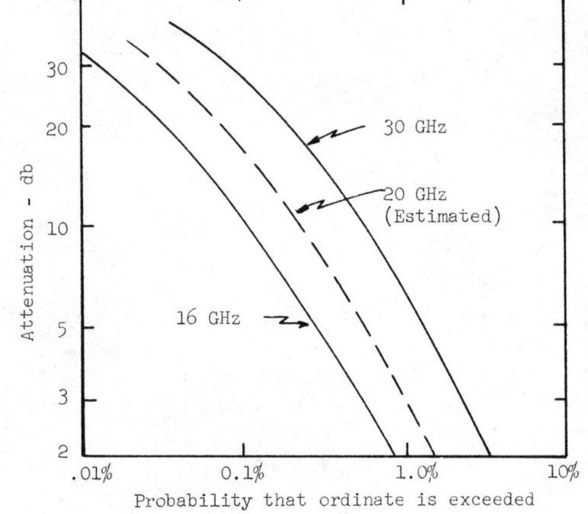

Fig. 4 Total attenuation through atmosphere measured at Crawford Hill, N. J., using the sun as a source.

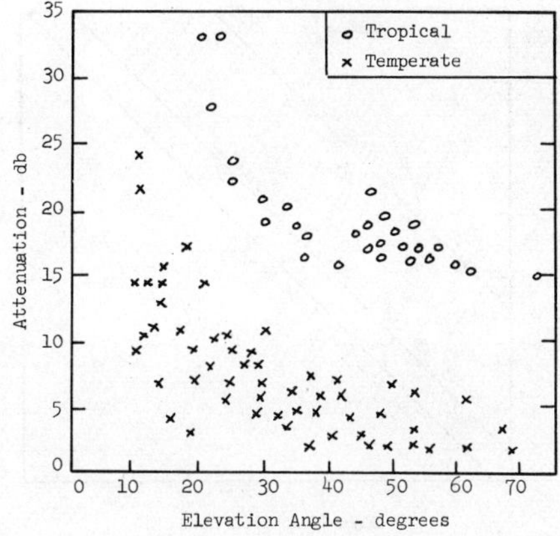

Fig. 5 Calculated rain and cloud attenuation vs elevation angle at 20 GHz for worldwide ground stations.

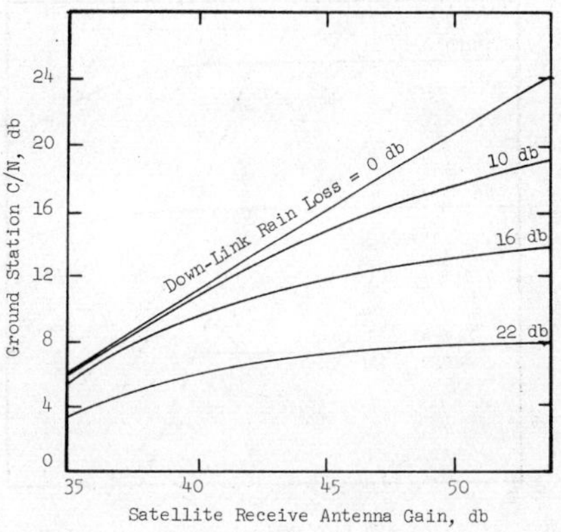

Fig. 7 Link performance vs satellite receive antenna gain for 40-db up-link rain loss.

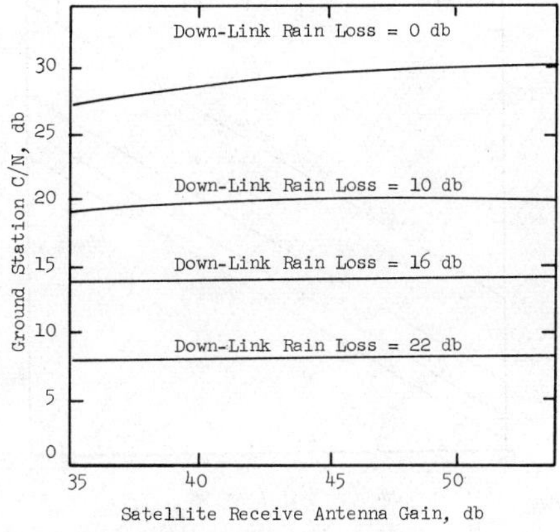

Fig. 6 Link performance vs satellite receive antenna gain for 16-db up-link rain loss.

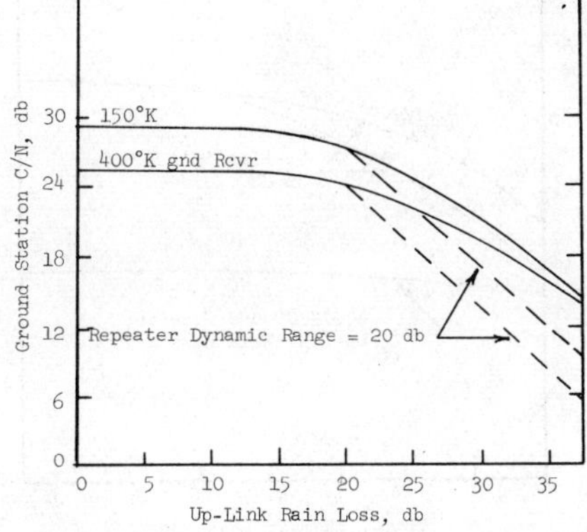

Fig. 8 Link performance vs up-link rain loss for 5-db down-link rain loss.

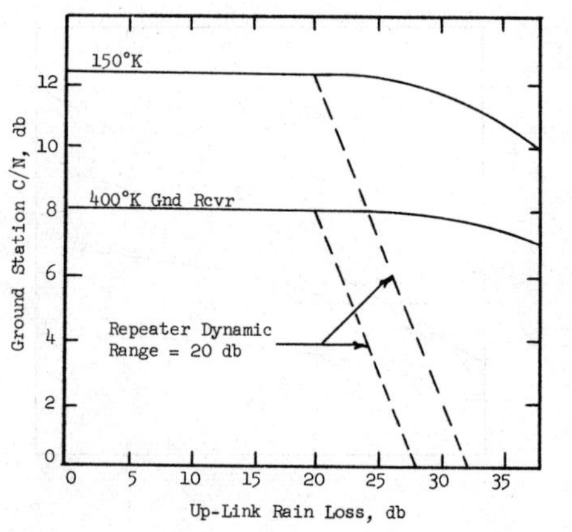

Fig. 9 Link performance vs up-link rain loss for 22-db down-link rain loss.

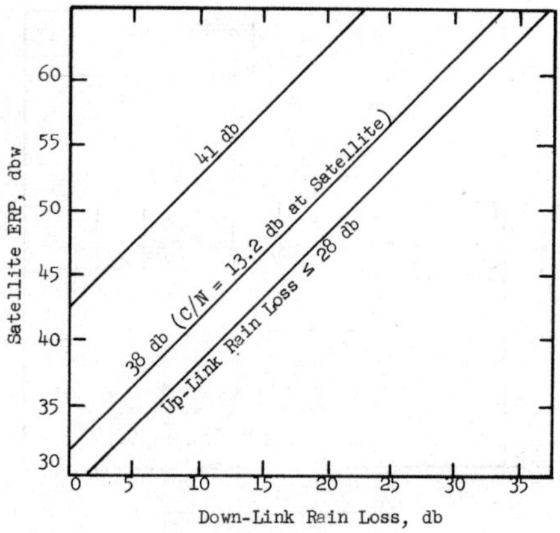

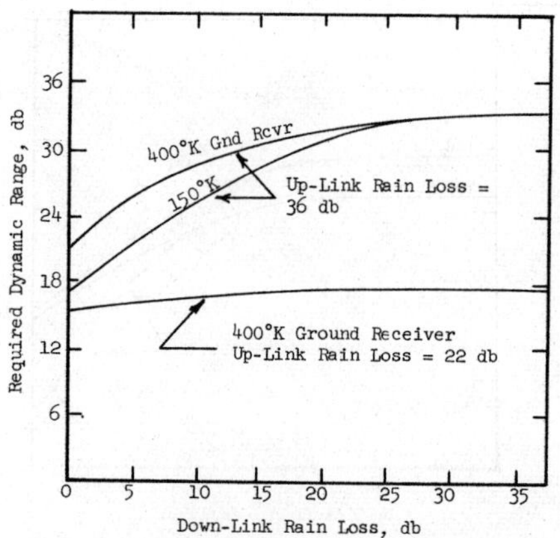

Fig. 10 Repeater dynamic range requirements as a function of maximum down-link rain loss.

Fig. 11 Repeater ERP requirements vs down-link rain loss.

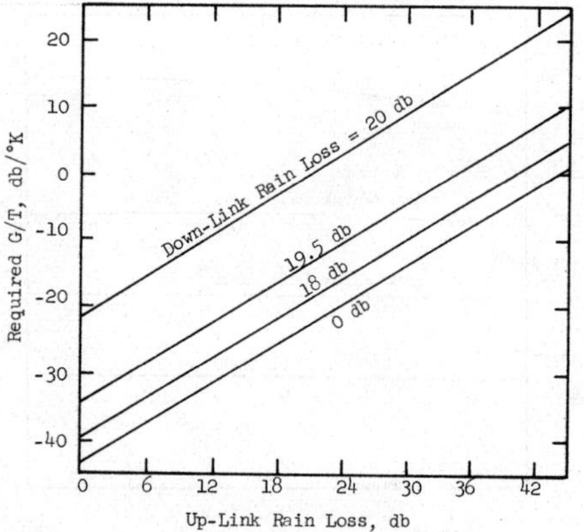

Fig. 12 Repeater G/T requirements vs up-link rain loss.

MEASUREMENT OF PRECIPITATION SCATTER EFFECTS ON PROPAGATION AT 6, 12, AND 18 GHz

Anthony Buige* and J.L. Levatich/

COMSAT Laboratories, Clarksburg, Md.

Abstract

New or expanded satellite systems must be compatible with existing terrestrial systems when both operate in the same frequency bands. COMSAT is measuring interference that might result from scattering by hydrometeors which are present in the common volume formed by the intersections or near-intersections of the main antenna beams of such systems. In March 1969, measurements were begun at 6 GHz, using the Etam, W. Va., satellite communications Earth station as the transmitting terminal and two 6-ft parabolic antennas as the receiving terminals. Signal statistics from this system have been obtained and correlated with predicted values based on rain-rate statistics. A third 6-GHz terminal is planned to investigate the angle dependence of the scattering and the effects of near-main beam intersections. Plans also call for similar measurements to be made at 12 and 18 GHz to obtain statistical data at these frequencies as an aid in determining their suitability for sharing.

I. Introduction

The possibility of interference between a satellite communications Earth station and a terrestrial radio-relay terminal is of concern to COMSAT. Interference can be caused by a number of phenomena, such as ducting, knife-edge diffraction, tropospheric scatter, and scattering by hydrometeors present in the common volume formed by the intersection of the antenna beams of the two systems. It

Presented as Paper 70-499 at the AIAA 3rd Communications Satellite Systems Conference, Los Angeles, Calif., April 6-8, 1970. This paper is based upon work performed in COMSAT Laboratories under Corporate sponsorship.

*Member of the Technical Staff, Propagation Branch.

/Propagation Branch Manager.

is often possible to satisfy the coordination criteria for most modes of propagation, but precipitation scatter can make site selection and coordination more difficult. The criteria utilized in avoiding precipitation-scatter interference are primarily based on theoretical studies since very little empirical data are available. The expected increase in the number of both communications satellite and terrestrial radio-relay systems in the next decade necessitates collection of these data.

COMSAT initiated a precipitation-scatter monitoring program to obtain data for the evaluation of the criteria required for sharing the same frequency bands with terrestrial radio-relay terminals. An initial experiment was conducted in the summer of 1968 in Rumford, Me., near the Andover satellite communications Earth station. It was a preliminary experiment designed to test some of the basic theory and to aid in the design of a more permanent system. Based on the results of this initial experiment, a 6-GHz experiment, simulating an operating system, was begun in March 1969. This paper presents the results obtained from this second experiment, discusses plans for enlarging the present experiment to evaluate other aspects of the scatter mechanism, and outlines a similar experiment which will be performed at 12 and 18 GHz.

II. Basic Assumptions and Equations

Theoretical treatment of precipitation scatter is based on the Mie theory of the scattering of a plane wave by a sphere. Simplifying assumptions can be introduced which show that in most cases the scattering can be considered to be Rayleigh in nature.[1] Those same assumptions are made here in the case of scattering at 6 GHz. A slightly different model is employed in the case of the 12- and 18-GHz problem, since the effects of attenuation as well as scattering must be considered. Further, at 6 GHz, the problems of both intersecting and nonintersecting beams are considered, while at the higher frequencies only the intersecting case is considered.

6-GHz Scattering

When the main antenna beams of the two systems intersect, the power received by the monitoring terminal can be expressed as

$$P_r = P_t \int \left(\frac{G_r}{4\pi d_r^{\,2}}\right) \left(\frac{\eta}{4\pi d_t^{\,2}}\right) \left(\frac{G_t\lambda^2}{4\pi}\right) dv \qquad (1)$$

where P_t is the transmitted power; G_t and G_r are the gains of the transmitting and receiving station antennas respectively; d_t and d_r are the distances from the transmitting and receiving stations to the common volume; dv is an element of the common volume; λ is the wavelength; and η is the scattering cross section per unit volume.

The transmitting station is assumed to have an antenna considerably larger than that of the receiving station, with the beam pointing at angles of 5° or more above the horizon and with the volume of intersection defined as[2]

$$\int dv = \frac{(\pi/4)\ (\Omega_t d_t)^2\ (\Omega_r d_r)}{\sin\Theta} \qquad (2)$$

where Ω is the antenna beam width in radians, and Θ is the angle between the antenna beams.

For Rayleigh-scattering, where the scatters are small water droplets, the scattering cross section/unit volume η generally is expressed as

$$\eta = Z/(3.6 \times 10^{15}\lambda^4) \text{ per meter} \qquad (3)$$

where Z is the equivalent radar reflectivity, expressed in mm^6/m^3.

By use of Eq. (3) and the relationship, $G\Omega^2 = \pi^2$, Eq. (1) reduces to

$$P_r = P_t Z\sqrt{G_r}/(\pi d_r \lambda^2 10^{17} \sin\Theta) \qquad (4)$$

In the case of nonintersecting beams, Eq. (1) must be modified to enable the integration along each of the two beams where the gains are the highest. By using the value for η given in Eq. (3) and defining the elemental volume as

$$dv = (\pi/4)\ (b_t\Omega_t)^2\ ds \qquad (5)$$

Eq. (1) can be reformulated as

$$P_r = \frac{P_t}{10^{18}\lambda^2} \left[\int_t \frac{G_r Z_t}{d_r^{\,2}} \, ds + \int_r \frac{G_t Z_r}{d_t^{\,2}} \, ds' \right] \qquad (6)$$

This case has been treated for widespread precipitation, where the gains of the antennas are assumed to follow a law, such as $G = 0.3/\alpha^2$ (where α is the off-beam angle in radians), and has been expressed as

$$P_r = P_t(Z_t + Z_r)/[(y \sin \Theta) 10^{18}\lambda^2] \qquad (7)$$

where y is the perpendicular distance between the two beams.

In computing the expected values for the experiment, the radar reflectivity factor used was[3]

$$Z = 200\ R^{1.6} \qquad (8)$$

where R is the surface rainfall rate in mm/hr.

Scattering at 12 and 18 GHz

The consideration of the effects of precipitation scatter at frequencies above 10 GHz must include consideration of attenuation caused by precipitation, especially for low-elevation paths. Two extreme-case weather models can be used to establish the limiting conditions of the problem. These models are: 1) precipitation in the common volume only (for all rainfall rates) and 2) precipitation along the entire path for rain rates of 1 mm/hr; 10 mm/hr (up to 50-km path length) and 100 mm/hr in the common volume, with 10-mm/hr rain rate up to 50 km over the path.

The worst interference resulting from precipitation scattering is expected to come from convective storms, where the high-intensity rainfall is concentrated over relatively small areas. In this case, a large amount of energy is essentially scattered isotropically from the precipitation, while little attenuation takes place outside the well-defined convective cell. Less interference is expected from heavy precipitation, spread over a large area, where the signal is attenuated as well as scattered.

Obviously, these extremes are only the limiting values to be considered, since neither of these oversimplified models will be encountered exactly as we have described. These models can be used, however, to establish the limiting conditions for the problem.

The basic geometry for the scattering problem is shown in Fig. 1. Distances and angles defined in the figure are not further defined.

The flux density P_σ, illuminating the common volume, can be approximated by[4]

$$P_\sigma = (P_t/A_t)e^{-\beta_{ex}L_1} \tag{9}$$

where P_t is the power transmitted, in watts; A_t is the effective aperture of transmitting antenna, in km^2; and β_{ex} is the extinction coefficient per km.

The total power scattered from the common volume can be obtained from

$$P_{sc} = \int_0^\ell P_\sigma \, e^{-\beta_{ex}x} \, \beta_{sc} A_t \, dx \tag{10}$$

where β_{sc} is the scattering coefficient per km.

$$P_{sc} = P_\sigma(\beta_{sc}/\beta_{ex})(1 - e^{-\beta_{ex}\ell}) \, A_t \tag{11}$$

The received interfering power P_R is

$$P_R = (P_{sc}e^{-\beta_{ex}L_2})/(4\pi d_{xt}^{\ 2}) \, A_R \tag{12}$$

where A_R is the effective area of receiving antenna, in km^2. Equations (9, 11, and 12) can now be combined, resulting in

$$P_R = \frac{P_t A_r}{4\pi \, d_{xt}^{\ 2}} \, \frac{\beta_{sc}}{\beta_{ex}} \, e^{-\beta_{ex}L_1} \, e^{-\beta_{ex}L_2} \left(1 - e^{-\beta_{ex}\ell}\right) \tag{13}$$

From Eq. (13), it can be seen that the worst case of interference occurs when the precipitation is confined to the common volume, resulting in $L_1 = L_2 = 0$. The least interference will be present when both L_1 and L_2 are large, resulting in large attenuation.

The scattering and extinction coefficients β_{sc} and β_{ex} are a function of the rainfall rate and operating frequency.[5] Table 1 shows the extinction coefficiencts, β_{ex}, and albedo, β_{sc}/β_{ex}, for 12 GHz and 18 GHz, for three rainfall rates.

Table 1 Extinction coefficients and albedo

R mm/hr	12 GHz		18 GHz	
	β_{ex} km^{-1}	Albedo	β_{ex} km^{-1}	Albedo
1	0.0068	0.054	0.015	0.10
10	0.097	0.11	0.20	0.20
100	1.1	0.19	2.0	0.34

In the example, $A_R = 1.57 \times 10^{-6}$ km^2 and $P_t = 1$ w were used. To solve Eq. (13), a realistic range of d_{xt} values is needed. The following assumptions will be used to arrive at a range of distances: a) the elevation angle at an Earth station (transmitting) terminal will be 20° or greater; b) the elevation angle at a terrestrial receiving terminal will be less than 5°; and c) no precipitation will occur above 12 km.

The results of these theoretical calculations are shown in Fig. 2 for 12 GHz and Fig. 3 for 18 GHz. The solid lines are for the worst case, with rain of given intensity in the common volume only, and the dashed lines are for rain over the entire path at 1-mm/hr rate, up to 50 km in extent over the path at 10 mm/hr, and 100-mm/hr in the common volume, with 10-mm/hr rain rate up to 50 km over the path.

III. Measurements at 6 GHz

The 6-GHz measurement system was initiated in March 1969. The site selected for monitoring purposes was the roof of a high school in Kingwood, W. Va., which is located about 13 miles from the Etam Earth station. Figure 4 is the great circle path profile between the two sites, showing the excellent site shielding. Both computations and experience have shown that the path loss for propagation modes other than scattering, such as tropospheric and knife-edge diffraction, was greater than 240 db 99.99% of the time.

Figure 5 shows the common-volume intersection geometry. The Etam Earth station is operating with an INTELSAT III satellite and is pointing along an azimuth of 119° and an elevation angle of 23°. Common volumes were selected to be

2000 and 6000 ft above the local terrain, giving heights of 4000 and 8000 ft above mean sea level. The intersection angles were as near perpendicular to the beam as could be obtained under the constraints involved.

The system block diagram is shown in Fig. 6. The transmitting system consists of a 6.460-GHz crystal-controlled oscillator, amplitude-modulated at a 1-kHz rate, whose signal is inserted into the Earth station transmitter at a level sufficient to produce 10 w at the antenna feed. With the antenna gain, this yields a transmitted e.i.r.p. of 103 dbm. The receive antennas consist of two 6-ft parabolic antennas, each having an approximate gain of 39 db. The system sensitivity threshold was set at -120 dbm for data reduction purposes.

Figure 7 gives the cumulative distributions of the scatter signals received for the two channels to date. About 98% of the time the signal was below threshold, and better than 99.99% of the time the signal level was less than -80 dbm. These distributions are based on 5-min. averages of the data obtained. The maximum signal level received on either channel was -76 dbm. This was a single isolated event which corresponds to an instantaneous rain rate of better than 200 mm/hr if one assumes the relationship $Z = 200 R^{1.6}$.

Correlation with Observed Rain Rates

To perform a correlation analysis between observed rain rates and measured signal levels, Weather Bureau statistics were obtained from the two nearest monitoring stations. Elkins, W. Va., is about 20 miles southeast of Etam, while Terra Alta is about 10 miles east of Kingwood. Figure 8 illustrates the cumulative distributions for rain at these sites, based on hourly averages. These distributions are considered to be representative rain rates for the area.

The rain distributions shown in Fig. 8 were used to predict the expected scatter distributions plotted in Fig. 9. These predicted values were compared with the channel II data received during the monitoring period. There is relatively good correlation between the distributions 99.95% of the time. The crossovers are attributable to the fact that instantaneous rain-rate data is not available.

Extended Experiment

Two areas of interest which need further attention are the determination of the statistics for near-intersections and the testing of the validity of the assumption that the scattering is essentially Rayleigh. A limited amount of data regarding near-intersections were obtained during the first three months of experiment operation. The distributions for the two off-beam angles observed are given in Fig. 10. It can be seen that the distributions fall off quite rapidly as the off-beam angle increases.

To obtain further information on this phenomenon and to investigate the angle dependence of the scattering mechanism itself, COMSAT is implementing a third receiving system. It is expected that this system will be installed early in 1970, in time for the spring and summer rains.

IV. Planned 12- and 18- GHz Measurements

COMSAT is purchasing the components for transmitting and receiving terminals at 12 and 18 GHz. Ideally, the system should be located in an area which experiences considerable uniformly structured, continuous rainfall over a relatively large area, as well as convective rain that is normally associated with thunderstorms. The present location in West Virginia meets this criteria; thus, current plans call for the 12- and 18-GHz experiment to be colocated with the 6-GHz system. This location has the additional feature that cross-correlation between the signals received at 6, 12, and 18 GHz can be performed and the frequency dependence of the scattering mechanism can be investigated.

The common-volume geometry will be essentially identical to that shown in Figs. 1 and 5, with the added flexibility that the azimuth and elevation angle of the transmitting sites will not be dependent upon the position of the satellite. With this flexibility, the angle dependence of the scattering mechanism can be investigated with a single receiving terminal.

A system block diagram is shown in Fig. 11. The transmitters will consist of solid-state, phase-locked oscillators, that will be square-wave modulated at a 1-kHz rate and that will be used to drive a TWT amplifier. The 10-w output power of the TWT will then be fed to a 10-ft parabolic dish having a gain of about 47 db at 12 GHz and 51 db at 18 GHz. The receiving terminal for each system will consist of a 10-ft parabola identical to the transmitting antenna, a tunnel-diode amplifier, a mixer-preamplifier, an i.f. receiver-detector, and a recorder. The terminals are being designed for unmanned operation with periodic visits for maintenance and pick up of recorded data planned.

Summary

Possible interference between communications satellite Earth stations and terrestrial radio-relay systems is being investigated by COMSAT. The specific mode of interference considered is scattering by hydrometeors within and near the common volume formed by the intersection of the main antenna beams of the two systems. Theoretical calculations indicate that the worst interference will be encountered when the common volume is filled by high-intensity convective precipitation.

Measurements at 6 GHz near a COMSAT Earth station indicate that 98% of the time the signal was below the receiving system sensitivity and 99.99% of the time it was less than -80 dbm. Measurements are continuing at 6 GHz, with one more terminal planned. In addition, measurements at 12 and 18 GHz are planned for the same location, to be started in 1970.

References

[1]Dennis, A. S., "Precipitation Scatter as an Interference Source in Communication Systems," I.R.E. Convention Record, Pt. I, Antennas and Propagation, 1962, p. 145.

[2]Altman, F. J., "Precipitation Scatter Interference in the 4 and 6 GHz Bands," Proceedings of the 1968 IEEE Internal Communications Convention, submitted to the IEEE Transactions of Communications Technology Group, 1968.

[3]Atlas, D., "Model Atmospheres For Precipitation," Handbook of Geophysics and Space Environments, Sec. 5.2, 1965, pp. 5-6.

[4]Hult, J. L., et al., "The Technology Potentials For Satellite Spacing and Frequency Sharing," Memorandum RM-5785-NASA, Oct. 1968, The Rand Corporation, Santa Monica, Calif.

[5]"Precipitation Scattering Considerations at Frequencies above 10 GHz," CCIR [International Radio Consultative Committee, Geneva] Document V/117, June 16, 1969.

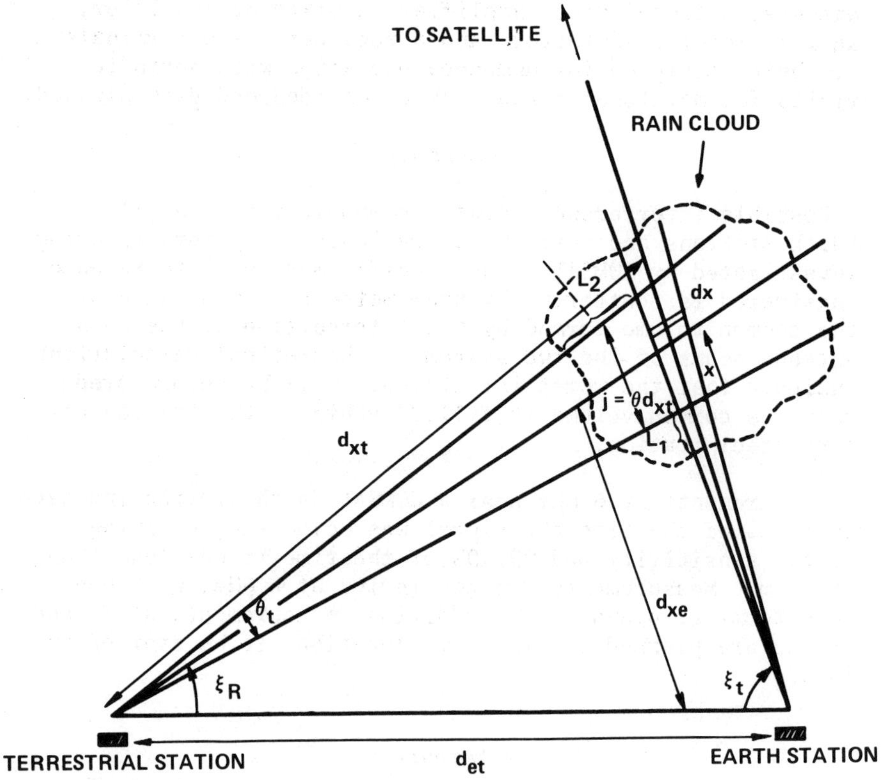

Fig. 1 Geometry of scattering problem.

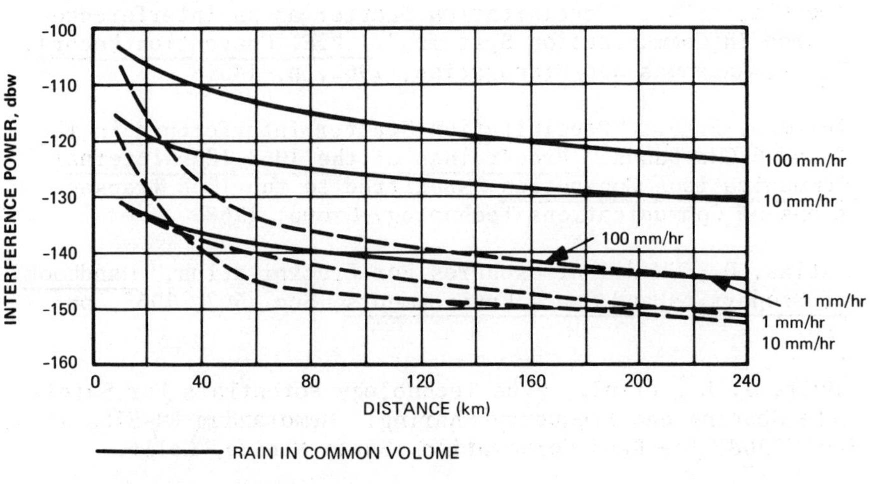

Fig. 2 Interference power vs distance for given rain rates at 12 GHz.

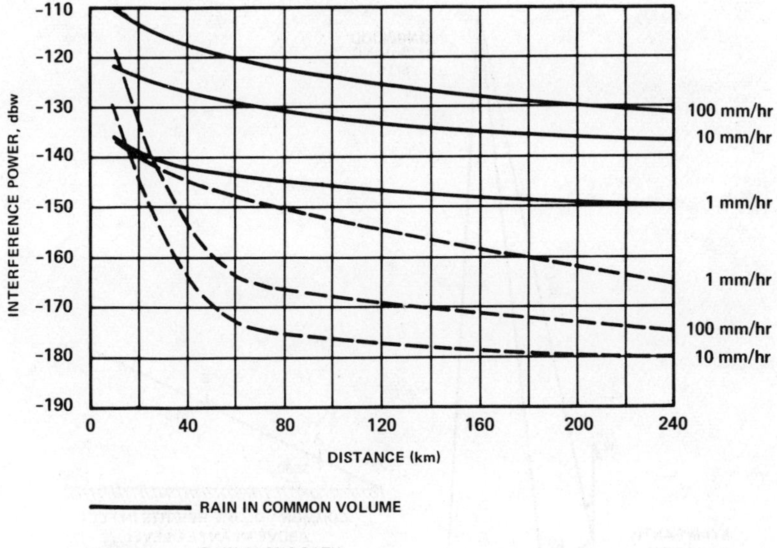

Fig. 3 Interference power vs distance for given rain rates at 18 GHz.

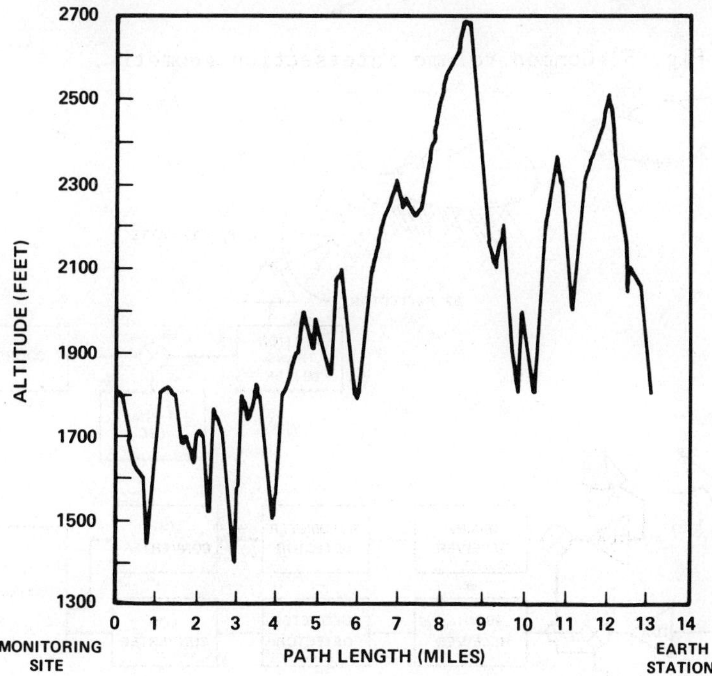

Fig. 4 Great circle path profile between the Kingwood monitoring site and the Etam Earth station.

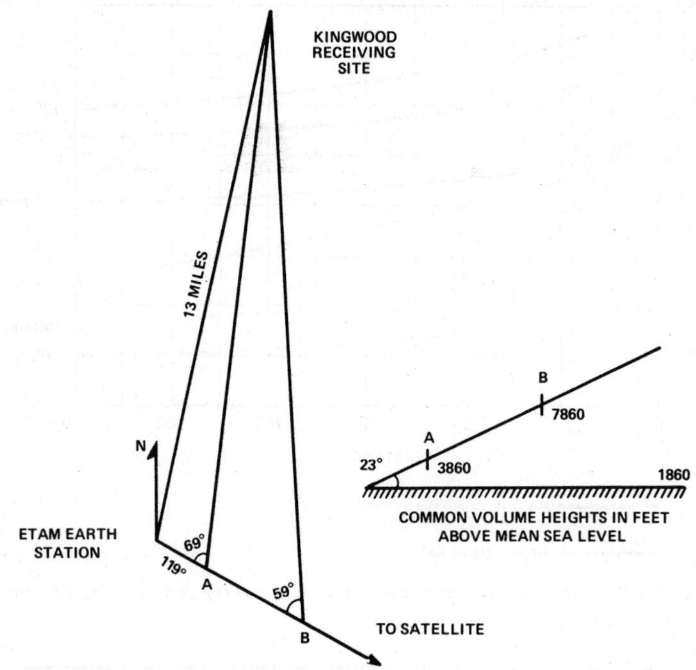

Fig. 5 Common volume intersection geometry.

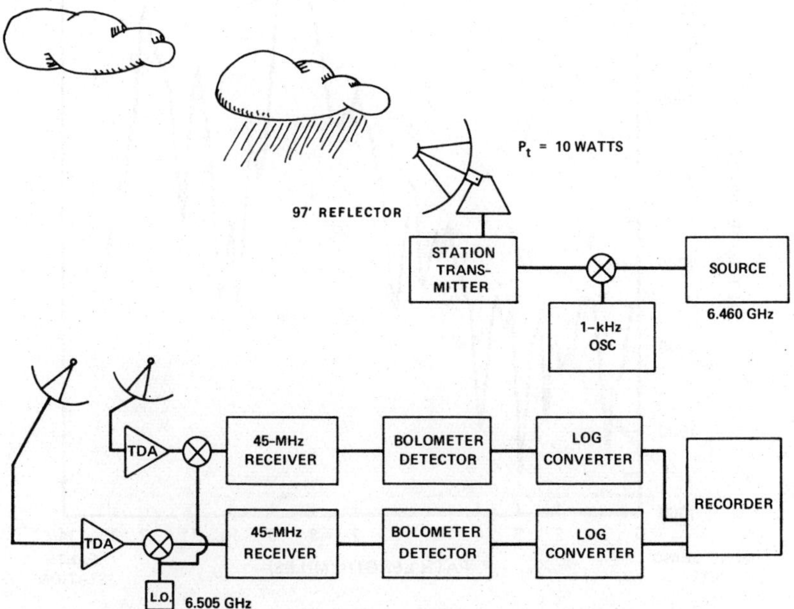

Fig. 6 Precipitation scatter system block diagram.

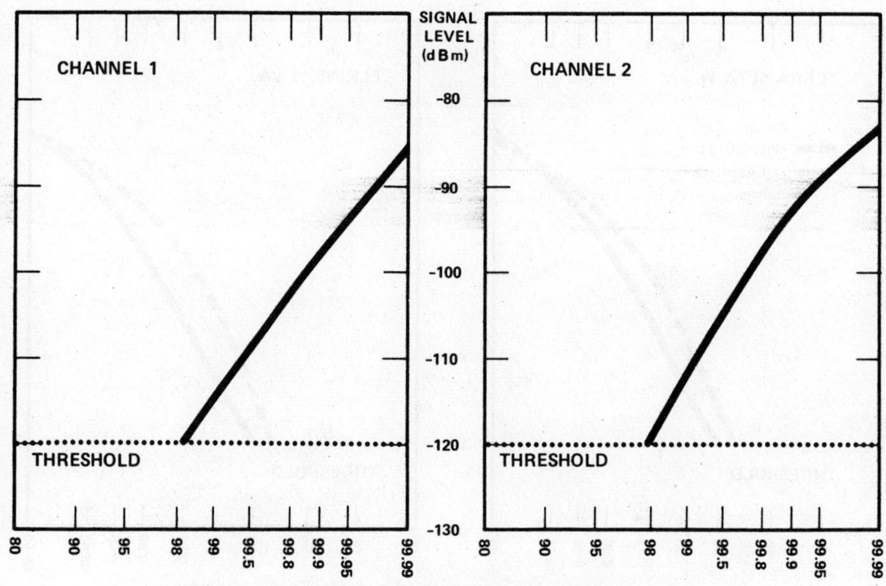

Fig. 7 Cumulative distributions of precipitation scatter signal levels received at the Kingwood monitoring site from March 21 to Oct. 31, 1969.

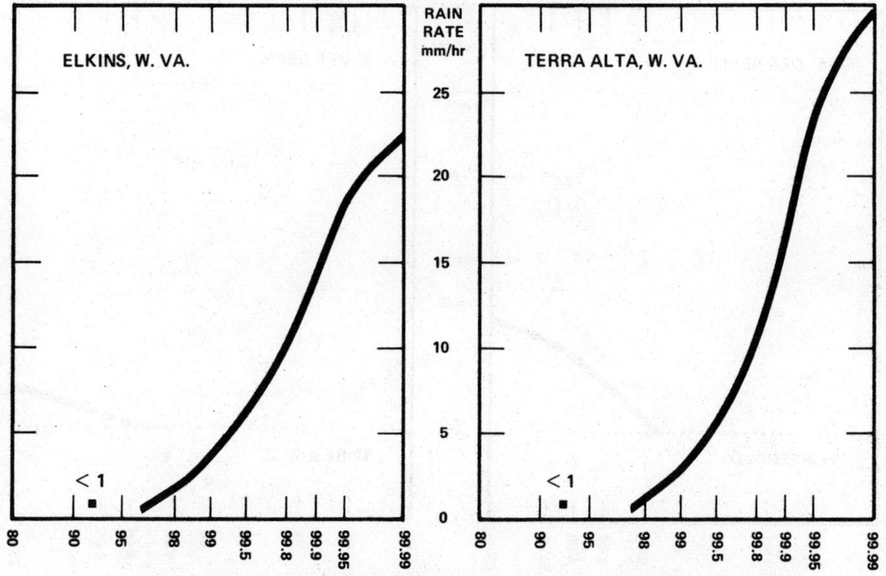

Fig. 8 Cumulative distributions of rain rates observed at Elkins and Terra Alta, W. Va., from March 1 through Sept. 30, 1969 (U.S. Weather Bureau statistics).

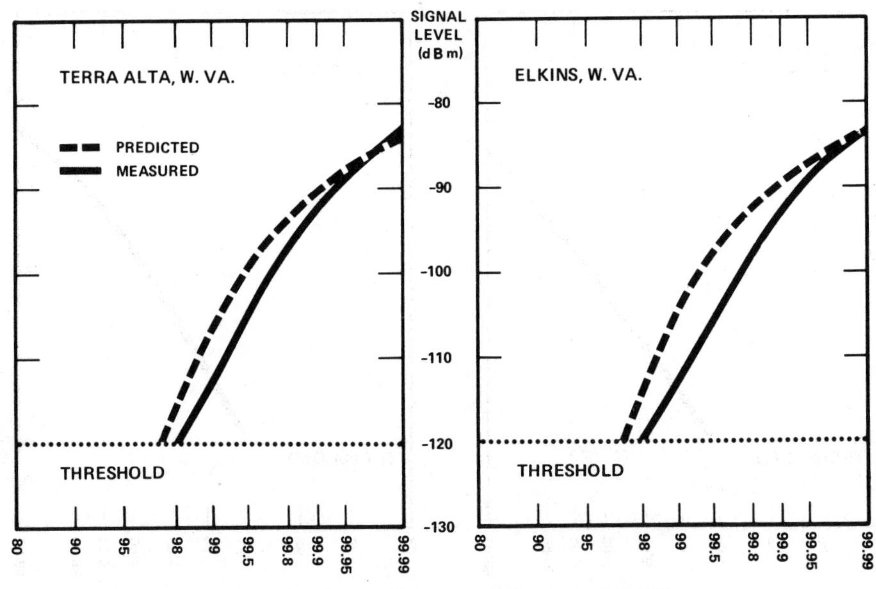

Fig. 9 Measured vs predicted cumulative distributions of received signal levels.

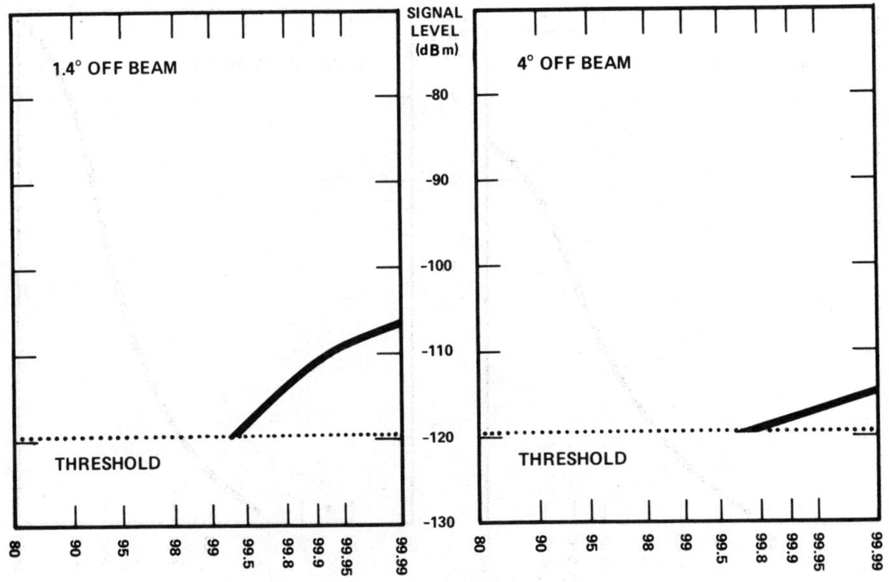

Fig. 10 Cumulative distribution of off-beam data obtained March 21 through June 10, 1969.

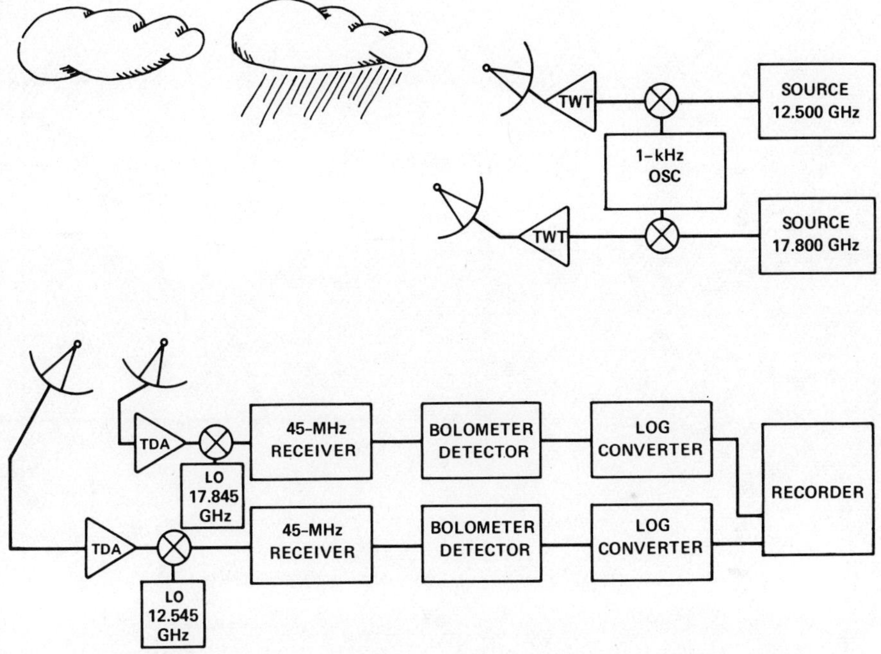

Fig. 11 12- and 18-GHz precipitation scatter system block diagram.

SPECTRUM FOR SPACE - PRESENT ALLOCATIONS

Richard G. Gould*

Office of Telecommunications Policy, Washington, D.C.

Abstract

Present allocations for space services and radio astronomy were made at the Extraordinary Administrative Radio Conference of 1963. This paper describes those allocations, the procedure by which international allocations are arrived at, and the further allocation of those frequencies to Government and non-Government users and for specific purposes within the United States. The paper lists those services which have been proposed but for which no frequency allocations have yet been made. A Table is included that lists current allocations in the United States for space services.

Introduction

Frequencies are allocated internationally at Administrative Radio Conferences convened under the auspices of the International Telecommunication Union (ITU). The agreements reached at these conferences comprise the Radio Regulations. Following ratification by a country, these Regulations have the force of a treaty to which each signatory is bound under international law.

Frequencies are allocated for specific services; for example, the communication-satellite service, the broadcasting service, and broadcasting-satellite service. At present, frequencies have been allocated to the first two services, but not to the third. Similarly, no frequencies have been allocated for Earth Sciences Satellites, and, in fact, such a service has not even been defined. Current allocations for space services in the United States are given in Table 1.

Presented at the AIAA 3rd Communications Satellite Systems Conference, Los Angeles, Calif., April 6-8, 1970 (not preprinted).

* Chief, Advanced Technology Division, Office of Telecommunications Policy. Now with the Federal Communications Commission. Any opinions expressed herein do not necessarily represent those of OTP or the FCC.

Table 1 Spectrum allocated for space services in the USA
(based on EARC, 1963, Geneva)

FREQUENCY (MHz) FROM	TO	BANDWIDTH (MHz)	SERVICE	SHARED WITH	COMMENTS*
2.495	2.505	0.010	RA	SF	U.S. 68,74
4.995	5.005	0.010	RA	SF	U.S. 68,74
9.995	10.005	0.010	RA,SR	SF	U.S. 74,83 ITU 215
14.990	15.010	0.020	RA	SF	U.S. 68,74
19.990	20.000	0.010	SR,RA	SF	U.S. 68,74
20.000	20.010	0.010	SR	SF	U.S. 84
24.990	25.010	0.020	RA	SF	U.S. 68,74
30.005	30.015	0.010	S,SR(s-e)	F,M	U.S. 94
38.000	38.160	0.160	RA	F,M,	U.S. 81
39.986	40.020	0.034	SR(s-e)	LM	U.S. 94
73.000	74.600	1.600	RA	--	U.S. 21,74

* U.S., followed by a number, refers to FCC footnotes to the Allocation Table. ITU followed by a number refers to footnotes to the International Radio Regulations. Only footnotes relating to space use are shown. Notes are given at the end of this Table.

Table 1 (continued)

FREQUENCY (MHz) FROM	TO	BANDWIDTH (MHz)	SERVICE	SHARED WITH	COMMENTS
117.975	123.075	5.100	AM(st)	AM	U.S. 85
123.575	136.000	12.425	AM(st)	F,M	U.S. 85
136.000	137.000	1.000	SR(tm & tk)	--	
137.000	138.000	1.000	MS,S(tm & tk)	--	
144.000	146.000	2.000	A(st)	A	ITU 284A
148.250 $\pm$ 15kHz		0.030	S(tc)	F,M	U.S. 86
149.900	150.050	0.150	RS	--	
154.200 $\pm$ 15kHz		0.030	S(tc)	LM	U.S. 86
399.900	400.050	0.150	RS	--	
400.050	401.000	0.950	SR(tm & tk)	MA	
401.000	402.000	1.000	S(tm)	MA	
404.000	406.000	2.000	RA	MA	
450.000 $\pm$ 0.25 MHz		0.500	S(tc)	A,F,M,LM	U.S. 87
608.000	614.000	6.000	RA(B after 1/1/74)	--	U.S. 88
1400	1427	27	RA	--	
1427	1429	2	S,E(tc)	F,M	
1525	1535	10	S(tm)	AM	
1535	1540	5	S(tm)	--	
1540	1660	120	AM(st),AR(st)	AR	Notes 1,2.
1660	1670	10	MS,RA	MA	U.S. 74,99,101
1690	1700	10	MS	MA	U.S. 99
1700	1710	10	SR(tm & tk)	--	
2110	2120	10	SR(tc)	F,M	U.S. 90
2290	2300	10	SR(tm & tk)	--	
2690	2700	10	RA	--	

Table 1 (continued)

FREQUENCY (MHz) FROM	TO	BANDWIDTH	SERVICE	SHARED WITH	COMMENTS
3700	4200	500	CS(s-e)	F	U.S. 91
4200	4400	200	AR(st)	AR	Note 1.
4990	5000	10	RA	--	
5000	5250	250	AM(st),AR(st)	AR	Notes 1,2.
5925	6425	500	CS(e-s)	F	U.S. 91
7250	7300	50	CS(s-e)	--	U.S. 91
7300	7750	450	CS(s-e)	F,M	U.S. 91,92 Note 3.
7900	7975	75	CS(e-s)	F,M	U.S. 91
7975	8025	50	CS(e-s)	--	U.S. 91
8025	8400	375	CS(e-s)	F,M	U.S. 91
8400	8500	100	SR	F,M	U.S. 62
10,680	10,700	20	RA	--	
14,300	14,400	100	RS	--	
15,250	15,350	100	SR	--	
15,350	15,400	50	RA	--	
15,400	15,700	300	AR	--	Notes 1,2.
19,300	19,400	100	RA	--	
31,300	31,500	200	RA	--	
31,500	31,800	300	SR	--	
34,400	34,500	100	MS	R	Note 4
88,000	90,000	2000	RA	--	
90,000 and above		Amateur, Experimental & Government			

ABBREVIATIONS USED IN TABLE

A	Amateur	RA	Radio Astronomy
AM	Aeronautical Mobile	RS	Radionavigation Satellite
AR	Aeronautical Radionavigation	S	Space
B	Broadcasting	SF	Standard Frequency
CS	Communication Satellite	SR	Space Research
E	Earth	e-s	earth-to-space
F	Fixed	s-e	space-to-earth
LM	Land Mobile	st	using space techniques
M	Mobile	tc	telecommand
MA	Meteorological Aids	tk	tracking
MS	Meteorological Satellite	tm	telemetry
R	Radiolocation		

NOTES

1. reserved for use and development of airborne electronic aids to air navigation and any directly associated ground-based or satellite borne facilities (ITU 352A)

2. . . . for use & development of systems using space communication techniques . . . (ITU 352B)

3. also 100 MHz for MS(tm & tk)

4. for cloud detection radars (ITU 412D)

However, an Administration may use any frequency so long as its use does not cause interference to stations operating in accord with the Radio Regulations. This provision is used primarily for experimental operations.

The last conference to deal with allocations for space services was the Extraordinary Administrative Radio Conference (EARC) held in 1963. The next one will be the World Administrative Radio Conference for Space Telecommunications (WARC-ST) scheduled for 1971. Thus, eight years will have elapsed since the previous space conference.

Within the U.S., nongovernment frequency usage is governed by Part 2 of the FCC Rules and Regulations while government use is developed by the Interdepartment Radio Advisory Committee (IRAC). Not all frequencies allocated internationally for a given service are included in our national allocations for that service. Moreover, U.S. national allocations provide that some frequencies may be used only by government users, others only by nongovernment, and some by both.

Specifically, within the U.S., the Communication-Satellite Service has 500 MHz (down) between 3700 and 4200 MHz, 500 MHz (up) between 5925 and 6425 MHz, 500 MHz (down) between 7250 and 7750 MHz, and 500 MHz (up) between 7900 and 8400 MHz. Although the ultimate disposition of this band as between government and nongovernment is deferred, the first two are now being used by nongovernment and the last two by government users.

Regarding aeronautical services, the bands 118-132 MHz and 132-136 MHz may employ space techniques for the aeronautical mobile service (subject "to coordination between administrations concerned... and affected."), but such systems are limited to communications as contrasted with radiodetermination, navigation, and location. Thus, a satellite for aeronautical communication in those bands could not supply navigation information except on an experimental noninterfering basis.

The 50-KHz band, 149.9-150.05 MHz, is now allocated for the Radionavigation Satellite Service and thus can be used for aeronautical radionavigation via satellite, but the difference in frequency from the aeronautical mobile band previously mentioned, 118-136 MHz, would complicate the system: a separate antenna for each band probably would be required on both the satellite and on the aircraft.

Four other bands, 1540-1660 MHz, 4200-4400 MHz, 5000-5250 MHz, and 15.4-15.7 GHz, are now allocated for aeronautical radionavigation and permit "the use and development of . . . satellite-borne facilities." All but the band 4200-4400 MHz are also allocated to the Aeronautical Mobile Service for the use of systems using space communications techniques, "subject to agreement of administrations concerned and affected."

No frequencies are now allocated for the Maritime Mobile Service using space techniques.

Bands also are allocated: to the Space Research Service for telemetry, telecommand, and tracking of manned and unmanned space missions; to the Meteorological Satellite Service; and to the Radio Astronomy Service.

Present Usage of Frequency Allocations for Space Services

A detailed listing of satellites that use, and have used, the spectrum is too voluminous to be presented here. What follows is a brief summary of these uses.

COMSAT, as System Manager for INTELSAT, has arranged for the launch of the nongovernment, commercial satellites now in orbit. These INTELSAT I, II, and III satellites use the 500-MHz wide bands 3.7-4.2 GHz down, and 5925-6425 GHz up, between about 40 Earth stations. It is estimated that this number of Earth stations will double by 1972. Eight INTELSAT IV's will be procured. Allowing for two failures and two spares, four will be operational over the Atlantic, Pacific, and Indian Oceans by mid-1974.

Government communication-satellite bands presently are used by about 26 IDCS satellites between about 6 Earth stations and by one TACSAT satellite (X-band plus UHF) and its predecessor LES satellites. Other users of these bands include two planned United Kingdom and two planned NATO satellites. A Phase II of the IDCS is also planned.

In the meteorological band around 1700 MHz, satellites include Nimbus 1, 2, 3, 4, E(1972) and F(1973). Satellites in the TIROS series and in the follow-on Tiros Operational Satellites (TOS) series (ESSA 1-9) also use this band. Forthcoming ERTS and SMS satellites also will employ 400 MHz for up-link transmissions from sensors and 460 MHz down. Navigation satellites in the 150-MHz and 400-MHz region have included Transit 1B.

Satellites using the Space Research bands have included OV 1-17, 18, 19, 17A; the biosatellites, APOLLO, EXPLORER, and MARINER missions, and the OSO, OGO satellites and IMP. Practically all bands allocated to this service have been employed, including 136 and 400 MHz, and, from time to time, nonallocated bands on a noninterfering basis. APOLLO spacecraft, for example, have used, and will use, the unified S-band at 2 GHz allocated for space research and frequencies below it outside the allocated band to satisfy their requirements.

Abroad, Canada flew the ISIS topside sounder that transmitted and received frequencies swept from 20 KHz to 12 MHz, as well as at 136 and 400 MHz. Russian space research and manned flights have used frequencies around 20 MHz and 960 MHz primarily.

The experimental ATS satellites have used a wide range of frequencies, including 4 and 6 GHz, 135 and 149 MHz, and 1540-1660 MHz for "Aeronautical Mobile using space techniques," as well as frequencies around 15 and 31 GHz. Typically, experimental spacecraft will use frequencies in the bands allocated for the service that a follow-on operational system will employ.

Regarding the radio-astronomy service, there are approximately 40 observatories active in the United States at present, with an additional 30 or so elsewhere in the world. These observe from time to time on all of the bands presently allocated, as well as on frequencies allocated for other services. For example, observations are being conducted at 22.23 GHz (a spectral line of water), although that frequency is allocated to another service. Radio astronomers would like to observe the formaldehyde line at 14.48 GHz, but this too is not allocated for radio-astronomy. Often, frequencies within their bands are unusable because of harmonic and other unintended interference.

We have already felt the pinch of frequency allocations in the Space Research Service. It is expected that the environmental systems, such as the METSATS and the Earth Resources systems, will soon place increased demands on the spectrum.

TECHNICAL FACTORS AND CRITERIA AFFECTING GEOSTATIONARY ORBIT UTILIZATION

D. M. Jansky* Executive Office of the President, Washington, D. C.

and

M. C. Jeruchim+ General Electric Company, Philadelphia, Pennsylvania

Abstract

A technical basis for sharing the geostationary orbit may be determined by 1) the variable factors of communication system design and 2) the criteria applied for utilization. This paper investigates the connection between the two, and forms some conclusions concerning the relationship between satellite system implementation and orbit utilization.

I. Introduction

The successful implementation of communication satellite systems has generated a sense of rising expectations in many countries that geostationary communication satellites offer a technology which may permit economical, efficient telecommunications for both inter- and intracountry information transfer requirements. Ready benefits are available as a result of the two prime related characteristics of geostationary communication satellites: a) a relatively inexpensive way to establish large channel capacities quickly; and b) accessibility between one point and many other points on the Earth.

Presented as Paper 70-441 at the AIAA 3rd Communications Satellite Systems Conference, Los Angeles, California, April 6-8, 1970.

*Chief, Spectrum Development Division, Office of Telecommunications Policy

+Engineer, Advanced Communications and Electronic Systems, Space Systems Organization

The actual number of satellites that can operate in the geostationary orbit in a particular frequency band is limited. Therefore, the fundamental question is to determine how close one (arbitrary) satellite may be placed to another without impairing its own or the other's effective operation. These considerations have prompted worldwide recognition of the need to study the various factors which influence the effective use of the geostationary orbit.

In addressing this problem consideration is first given to some of the many technical factors involved; curves are presented (for homogenous FDM/FM telephony systems) that permit tradeoffs to be made between orbital use measures and various system parameters. Second, the technical criteria that could be used to define "optimum" use of the orbit are considered. These criteria reflect different types of needs and seem representative of the standards on which sharing may be based. The results are then analyzed with a view toward a) isolating those factors which are most critical; and b) developing the guides to be applied to communication satellite systems sharing the limited orbit and spectrum.

II. Basic analytic model

The fundamental analytic problem is to determine how close several satellites may be placed to one another without impairing their own or the others' effective operation. There are many factors having a wide range of variability that influence the sharing possibilities: a) modulation method, b) modulating signal type, c) modulating signal baseband bandwidth, d) modulation index, e) RF bandwidth, f) satellite e.i.r.p., g) number of carriers per transponder, h) carrier frequencies, i) frequency spacing between carriers on different satellites, j) satellite antenna beamwidth, k) satellite drift and pointing accuracy, l) number of antenna beams per satellite m) required SNR or noise budget at the receiver output, n) Earth-station figure of merit (G/T), o) Earth-station antenna pointing accuracy, and p) propagation conditions.

Various combinations of these items lead to a large number of interesting possibilities. Consequently, it is advantageous to consider a simplified situation, which nevertheless retains many of the essential features of the problem. Accordingly, the basic system model considered is defined by the following characteristics: a) A full ring of equally spaced satellites are in geostationary orbit; b) All satellites have single-beam, earth-coverage antennas with identical polarization; c) All satellites have identical e.i.r.p.; d) All satellites transmit single-carrier FDM/FM multichannel telephony with the same baseband bandwidth and modulation index, or a single FM-TV channel with the same modulation index; e) All Earth stations are identical with antenna sidelobes per CCIR Report 391; f) A typical Earth-station is located at 45° latitude and 0° relative longitude with respect to its wanted satellite; g) Clear weather conditions pertain; and h) The down-link only is considered.

The foregoing model will be referred to, for brevity, as a "homogeneous system." Although this model may appear artificial, it reduces a situation of great complexity to a tractable one and it is sufficiently unconstrained to show the general trends in orbit/spectrum utilization as a function of changes in system parameters. In addition, some of the assumptions are not fundamentally constraining and can be changed with relative ease, at least conceptually.

Tradeoff Curves for the Homogeneous Telephony System

The performance of a multichannel telephony system will be specified by the total noise N_T in the top telephone channel at a point of zero relative level. The total noise has two constituents,‡ namely thermal noise N_t and interference noise N_i. Symbolically,

$$N_T = N_t + N_i \tag{1}$$

For fixed N_T there is a tradeoff between N_t and N_i. The former is a function only of the link parameters, i.e., modulation index, e.i.r.p., Earth station G/T, etc, whereas the latter is a function (among other things) of satellite spacing. This equation embodies the tradeoff between system cost and orbit utilization.

The cost factors are reflected in the thermal noise performance, which is given by

$$10 \log N_t = 64\text{-CNR-}10 \log \{M^2_{rms}(\sqrt{\Lambda}M_{rms} + 1)\} \tag{2}$$

where CNR = carrier-to-noise ratio (db), M_{rms} = rms modulation index of the multichannel baseband, Λ = peak-to-average power ratio of the multichannel baseband, and N_t = thermal noise, at a point of zero relative level, in picowatts psophometrically weighted (pW0p).

‡It is recognized that in practice due allowance must be given to other noise sources such as delay distortion, multiplex equipment, etc.

Equation (2) is based further on assuming a) 2.5 db psophometric weighting, b) 4 db preemphasis advantage, c) $n \geq 240$ channels.

Finally, N_t can be related to the link parameters through the CNR as follows:

$$CNR = e.i.r.p. - 10 \log W + 10 \log (A/T) + 66 \quad (3)$$

where W = RF bandwidth, Hz, T = earth station noise temperature, °K, and A = earth station antenna effective receiving area, meters^2.

Next, the interference noise depends on the relative frequencies of the wanted and unwanted carriers. It is convenient to treat separately the case of co-channel operation and the case where the various carriers may be offset from one another. For the co-channel case, it can be shown that[1]

$$10 \log N_i = 71.2 - CXR - 10 \log K \quad (4)$$

where CXR = carrier-to-interference ratio, (db), K = a constant depending on M_{rms}, and N_i = interference noise at baseband, (pW0p).

Equation (4) also implicitly depends on assumptions a and c above. Since CXR represents the carrier-to-total interference, Eq. (4) further assumes that the effect of multiple interferers is additive at baseband; this assumption seems well-founded for high CXR. The factor K may be thought of as an FM improvement factor with respect to interference. The numerical evaluation of K is, in general, quite complex. This paper uses the results of reference 2 which, for the co-channel case, and including CCIR preemphasis, may be fitted to the following equation

$$K \simeq 1 + 9.5 M_{rms}^3 \quad (5)$$

It remains to connect the CXR to satellite spacing. To a good approximation,

$$\frac{X}{C} = \frac{1}{G_m} \Sigma G(i\Delta\Theta) \quad (6)$$

where (X/C) = interference-to-carrier ratio, i.e., CXR = 10 log (C/X); G_m = on-axis gain of the wanted earth station (numeric) = $\eta(\pi D/\lambda)^2$; $G(i\Delta\Theta)$ = gain of the earth station $i\Delta\Theta$

degrees off axis; $\Delta\Theta$ = intersatellite spacing; D/λ = ratio of antenna diameter to wavelength; and η = antenna efficiency.

The summation in (6) is taken over all interfering satellites. The approximation results from the following assumptions: a) the e.i.r.p. of the interfering satellites in the direction of the wanted Earth station is the same as that of the wanted satellite; b) the slant range, hence the space loss, does not vary with satellite position; c) the angle subtended at the antenna is the same as the satellite longitudinal spacing. The assumptions above are all conservative and will tend to give slightly larger spacings than could be obtained. As mentioned earlier, the gain in the sidelobe region will be taken according to the model developed by the CCIR[3], namely

$$10 \log G(i\Delta\Theta) = 32 - 25 \log (i\Delta\Theta) \tag{7}$$

At this stage it is convenient to consider the effect of polarization. Accordingly, suppose alternate satellites are oppositely polarized so as to provide Y db of isolation. Equation (6) now becomes, using Eq. (7),

$$\frac{X}{C} = \frac{10^{3.2}}{G_m(\Delta\Theta)^{2.5}} \{10^{-Y/10} \Sigma(2i-1)^{-2.5} + \Sigma(2i)^{-2.5}\} \tag{8}$$

Using equations (4), (5), and (8) we can now obtain an important piece of information, namely the relationship between N_i and the satellite spacing. In particular, when Y=0, and assuming η= 0.5, a simple equation is obtained

$$\Delta\Theta \simeq \{\frac{11.45 \times 10^9}{N_i (D/\lambda)^2 (1+9.5 M_{rms}^3)}\}^{0.4} \tag{9}$$

Figure 1 shows this relationship for D/λ=300 and M_{rms} a parameter. A value for N_i of 1000 pWOp has been provisionally recommended[4]. However, the range shown for N_i gives an idea of the tradeoffs possible should further reconsideration seem desirable.

Spacing vs Interference Noise/Modulation. Although equation (9) contains all possible information concerning the effect of the major parameters on $\Delta\Theta$, it is nevertheless worthwhile to point out some specific features for the parameter range expected in practice. For the values of M_{rms} shown in Figure 1, which should nearly bracket the range of practical interest, and for a fixed value of N_i, it is found that the required spacings relative to M_{rms}= 0.3 are as given in the following table:

M_{rms}	$\Delta\Theta$	W
0.3	1	1
0.5	0.8	1.33
1.0	0.43	2.15
2.0	0.194	3.75

The Carson's rule bandwidth, W, relative to that for M_{rms}= 0.3 is also given in the table and it can be seen that the product $\Delta\Theta \cdot W$ is close to unity (.725 for M_{rms}= 2.0). Thus, for the range shown, spacing and bandwidth <u>potentially</u> interchange nearly on a one-to-one basis. It should be remembered, however, that for a given N_i, a different CNR will be required for each M_{rms} in order to meet the thermal noise budget.

On the other hand, for fixed M_{rms}, $\Delta\Theta$ varies as $N_i^{-0.4}$. As N_i varies from 250 to 7500 pW, a 30:1 change, $\Delta\Theta$ varies by about a factor of 4. It is observed that interference noise is quite sensitive to spacing and there are diminishing returns in $\Delta\Theta$ for allowing more and more interference noise. The foregoing is unaffected by the actual antenna size which acts only as a scale factor.

<u>Spacing vs antenna size</u>. The relationship of spacing to antenna size is evident from equation (9) which shows, for given N_i, that $\Delta\Theta$ is proportional to $(D/\lambda)^{-0.8}$. For (D/λ) varying from 30 to 1000, which should bracket the range of interest, it is found that $\Delta\Theta$ varies by a factor of 16.4. Thus, spacing is traded relatively inefficiently for antenna size.

<u>Spacing vs polarization</u>. The effect of polarization can be simply obtained from Eq. (8); one merely solves for that $\Delta\Theta$, for given γ, which keeps the left side constant. In this way one generates the ratio of spacings, with and without polarization, as a function Υ. The result is shown in Fig. 2. It can be seen that as $\Upsilon \to \infty$ a spacing reduction of two can be obtained. It can be also seen, however, that for $\Upsilon \geq 10$ db, the benefits diminish rapidly.

<u>Spacing vs carrier Interleaving</u>. The possibility of improving orbital spacing arises when the carrier frequencies of adjacent satellites are staggered. This is because the factor K is actually $K(f_D)$, i.e., a function of the carrier frequency difference between the wanted and unwanted signals. Consider for the moment a single interferer, at some frequency offset f_D.

We can then write

$$10 \log N_i = 71.2 - CXR - 10 \log K(0) - \{10 \log K(f_D) - 10 \log K(0)\} \quad (10)$$

where K(0) is the factor K at zero offset, i.e., cochannel. Since, for most cases of interest, $K(f_D) > K(0)$, the term in brackets on the right hand is positive. But Eq. (10) is precisely what would result if a cochannel interferer were provided with a polarization isolation of $Y = 10 \log K(f_D) - 10 \log K(0)$ Thus the effect of carrier offset can be reduced to an equivalent polarization isolation, which may be different for each interferer. A case of particular interest can be simply solved. Suppose that alternate satellites are offset by half the signal bandwidth. The arrangement is then symmetrical, with alternate satellites providing an (equivalent) amount of isolation, Y db. Eq. (9) and Fig. 2 can then be applied directly. Values of $K(f_D)$ may be found in reference 2.

<u>Spacing vs Varying above parameters.</u> So far we have examined only the relationship between the interference noise budget N_i and the satellite spacing, $\Delta\Theta$. Varying N_i, however, implies a change in the link parameters in order to change N_t so that the total noise is kept constant. Furthermore, the specific link parameters are dependent on M_{rms}, which, in turn, determines the bandwidth. It seems desirable, therefore, to display graphically all these interactions. To this end, we define a rather natural measure, channels/orbit degree/MHz, with which to evaluate system performance. This measure reflects the efficiency with which not only the orbit, but also the frequency spectrum is being used. To obtain channels/deg/MHz (which will be denoted $\ddot{n}$) we note that

$$\ddot{n} = \dot{n}/\Delta\Theta \quad \text{(ch./deg./MHz)} \quad (11)$$

where $\dot{n}$ is the number of channels/MHz. Using carson's rule for the FM bandwidth and assuming $\Lambda = 10$ we have

$$\dot{n} = 119/(\sqrt{10}\, M_{rms} + 1), \quad \text{(ch/MHz)} \quad (12)$$

To evaluate $\ddot{n}$ as M_{rms} varies we assume a fixed CNR; this permits a more equitable comparison between alternatives. Manipulation of the previous equations gives

$$\frac{1}{\Delta\Theta} = \left\{ \left(\frac{N_T}{10^9} - \frac{1}{(C/N)R_t} \right) \frac{(D/\lambda)^2 \; (1 + 9.5\, M_{rms}^3)}{11.45} \right\}^{0.4} \quad (13)$$

where $10 \log R_t = 90 - 10 \log N_t - CNR$ and is obtained from Eq. (2). Substituting Eq. (13) into Eq. (11) provides an explicit expression, for cochannel operation, interrelating all significant parameters. This expression was programmed on a computer and a typical result, for N_T = 7500 pW and D/λ = 300, is shown in Fig. 3. Also shown on the figure are lines of constant $B= N_t/N_i$.

There are a number of interesting points to be made from Fig. 3. First, it will be recalled that the abscissa is the number of channels/MHz (per satellite). If every satellite is allowed the same bandwidth, the abscissa gives the relative number of channels per satellite. Hence, for fixed $\ddot{n}$, the relative number of satellites is inversely proportional to the abscissa. At the same time, the relative e.i.r.p. is given by cutting across at constant $\ddot{n}$ if the link parameters are assumed the same. For example, suppose $\ddot{n}$ = 19. From the figure, this can be achieved for $\dot{n}$ = 64 and CNR = 39 db and also for $\dot{n}$ = 30 and CNR = 24 db. Thus, in the latter case over twice as many satellites are required but the e.i.r.p. can be 15 dB lower. The case which is preferable will depend on many factors including cost and possible limitation on the power flux-density.

Another fact evident from Figure 3 is that the larger $\dot{n}$ is, the more sensitive $\ddot{n}$ is to e.i.r.p. variations. This fact is worth noting in light of the various factors which may cause the e.i.r.p. of a satellite to degrade during its lifetime.

Another feature worth noting is that as the CNR decreases the possible range of $\dot{n}$ decreases. This is explained by the fact that, for each CNR, there is an $\dot{n}$ for which $N_t = N_T$, and hence permits no interference noise. This accounts for the downward bend and asymptotic appearance of the curves. The inference is that as the CNR is decreased one is automatically forced to satellites with higher modulation indices and hence lower capacity per unit bandwidth.

Thus Eq. (13) and Fig. 3 provide useful tools for trading off orbit/spectrum utilization with satellite capacity, as a function of the link parameters and the composition of the noise budget.

Tradeoffs for Homogeneous Television

The study of mutually interfering FM television systems can be approached in two ways, "subjectively" and "objectively." These two approaches refer to methods of characterizing interference. In the subjective approach[5] the effect of interference is determined in viewing tests. The carrier-to-interference radio (CXR) that produces a given reaction is in general a function of 1) the modulation index of wanted and unwanted signals, 2) their relative carrier frequencies, 3) the signal-to-thermal noise ratio in the wanted system. The value of CXR which results in the desired reaction for some set of 1), 2), and 3) is termed the protection ratio (P.R.). Thus, system performance can be characterized by stating the required P. R.

The "objective" approach[6] attempts to define an equivalent SNR resulting from interference.§ This SNR is some suitable function of relevant variables. It can be combined with the thermal SNR to yield an overall SNR. This approach is subject to the difficulty of defining the SNR resulting from interference in such a way that it is indeed equivalent to a given amount of thermal noise. On the other hand, once such an equivalence is found it becomes simpler to tradeoff interference noise and thermal noise.

We present only an example showing the orbit utilization potential of homogeneous TV systems. We assume cochannel operation and use the results of Ref. 5 which gives P.R. = 28.7 dB for M = 1 and P. R. = 19.9 dB for M = 3 where M is the peak deviation ratio.# Using Eq. (8) with Y = 0 and $D/\lambda = 300$, it is found that for the protection ratios given, $\Delta\Theta = 1.7°$ and 0.8° for M = 1 and 3, respectively. Since the Carson rule bandwidth ratio is 2:1 it is seen that the potential number of channels per degree MHz is about equal. It must be remembered however, that the system with M = 1 will require greater power.

General remarks on Change of Assumption

The previous discussion made several assumptions, namely those for homogeneous systems. Brief comment should be made on the effect of changing some of these assumptions.

---

§ Naturally, the "objective" approach must ultimately resort to subjective tests.

The protection ratios given in Ref. (5) correspond to the subjective evaluation of "just perceptible."

Effect of Earth-station antenna sidelobe characteristics. Utilization of the orbit can obviously be improved if the sidelobes of the receiving antenna can be decreased. Studies have been made assuming an "adjustable" CCIR model, 10A-10K logθ, and it appears that for increases in sidelobe decay rate from 2.5 to, say, 3.5, improvement in spacing by a factor of about 2-3 could be expected. Since the bulk of the interference arises from the nearest satellites, a more dramatic improvement in the near-in sidelobes would produce similar reductions in spacing. In fact, if one assumes a Taylor pattern, it can be shown [7,8] that spacing can be reduced materially. Thus, in principle, substantial improvement in orbit utilization is possible with sidelobe control.

Effect of satellite antenna beamwidth. The homogeneous system assumed Earth-coverage satellite antennas. If the spacecraft antenna is directional additional isolation can accrue, thus allowing closer satellite spacings. The exact amount of improvement, however, remains an open question since it evidently depends on specific assumptions such as beamwidth and location of beam centers.

Flux-density limits. In discussing the homogeneous system no particular constraints were supposed to limit the extent to which tradeoffs could be made. In practice, some of the regions of the tradeoff curves may be disallowed as a result of, for example, economic or regulatory constraints. One of the most important constraints which may apply is a limit on power flux-density at the earth. Basically this limits the options on satellite e.i.r.p. For the shared bands between 1 and 10 GHz the CCIR has currently a recommendation[9] on the allowable flux-density. The way in which this limits the satellite e.i.r.p. has been computed [1,10] for FM telephony.

System nonhomogeneity. Perhaps the "biggest" assumption made in the homogeneous system is the very postulation of homogeneity itself. What then can be said about the effect on orbit utilization of allowing dissimilar systems to operate in proximity? A truly general treatment should be able to encompass every conceivable orbital arrangement of satellites as well as every possible specification of technical characteristics, both of the satellites and the Earth stations. Such generality has so far eluded a workable treatment. However, some tentative comments can be made. Consider just two satellites, A and B. Satellite A will require some minimum spacing θ_A to meet its interference noise requirements. Similarly, satellite B will require some minimum spacing θ_B for its satisfactory operation, and in general $\theta_A \neq \theta_B$. Since the larger spacing must prevail, it can be said that an inefficiency results in the sense that one satellite is operating better than its design goal; conversely, the satellite requiring the larger spacing could be replaced by another which tends to make $\theta_A = \theta_B$. If the ratio θ_A/θ_B is much different from unity it is obvious that a large portion of the orbit is unused which could support several of the satellites requiring the smaller spacing. On the other hand, the statement that $\theta_A \neq \theta_B$ represents inefficient use of the orbit tends to be oversimplified: is there need of the "wasted" portion of the orbit, at least for the expected lifetime of the "worst" satellite; is there undue economic penalty to the user to force θ_A to equal θ_B? Such considerations, while not really technical, appear necessary for a realistic appraisal of the efficiency of orbit use by nonhomogeneous systems. It should not be inferred, however, that nonhomogeneity per se necessarily implies orbit wastage. It can be shown [7], in fact, that nonhomogeneous systems can operate compatibly (in the sense $\theta_A = \theta_B$) if sufficient control over the parameters of both systems can be assumed. Perhaps the fairest statement that can be made at this stage is that nonhomogeneity need not lead to appreciable orbital inefficiency but that the uncontrolled placement of nonhomogeneous systems almost certainly will.

III. Influence of criteria on technical characteristics

The preceding sections discussed the various technical characteristics as they related to use of the geostationary orbit. It is now of interest to discuss how various criteria influence the choice of technical characteristics. The criteria have been chosen to illustrate how these technical characteristics take on different importance for different satellite configurations. Each is analyzed to indicate how various parameters become important depending on what criteria are chosen. The criteria discussed below were all chosen with a view to optimization in the geostationary orbit.

Maximize the Number of Satellites

The criterion is, of course, equivalent to minimizing the intersatellite spacing. For systems where interference noise can generally be traded for thermal noise (e.g., telephony systems), this minimization tends to be satisfied when any combination of the following courses of action is taken: a) Operate in the interference-limited mode, i.e., assign as great a percentage of the noise budget as possible to interference noise; b) Use Earth-station antennas with high D/λ. Everything else being fixed, this will allow satellites to be placed closer together; c) Use a high (FM) modulation index. The immunity to interference increases with modulation index. Thus, for a given amount of baseband interference noise, the satellites can be placed closer as the modulation index increases; d) Use a comparatively high e.i.r.p. and/or high ratio of Earth-antenna area to system noise temperature. This strategy is actually a necessary consequence of item a). For if operation tends to be interference limited, the allowable thermal noise is very small. This in turn implies a high CNR which can be achieved with high e.i.r.p. and/or high G/T. It should be noted that increasing e.i.r.p. may eventually conflict with flux-density limits; e) Use polarization isolation. For a given amount of interference noise power this allows, of course, closer satellite spacings; f) Use carrier offset. When the carriers are appropriately offset between satellites there is increased immunity to interference, thus permitting closer spacings.

When systems are required to operate with a given protection ratio, the situation is somewhat different. One cannot then speak of operating in the interference limited mode. Hence, given a number of satellites with some fixed protection ratio, their number can be maximized with any or all of the following measures: a) Use earth station antennas with high D/λ; b) Use polarization isolation; c) Use frequency staggering.

The use of directional satellite antennas, to the degree that they provide additional isolation, will also permit closer satellite spacings.

When systems with different technical characteristics coexist in the geostationary orbit, the last three measures, namely polarization-isolation carrier-offset, and increasing D/λ will always tend to decrease the spacings.

In general, however, when all parameters are free, the relationship between these parameters and satellite spacing is neither straightforward nor unique. If all the system characteristics are fixed, however, the only way to maximize the number of satellites is to have the flexibility of ordering these satellites in the orbit so as to minimize the overall subtended arc.

Maximize the number of Earth-satellite links.

This criterion tends to be satisfied when any or all of the following measures are taken: a) Maximize the number of satellites in orbit. The various factors and tradeoffs mentioned in that connection apply here equally. b) Increase the number of antenna beams per satellite. This evidently will increase the number of links. It will be noted, however, that this measure can be used equally well to increase the number of satellites by providing additional isolation. c) Increase the number of transponders per satellite; this can be done by increasing the number of antenna beams, one for each transponder, in which case this measure is not independent of the previous one. Alternatively, the number of transponders can be increased by simply giving a different frequency assignment for each one; of course, this is simply using more spectrum. d) Increase the number of Earth-stations; to the extent that more than one Earth-station receives the same signal, the number of space-Earth links is increased. This is primarily a matter, however, of the logistics and economics of the situation and has no direct effect on the link parameters.

Maximize the Total Data Rate

This criterion represents a literal interpretation of orbit/spectrum utilization. That is, the concern for conservation of orbit and spectrum is not attributable to these quantities, per se, but for their communication-carrying capacity. Maximizing such capacity, therefore, is really what is of interest. It is more realistic to maximize total data rate under some reasonable set of constraints. In particular, therefore, it is assumed that there are a number of satellites spaced along the geostationary orbit. In this situation, total data rate is simply the "integral" of the data rate "density," namely data rate per degree per megahertz. Maximizing the latter also maximizes the former. The conditions which lead to this maximization are described in the next paragraph.

Maximize Data Rate per Degree per Megahertz

The conditions that tend to satisfy this criterion also tend to satisfy the first and third criteria discussed above. These conditions are as follows: a) Increase D/λ; this allows closer spacings for given interference noise and hence increases the data rate density in the orbital dimension. b) Increase modulation index; the data rate density is generally an increasing function of modulation index when the satellites operate cochannel. For carrier-offset on alternate satellites, data rate density is always a monotonic function of modulation index. Increasing the modulation index improves (generally) the immunity to interference and may be viewed as increasing the data rate density in the frequency dimension. c) Operate in the interference-limited mode. This allows closer spacing of satellites. d) Increase e.i.r.p.; this is actually a consequence of the preceding measure since a relatively small allocation to thermal noise implies a compensation in higher e.i.r.p. and/or earth-terminal G/T. e) Use carrier-offset; this measure essentially improves the immunity to interference and therefore allows greater satellite packing. f) Use polarization isolation; this produces a lower interference level which implies the option of closer satellite spacing.

Optimize the Number of Channels per Transponder

The technical characteristics that tend to satisfy this criterion depend on the interpretation of "optimize." The basic situation involves the tradeoffs between the transponder** (or satellite) parameters on the one hand and the number of transponders (or satellites) on the other; either may be "optimized" depending on one's viewpoint and operating constraints.

Basically, for a given communication capacity, $\ddot{n}$, there exists a tradeoff between the number of satellites required to achieve this capacity and the number of channels per satellite. Obviously, for fixed n, the number of satellites decreases as the number of channels per satellite increases, and vice versa. A large number of channels per unit bandwidth implies a low modulation index and a high e.i.r.p. for a given performance. The "optimum" point in the tradeoffs, then, will depend on which of the following constraints is controlling: a) Economic constraints - if these constraints are paramount, the optimum point depends on the cost of building and launching a relatively large number of "small" satellites compared to the same costs for a smaller number of "large" satellites. b) Operational constraints - if the optimum economic solution is for fewer large satellites, this solution might be short of realization since such satellites require higher e.i.r.p.'s which

**We have assumed a single transponder per satellite so that, in this section, the two words are used synonymously.

may run into flux-density limitations. On the other hand, there may be a minimum number of satellites required in a given arc of the geostationary orbit, depending on the earth-coverage area of this arc.

Limit Interference between Systems to Tolerate Levels

There are two types of systems with which satellites may interfere: a) Other satellite systems and b) Terrestrial systems.

In the first instance, the "tolerable" level would most likely be specified as an upper limit on the noise contributed by the interference in a telephone channel. To meet prescribed limits on the tolerable interference noise, one or more of the following measures can be taken: a) For given modulation index and desired minimum spacing, one must adjust D/λ to the requisite value. b) For given modulation index and antenna size, one must adjust the satellite spacing. c) For given D/λ and $\Delta\Theta$, one must adjust the modulation index to the proper value. d) Everything else being fixed, the relative e.i.r.p. of adjacent satellites may be varied although this course of action, by itself, may not be sufficient. e) The carrier-offsetting of spatially adjacent signals can significantly reduce the interference noise. This measure, therefore, can be used in conjunction with the ones above to attain the specified output noise level.

The preceding has referred to telephone channels. In some situations, such as TV interfering with TV, the tolerable interference level is specified in terms of the protection ratio. The required protection ratio is a function of the modulation indices of the wanted and unwanted signals and their relative carrier frequencies, while the achievement of a given protection ratio depends on the proper adjustment of relative e.i.r.p.'s, satellite spacing, and ground antenna size. Thus, although the tolerable interference level is specified in different terms than for the telephony case, the same parameters and tradeoffs are involved as described above.

With respect to terrestrial systems, the tolerable level of interference to these systems is specified, as a function of angle of arrival, in terms of the power flux-density per 4 kHz. The resultant constraint on satellite systems essentially is to provide an upper bound on the e.i.r.p. of satellites. This constraint, in general, will tend to conflict with other criteria, particularly the first and fourth criteria discussed above.

Permit Use of Systems with Different Characteristics

This criterion depends on the coordination between neighboring satellite systems so that each induces in the other no more than a specified and tolerable interference.

For systems with different but fixed characteristics, their ability to coexist depends strictly on the geometry and the relative frequency assignments. In other words, everything else being fixed, there is always some spacing that will permit different systems to share the orbit. It is, of course, possible that the spacing required precludes coverage of intended areas, or that it is much too large for a reasonably effective use of the orbit. In such instances, it may become necessary to impose minimum specifications on system characteristics.

Permit Reallocation of Orbit/Spectrum Assignments

This criterion, like the previous one, does not imply a particular trend in system characteristics for its satisfaction. Rather, it represents the viewpoint that sufficient flexibility should be incorporated in any regulatory scheme, so that, as technology and traffic requirements change, new satellites with possibly different characteristics can replace outdated or inoperative ones. This kind of situation can occur, for example, when the traffic capacity of a given satellite is exceeded by the demand and the system operator may whish to replace it by a higher capacity satellite. The characteristics of the new satellite will, in general, be different from that of the old ones. To accommodate these new satellites, and to maintain the noise budgets of the ones already existing and remaining in operation, generally implies that one or more of the following courses of action be taken: a) the orbital position of the new satellite(s) may be different from the one(s) it replaces; b) the frequency assignment of the new satellite(s) may be different from the one(s) it replaces; and c) the orbital position of the existing satellites may have to be changed.

The last item implies that the existing systems have station-changing capability. It may also be the case that if the parameters of existing systems under the operator's control, such as the Earth-station G/T, could be changed, the accommodation of new and old systems would be facilitated. However, it appears realistic to suppose that the characteristics of existing systems, except for satellite position, would remain unchanged. It seems reasonable that some degree of control can be exercised over the characteristics of the new systems so that all systems operate compatibly. Nevertheless, it is not possible to say what these new characteristics should be without knowing the ones of the existing systems. The most that can be said is that the situation prior to and after reallocation is still governed by the applicable equations.

Conserve rf Bandwidth

If this criterion is of prime importance, there are essentially two ways of achieving it. The first is simply to reuse the same frequency band as much as possible. This can be done by providing the greatest isolation possible between systems which implies the use of (narrow) multibeam satellite antennas and polarization isolation; of course, narrow beam earth station antennas are also indicated. Once these measures have been taken to the fullest extent, the only obvious step left is to use as small a modulation index as possible -- however, the price one pays for narrowband systems will usually be one of the following: a) higher satellite e.i.r.p., b) higher earth station G/T, c) larger satellite spacings.

In the bands shared with terrestrial systems, item a) is, of course, in conflict with the requirements of the sixth criterion, whereas item c) conflicts with the first criterion.

Protect Existing Satellite Services

This criterion again does not imply an unambigous trend in system characteristics. Rather, it is the expression of a view which considers that systems already in place and operating be disturbed to the minimum extent possible. It is meant to imply protection only to the degree that further development of the orbit or spectrum is not severely inhibited. In the latter eventuality, existing systems may be required to accept some degradation in one or more performance parameters or (more likely) accept some reallocation in orbit or frequency assignment. Thus, the mutual protection of existing and new systems is more or less synonymous with the sixth and eighth criteria discussed above.

Comparative Analysis of Criteria

The interrelationships among the criteria are relatively subtle interplays of economic, nationalistic, political, and technical factors, not easily amenable to quantification, or even ordering. Nevertheless, to the extent that criteria conflict or overlap, it is expected that the technical characteristics implied for their satisfaction will also conflict or overlap to more or less the same degree. From the preceding discussion, one can find which system characteristics tend to satisfy each criterion. By observing these connections it is possible to infer the interrelationships among the criteria to the extent dictated by technical parameters. This process is embodied in Table 1.

The entries marked N.A. (Not Applicable) imply that the parameter in question is not particularly relevant to or is not unambiguously specified by a criterion. For example, criterion 7 inherently leaves the e.i.r.p. unspecified. Another point is that criteria 8 and 10 implicitly refer to two types of systems, those in existence and those to be installed. The characteristics of the former are assumed fixed except, possibly, for orbital position. These criteria apply primarily to projected systems. (In the table the top line next to criterion 8 refers to old systems and the bottom line to new systems.) In reference to criterion 6 the top line refers to intrasatellite system interference and the second line to satellite interference to terrestrial systems. Finally, the term "adjust" in the table implies that the characteristic under consideration is not forced in a unique way by a given criterion, but that if enough other variables were assumed, one could compute how this parameter ought to be adjusted.

Table 1 Relationships Between Criteria and Technical Characteristics

Parameters / Criteria	Satellite eirp A	Satellite Antenna Size B	Number of Trans-ponders C	Satellite Spacing D	Modula-tion Index E	Carrier Fre-quency Offset F	Polar-iza-tion Isola-tion G	Earth-Station Antenna Size H	Earth-Station Receiver Temp. I
1. Maximize Number of Satellites	Increase	Increase	N.A.	Decrease	Increase	Use	Use	Increase	Decrease
2. Maximize Number of Earth-satellite Links	Increase	Increase	Increase	Decrease	Increase	Use	Use	Increase	Decrease
3. Maximize Total Data Rate	Increase	Increase	Increase	Decrease	Increase	Use	Use	Increase	Decrease
4. Maximize Data Rate per Degree per MHz	Increase	Increase	N.A.	Decrease	Increase	Use	Use	Increase	Decrease
5. Optimize No. of Telephone Channels per Transponder	Adjust	Adjust	Adjust	Adjust	Adjust	Use	Use	Adjust	Adjust
6. Limit Interference to Tolerable Levels	Constrain Constrain	Increase N.A.	N.A. N.A.	Increase N.A.	Increase Increase	Use N.A.	Use N.A.	Increase N.A.	Decrease N.A.
7. Permit Systems with Different Characteristics	N.A.	N.A.	N.A.	Adjust	N.A.	Use	Use	N.A.	N.A.
8. Permit Re-allocation of Orbit/Spectrum Assignments	N.A. Adjust	N.A. Adjust	N.A. Adjust	Adjust Adjust	N.A. Adjust	Adjust Adjust	N.A. Use	N.A. Adjust	N.A. Adjust
9. Conserve RF Bandwidth	Increase	Increase	N.A.	Increase	Decrease	May be Useful	Use	Increase	Decrease
10. Accommodate Existing Comsat Services	Adjust	Adjust	N.A.	Adjust	Adjust	Use	Use	Adjust	Adjust

IV. Discussion and Interpretation of Results

The previous sections considered technical factors, criteria, and their interrelationships. The results presented permit the evaluation of technical tradeoffs and the determination of those factors that tend to satisfy any particular criterion. Which criterion is uppermost depends, of course, on the administrations and users involved. In practice, however, no one criterion is likely to prevail and general use of the orbit/spectrum space is most likely to develop on some sort of compromise basis. In fact, development of generally acceptable sharing principles will not only aim at a reasonably efficient and effective use of orbit/spectrum, but will very probably represent an accommodation of competing economic, political and national interests. While we do not discuss these less tangible aspects they frame the realworld context within which one operates. We should like, therefore, to re-evaluate our technical results in the light of these realities, with a view toward 1) isolating those factors which are most critical, most generally applicable, and most practical of implementation; and 2) developing general guides to be applied to communication satellite systems sharing limited orbit and spectrum.

Critical Factors for Orbit/Spectrum Use

From the previous work it can be seen that the factors most critical to orbit utilization are, naturally enough, those which provide isolation between satellites, whether arising from geometry or from signal structure. Specifically, these factors are 1) earth station antenna D/λ and sidelobe characteristics, 2) satellite antenna pattern and coverage, 3) modulation index or satellite capacity, 4) ratio of thermal-to-interference noise budgets, and 5) isolation methods such as polarization of carrier offset. All of these factors are not equally amenable to application as they imply varying degrees of control over system parameters and a certain amount of intersystem coordination or cooperation.

Consider, for example, the Earth station antenna D/λ. From the point of view of orbit utilization it is desirable to make D/λ as large as possible. On the other hand, the Earth station being a substantial portion of system cost, the system designer will very likely want to minimize its size. Thus, orbit utilization and economics are here in conflict and one should not realistically expect uniformly large earth stations without some external imposition of a minimum size. On the other hand, for a given antenna size, appreciable spacing decrease can result from improvement of the sidelobe characteristics. Apart from developmental costs of sidelobe reduction techniques, it seems that cost of Earth stations should be negligibly affected. Furthermore, such techniques are not particularly criteria oriented. Thus, it would seem that investigation of reliable, easily implemented sidelobe reduction techniques would have a big "payoff" in terms of potential application.

Narrow beam satellite antennas can have a highly beneficial effect on orbit utilization. It would seem, however, that this effect is most likely to take place to the degree that coverage areas can be substantially separated. This is as likely to be the case as not, depending on the particular arc of the orbit considered. Where applicable, however, spot coverage is desirable and associated techniques appear to be another "payoff" area for study.

Generally speaking, increasing the modulation index improves orbit utilization. However, for fixed bandwidth allocation this reduces satellite capacity. Furthermore, modulation index is one of those fundamental parameters likely to remain under the system designer's control. Thus, while universal use of high modulation indices is desirable, it appears debatable at this point to what degree this can be imposed.

Operating in the interference-limited mode, all other things being fixed, provides the ultimate in orbit utilization. Thus, a low ratio of thermal-to-interference noise is desirable. The penalty, obviously, is economic. Provisional recommendations exist, but further study is needed to find the point where economic penalty becomes disproportionate to orbit utilization improvement.

Isolation methods such as different polarization and/or carrier offset on alternate satellites appear especially desirable since they are criteria-independent, do not otherwise affect system design, and thus permit as great a diversity of systems as may be required. The application of these methods requires only coordination and cooperation among adjacent systems. The degree of polarization isolation to be practially expected is still a matter under study. It appears that great returns for such studies are possible in terms of potential benefits.

In view of the multitudinous factors involved,it appears that there exist no unequivocal recommendations for "efficient" sharing of the orbit. However the following conclusions can be used as a basis for developing orbital use guidelines: a) For all, except very high,carrier-to-noise ratios (CNR) increasing the number of channels per transponder imposes a reduction in channels/deg/MHz when the CNR is held constant; b) In general, there is incompatibility between maximizing the number of channels in a single satellite and maximizing the total number of channels for the whole orbital arc; c) It is possible to derive a set of criteria which may be applied to find the technical characteristics that will satisfy them; the ranking of criteria in order of importance depends basically on the requirement set by users and administrations; d) Satellite system development is still in a stage of rapid change, and maximization of orbital utilization according to any one of these criteria would be an unrealistic objective; e) Coordination between neighboring satellite systems, with respect to such things as polarization diversity or carrier frequency offset, can materially improve orbital utilization irrespective of any criterion; f) Any coordination procedure between satellite systems, to the extent that it incorporates orbital use criteria, should be such as to permit both high-capacity and low-capacity communication satellites; g) In view of generally conflicting requirements of the various criteria, and in view of changing traffic demands and patterns, it seems that any policy for orbit/spectrum assignments ought to be kept flexible.

Orbital Use Guidelines

The above has pointed the direction toward a systematic approach to establishing a viable technical base upon which to pursue use of the geostationary orbit. It is not sufficient to use the assumptions of a homogeneous system to guide the design of future systems. However, the investigations resulting from analyses based on homogeneous systems do provide insight into the sensitivity of various satellite parameters to change. From an examination of the factors as they vary for different criteria, it is clear that they have different degrees of importance for different optimizations. It is incumbent to make preliminary determinations with respect to which parts of the orbit should be used for what type of satellite. Since it is advantageous to put similar satellites adjacent to one another, orbital location should be considered in light of the type of traffic that is anticipated. Next, some thought should be given to the lifetime of the satellite and the type of techniques which a particular generation of satellites might employ, whether or not it is one of a kind or whether there will be many others like it. Finally, it is apparent that there are many spectrum reuse techniques, which will permit enhanced use of the orbit if properly utilized. However, certain of these techniques are more usefully employed in one part of the orbit than another. Furthermore, care should be exercised in applying the various techniques in a fashion that does not deny the use of others.

It is seen from the above discussion that it should be possible to construct some general rules or guidelines, which should be applied prior to satellite system implementation. These guidelines as a minimum should include the answers to the following questions: a) What are the particular existing satellites which would be affected by the placing in orbit of new satellites,or by the relocation of an existing satellite? b) What is the nature of compatibility between the two systems - what is the minimum distance to which the proposed satellite may be moved to an adjacent satellite? c) What is the range of orbital positions for which the satellite coverage would be satisfactory? d) Is this satellite operation to be conducted in tandem with any existing or proposed satellite? e) What is the estimated lifetime of the satellite - is it an experimental or operational satellite? If experimental, is it intended to replace this satellite with a satellite of similar characteristics? f) How can future satellites be designed so as to minimize the distance between them and existing satellites? g) What cost/effective options are available on the proposed satellite to further minimize the distance between it and adjacent satellites? h) What type of traffic, i.e., telephony - TV, etc., is for this satellite's lifetime? i) Is it envisioned moving the location of this satellite in orbit, and if so in what way? What operational orbital positioning tolerances are envisioned, and what are the tightest tolerances possible - for what periods of time?

The answers to these questions should provide sufficient information to determine how effectively proposed satellites would utilize the geostationary orbit, and begin to furnish a heuristic methodology for providing flexibility in system implementation.

Topics for further study

There are a number of areas that require further study before a more complete picture of orbit/spectrum utilization is developed. We list here some of the more important of these topics: a) Develop systematic procedures for handling nonhomogeneous systems; b) Study in detail the orbit utilization of digital systems; c) Obtain more information on the effects of arbitrary, mutually interfering signals; d) Evaluate the realistic possibilities for improving earth station sidelobe characteristics; e) Study the effects of multibeam, directional satellite antennas. We also mention the need to study a topic of great importance namely economics. It appears likely that realistic system-sharing criteria cannot be evolved without at least some evaluation of the economic implications.

References

[1]Jeruchim, M. C. and Sayer, T. C., "Orbit/Spectrum Utilization Study," Interim Rept. (Vol.I), Document No. 69SD4270, May 15, 1969, General Electric Co., Philadelphia, Pa.

[2]CCIR Study Groups, Document USSG IV/1032, August 1, 1970.

[3]CCIR, Documents of the Xlth Plenary Assembly, Vol. 4, Pt.2, Rept. 391, Oslo, Norway, 1966.

[4]CCIR Document IV/427-E, Draft Recommendation, Sept. 30,1969.

[5]CCIR Study Groups, Document IV/299-E, June 10, 1969.

[6]CCIR Study Groups, Document IV/224-E, May 19, 1969.

[7]Jeruchim, M. C. and Moore, F. D., "Orbit/Spectrum Utilization Study, " Vol. II, Document No. 69SD4348, Sept. 12, 1969, General Electric Co., Philadelphia, Pa.

[8]Hult, J. L. et. al., "The Technology Potentials for Satellite Spacing and Frequency Sharing," Memo. RM-5785-NASA, Oct. 1968, The RAND Corp., Santa Monica, Calif.

[9]CCIR, Documents of the Xlth Plenary Assembly, Vol. 4, Pt. 2, Rec. 358-1, Oslo, Norway, 1966.

[10]Kane, D. A. and Jeruchim, M. C., "Orbit/Spectrum Utilization Study," Vol. III, Document No. 70SD4246, June 30, 1970, General Electric Co., Philadelphia, Pa.

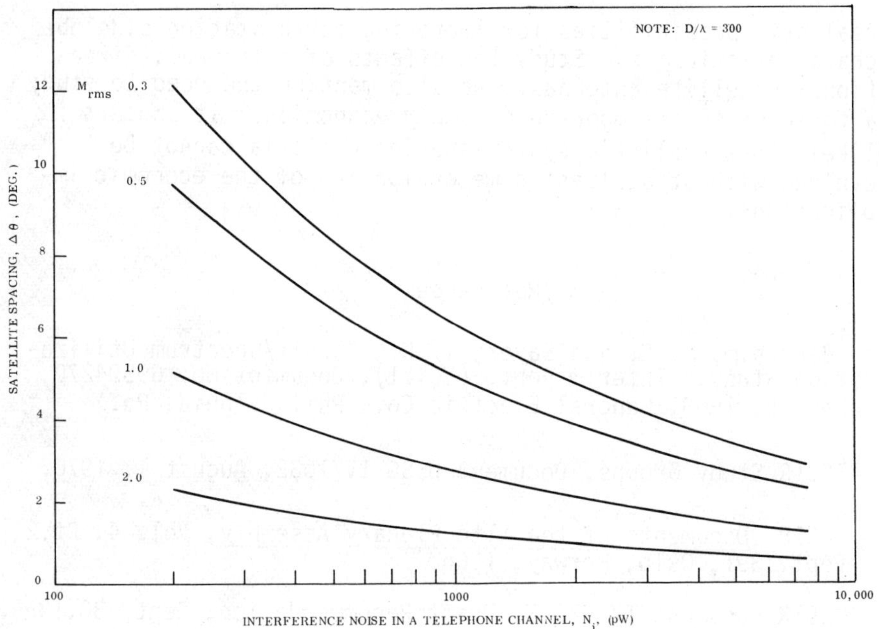

Fig. 1. Satellite Spacing as a Function of Interference Noise, Cochannel Case

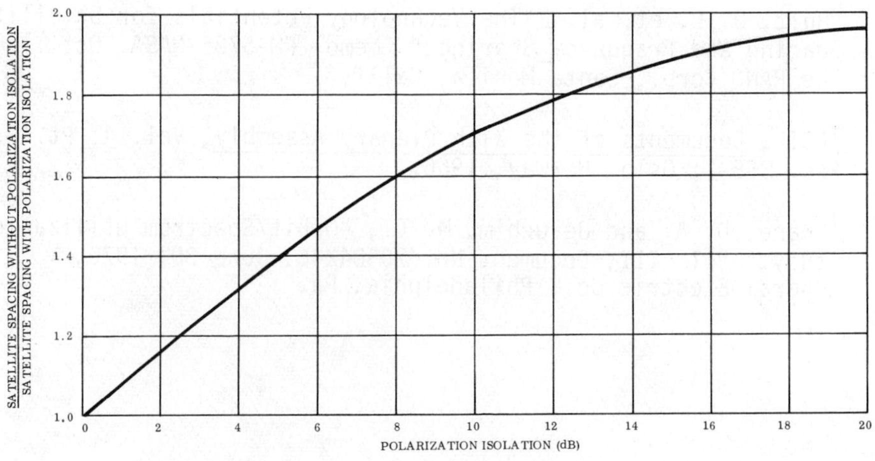

Fig. 2. Decrease in Satellite Spacing as a Function of Polarization Isolation.

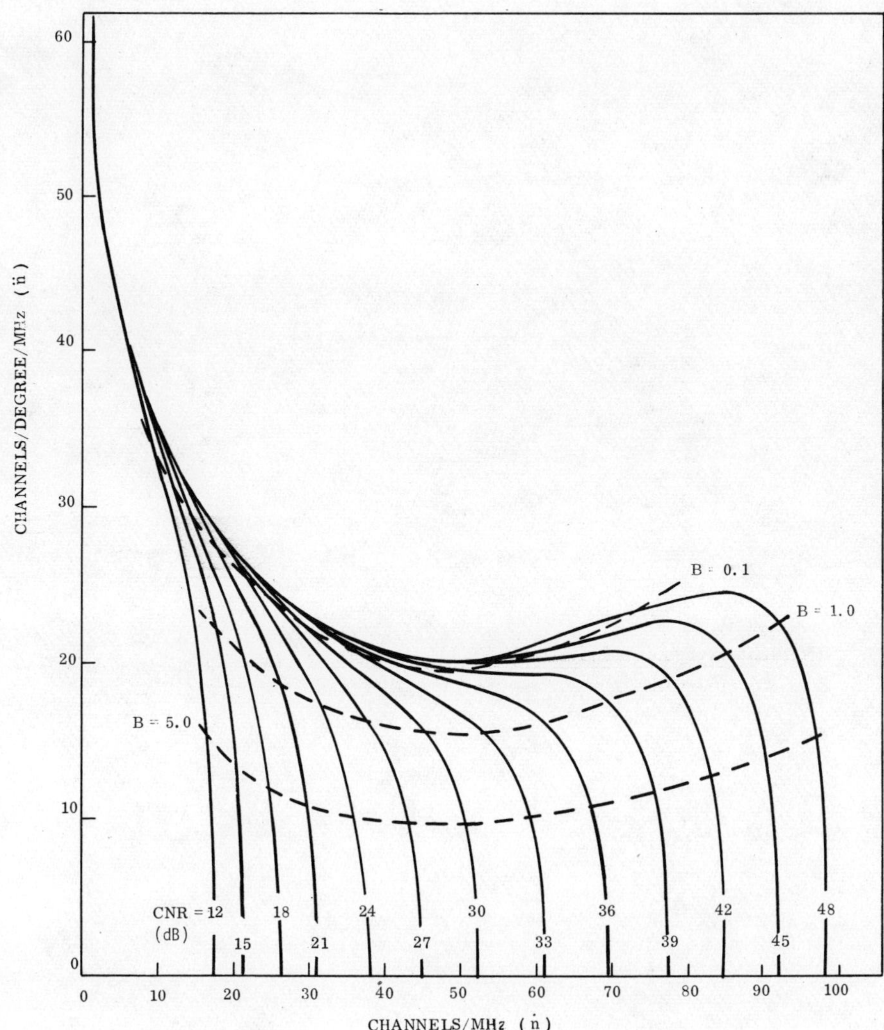

Fig. 3. Orbit/Spectrum Utilization as a Function of Channel per Satellite in one MHz, Cochannel Case.

REUSE OF THE FREQUENCY SPECTRUM AT THE SATELLITE

J. C. Fuenzalida* and E. Podraczky/

COMSAT Laboratories, Clarksburg, Md.

Abstract

The total bandwidth available at a geostationary satellite could be increased by reusing the same frequency band in multiple independent Earthward beams. Such beams could be utilized either for distribution services in nonoverlapping areas or for point-to-point communications services. This paper addresses the technique of frequency reuse at the satellite with the objective of obtaining an indication of the relative importance of the various parameters. The interference aggregate caused by unwanted cochannel transmissions is estimated for several satellite antenna side-lobe models. Using these results, calculations are performed of the maximum number of reuses of the frequency spectrum. The effects of both the modulation methods as well as the interference noise allocation are discussed. Finally, some geographical and traffic problem considerations are given.

I. Introduction

There is growing interest in studying the feasibility of increasing the total bandwidth available at a geostationary satellite by reusing the same frequency band in multiple independent Earthward beams. This technique would increase the usefulness of the geostationary satellite orbit, since a number of beams utilizing the same frequency band could carry independent information. A limited-area distribution system

Presented as Paper 70-442 at the AIAA 3rd Communications Satellite Systems Conference, Los Angeles, Calif., April 6-8, 1970. This paper is based upon work performed in COMSAT Laboratories under the sponsorship of the International Telecommunications Satellite Consortium (INTELSAT). Any views expressed in the paper are not necessarily those of INTELSAT.

*Member of the Technical Staff, Systems Simulation Branch.

/Manager, Systems Integration Laboratory.

would most probably reuse the frequency spectrum only a few times, whereas point-to-point communication systems could conceivably reuse it in a large number of beams.

An investigation of the potentials of frequency reuse at the satellite requires the consideration of a large number of factors. It is difficult to predict how technology will develop and single out those factors that will ultimately limit the performance of multibeam satellites. For this reason, the general approach chosen in this paper is an attempt to gain insight into possible tradeoffs, with the objective of obtaining an indication of the relative importance of the various parameters.

The total number of independent Earthward beams that could operate from a single orbital location is first derived. This number depends basically on the characteristics of the satellite antennas and on the carrier-to-interference ratio, which is required to meet the desired performance objectives. The interference aggregate caused by the unwanted cochannel transmissions is determined for several antenna side-lobe models. Using these results, calculations are performed to determine the maximum number of reuses of the frequency spectrum that are possible from a satellite. This problem is somewhat analogous to the reuse of the frequency spectrum at the Earth station, using a multiplicity of satellites. Since Earthward beams can be accommodated in two dimensions (as opposed to the one-dimensional arrangement of satellites in the geostationary satellite orbit), some interesting differences are found.

The modulation technique utilized has an important effect on the feasibility of frequency reuse. A comparison of several modulations is presented. A derivation is also given for the bandwidth expansion ratio, based on Shannon's bandwidth tradeoff law, which maximizes the total capacity of a multibeam satellite.

Next, the effects on frequency reuse of two interbeam isolation mechanisms are investigated: polarization discrimination, and interleaved frequency plans (for frequency modulation). It is concluded that this additional isolation is particularly important for distribution systems where the required interbeam spacing could be just a few beam widths.

A discussion of the interbeam interference allocation in the total noise budget and its effects on frequency reuse is then presented. Geographical and traffic pattern considerations affecting the implementation of multibeam satellites are discussed; and, finally, some conclusions are given.

II. Maximum Number of Independent Earthward Beams

The maximum number of frequency reuses possible at a geostationary orbit location is considered here. Essentially, this number is determined by the minimum interbeam angular separation. The model considered for the average side-lobe level plays an important role in determining the minimum angular separation.

Equation (1) represents a model for the average side-lobe level of an antenna.

$$G(0)/G(\theta) = 1 + (\theta/\theta_0)^m \tag{1}$$

where $G(\theta)$ = gain function; θ_0 = 1/2 of 3-db beam width; and m = index of antenna pattern.

This equation predicts that the isolation (defined as the gain normalized to the main beam gain) is a function of the angle to the axis, normalized to the half-power half-beam width. The required isolation from an adjacent cochannel beam determines the minimum ratio of θ/θ_0. This desired isolation is clearly higher than the acceptable carrier-to-total interference, since there are many interference entries.

The carrier-to-total interference ratio required to meet the desired objectives depends upon a variety of factors, such as modulation method, interference noise allocation in the budget, etc. To a good approximation, the total interference caused by equidistant interfering beams can be expressed as the product of the interference caused by one adjacent beam and a function of the parameter m.

Expressing this relationship in db, it may be seen that

$$(C/I)_{db} = [G(0)/G(\theta)]_{db} - F(m) \tag{2}$$

Considering a coverage area determined by the projection of the 3-db contour of the satellite antenna, and considering the case of closely packed beams, it can be shown that F(m) is typically 12 to 14 db.

When Eq. (2) is introduced into Eq. (1) and the ratio of interference angle to coverage angle is equal to $\theta_i/\theta_0 = k$, then

$$(C/I)_{db} = 10 \log (k^m) - F(m) \qquad (3)$$

Figure 1 shows the ratio k as a function of the total required carrier-to-interference ratio where the parameter m is given the values 2.5, 3.0, and 4.0.

It can be observed that the ratio of interference angle to coverage angle depends strongly on the parameter m of the satellite antenna. Furthermore, it can be observed that operating conditions requiring a high carrier-to-total-interference ratio will require interbeam separations equivalent to an unreasonably large number of beam widths.

Since the dependence of the parameter k on the desired carrier-to-interference objective and on the model for the satellite antenna directivity has been discussed, the total number of possible frequency reuses at one orbital location can be determined. For this purpose, it is helpful to consider a satellite-centered coordinate system which permits the satellite beams to retain their shapes. The Earth appears as a circle, and if circular satellite beams are considered, the problem reduces to arranging these beams in the Earth's projection. This was done, and the results are summarized in Fig. 2 for k = 10, 20, and 30. The abrupt discontinuities shown here are actually an artifact caused by the requirement that all beams be equally spaced. If the model is changed so that $k\theta_0$ is the minimum beam separation, and so that all beams need not be equally spaced, a smoother function would be obtained. The straight lines in this figure represent good asymptotic approximations for the case in which a large number of beams are used. It should be noted that significant frequency reuse is possible with satellite antenna 3-db beam widths of up to 2°.

Since Eq. (1) is believed to be too conservative for small (θ/θ_0), it is possible that systems with k ratios of less than 10 might be implemented. This would permit frequency reuse at the satellite with even larger satellite beam widths.

III. Cross-Polarization Isolation and Frequency-Staggering

Isolation which is achievable with the use of orthogonal linear polarizations has been estimated to be of the order of 10 db between adjacent beams. When the frequency spectrum is reused at the same beam, the resulting isolation would be considerably greater.

Frequency-staggering decreases the required isolation from interference and can, therefore, be treated similarly to an equivalent increase in the isolation. This technique is expected to be of some interest when FM is used.

Although it would be desirable to be able to reuse the frequencies at interbeam distances of just a few beam widths for all types of services, this is especially important for area distribution systems for which the size of the satellite beams cannot be reduced arbitrarily. These systems have a small number of beams that would overlap in some areas. The isolation between beams obtained by the antenna directivity might or might not permit the reuse of frequencies between some of the beams. To increase the total bandwidth available to such systems, cross-polarization and frequency-staggering can be utilized at each beam, effectively doubling the bandwidth available, or at neighboring beams, thus increasing the number of beams sharing the frequency.

Another consideration pertinent to a modest number of reuses is that, since actual antenna patterns do not have the symmetry of a figure of revolution, a simple rotation of the antennas can increase the interbeam isolation.

When a satellite model with a large number of Earthward beams is considered, the question arises as to how many of these beams can carry independent information when, in addition to the simple antenna directivity, two other independent isolation mechanisms are available.

Figure 3 shows one of the many possible arrangements of equidistant beams. This particular arrangement was selected because it is highly symmetric. It can be shown that when p and f represent the corresponding isolation ratio advantages, the interference is equal to a fraction of the corresponding interference, which occurs in the absence of additional isolation and which is given by Eq. (4):

$$I_{P,F}/I_{0,0} = 2^{-m} + 1/3(1 - 2^{-m})\ (p + f + pf) \qquad (4)$$

In other words, this ratio is always greater than 2^{-m}. The hypothetical case of perfect isolation would reduce the interference to 2^{-m}, the equivalent effect of increasing all angular spacing by a factor of 2.

Figure 4 shows the increase in the number of Earthward beams that can reuse the frequency (γ) as a function of P and F, the isolation advantages in db. An increase in the number of beams by a factor of 2 would be the asymptotical solution when one of the isolations is -∞ db and the other is negligible.

IV. Modulation Techniques Suitable for Spectrum Reuse at the Satellite

The choice of modulation technique determines the minimum acceptable carrier-to-interference ratio. It is well known that modulation techniques which can be used with a low-carrier-to-interference ratio require a large ratio bandwidth-to-information rate. The potential number of reuses of the spectrum by independent beams from the satellite is inversely proportional to the square of the interbeam spacing. The total bandwidth available (the number of reuses times the bandwidth per beam) increases, therefore, very rapidly as smaller C/I ratios are required. A somewhat similar situation exists when the spectrum is reused at the Earth station, thus exploiting the Earth station antenna directivity. In that case, however, the number of reuses is inversely proportional to the interbeam spacing only. For this reason, spectrum reuse from the satellite will strongly favor modulation techniques requiring low C/I.

The traffic capacity from an orbital location, normalized to the square of the angle associated with each independent beam, can be expressed as

$$\dot{n}_s = A_1 \cdot n/(\theta_i/2)^2 B_{RF} = A_2/\theta_i^2 \cdot R \qquad (5)$$

where $\dot{n}_s$ = traffic capacity/MHz, normalized to the square of the angle associated with each beam; B_{RF} = total RF bandwidth occupied; θ_i = minimum separation angle between independent beams; A_1, A_2 = constants; and R = bandwidth-expansion ratio = $B_{RF}/4200\ n = B_{RF}/f_m$. By recalling Eq. (2) and letting the interference be a constant fraction of the noise budget,

$$C/I = \beta C/N, \tag{6}$$

it may be seen that

$$(\theta_i/\theta_0) = [(\beta C/N)/F(m)]^{1/m} \tag{7}$$

and finally that

$$\dot{n}_s = A_2[F(m)]^{1/m}/(\beta \cdot C/N)^{2/m} \cdot \theta_0^2 \cdot R \tag{8}$$

Equation (8) is based on the assumption that the interference aggregate can be treated as white Gaussian noise.

It can be seen that the capacity is inversely proportional to the product of $(C/N)^{2/m}$ and R. For each modulation technique, the combinations of C/N and R required to meet the channel objectives can be determined. When these values are introduced into Eq. (8), a comparison of modulations for frequency reuse at the satellite may be obtained.

Reuse of the Frequency Spectrum with Analog and Digital Modulations

Analog modulation methods. Single-sideband was used as a reference in Fig. 5, and it is by far the least attractive modulation method in this context. The total capacity of a satellite using FM depends on the bandwidth-expansion ratio. As a smaller bandwidth-expansion ratio is considered, more channels can be transmitted in every MHz of bandwidth. Therefore, the potential traffic capacity and the isolation requirements of every beam will increase.

The total effect of using small bandwidth-expansion ratios, as shown by Fig. 5, is to reduce the capacity available at an orbital location. The advantages from frequency-staggering can be considerable for FM carriers.

Digital modulation methods. PCM-PSK is the most likely digital candidate for this application. This modulation technique shows a somewhat similar behavior to FM. That is, the total effect of using small bandwidth-expansion ratios (i.e., high-order PSK) is to reduce the potential capacity available at an orbital location. This is also shown in Fig. 5.

In the case of a few interference entries, the interference effect has been shown to be considerably smaller than that caused by white Gaussian noise of the same power. This is of special interest for a distribution system which involves a small number of frequency reuses. Unfortunately, the advantage of frequency interleaving is not expected to be significant for this modulation method.

Reuse of the Frequency Spectrum Using the Optimum Power-Bandwidth Tradeoff

When the carrier-to-interference ratio in the multibeam satellite model is introduced into Shannon's power-bandwidth tradeoff law, and normalized by the square of the angle associated with each independent beam, the following relationship is obtained

$$\dot{n}_s = [\alpha/(k\theta_0)^2 R] \log_2 (1 + C/I) \tag{9}$$

When α is a constant, the C/I ratio from Eq. (2) is introduced

$$\dot{n}_s \doteq [\alpha/(k\theta_0)^2 R] \log_2 \left\{1 + (\theta_i/\theta_0)^m/f(m)\right\} \tag{10}$$

$$= [\alpha/(k\theta_0)^2 R] \left[\log_2 \left\{f(m) + (0_i/\theta_0)m\right\} - \log_2 f(m)\right] \tag{11}$$

For m = 2.5, f(m) can be shown to be equal to 24 and, therefore

$$\dot{n}_s \doteq (\alpha'/k^2) \left\{\log_2 (24 + k^m) - 4.6\right\} \tag{12}$$

where $\alpha' = \alpha/(\theta_0^2 R)$.

Equation (12) is valid for closely-packed satellite beams in a location where the number of beams is not reduced by the geographical distribution of the traffic requirements. Figure 6 shows this function arbitrarily normalized to 100 for k = 4. It can easily be observed that considerable penalty would occur when using a large k.

V. Interbeam Interference Noise Allowance

A prerequisite for the reuse of the same frequency band at the satellite is the acceptance of a certain amount of interbeam interference. The amount of interference noise budgeted for this purpose determines the required isolation and ultimately affects the number of possible beams reusing the frequency spectrum.

In Eq. (8), it was shown that the total capacity of a multibeam satellite is inversely proportional to the carrier-to-interference ratio, to the power 2/m. Assuming that the effect of the interference aggregate is the same as an equivalent amount of white Gaussian noise, then the carrier-to-interference ratio is proportional to the carrier-to-total-noise ratio. The constant of proportionality β is given by

$$\beta = 1 + N_I/N_{Th} \tag{13}$$

where N_I/N_{Th} = interference-to-thermal-noise ratio.

For a given modulation technique, the required carrier-to-total-noise ratio is constant; therefore, the multibeam satellite capacity is inversely proportional to $(1 + N_I/N_{Th})^{2m}$. This relationship is shown in Fig. 7, curve A, where m = 2.5.

An equivalent tradeoff curve for one-dimensional spectrum reuse is shown in the same figure, curve B. (This case corresponds to the spectrum reuse at the Earth stations.) A comparison of these curves shows that, in order to achieve the same fraction of reuses of the interference-limited case, a larger fraction of the noise budget should be allocated for interbeam interference at the satellite than at the Earth station.

VI. Geographical and Traffic Pattern Considerations

The calculations of the number of possible reuses of the frequency spectrum presented earlier considered a purely geometrical problem and have thus disregarded other important elements. First, only a fraction of the Earth is covered with land masses, and second, the potential communication traffic requirements are certainly not uniformly distributed. The communication requirements generally decrease with distance and, therefore, traffic considerations might dictate the implementation of a small number of very close beams carrying large traffic capacities, as in a typical domestic satellite system.

Another consideration is that the size of the Earth surface projection of identical and equally spaced satellite beams depends strongly on the elevation angle of each beam. Beams pointing close to the horizon illuminate larger areas than beams pointing near the subsatellite point. This effect is expected to restrict the orbital arcs, which are suitable for the different multibeam communication satellite services.

Frequency reuse, in which two beams serve closely located traffic nodal points, requires orbital locations from which these two points subtend the largest possible angle at the satellite. In general, for coverage areas at low latitudes, this is the case for the orbital locations at approximately the same longitude.

For service at higher latitudes, however, this is not necessarily the case. As an example, two points on the surface of the Earth located at latitudes of 50° and 55°, respectively, with a 5° difference in their longitudes, subtend an angle at the satellite varying from about 1° to about 0.4°, depending on the satellite location. The largest angle is associated with a longitude difference of about 50°.

The size of the projections on the surface of the Earth depends on the elevation. Figure 8 shows the geostationary satellite geometry, where $\emptyset$ is the great circle angle and T is the tilt angle at the satellite. The derivative $d\emptyset/dT$ is plotted against the elevation angle in Fig. 9. One degree at the satellite is subtended by a distance of about 620 km at $\varepsilon = 90°$, whereas as much as 2000 km are required to subtend the same angle at low elevations such as 5°.

It can be concluded from these considerations that the relationship of the satellite location to the coverage areas plays an important role in the potential reuse of the frequency spectrum by independent beams.

VII. Conclusions

The importance of reusing the frequency spectrum in a satellite is considerable. This technique could increase the total available spectrum at a geostationary location, thus permitting the implementation of satellites with large communication capacities, which operate in a power-limited mode.

One of the critical elements in frequency reuse is the control of the side-lobe level of the satellite antenna. It is, therefore, necessary to obtain data on how successfully this can be accomplished with multibeam spacecraft antennas. It has been shown here that a large number of frequency reuses would be possible for point-to-point communication systems if narrow-beam satellite antennas become available.

The largest potential multibeam satellite capacities are associated with operation at low-carrier-to-interference ratios. For this reason, large bandwidth-expansion ratios are likely to be favored, and a significant portion of the noise budget should be set aside for interbeam interference.

Distribution systems, such as a possible time-zone coverage in the U.S., which reuse the frequency in nonadjacent time zones are conceivable. Nevertheless, frequency reuse at the satellite is expected to be particularly attractive for point-to-point communications service.

It is expected that multibeam satellites reusing the spectrum will be more restrictive, with respect to the optimum orbital locations for each service, than Earth-coverage satellites. This will add new complexity to the problem of coordinating a variety of systems in the geostationary satellite orbit.

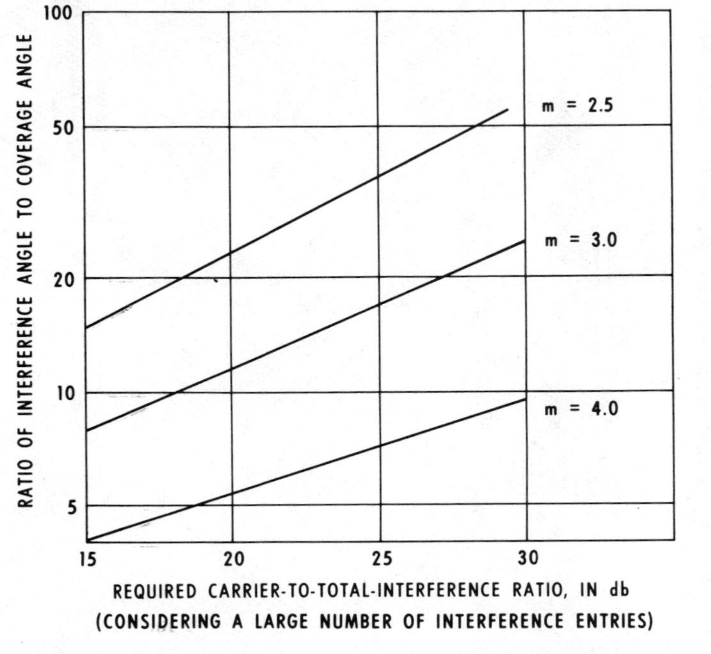

Fig. 1 Required ratio of interference angle to coverage angle.

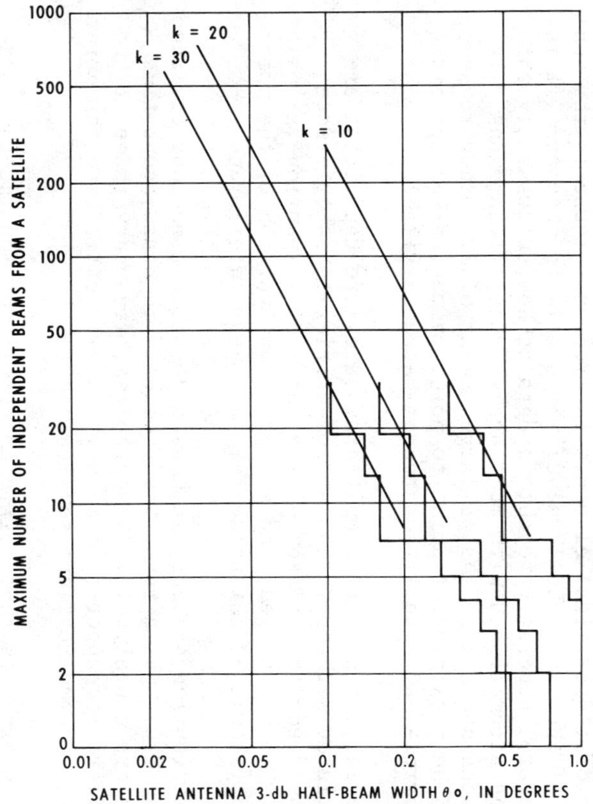

Fig. 2 Maximum number of independent beams.

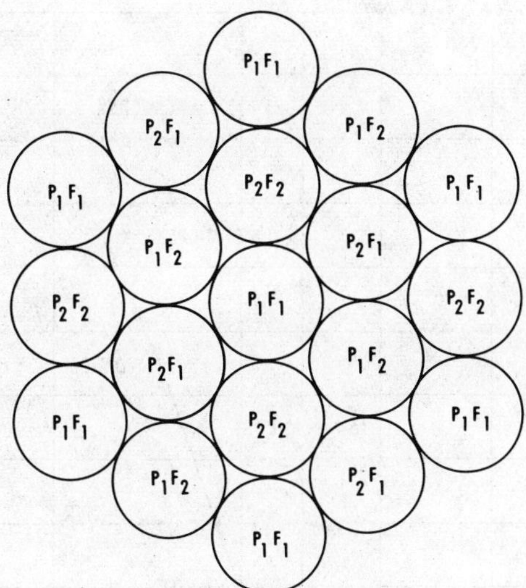

Fig. 3 Arrangement of polarizations and frequency plans.

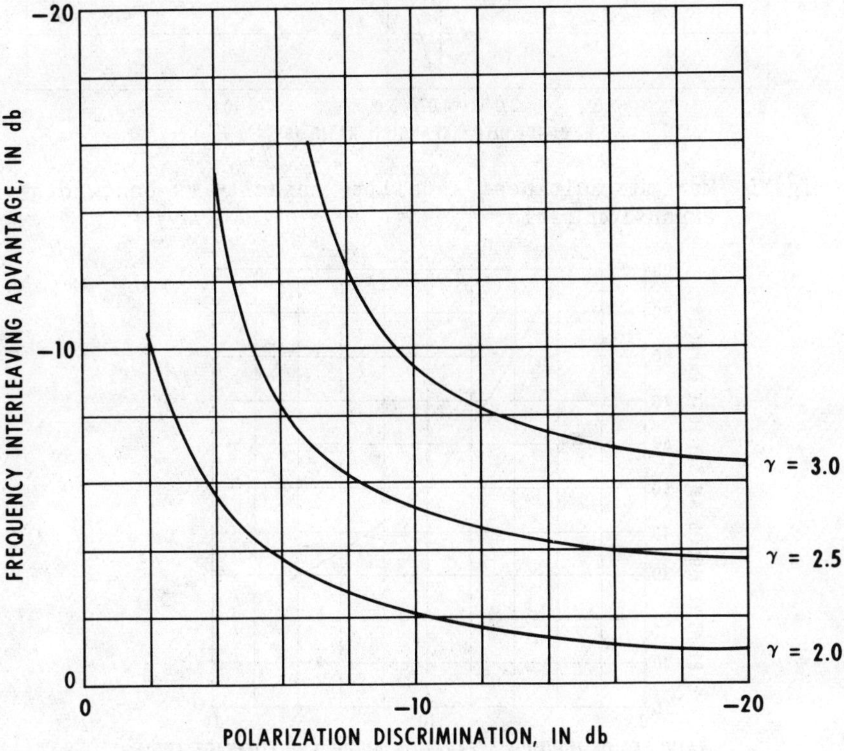

Fig. 4 Increase in beam density due to polarization and frequency interleaving advantages.

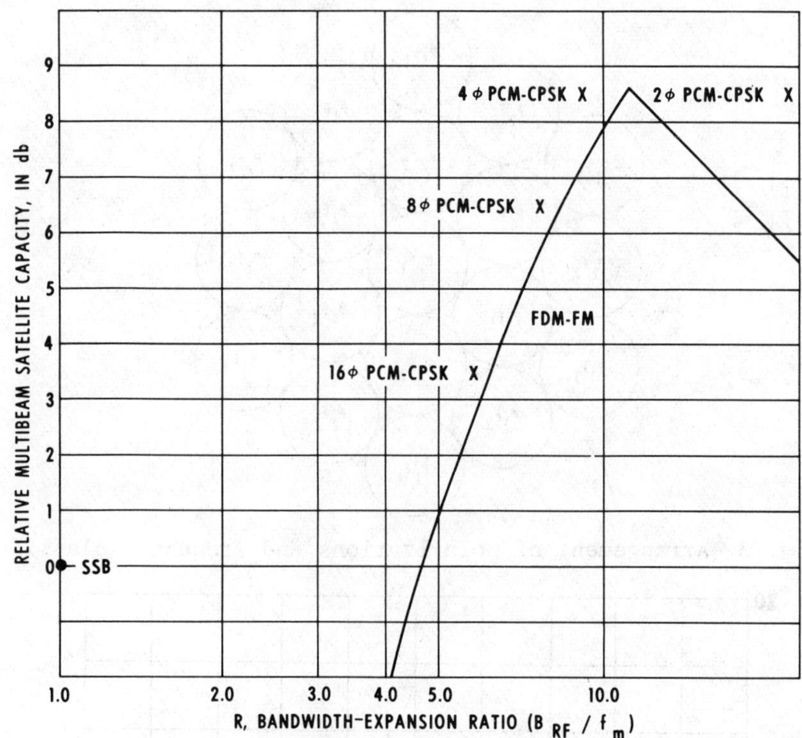

Fig. 5 Maximum multibeam satellite capacity vs bandwidth-expansion ratio.

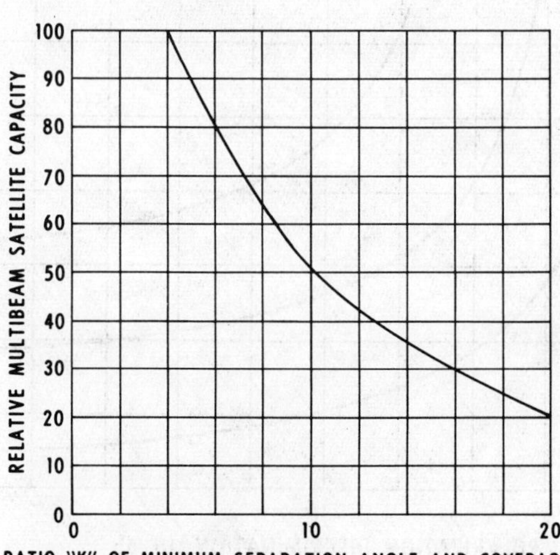

Fig. 6 Relative multibeam satellite capacity using ideal bandwidth-to-C/I tradeoff.

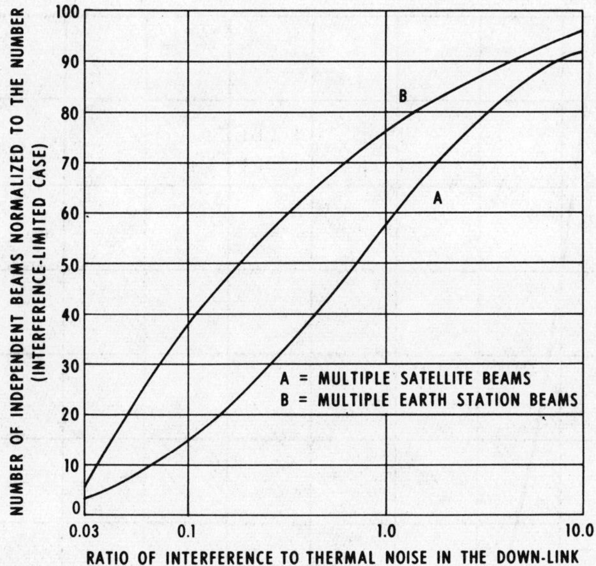

Fig. 7 Number of independent beams vs the interference-to-thermal-noise ratio.

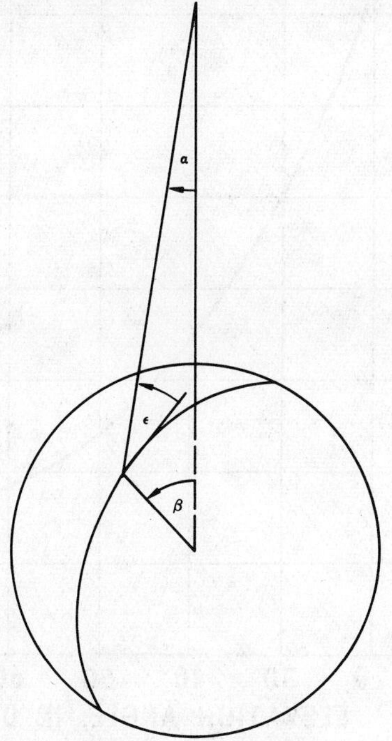

Fig. 8 Geostationary satellite geometry.

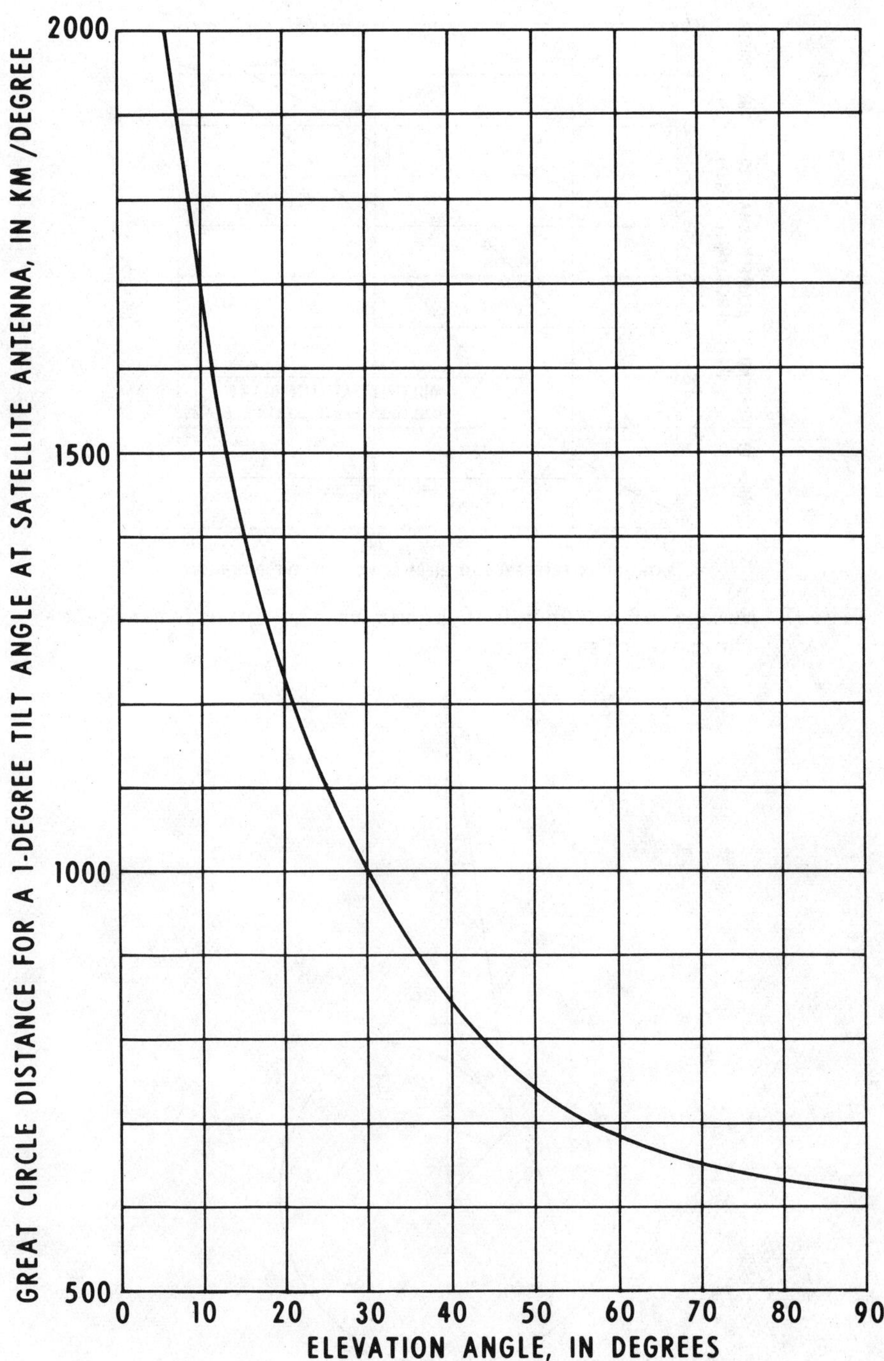

Fig. 9 Great circle arc for a 1° tilt at the satellite vs elevation angle.

IX. THE RELATIVE MERITS OF THREE-AXIS AND DUAL-SPIN STABILIZATION SYSTEMS FOR FUTURE SYNCHRONOUS COMMUNICATION SATELLITES*

* The papers in this chapter are based on material presented at a panel session of the AIAA 3rd Communications Satellite Systems Conference. They examine, on an over-all systems basis, the influence of antenna pointing accuracy and prime power requirements upon stabilization system mechanization for future communication satellites. The relative advantages of three-axis and dual-spin stabilization systems insofar as pointing accuracy, reliability, weight, power requirements, and dimensional constraints are concerned are examined, and possible tradeoffs between these and other relevant factors are considered. The last paper of the chapter is a summary of the panel discussion as viewed by the moderator.

T. B. Garber

DUAL-SPIN VERSUS THREE-AXIS STABILIZATION FOR FUTURE SYNCHRONOUS COMMUNICATION SATELLITES

D. C. Bakeman*

TRW Systems, Redondo Beach, Calif.

Abstract

The present generation of synchronous communication satellites utilize dual spin stabilization. However, the future need for satellites with an expanded communication capability (larger antenna array, increased rf power and improved antenna pointing accuracy) leads to further study of the relative merits of dual spin as compared to 3-axis stabilization. The study is made by examining both a dual spin and a 3-axis stabilized spacecraft design for two sets of future communication requirements. The results show increasing 3-axis design advantages as the requirements expand. However, full realization of these advantages requires the utilization of yet unproven lightweight extendible solar arrays and attitude control and propulsion hardware. It is concluded that the next generation of comsats will include 3-axis stabilized spacecraft.

Introduction

The relative merits of dual-spin with respect to 3-axis stabilization and control approaches† for synchronous communication satellites is a subject broadly debated, and each new spacecraft system design renews the debate and stimulates further investigation. The "present generation" of operational satellites utilize the dual spin approach. The "next

---

Presented at the AIAA 3rd Communications Satellite Systems Conference, Los Angeles, Calif., April 6-8, 1970 (not preprinted).

*Sr. Systems Engineer; now with Aerospace Corporation, Los Angeles, Calif.

†For purposes of this discussion, dual-spin systems are those in which the antenna platform is despun about the spin axis (normal to the orbit plane) of a spin-stabilized spacecraft, whereas 3-axis systems are those in which the spacecraft itself is utilized as the antenna platform.

generation" of operational satellites must satisfy demands for an expanded communications capability, thus requiring larger antennas and antenna arrays, greater power, and increased antenna pointing accuracy. These more severe requirements will have a major impact on the stabilization and control configuration chosen, and it is this relationship between expanded communication capability and the choice of dual-spin or 3-axis stabilization that will be discussed here.

Present Generation Satellites

The present generation of operational synchronous communication satellites will be defined as including Intelsat III, Skynet I, TACSAT I, DSCS-II, and Intelsat IV (see Table 1†). All of these satellites are of the dual-spin design with a spinning drum covered with solar cells and a despun antenna. The spinning and despun sections are connected by a mechanical despin mechanism (MDA) that can transfer signals or power (dc or rf) between the sections as well as provide the relative motion.

The communication requirements for the present generation satellites are approximately bounded as follows:

Antenna Configuration	
Antenna Envelope	10 × 12 × 12 ft
No. of Cantilevered Antennas and/or Elements	5
Solar Array Power	1 kw
Required Antenna Pointing Accuracy	0.25°
Lifetime	7 yr

Intelsat III was the first operational communication satellite to utilize a spinner with a mechanically despun antenna. The decision to do so was made during 1965, in a technical atmosphere where 3-axis stabilization seemed relatively complex, gravity gradient spacecraft had yet to be developed for synchronous altitude, spinning spacecraft had been successful, some confidence had been gained for the reliability of

---

†All tables and figures mentioned in this article will be found at the end of the Chapter.

nonhermetically sealed bearings in outer space, and electrically despun antennas appeared to be relatively inefficient.

The dual-spin concept proved to have a number of advantages relative to 3-axis stabilization, namely:

1) A simpler attitude sensing system. Scanning is provided by the spinner, and the spin momentum obviates the need for direct yaw angle measurements.

2) The propulsion system obtains ullage control from the spinner centrifugal force, a minimum number of jet thrusters are required, the same relatively high thrust level can be used for both attitude control and station keeping, and torques induced by leaks are averaged out.

3) The attitude stiffness resulting from the spinning momentum reduces the effects of torques created within the spacecraft and prevents any rapid buildup of attitude error resulting from environmental torques. Thus there is sufficient time for compensation by ground command to the spacecraft or to a gimbaled antenna. Also, the attitude stiffness can be utilized for attitude control during an apogee motor burn. (A 3-axis system using a momentum wheel also possesses this stiffness, but normally to a lesser degree.)

On the other hand, certain disadvantages of dual spin became clear.

1) The necessity for the MDA with its associated weight, power requirement, and cost.

2) A need for mechanical damping on the despun platform to compensate for nutational instability caused by an unfavorable ratio of spin to lateral moments of inertia and by energy dissipation in the spinning portion of the spacecraft.

3) Weight and cost penalty resulting from the added number of solar cells needed for a given power when mounted on a rotating drum.

Communication Requirements for Future Satellites

Trends in requirements for the communication payload of synchronous satellites include larger and more complex antenna configurations, more accurately pointed beams, and higher power.

Single unfurable parabolic antennas with apertures of 30 ft will be flown within the next few years, and even larger sizes are being considered. Large phased arrays are also under consideration. These large antennas can be used to reduce spacecraft dc power requirements, and to maintain a narrow, thus noninterfering and relatively secure, antenna beam.

Multiple antenna configurations result from requirements to relay information, either ground to ground or another spacecraft to ground, for a number of users and a range of frequencies. Configurations with as many as five gimbaled parabolic antennas have been seriously considered.

The increasing beam pointing accuracy requirement results from the use of narrower beams. Pointing errors will in some cases, have to be decreased to 0.1° or better in order to utilize contemplated beamwidths measured in tenths of a degree.

Higher dc power requirements at the spacecraft result from a number of factors, including the need for more rf power from any single antenna, the use of more communication links and channels, the use of higher frequencies with attendant lower efficiency transmitters, and the increase in onboard automation and data processing.

Design Examples for Future Communication Requirements

The relative merits of dual-spin vs 3-axis stabilization for future communication requirements will now be studied by examining spacecraft design examples for two sets of possible expanded requirements. The sets of requirements were chosen to illustrate both a reasonable extension of the present generation requirements in order to perceive trends (Design Example 1), and a major increase in power and antenna size requirements in order to examine more of a limiting case (Design Example 2). Only solar array power systems are considered in these design examples.

DESIGN EXAMPLE 1

Requirements

Antennas	Two 4-ft parabolic gimbaled
	Two 7-ft parabolic gimbaled
	Four 9-ft yagis
	One 2-ft horn

Array Power	1200 w
Gimbaled Antenna Beam Pointing Accuracy	0.15°
Lifetime	7 yr
Launch Fairing Diameter	10 ft

Spacecraft Design

A dual-spin spacecraft configuration for the above communication requirements is shown in Fig. 1, and a 3-axis-stabilized configuration is shown in Fig. 2. Important characteristics of the two configurations are presented for comparison in Table 2. These configurations are the result of first iteration "feasibility" designs, and no detailed computations were made of solar torque, cost, and reliability.

Discussion

The dual-spin vs 3-axis stabilization tradeoffs related to the spacecraft characteristics of Table 2 will now be discussed.

1) Weight. The weight is less for the 3-axis configuration because of the use of arrays that are mechanically sun-oriented for maximum efficiency and that are the lightweight extendible type which are still in the development phase although flight models are being readied. Their weight is less than half the weight per unit area of a normal spinner array, since the heavier substrate used on a spinner or rigid paddle array for physical integrity during launch is not needed when the array is folded or rolled up. The weight difference can be very important since the predicted spacecraft weight is near the maximum that can be carried on an Atlas Centaur, and a requirement for the greater lifting capability of the Titan IIIC would increase launch costs by at least 5 million dollars. Therefore some spacecraft capability may have to be sacrificed in order to limit the launch weight of the dual spin configuration.

Although not considered in the designs of Table 2, an additional weight advantage for a 3-axis design can result from a north-south station keeping requirement. The potential advantage is realized when light weight (high ISP) electric propulsion can be utilized because its power consumption is provided by relatively light weight solar arrays. However, electric propulsion is still in the development phase.

2) Size of Launch Configuration. The smaller size of the 3-axis stabilized launch (stowed) configuration reflects its greater design flexibility. The height of the stowed configuration for the dual-spin design is approaching the limit for existing Atlas-Centaur fairings (approximately 24-ft cylindrical length topped by 12-ft conical). The use of a shortened spinner with radial paddles, as shown in Fig. 3, decreases the stowed height to 18 ft but does not decrease weight. The use of a shortened spinner plus a directed array would reduce weight, particularly if an extendible array is used, and it would reduce stowed height. However, if the directed array is attached at the top of the spinner, antenna beam interference may result; and if the array is mounted at the bottom end, then mechanical complexity is increased and an apogee kick motor would be difficult to integrate. In either case, the resulting solar torque may be unacceptably large. The stowed configuration for the 3-axis design is shown in Fig. 4. With this design, two spacecraft can be carried in a 10- × 25-ft fairing, and a projected Titan III booster could launch the two simultaneously for perhaps 5 million dollars less than launching one each on Atlas-Centaurs.

The dual-spin design also is constrained to some extent by the practical MDA design limitation on the number of af and rf signals that can be transferred between the spinner and the despun platform, and by the power losses associated with transferring rf signals (the losses increase with frequency). As a result, redundant encoders/decoders may have to be used on both sides of the MDA, and the transmitters and receivers usually must be located on the despun platform.

3) Length of Antenna Structural Supports. The total length of the structural members which attach the antennas to the spacecraft body is of interest because, as those lengths increase, so does structural bending resulting from inertial forces and changing thermal gradients -- thus tending to degrade control system stability and antenna pointing accuracy. Therefore the addition of structural weight and/or complexity may be necessary to compensate for the longer appendage support members of the dual-spin configuration.

4) Heat Rejection Area. Heat rejection is normally accomplished from surfaces parallel to the orbital plane. The dual-spin design loses the use of some of that surface when an apogee motor partially covers one end of the spinner, but heat can be transferred to the solar cell surface if increased internal temperature changes during eclipse are acceptable.

The 3-axis design requires more insulation than does the dual-spin because the 3-axis spacecraft rotates relatively slowly with respect to the sun line. For the power requirements of this example, the available heat rejection areas for the two spacecraft designs are considered adequate, and both should require thermal control systems of comparable weight and complexity.

5) Solar Pressure Torque. The solar pressure torque will tend to be larger for the dual-spin configuration because the spacecraft is not symmetrical about the center of gravity, and a larger solar torque will increase the amount of propellant used for attitude control. Also, for a dual-spin design with attitude control from the ground, more frequent ground commands will be necessary to maintain the same attitude accuracy. The torque is minimized by locating components such that the center of gravity approaches the center of pressure, but undesirable tradeoffs in component location may be necessary.

6) Development Risk. A major development risk results from the difficult gimbaled antenna beam pointing accuracy of 0.15° relative to the line to a point on Earth, without automatic tracking. To achieve this accuracy, the antenna gimbaling capability would be utilized to avoid the propellant weight penalty associated with pointing the spacecraft and the difficulty in minimizing spacecraft attitude errors resulting from body dynamics. Pointing errors caused by orbital effects then can be removed by programed antenna commands, and those resulting from changes in spacecraft attitude can be compensated by antenna commands based on error signals from spacecraft attitude sensors or from separate attitude sensors located on the antenna. Those pointing errors which are not detected by attitude sensors can be minimized by electronic smoothing or careful mechanical design but cannot be explicitly compensated for by the control systems. Such errors include sensor noise, antenna structural support dynamics, and antenna thermal distortion. Potential error sources that are unique to the dual-spin configuration include spinner dynamic unbalance and MDA torque ripple resulting from variations in bearing friction.

A 0.15° pointing accuracy can be achieved for a dual spin system by the sole use of state-of-the-art Earth sensors if alignment bias errors can be calibrated in flight by pointing an antenna to a receiver with a known location. Earth sensors alone can also be used for a 3-axis system if the dual spin type of attitude stiffness can be achieved by a relatively large momentum wheel spinning about the pitch axis (perpendicular to orbital plane) or by a smaller wheel utilizing control-moment-gyro principles. The latter would be a development item. For a 3-axis system without inertial attitude stiffness, an attitude error about the yaw axis (vehicle to Earth) does not become an error about the roll axis (in orbital plane) for part of an orbit. Therefore, the yaw error cannot be measured by an Earth sensor, and devices such as a Polaris sensor or an rf interferometer would have to be utilized. A Polaris sensor with an electronic scan is a state-of-the-art development; and, although it may have to be deployed to achieve its required field of view, the resulting accuracy would still be adequate for the relatively relaxed requirements in yaw.

As pointing accuracy is further improved to reduce errors to less than 0.1°, Earth sensor accuracy becomes inadequate, and a sensing system using devices such as star sensors or rf interferometers would be needed for control about all three axes. For such a system the dual-spin spacecraft could utilize a state-of-the-art star mapper located on the spinner.

Pointing accuracy also requires a stable spacecraft and well-damped resonant modes. The relative difficulty of achieving these for a dual-spin or a 3-axis stabilized spacecraft must be considered for both the rigid and the flexible body dynamics.

Rigid body control is effected by gimbaling the antennas, and the effects increase with gimbaling requirements (angle, rate, and acceleration), antenna weight and size, and antenna mass unbalance about its gimbal point. A change in gimbal angle of an unbalanced antenna, for example, can produce despun platform mass unbalance that results in cross products of inertia and coupling between control axes. Such coupling increases the difficulty of control system design, particularly for a dual-spin spacecraft where stability can be seriously affected and the flexibility of control system design is more limited than for 3-axis stabilization. The momentum of a slewing antenna also can affect attitude control; however, if necessary, a momentum wheel can be attached to the antenna for compensation.

Adequate damping of flexible body dynamics normally requires that the bending mode frequencies be maintained well above the frequency of rigid body modes. This becomes more difficult as the bending frequencies tend to decrease with increasing antenna array size, and the dual-spin design is at a relative disadvantage since its nutation frequency is higher than the rigid body mode frequency for a 3-axis design. On the other hand the 3-axis design must contend with the bending of the cantilevered solar arrays.

In evaluating total development risk, it appears that the risk associated with the dual spin design is somewhat less than with a 3-axis design. The dual spin design need mainly contend with the mechanical/structural difficulty of obtaining acceptable dynamics. However, if full advantage is to be gained from the 3-axis design, then subsystem development is required in the areas of yaw attitude controls, lightweight extendible solar arrays, and probably electric propulsion.

7) Control during Apogee Kick Motor Burn. If an Atlas Centaur is to be the launch vehicle, then an apogee kick motor (AKM) must be added to inject the spacecraft into circular synchronous orbit. For the dual-spin spacecraft, a solid propellant engine can be integrated into the spacecraft, and attitude control during injection burn is provided by the momentum of the spinning drum. The 3-axis stabilized spacecraft also could be spun up to achieve attitude control during injection and then later despun (probably by the use of two identical sets of solid rockets), but the spacecraft configuration then has to be designed to withstand centrifugal force loads and to be adequately spin-stable. Another possibility for the 3-axis design is to control attitude with either a gimbaled motor or added high-thrust attitude control jets controlled by inertial sensors, but the associated cost is appreciable. For example, the use of a Burner II stage, which combines an AKM with attitude control, may cost on the order of a million dollars.

8) Reliability and Cost. No detailed spacecraft design work, particularly at the subsystem level, was performed for this design example, and therefore no detailed reliability or cost computations were made. However, from an over-all comparative examination, it appears that the reliability for the 3-axis design is reduced by the more complex attitude sensing system (which can be made redundant) while the dual spin design reliability is degraded by the platform despin system (the MDA cannot be made redundant). Also, the 3-axis design weight advantage possibly could be utilized to improve reliability by additional redundancy. Viewing this comparison and reviewing the results of previous design studies, it is judged that the reliability is comparable and would not be a major factor in a choice of configuration. Based on previous design studies, it also is judged that the spacecraft costs

for the dual-spin and 3-axis designs would be comparable. However, launch costs could differ as discussed previously.

Conclusions

Based on the preceding discussion, the advantages of a 3-axis stabilized spacecraft design for this design example are:

1) Lower weight reduces the probability that the final spacecraft design will exceed the launch capability of the Atlas-Centaur and thus require that the design capability be compromised or that a more expensive launch vehicle be used. In any case, the weight advantage can be used for greater reliability or radiated power.

2) A shorter shroud is needed to cover the spacecraft during launch, and the possibility exists of launching two spacecraft on a single launch vehicle.

3) The attitude control system is less sensitive to relatively low-frequency structural bending modes, to energy dissipation in the spacecraft, and to the increasing weight and size of the antennas and solar array.

4) Design modifications are easier to implement because the spacecraft configuration is less constrained.

Advantages of the dual spin design are: 1) simpler and flight-proven attitude sensing/control and solar array subsystems; 2) ease of implementing attitude control during an apogee kick motor burn, and 3) sufficient attitude stiffness that spacecraft attitude can be controlled by only ground commands at intervals of days.

Thus the dual-spin advantages previously listed for the present generation satellites are still applicable for this design example. On the other hand, some disadvantages that are inherent to the dual-spin design are becoming more important with the expanded communication requirements. In particular, the weight and stowed size are becoming less competitive. Also, confidence in achieving adequate dynamic stability and damping decreases more rapidly for a dual-spin design as communication requirements increase.

DESIGN EXAMPLE 2

Requirements

Antennas	One 30-ft parabolic
	One 2-ft horn
Array Power	12 kw
Beam Pointing Accuracy	0.25°
Lifetime	3 yr
Launch Fairing Diameter	10 ft

Spacecraft Design

A spacecraft configuration that satisfies the requirements is shown in Fig. 5. This is a 3-axis stabilized configuration with an unfurlable antenna and extendible solar arrays.

Discussion

The requirements for this design example are extreme but have been considered for a TV broadcast satellite. Only a 3-axis stabilized spacecraft design is shown (Fig. 5) because the design study failed to produce a feasible dual-spin design.

Since a solar array for this power cannot be mounted on the spinner of a dual-spin configuration, the design study sized the spinner to produce only the minimum momentum necessary for attitude control. Even then a dual-spin design exceeded constraints in the areas of launch size, weight, and mechanical complexity.

Conclusions

As the dc power requirements and the antennas become very large, the classical dual spin design and characteristics are no longer applicable. The reason is that the solar array must then be an extendible directed array, and a spinner, if used, would be essentially just a momentum wheel probably carrying little or no electrical equipment and rotating at speeds considerably higher than the normal dual-spin rate.

General Conclusions

A great deal of experience and precedence has been accumulated for dual-spin stabilization by the present generation of synchronous communication satellites. Therefore an operational 3-axis stabilized satellite will be built only when a 3-axis design clearly can be shown to be both feasible and advantageous. This will probably require that improved attitude control, solar array and propulsion subsystems be developed for a 3-axis design and that future communication requirements increase solar array size, antenna size, and beam-pointing accuracy to the degree that either 1) the weight and configuration advantages of a 3-axis stabilized design provide a clear cost or capability advantage, or 2) the greater structural/mechanical difficulties in achieving acceptable dynamic characteristics for a dual-spin design are judged to involve comparatively high risk.

There is a level of communication requirements at which the above will occur, but that level cannot be defined as a general criteria because of the complex factors and value judgments that are involved in any specific dual-spin vs 3-axis decision. However, since improved subsystems for 3-axis control will be developed and proven, and since future communication requirements will expand, it can be concluded that the next generation of synchronous communication satellites will include some that are 3-axis stabilized.

RELATIVE MERITS OF THREE-AXIS AND DUAL-SPIN STABILIZATION SYSTEMS FOR FUTURE COMMUNICATION SATELLITES

K.W. Jenkins*

General Electric Company, Valley Forge, Pa.

Abstract

An analysis of the future choice of attitude control sub - system design for communications satellites is presented. The previous success of the dual spin design is primarily due to its design simplicity. (No oriented solar array, passive thermal design, and simple propulsion design.) In the future, as power requirements extend beyond a kilowatt, the need for lightweight roll-out arrays will appear. This requirement, plus the appearance of surface tension propellent tanks, requiring no bladders and the greater possible reliability of three axis designs make them appear more attractive. A comparison table of the dual spin design, the three axis momentum wheel design, and the flywheel design show that the next generation communications satellite will utilize either a three axis or flywheel stabilized attitude control subsystem instead of the present dual spinner type of design.

Development History

Early efforts to establish a military communications satellite system included the Score, Courier, and Advent Programs. The Advent Program was an ambitious three-axis stabilized satellite having oriented solar arrays. As a result of weight limitations of the then available boosters, the Advent Program was canceled in 1962, and it was decided to rely upon very simple satellites to increase reliability. The result was the Initial Defense Satellite Communications System (IDSCS), which was a spin-stabilized design.

Presented at the AIAA 3rd Communications Satellite Systems Conference, Los Angeles, Calif., April 6-8, 1970 (not pre - printed).

*Manager, ATS F and G Attitude Control; now Manager, Process Control Technology, General Electric Company, West Lynn, Mass.

At the same time, nonmilitary synchronous communication satellites, such as SYNCOM, ATS-1 and Early Bird, were appearing. These vehicles were also spin-stabilized. As a result of their success, the next series of spacecraft designs attempted to increase the effective radiated power (ERP) by both increasing the spacecraft size and by despinning the antenna to provide directional antenna gain. Continued success with this design approach has led to larger and larger vehicles and also to a greater percentage of the total weight in the despun section. As a result, the present designs are approach ing a case of "Tail Wags Dog".

In order to predict the future design for synchronous COMSATS, it is first necessary to review some of the previous designs. Table 1 presents some of the parameters for four different spacecraft to illustrate the evolution of the dual spinner.

The "Tacsat" vehicle is meant to be a design similar to the Hughes TACSAT and represents the class of vehicles launched by a Titan IIIC,using the Gyrostat principle. The programs listed are chosen as representative of the growth of the dual-spin design. This is clearly shown by the percentage despun weight column.

ATS-1 was a spin-stabilized design. The usual cited advantages of spin stabilization are 1) passive control during orbit inject, 2) only two thrusters required, 3) scanning sensors require no moving parts, 4) no solar array orientation required, 5) passive thermal design 6) insensitive to thruster leakage, 7) no active control loops, and 8) positive propellant expulsion.

The success of the OSO dual-spin design and the desire to increase ERP led to the adoption of dual spin for Intelsat-3. Note the more efficient conversion of array power to ERP. However, design of the despun bearing is a new problem. On Intelsat 3,the bearing seized, evidently because of thermal problems. Similarly,the TACSAT vehicle has been experiencing an unexpected wobble with an amplitude of about 1°. This has been attributed to an unexpected flexibility in the bearing. The advantages of the spinner also apply to the dual-spin design. However, there are disadvantages to this design. First, it depends on the rotating bearing design which, as we have seen, can create problems. Second, the pointing accuracy is relatively low since absorption of solar-pressure-induced momentum causes precession. Typically, to hold a TACSAT type design to $\pm 0.2^{o}$ requires daily corrections. Last, and most important, the dual-spin design is power limited by the diameter of the available shroud. At present, the 10-ft shroud limits the drum diameter to approximately 108 in. This results in an effective limit of 1kw for the dual-spin design. In addition, at high array powers the cost of a non-oriented array is considerable. For a 5kw array, assuming 500 sq ft of oriented array and a cost of $\$2700/ft^2$, the cost penalty for a drum array is \$3.4 million.

Future Requirements

Future communication satellites will require greater pointing accuracy and higher power. Figure 6† shows the way in which the ERP has grown, The higher ERP will be achieved both through higher transmitter power and narrower beamwidth, which in turn will require greater pointing accuracy. Sabroff[1] has pointed out that increased accuracy significantly increases power at the half-power points. He notes that, by increasing accuracy from 0.5 to 0.1^{o}, the half-power point of a 2° beamwidth antenna is increased 3.6 db.

Recent studies at General Electric have shown that a spacecraft capable of providing TV and some communications capability for an emerging nation would require 2.5 to 5.0 kw of array power. Similarly, to perform an operational version of the PLACE air traffic control experiment will require an array power of 2-3 kw based upon a 10-channel system. Studies also are underway of a 1.3 to 1.5kw NAVSAT using L-Band and having 350 w of rf power, consisting of one voice, one data, and one surveillance channel. Finally, to implement a Direct Broadcast Satellite, a power of 20 kw is required.

At these higher power levels the lightweight roll-out array is preferable since conventional arrays at 10w/lb impose a considerable weight constraint over the 30w/lb achievable with the roll-outs. For instance, the weight penalty at 5kw is 300 lb.

In general, the trend seems fairly clear. The next generation communication satellite will have a 2kw or higher array output and pointing accuracies of 0.1^{o} or better. They will feature ERP's of 70 and higher, using large ($\sim$20 ft) reflectors and lightweight solar arrays.

ATS F and G Attitude Control

General Electric's design for ATS F and G is an example of a 3-axis vehicle. ATS F and G features a radiation balance type

† All figures and tables are at the end of the Chapter.

Earth sensor, yaw control via a Polaris tracker, and torquing via three momentum wheels. The propulsion subsystem is a hydrazine monopropellant system operated in a blow-down mode. The tank uses a surface tension device for propellant expulsion. Full thruster redundancy is achieved by backup pitch thrusters plus roll and yaw being backups for each other. The attitude control subsystem is capable of pointing the 30 ft reflector to any spot on the Earth's surface to an accuracy of $\pm 0.1^{\circ}$. The momentum wheels are used for all pointing and slewing maneuvers with secular momentum dump by the thrusters occurring every two days at a maximum. Reliability is enhanced by using dual controllers and gyros/ sun sensors as a backup for Earth sensor/polaris tracker combination. Subsystem weight is 105 lb, and the base power is 41.5 w plus an additional 20 w to keep the gyros on standby. Other vehicle parameters of interest are

Total vehicle weight	2000 lb
Fixed solar array	160 ft^2
EOL power	300 w
Antenna beamwidth	
X-band	0.3°
S-band	1.0°
L-band(FM)	$1.0^{\circ} \times 7.5^{\circ}$
L-band (pencil)	2.6°
UHF	2.6°

Except for the array power, which will grow, these parameters, and in particular those of the antenna, are typical of the next generation communication satellite.

Comparison of Attitude Control Systems

Table 3 presents a tradeoff matrix between the conventional dual-spin design, such as TACSAT; the 3-axis design, such as ATS F and G; and a hybrid system using a flywheel rather than the drum to produce a momentum bias. The flywheel design was included to explore the possibility of including the good portions of both designs. Much of the table is self-explanatory; however, a few comments are in order.

The reliability of the dual spinner is no better than that of the bearing and power transfer assembly. Although the slip ring assembly is usually made redundant, it is difficult to make a redundant bearing. On the other hand, the reliability of momentum wheels can be enhanced by additional wheels. Likewise, the dual spinner can employ redundant flywheels or a backup 3-axis system.

The use of an apogee kick stage is a definite advantage of dual spinners; however, as satellite weight increases the use of the Titan class launch vehicles eliminates the need for the kick stage feature. Similarly, the advent of radiation balance type Earth sensors and image dissector star trackers has lessened the need for scanning sensors.

As discussed, one of the biggest limitations of the dual-spin design is the power available. With the 10-ft Titan shroud, this results in a 1kw limit. Even assuming the 14-ft shroud were available, the power limit would be approximately 2kw, and the weight penalty in drum weight would be 530 lb.

Testing is another area in which the 3-axis type design or the flywheel design has the advantage of simplicity. The Gyrostat principle requires determination of all damping sources in the spinning body. This means that it is necessary to retest after adding any components to the spacecraft. Failure to retest after adding heat pipes on ATS-E is thought to be the reason for its failure. The dual spin design may be difficult to balance dynamically if the inertia ratios are small. It is much easier to balance a rigid flywheel, and of course the 3-axis design does not need accurate inertia or damping measurements. Therefore, on a production basis, component tests of the attitude control are sufficient.

The dual-spinner design typically has a large cp-cg offset resulting in large solar pressure momentum buildup. For instance, a dual-spin version of ATS F and G would have a peak solar torque of 263×10^{-6} ft -lb vs 30×10^{-6} for the 3-axis design. In addition, the dual spinner responds to solar pressure with attitude drift, and the 3-axis design absorbs the momentum in the reaction wheels.

An often-cited advantage of the dual spinner is the positive propellant expulsion, eliminating the need for bladders, and the fewer number of thruster and valve assemblies resulting in greater reliability. The introduction of surface-tension devices, such as those used by Lockheed on the Agena, eliminates the first advantage. The second one still applies. However, in the case of hydrazine, valve reliability is primarily a function of the number of cycles. Because the thruster must fire at the proper point in the spin cycle, the two thrusters on a dual-spin spacecraft must cycle 40 times as frequently per year.

The last item on the matrix may be used to summarize. The basic problem with a dual spinner is its relatively rigid configuration. As power requirement go up, so do the cost and weight penalties of the drum-mounted array. Further the dual spinner has high solar pressure. If the arrays are removed from the drum, then two sets of power slip rings are needed to get power on and off the spinning section, plus another set of slip rings if the array is oriented. With surface tension devices available, the design flexibility of the flywheel or 3-axis system seems irresistible. The flywheel can be packaged anywhere. The rollout arrays can be placed on one side of the cg with the antennas on the other to minimize solar - pressure effects. Further, precision design of flywheel bearings whould be much easier than despun bearings. Reliability of momentum wheels is excellent. The Nimbus wheel was still operating properly after 32 months in orbit.

The choice between a flywheel and a 3-axis design is more a function of the particular mission requirements, since these two systems are very similar. For missions requiring frequent reorientation of the vehicle, the 3-axis design is preferable. But this also can be done by gimbaling the flywheel, in which case it becomes a CMG. The reverse procedure is also possible. The rotor section of the Apollo Telescope Mount CMG generates 2000 ft-lb sec of momentum and is therefore equivalent to an Intelsat IV spinner.

Summary

The next generation communications satellite will be a 3-axis or flywheel stabilized spacecraft. It will utilize rollout arrays to generate from 2 to 5 kw of power, and it will be launced from a Tital IIIC. For those systems where the entire vehicle is not rapidly reoriented, the system will most certainly feature a large (2000-ft-lb-sec) flywheel to take advantage of the present features of a dual spin design.

Reference

[1] Sabroff, A.E., "Advanced Spacecraft Stabilization and Control Techniques," Journal of Spacecraft and Rockets, Vol.5, no. 12, Dec. 1968, pp. 1377-1393.

RELATIVE MERITS OF THREE AXIS AND DUAL-SPIN STABILIZATION SYSTEMS FOR FUTURE SYNCHRONOUS COMMUNICATION SATELLITES

J. J. Rodden*

Lockheed Missiles and Space Company, Sunnyvale, Calif.

Abstract

The design practice for future large, complex communication satellites is expected to prefer three-axis attitude stabilized configurations with sun tracking solar arrays over spin stabilized systems. A comparative evaluation of subsystem considerations shows that the use of three-axis stabilization results in better design flexibility in terms of equipment selection, placement and interconnection. The integration of subsystems is less demanding in a non-rotating system than one spin stabilized for spacecraft with large channel capacity and multipurpose communication applications.

Introduction

Future communication missions will require increased rf-channel capacity together with extended operational life and multipurpose applications. Two competing satellite design philosopies lead to systems with de-spun Earth pointing antennas and three-axis Earth pointing systems. Current communication satellites are spin stabilized. Considering the electrical power requirements to sustain increased communications capacity and the physical limitation on the amount of solar cells that can be mounted on a spinning body drum, coupled with physical complexity of multipurpose antennas, it is expected that three-axis stabilized satellites with sun tracking solar arrays will be the prime candidate for future missions. The stabilization systems using momentum and reaction wheels as control torque sources are considered best choice on the basis of meeting

---

Presented at the AIAA 3rd Communications Satellite Systems Conference, Los Angeles, Calif., April 6-8, 1970 (not preprinted).

*Manager, Attitude Stability & Controls Systems, Space Systems Division.

life and accuracy requirements with proven performance. A comparison of dual-spin and wheel-stabilized configuration involves consideration of each spacecraft subsystem.

Electrical Power

The wheel-stabilized satellite has significant flexibility in array sizing to accommodate varying loads, and it features either hardwiring of all power to the load or through slip-rings across a slow array drive system. The body-spin satellite has a solar array area, limited in diameter and length with a separate battery-charging section, and requires a high-speed slip-ring/brush assembly between the array and the despun load.

Flexibility of power requirements is inherent in the wheel-stabilized satellite. The extendible array can be designed to provide extra area per extension without further spacecraft modification except for matching the battery amperehours to the load. This arrangement enables easy adjustment of solar power to accommodate design changes. The body-spin satellite is severely limited in solar power flexibility, especially to increases in drum length. Since the drum and the enclosed equipment constitute the flywheel of a momentum device, any change in mass distribution severely affects the dynamic balance, center-of-gravity location, and bearing loads; therefore, redesign would be necessary to regain antenna pointing accuracy.

Power source weights for solar array and NiCd batteries as a function of the power requirement are shown in Fig. 7.† For multi-kilowatt systems, the weight advantage of suntracking flexible substrate solar arrays is more than 10 to 1 over a cylindrical spin-stabilized system giving the same power. The brush and slip ring problem is significantly more severe for spinner systems. For a typical rotation of 60 rpm, there are more than 86,000 slip-ring rotations required against one revolution of a sun tracking system. Qualification testing has shown that slip-ring wear and resulting debris can be a major problem. Redundancy is more readily applied to the slower rotation system. A hardwire recycling system can be used to eliminate slip-rings altogether if desired.

Attitude Stabilization

The wheel-stabilization approach to attitude control features simple on-board equipment, on-board data processing, and

† The figure is at the end of the Chapter.

gives accurate pointing without reliance on continuous ground station corrections. The wheel-stabilized satellite can be completely fail-safe through redundancy and backup systems, whereas the body-spin system has bearing failure modes leading to interruptions in communications and even permanent outage.

Both the wheel-stabilized and body-spin satellites maintain antenna beam pointing by means of gyroscopic stiffness and a nonspinning communication assembly. In the wheel-stabilized system, the entire spacecraft is despun with a hermetically sealed single or multiple flywheel, rather than an exposed despin joint. With the possible exception of any horizon sensor's scanning mirrors, the wheel-stabilized attitude system can be, part-for-part, less complex than the body-spin system both in terms of equipment required and in electronic parts count.

The wheel-stabilized systems are controlled by wheel speed changes. A reaction wheel control loop commands the wheel speed to change in proportion to the attitude error signal. Knowledge of wheel speed and position is not required. In the body-spin system, both the speed and position of the despun antenna must be continuously and precisely controlled with respect to the attitude reference horizon pippers located on the spinning drum. In addition, the servo system in the body-spin system requires on-board electronic data processing to capture, synchronize, and precisely orient the stationary section relative to its spinning section.

Pointing accuracy is inherent in the wheel satellite. Alignment accuracies can be directly measured and controlled; and correction, if required, is direct. Contrasted with this simplicity is the spin system's requirement for stringent control of the dynamic balance and alignment of the rotating drum. Since the momentum stored in this drum provides the attitude orientation stiffness, any unbalance or misalignment produces a wobble in the despun antenna. Furthermore, accuracy degrades with bearing wear, thermal and load distortion of the structure, and spin-rate variation.

Two considerations are important in bearing operation -- preventing temporary outages and preventing catastrophic failure. In both systems, spurious thrusting from leakage of the mass expulsion system can be avoided by the use of series valves. In the wheel-stabilized system, an inactive standby thruster control system can insure antenna orientation in the event of failure of the wheel servo until control is switched to a redundant wheel. With wheel redundancy, no single failure can lead to communications outage. Any failure of the despin joint of the body-spin system, with its critical bearing design, is catastrophic.

Positioning and Orientation

The wheel and body-spin satellites can both feature multiple redundancy of hardware and meet the communication system specification with one thruster out. By employing sufficient thrusters, the wheel-stabilized satellite provides multiple redundancy in the required functions of despin, reorientation, roll, pitch, stationkeeping, and wheel desaturation. Such a system performs normally with any thruster failed closed, and in some cases with two thrusters failed closed.

Propellant management of the wheel-stabilized satellite in the zero-gravity environment may be obtained by use of an aluminum perforated sheet ball, installed inside the propellant tank 0.1 in. from the tank wall. Liquid surface tension entraps the propellant between the ball and the tank wall. This method insures that liquid will fill the tank outlet under any orientation and at externally applied forces up to 1 g. The forces imposed on the propellant by the thrusters does not exceed 0.01 g. This tank system requires only two tanks for complete redundancy and affords the further advantages of having no pressure balance lines and having all metal surfaces wetted with propellant. In addition, it does not require precise installation location. Tank design in the wheel-stabilized vehicle is simplified by the absence of strict mass distribution requirements during propellant utilization.

The body-spin satellite generates an artificial gravity by spinning the tanks. This establishes the pressurizing gas on the inner hemispheres of the four tanks and in the two tank pressure balance lines. These gas-to-metal surface contacts increase the possibility of leaks from pressure with resulting unavailability of the propellant.

Structures

The wheel-stabilized satellite structure can be more efficient and lighter than that of the body-spin system. The wheel-stabilized satellite has a control load path with compactly stowed solar arrays. This simple, lightweight structure can fit within a flight-qualified shroud. The body-spin stabilized satellite also has a central load path; but, with this satellite, it must support the despin assembly, the cantilevered equipment rack, and the solar-array drum. In addition, the despin bearings must support a cantilevered despun platform through the ascent environment.

Telemetry, Command, Communications, and Antennas

The telemetry and command subsystems are more complicated in the body-spin satellite because of the slip rings between the spacecraft and the transmitter and receiver. The communications packages are the same, and the antennas are equivalent in complexity on the body-spin satellite.

In the wheel-stabilized satellites, all telemetry and command signals are hardwired to the redundant telemetry encoders. In the body-spin satellite, a method of transferring information across the despin joint is to multiplex it on each side of the joint, cross the joint on two lines per telemetry system, and combine the multiplexed signals on the transponder side. The command link similarly transmits the coded command message across the interface and decodes the proper commands on each side, requiring four slip rings. The body-spin satellite is thus burdened by the complexity of an extra synchronized multiplexer, an extra signal conditioning box, and an extra coder over that needed for the wheel-stabilized satellite in addition to the reliability degradation from the eight slip-rings.

Thermal Control

Passive thermal control of the wheel satellite equipment rack is provided by multilayer insulation and thermal control surfaces. Thermal control of the equipment in the solar array cylinder of the body-spin satellite is maintained by cyclic exposure to solar heating during sunlight operation. During eclipse, insulation and heaters are required in the spinning vehicle.

The insulated equipment compartment with North-South facing radiating surfaces used in the wheel-stabilized satellite isolates all spacecraft equipment from solar radiation flux to keep a uniform ambient temperature. A mosaic of optical solar reflector surfaces on the ends controls outflow of heat to space. This heat rejection system is unaffected by eclipse.

The solar-array drum of the body-spin system radiates an even temperature to all spacecraft equipment during sunlight operation, typically about 75° F. In eclipse, however, the solar array becomes an efficient radiator to space, rapidly drawing heat away from the spacecraft equipment. Temperature-sensitive equipment, such as batteries and the hydrazine tanks and thrusters, must be insulated, making the thermal design problems of the two systems the same. The wheel satellite's solar-oriented array arms experience no diurnal temperature variations but operate at a temperature of about 130° F except and during eclipse periods. The result is that the wheel-stabilized satellite solar arrays experience the same temperature cycles as those of the body-spin system but operate at a higher temperature.

The hydrazine thrusters of the wheel-stabilized satellite are mounted inside the insulated equipment compartment and probably require no heaters. For the body-spin satellite, insulation and heaters are needed during eclipse, since the thrusters are tied to the solar array drum. The wheel satellite apogee motor is exposed to sunlight during transfer orbit so it receives adequate solar energy to exceed a 40° F temperature. The body-spin apogee motor is shielded from the sun by the solar-array drum, so it requires insulation blankets and a heater. Larger view factors on the wheel-stabilized communications electronic assemblies make the thermal control problems and solutions more tractable because of the larger radiation areas available. The communications package on the spinning vehicle must be thermally decoupled to protect the despin bearing assembly from high temperatures.

Orbit Operations

Attitude and orbit control of both wheel or spin-stabilized satellites are conducted in almost identical fashion in final orbit. In both satellites, East-West antenna pointing is normally controlled continuously by on board equipment. North-South pointing drift, resulting from solar pressure torque, is corrected by ground command about once every 2 or 3 weeks, and orbit correction maneuvers in both systems also are commanded from the ground about once every 2 weeks. A pair of jets are used for North-South stationkeeping to keep the satellite within the antenna pointing specification.

Ground Handling Comparisons

The wheel-stabilized satellite offers the advantages of service access to equipment without removal of fragile solar-array panels, elimination of high-precision dynamic balancing, static system testing instead of an active spinning operation, transportability in smaller aircraft, and simplified spacecraft storage. The complete integrated systems run on the wheel-stabilized satellite can be accomplished with no spinning motion except for the stabilization wheels running inside hermetically sealed cases, no spin tables with test table slip rings, no windage or nutation damper test uncertainties, no fragile exposed solar cells during system tests, and no dynamic balancing in the orbit condition. The stabilization wheels are permanently balanced separately at the vendor's factory. The spacecraft is balanced in a single-step process at the launch base to the ascent requirements. These requirements are based on the accuracy of the final orbit injection, which is less stringent than the requirements for stabilizing a satellite to an accurate pointing reference in final orbit.

The body-spin satellite must be balanced in a two step process -- for the transfer orbit and for the final orbit configurations. This latter balancing has to be conducted repeatedly throughout the manufacture and test cycle, whenever the configuration changes, because of test instrumentation or component replacement. Test facilities for the body-spin satellite must include an accurate spin table for dynamic operation during system test. This facility also must contain highly accurate despin platform alignment equipment to monitor the operation of the despin servo. None of this equipment is necessary for system testing of the wheel-stabilized satellite.

Conclusions

Simplicity. The wheel-stabilized satellite is at least as simple as the body-spin system in spacecraft design and operations and much simpler to assemble, test, and transport.

Confidence in design. Both the wheel-stabilized satellite and the dual-spin designs are based on qualified hardware and approaches that have been applied successfully in major military and NASA programs.

Reliability. Of utmost importance, the wheel-stabilized satellite can be immune to any single catastrophic bearing failure mode that could cause telecommunications outage. Reliability of the body-spin system is compromised by the vulnerable slip-rings and brushes, and binding of the despin bearings could paralyze the satellite and cut off communications. There is no practical backup for these bearings.

Flexibility/growth. In addition to its greater dependability, the wheel-stabilized satellite has inherent growth potential. Simply adding or subtracting solar-array area equips it for different missions, and updating it for longer life only requires larger propellant capacity. By contrast, the body-spin system is limited in growth potential. Any changes in the length of the drum or even the propellant capacity disrupts mass properties control and consequently the dynamic balance crucial for pointing accuracy. Therefore, adaptation to a different mission could require major redesign.

In summary, the wheel-stabilized satellite is comparable to the body-spin system in simplicity. In design experience it can eliminate single path failure modes and afford mission flexibility and growth potential.

STABILIZATION OF FUTURE SYNCHRONOUS COMMUNICATION SATELLITES

A. J. Iorillo*

Hughes Aircraft Company, El Segundo, Calif.

Abstract

Synchronous communication mission requirements projected by various system designers for implementation in the 1970 to 1975 time frame are examined to determine appropriate satellite design characteristics including antenna size, power system capacity, and vehicle pointing accuracy. The resultant characteristics are compared with current vehicle systems. It is shown that the required spacecraft capabilities are well within the current state of the art.

Nearly a decade has passed since the first successful synchronous communication satellite program was started — NASA's SYNCOM program in 1961. We have witnessed dramatic advances in the technology, which has evolved from the 80-lb SYNCOM with 5 voice circuits to the 1300-lb INTELSAT IV with 9000 circuits or 12 television channels. Figures 6 and 8† illustrate the historical progression in terms of radiated power and satellite weight. The technological reasons for this growth were many-fold. The single most important, of course, was the increase in launch vehicle weight and volume capacity from the early THOR-DELTA to the ATLAS-CENTAUR and TITAN III. With all this marvelous progress and experience, one would imagine that by now we would have a more or less clear picture of the future of communication satellites. Unfortunately this is not the case. The technology developed so fast that our institutional structure has not been able to develop adequate policy for its use. Though we may envision a broad range of

---

Presented at the AIAA 3rd Communications Satellite Systems Conference, Los Angeles, Calif., April 6-8, 1970 (not preprinted).

*Manager, System Studies, Space and Communications Group.

†All figures and tables are at the end of the Chapter.

possibilities for satellites, it is not yet clear how such services would be implemented nor by whom.

For example, current nonmilitary communication satellite policy in the form of the 1962 COMSAT Act does not yet reflect the possibility of stationary satellite services. Formulated in the presynchronous TELSTAR era, the Act directs the COMSAT Corporation to establish a global communication satellite system as an international joint venture with other nations of the world in INTELSAT and does not provide for domestic service. For years, debate regarding the domestic issue has been under way. Recent White House position papers indicate that formal policy may be forthcoming. Nonetheless, the issue is not resolved. This nontechnical situation affects us because policy will influence future system trends greatly. Should exploitation of this resource be left solely to free enterprise, the characteristics of future systems could differ markedly from those that might be deployed by a government-subsidized agency charged with eleemosynary objectives. Thus, after a remarkable decade, the only nonmilitary operational use of synchronous technology is the international trunking task assigned COMSAT. As such, their INTELSAT vehicles represent alternates to and expansion of existing cable and long-line installations.

We are familiar with the various new satellite services which have been studied and proposed in the last few years. The most durable are listed in Table 4. A number of these have been demonstrated in NASA's ATS program. In particular, mobile communications involving aircraft and ships were accomplished with ATS I and III, including television transmission from the APOLLO recovery ships. The remaining missions will be performed in limited ways on NASA's ATS F/G satellites. By 1974 most of the conceivable uses of synchronous satellites will have been demonstrated to a greater or lesser extent. Few seriously doubt the technical feasibility of any of these.

The real challenge of the next decade is that of creating business structures and governmental regulatory policies that will integrate satellite technology into the routine operation of the private sector. This will not be easy. What we need most is the merging of daring entrepreneurs, creative application engineers, and enlightened regulators.

This writer believes that the near future of communication satellites will be characterized by application of "current" technology to missions such as those in Table 4. We do not expect great emphasis on developing "new" technology. If we are not successful at applying current technology, it is difficult to believe that we will be given an opportunity to develop new technology.

The foregoing summary of our views is intended to provide a backdrop for our conclusions. Rather that discuss stabilization systems merits for abstract mission parameters, we felt compelled to attempt to bound the probable future. Our approach has been to examine available recent studies which relate to the application missions in Table 4 and to extract the satellite parameters suggested. We have not attempted to project beyond these applications to more "advanced" systems. We recognize that these studies may never result in any action and that controversies continue over frequency selections and the like. Nonetheless, on the whole, the list of missions and typical parameters represents a fair cross section of community thought in the area of future requirements.

An interesting characteristic common to all the systems - except the data relay — is that the selected antenna sizes were limited by the requirement to cover specified geographic areas. For example, the L-band communication and navigation antennas on the navigation and traffic control satellite are required to cover the northern hemisphere and the entire globe, respectively. The domestic satellite antennas are sized to cover time zones in the United States and all of Canada and India for their domestic services. The common impression that the trend in communication satellites is to larger and larger antennas is not true. In fact, the only "large" antenna on the list is the Indian system antenna. Even this is smaller that the 30-ft design to be flown on ATS F/G and is not much larger than the 17-ft high antenna tower on TACSAT. Moreover, a serious debate rages as to whether domestic satellites should be allocated 860-MHz spectrum at all. Arguments have been put forth to confine domestic service to somewhere about 8 GHz. Should these arguments prevail, even the small country coverage antennas may shrink to values such as 2 ft. (The data relay antennas are limited by manufacturing tolerance considerations since they operate at a rather high frequency.)

The prime power values developed in the various studies arise from radiated power requirements to achieve adequate communication link parameters (including a tradeoff on ground or airborne antenna sizes) and the normal satellite housekeeping power drain. Degradation with life also has been taken into account. Note that the prime power values required are all within the general range (0.5 to 1 kw) of current technology as represented by TACSAT and INTELSAT IV. Large unfurlable solar arrays to generate many kilowatts of power do not appear to be needed in these future missions (we will return later to the abstract question of how large arrays would be implemented).

Examining pointing accuracy requirements, we find that they too can be easily satisfied by the 0.03° capability demonstrated by TACSAT. Except for the remote case of a "Los Angeles" satellite† which would be pointed to about 0.05° (or 15 miles on the ground), communication satellites of the reasonable future will not require precision pointing. In fact, this writer takes issue with the reference conclusion that the 4° Indian antenna beam should be pointed much more accurately than 0.25°.

At first hand then, this selected list of requirements and derived parameters may look suspect in that they all lie within current communication satellite technology. It may appear that this is the result of shortsighted decisions made by the system synthesizers merely to sell what they have. Certainly to some extent this is the case, but, as stated earlier, near future advances in communication satellites must come in the form of application of what we have. Further, the applications must be made within the constraints of price competition with improving terrestrial technologies such as LORAN, inertial guidance systems, microwave and cable systems, and the like. These practical considerations play an important role in application engineering.

However, there are some equally fundamental technical considerations. Figures 6 and 8 shed some light. Consider that there is about a 3000-fold increase in capacity from SYNCOM to INTELSAT IV. Yet their weight ratio is only 12. Their prime power ratio is about 15. (Another interesting ratio is that of price, which is about 6.) The big leverage is antenna gain. Here their ratio is several hundred. As indicated earlier, for most missions further increases in antenna gain ratios are limited by coverage requirements. Brute force prime power increase is the only degree of freedom left. The next factor of ten comes at great expense. An increase to 10 kw requires a weight increase of about 2000 lb in solar panel, batteries for eclipse operation, velocity control

---

†A pet idea of a colleague whose Malibu, Calif., residence does not have good TV service.

propulsion, and structure (not to mention severe thermal control problems and the like). Moreover, there may be no economical boosters to launch the resultant 3000-lb satellite in the next decade. Table 5 lists typical launch vehicle characteristics that are destined to remain static for a long time because of curtailment of booster development. The result is that about 1 kw is a fairly natural breakpoint. Future satellites probably will not exceed this level by more than a factor of two or so.

The stabilization systems recommended by the study groups for the selected future missions also are shown in Table 4. We see no reason why dual-spin systems were not recommended for all the missions. Each required parameter is within those achievable with current dual-spin designs, both in power and antenna size carrying ability. We need not repeat here the reasons for preferring dual-spin systems to three-axis systems where practical.[7] It is sufficient to point out that every operational synchronous communication satellite system selected in the last 5 years by the USAF, COMSAT, and the Canadian government has been a dual-spin system. (The ATS F/G configuration was selected long before the dual-spin gyrostat system was developed.) Their record of performance is still outstanding. We now predict lifetimes of 5 to 10 years for INTELSAT IV-type vehicles. To date, TACSAT has functioned for more than a year without evidence of spin control, spin bearing, or power and slip ring subsystem deterioration.

For missions that may require power levels of more than 1 kw, we submit the dual-spin configuration shown in Fig.9. The configuration differs from current craft in that a single-axis sun tracking rollout array is deployed on the oriented platform. The rotor still carries apogee kick motor, velocity and attitude control, and attitude sensor subsystems. Power for these systems as well as payload power before deployment of the large array is provided by a small panel on the rotor. This feature is particularly important for transfer orbit operation and post-apogee kick during orbit establishment prior to beginning the on-orbit mission. The penalty for a single axis array point system relative to a panel held normal to the sunline in all seasons is about 25 lb for a 2-kw array. The attendant simplicity is well worth this cost.

We feel it is incumbent on the three-axis system designer to demonstrate the rationale for deviating from the successful spin and dual-spin technology path developed primarily for synchronous operation.

References

[1]"Navigation/Traffic Control Satellite Missions Study", June 1969, (for NASA ERC), TRW No.09778-6008-RO-00.

[2]"Hughes Multipurpose Synchronous Communications Satellite Study", August 1967 (for COMSAT Corporation), HAC SSD 70300R-2.

[3]"Canadian Domestic Satellite Study", Canadian Government contract file IRA 9122-03-3, Nov. 1968.

[4]Hesselbacher, R.W., "An Evaluation of Television Broadcast Satellite Systems", Journal of Spacecraft and Rockets, Vol.6, No.10, Oct.1969, pp. 1124-1132.

[5]"Useful Applications of Earth-Oriented Satellites", National Academy of Sciences, Summer Study on Space Applications for NASA, Vol.7, 1969.

[6]"Study of Low Altitude Satellite Utilizing a Data Relay Satellite System - Final Report", HAC SSD 90283R, Hughes Aircraft Company, Aug. 1969.

[7]Iorillo, A.J., "Analyses Related to Hughes Gyrostat System", HAC SSD 70012R.

THREE-AXIS OR DUAL-SPIN STABILIZATION FOR FUTURE SYNCHRONOUS COMMUNICATION SATELLITES - A SUMMARY

T. B. Garber*

The Rand Corporation, Santa Monica, California

Abstract

Synchronous communication satellites have, since their inception, utilized some variation of spin stabilization to achieve antenna pointing requirements. As prime power and antenna pointing accuracy increase, the problem of adequately balancing solar disturbing torques becomes difficult. If the disturbances acting on the satellite are such that frequent control is necessary to maintain the pointing accuracy dictated by mission requirements, then attitude stiffness derived by spinning the vehicle or a portion of the vehicle is no longer advantageous and full three-axis stabilization is necessary. The projected requirements for the next generation of commercial communication satellites can be met by either dual-spin systems or by stabilization systems which employ an internal spinning wheel to enhance attitude stiffness. However the trend toward higher power, larger antennas, and increased pointing accuracy will necessitate the eventual use of three-axis stabilization for at least some communication satellites.

Summary and Conclusions

A communication satellite in a synchronous, stationary orbit has no need to make frequent, large-amplitude adjustments in body attitude. In addition, external disturbing torques tend to be periodic in nature. As a consequence of these two factors, "attitude stiffness" with respect to the orbital plane is desirable and can be achieved most easily by enhancing or augmenting the satellite's angular momentum vector.

As indicated in the preceding 4 papers, the first synchronous communication satellites achieved attitude stiffness by spin stabilization, with the angular momentum vector perpendicular to the orbital plane. In order to employ Earth-pointing antennas

---

*Physical Scientist, Engineering Sciences Department.

on such satellites, dual-spin stabilization was developed, as currently represented by the Intelsat 3 and 4 designs.

A different approach to the problem of achieving attitude stiffness is possible. Rather than spinning the entire vehicle or at least a substantial fraction of the vehicle, momentum augmentation can be achieved by using a pitch wheel mounted inside the satellite. In this case the entire satellite is the despun portion.

The relative merits of these two methods of increasing satellite attitude stiffness about the roll and yaw axes have been considered in detail in the preceding 4 papers. However, as pointed out by Jenkins, the trend in dual-spin designs is toward a larger percentage of the satellite's mass to be despun (Table 1). The question arises, as this trend continues, as to when the stabilization system is no longer to be classified dual-spin but rather three-axis with angular momentum augmentation. One possible definition of dual-spin stabilization is that the spinning portion of the satellite performs functions other than angular momentum augmentation or modulation. This definition would encompass a wide variety of dual-spin satellite configurations, and in limiting cases, such as when the spinning portion of the satellite contains only horizon scanners and associated electronics, they would hardly differ from systems with an internal pitch wheel as the momentum source.

In any case, the required angular momentum, and thus the moment of inertia and spin rate of the rotating part of the satellite, is a function of the disturbing torques about the satellite's roll and yaw axes and the angular errors that can be tolerated about those axes. If, for a given level of disturbing torques, attitude accuracy requirements are so stringent that frequent roll and yaw corrections are necessary, then many of the advantages of achieving roll-yaw stiffness by momentum augmentation disappear.

What sort of communication requirements are on the immediate horizon? Table 4 presents a list of missions, together with the required satellite parameters, that might have commercial significance within the next two to five years. Of the five missions listed, only two, the NASA Data Relay and the Indian Domestic satellite, have pointing accuracy requirements that exceed what can be achieved by state-of-the-art Earth sensors. An antenna pointing accuracy requirement of 0.05° would necessitate a three-axis sensing system, and, in the case of the Indian Domestic satellite, a 21-ft-diam antenna would present difficulties in balancing solar disturbing torques. Both of these considerations tend to make dual-spin or momentum wheel systems less attractive as compared to three-axis systems with reaction wheels.

The remaining three missions of Table 4 are, performance-wise, well within the current state-of-the-art of dual spin or wheel-stabilized satellites. As prime power requirements increase beyond 1 kw, a drum configuration for dual-spin satellites cannot provide sufficient solar cell area because of size constraints imposed by current launch vehicles. In addition, the weight and cost of drum-mounted solar cells increase rapidly beyond 1 kw relative to directed solar arrays. However, as indicated by Iorillo, a directed solar array, if required, can be mounted on the despun portion of a dual-spin satellite.

Thus it would appear that the next generation of communication satellites will most likely utilize dual-spin or momentum-wheel stabilization to achieve attitude stiffness about the roll and yaw axes. Either dual-spin or wheel-stabilized satellites can satisfy the probable mission requirements, and the "preferred" approach is, to a great degree, dependent upon the manner in which a particular design group assesses the relative risks and tradeoffs involved. However, as solar arrays and antennas become very large, the problem of adequately balancing solar disturbing torques becomes difficult, and full three-axis stabilization systems become a necessity.

Table 1 The evolution of synchronous communication satellites

Vehicle	Weight (lb)	Size (in.)	Array power (w)	Weight despun (%)	Design life (yrs)	ERP (w)
ATS-1	775	56 × 57	185	0	3	26
Intelsat 3	322	56 × 41	130	8	5	27
Intelsat 4	1019	106 × 110	780	32	7	35-45
TACSAT	1600	108 × 300	900	40	7	-

Table 2 Characteristics of Design Example 2 spacecraft

Parameter	Dual spin	3-axis
Weight (without apogee kick motor)	1750	1600[a]
Size of launch (stowed) configuration	9 × 26 ft	9 × 9 ft
Length of structural support members between antennas and spacecraft body		
4-ft antennas	9 ft	2 ft
7-ft antennas	8.5 ft	4 ft
Heat rejection area[b]		
Without apogee kick motor	125 ft^2	100 ft^2
With apogee kick motor	110 ft^2	100 ft^2
Solar pressure torque	Easier to minimize for 3-axis	
Development risk	Higher for 3-axis	
Control during apogee kick motor burn	Easier for dual-spin	
Reliability	Comparable	
Cost	Comparable	

[a]Assumes use of extendible arrays.

[b]Considers only surfaces parallel to the orbital plane.

Table 3 A comparison of dual-spin, flywheel and three-axis stabilization system characteristics

Item	Dual spinner	Flywheel	3-axis stabilized
Reliability	Cannot be made better than that of bearing and power transfer assembly	Unlimited, using redundant flywheels or backup 3 axis	Unlimited, at expense of size, weight, and power
Passive stabilized apogee kick (not required on TIIIC launches)	Yes	Yes - depending on flywheel size	No - use Burner II Penalty - 100 lb
Sensor selection	Scanning Earth sensor slit-type star tracker	(Depends upon flywheel design); scanning Earth sensor. Image dissector star tracker	Radiation balance Earth sensor image dissector star tracker (no mechanical motion)
Scanning payload sensors (not generally used on COMSATS)	Easy	Difficult (depending upon design)	Requires spinning platform
Thermal control	Simplified by rotating drum; may require louvers or heat pipes, particularly in despun part	Same as 3-axis	Louvers and heat pipes
Power capacity	1 kw	Not limited	Not limited

Table 3 (continued)

Item	Dual spinner	Flywheel	3-axis stabilized
Testing	May be difficult; must determine unbalance and all damping sources	Not difficult	Accuracy inertia and damping measurements not needed
Attitude drift (based on once per day precession correction)	±0.3°	±0.3° - simple 0.1° - hybrid	0.1°
Roll maneuvers	Requires 1/30 lb/° of reorientation	1/30 lb/°	Electric power
Propellant expulsion	g forces	Surface tension device	Surface tension device
Attitude control thrusters required	2	6	6
Thrusters cycles/yr	14,600	365	365
Configuration freedom	Limited - causes high solar pressure	Relatively unlimited	Relatively unlimited

Table 4 Projected communication satellite mission requirements

		Satellite parameters				
Possible missions	Reference[a]	Frequencies	Required antennas (diam)	Prime power (w)	Pointing accuracy (deg)	Suggested stabilization
Communication/ navigation and traffic control (Aeronautical and marine)	TRW study[1]	1600 MHz	5.5 ft and 3.0 ft	760	0.5	Three-axis
United States domestic (TV and voice)	Hughes study[2]	4 and 6 GHz	8 ft (multi-feeds)	500	0.25	Dual-spin
Canadian domestic (TV and voice)	Canadian Government study[3]	4 and 6 GHz	5 × 2 ft	250	0.5	Dual-spin
Indian domestic (TV and voice)	GE study[4]	860 MHz and 8.4 GHz	21 ft and 2 ft	1100	0.05	Three-axis
NASA data relay	National Academy of Science study[5] and Hughes study[6]	15 GHz and 8.4 GHz	9 ft and 9 ft	500	0.05	Not defined

[a]References are listed at the end of the Iorillo paper in this chapter.

Table 5 Typical launch vehicle characteristics

	Thor-Delta and apogee kick motor	Atlas-Centaur and apogee kick motor	Titan IIIC
Cost to nongovernment user (millions)	7.0	16	26
Payload into synchronous orbit (lb)	600	2000	2600
Available shroud diam (ft)	5.4	10	10
Available shroud lengths (ft)	19.0	39.0	50

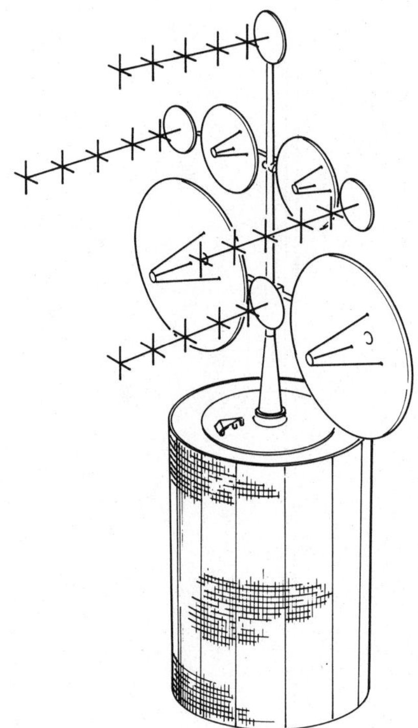

Fig. 1 Dual-spin stabilized spacecraft for Design Example 1.

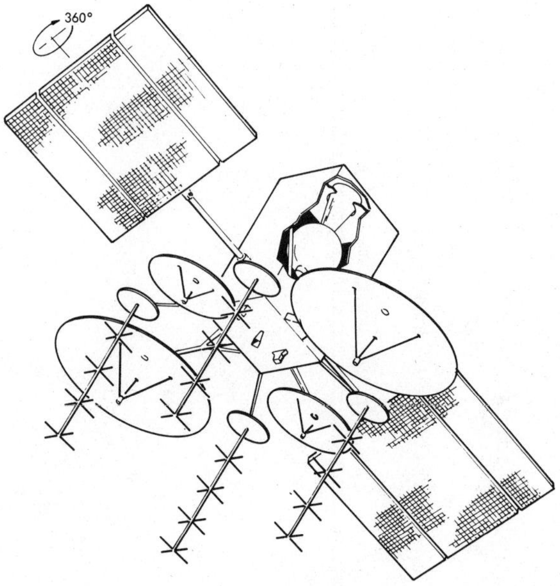

Fig. 2 3-axis stabilized spacecraft for Design Example 1.

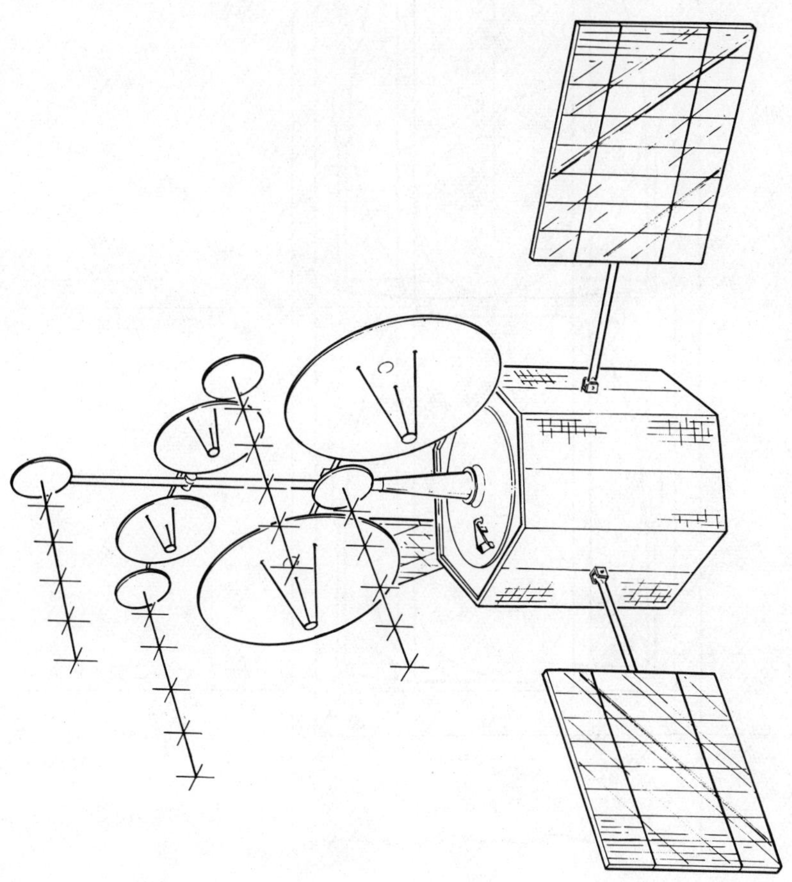

Fig. 3 Dual-spin stabilized spacecraft with solar paddles.

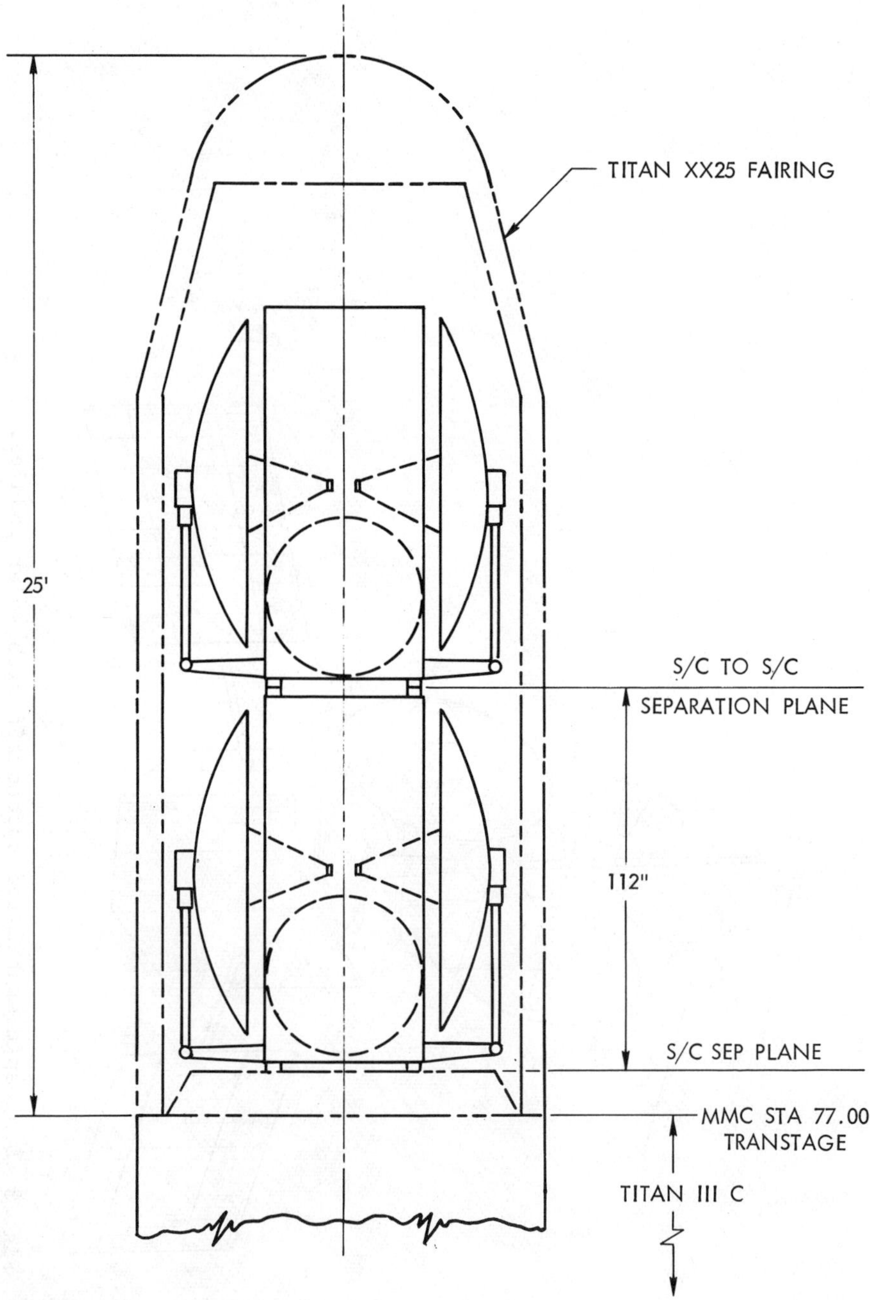

Fig. 4 Two 3-axis stabilized spacecraft for Design Example 1 in launch (stowed) configuration.

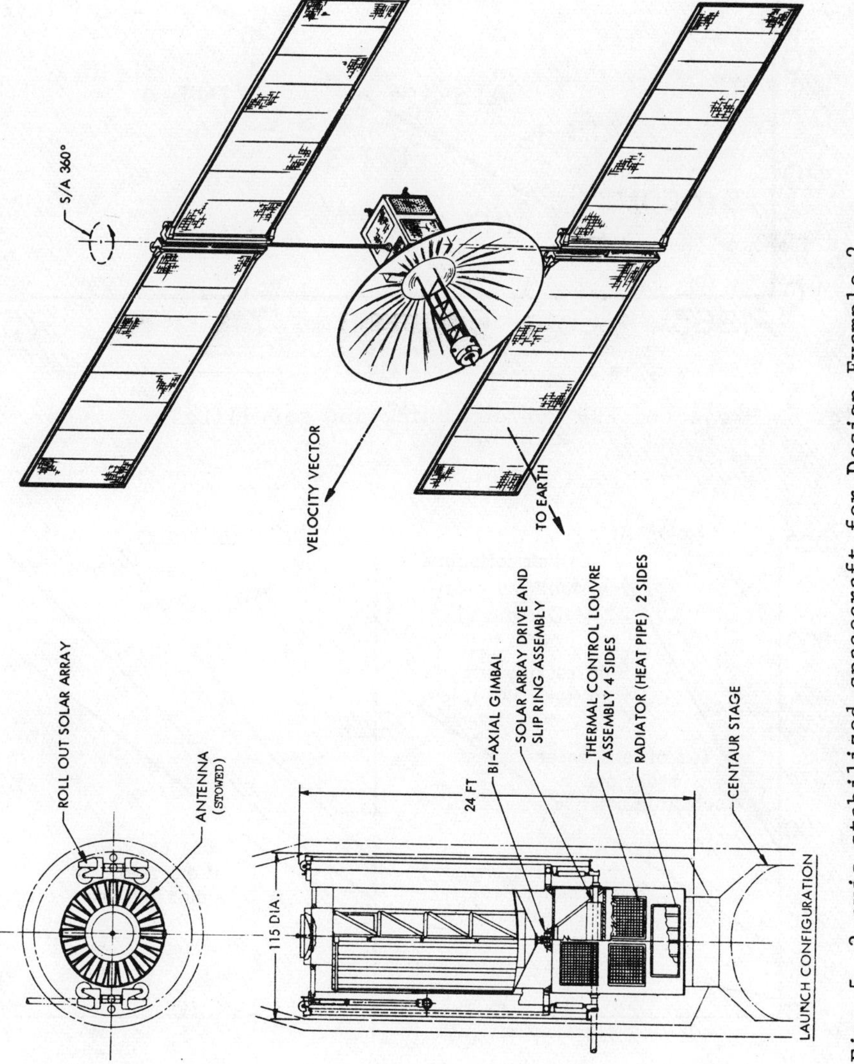

Fig. 5 3-axis stabilized spacecraft for Design Example 2.

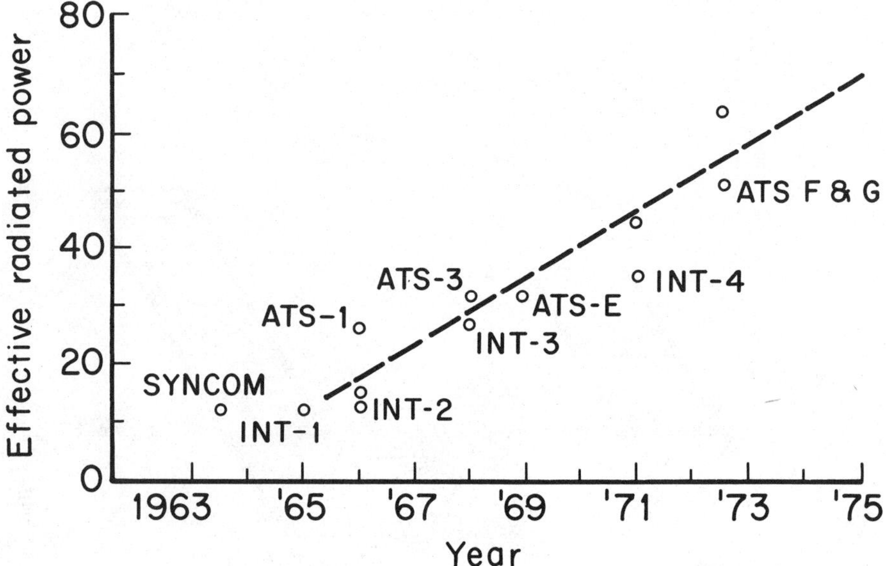

Fig. 6 Projected ERP for communication satellites.

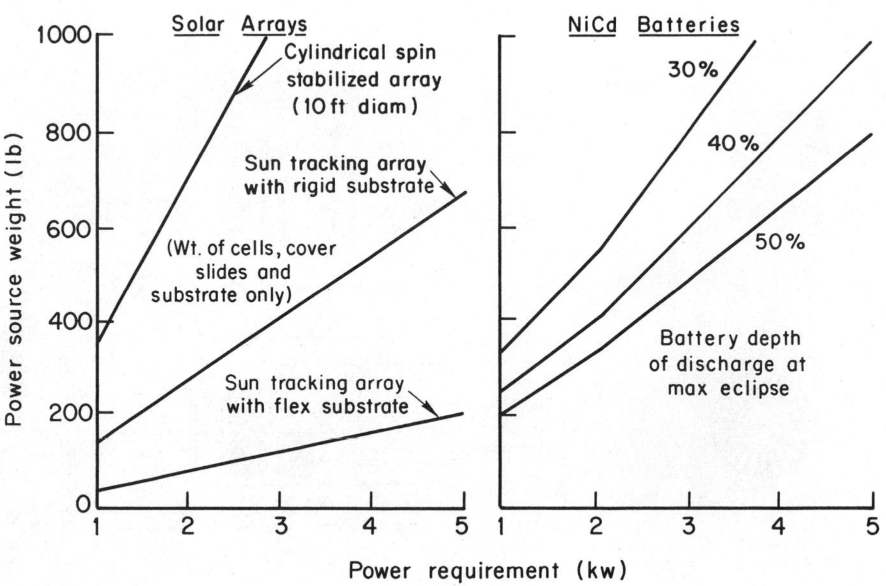

Fig. 7 Synchronous satellite electrical power system.

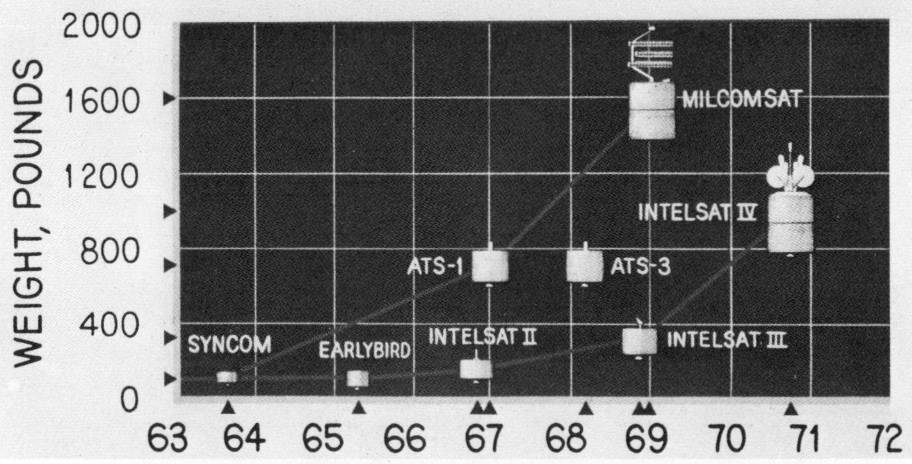

Fig. 8 Growth of synchronous communication satellites.

Fig. 9 Dual-spin satellite with single-axis sun tracking array.

Index to Contributors to Volume 26